Essential Computational Modeling for the Human Body

Essential Computational Modeling for the Human Body

A derivative of Handbook of Numerical Analysis, Special Volume: Computational Models for the Human Body, Vol XII

General Editor:

P.G. Ciarlet

Laboratoire Jacques-Louis Lions
Université Pierre et Marie Curie
4 Place Jussieu
75005 PARIS, France

and

Department of Mathematics
City University of Hong Kong
Tat Chee Avenue
KOWLOON, Hong Kong

Guest Editor:

N. Ayache

INRIA
2004 Route des Lucioles
06902 Sophia–Antipolis
France

Amsterdam • Boston • Heidelberg • London • New York
Oxford • Paris • San Diego • San Francisco • Singapore
Sydney • Tokyo

Elsevier
30 Corporate Drive, Suite 400, Burlington, MA 01803, USA
Linacre House, Jordan Hill, Oxford OX2 8DP, UK
Radarweg 29, PO Box 211, 1000 AE Amsterdam, The Netherlands

First edition 2011

Material in the text originally appeared in the Handbook of Numerical Analysis, Special Volume: Computational Models for the Human Body, Vol XII edited by P.G. Ciarlet (editor) and N. Ayache (guest editor) (Elsevier B.V. 2004)

British Library Cataloguing in Publication Data
A catalogue record for this book is available from the British Library

Library of Congress Cataloging-in-Publication Data
A catalog record for this book is available from the Library of Congress

ISBN: 978-0-444-53755-3

For information on all Elsevier publications
visit our web site at *books.elsevier.com*

Typeset by: diacriTech, India

Printed and bound in Great Britain
11 12 13   10 9 8 7 6 5 4 3 2 1

General Preface

In the early eighties, when Jacques-Louis Lions and I considered the idea of a *Handbook of Numerical Analysis*, we carefully laid out specific objectives, outlined in the following excerpts from the "General Preface" which has appeared at the beginning of each of the volumes published so far:

> During the past decades, giant needs for ever more sophisticated mathematical models and increasingly complex and extensive computer simulations have arisen. In this fashion, two indissociable activities, *mathematical modeling* and *computer simulation*, have gained a major status in all aspects of science, technology and industry.
>
> In order that these two sciences be established on the safest possible grounds, mathematical rigor is indispensable. For this reason, two companion sciences, *Numerical Analysis* and *Scientific Software*, have emerged as essential steps for validating the mathematical models and the computer simulations that are based on them.
>
> *Numerical Analysis* is here understood as the part of *Mathematics* that describes and analyzes all the numerical schemes that are used on computers; its objective consists in obtaining a clear, precise, and faithful, representation of all the "information" contained in a mathematical model; as such, it is the natural extension of more classical tools, such as analytic solutions, special transforms, functional analysis, as well as stability and asymptotic analysis.
>
> The various volumes comprising the *Handbook of Numerical Analysis* will thoroughly cover all the major aspects of Numerical Analysis, by presenting accessible and in-depth surveys, which include the most recent trends.
>
> More precisely, the Handbook will cover the *basic methods of Numerical Analysis*, gathered under the following general headings:
>
> – Solution of Equations in $\mathbb{R}^n$,
> – Finite Difference Methods,
> – Finite Element Methods,
> – Techniques of Scientific Computing.

It will also cover the *numerical solution of actual problems of contemporary interest in Applied Mathematics*, gathered under the following general headings:

- Numerical Methods for Fluids,
- Numerical Methods for Solids.

In retrospect, it can be safely asserted that Volumes I to IX, which were edited by both of us, fulfilled most of these objectives, thanks to the eminence of the authors and the quality of their contributions.

After Jacques-Louis Lions' tragic loss in 2001, it became clear that Volume IX would be the last one of the type published so far, i.e., edited by both of us and devoted to some of the general headings defined above. It was then decided, in consultation with the publisher, that each future volume will instead be devoted to a single "*specific application*" and called for this reason a "*Special Volume*". "*Specific applications*" will include Mathematical Finance, Meteorology, Celestial Mechanics, Computational Chemistry, Living Systems, Electromagnetism, Computational Mathematics etc. It is worth noting that the inclusion of such "specific applications" in the *Handbook of Numerical Analysis* was part of our initial project.

To ensure the continuity of this enterprise, I will continue to act as Editor of each Special Volume, whose conception will be jointly coordinated and supervised by a Guest Editor.

P.G. CIARLET
July 2002

List of Contributors

Alfio Quarteroni, *Institute of Mathematics, EPFL, Lausanne, Switzerland, MOX, Department of Mathematics, Politecnico di Milano, Milano, Italy, e-mail: alfio.quarteroni@epfl.ch* (Ch. 1).

Eberhard Haug, *ESI Software S.A., 99, rue des Solets, BP 80112, 94513 Rungis Cedex, France, e-mail: eha@esi-group.com, URL: http://www.esi-group.com* (Ch. 2).

Hervé Delingette, *INRIA Sophia–Antipolis, 2004, route des Lucioles, BP 93, 06902 Sophia–Antipolis, France, e-mail: Herve.Delingette@sophia.inria.fr* (Ch. 3).

Hyung-Yun Choi, *Hong-Ik University, Seoul, South Korea, e-mail: hychoi@hongik.ac.kr* (Ch. 2).

Luca Formaggia, *MOX, Department of Mathematics, Politecnico di Milano, Milano, Italy, e-mail: luca.formaggia@mate.polimi.it* (Ch. 1).

Muriel Beaugonin, *ESI Software S.A., Paris, France, e-mail: mbe@esi-group.com* (Ch. 2).

Nicholas Ayache, *INRIA Sophia–Antipolis, 2004, route des Lucioles, BP 93, 06902 Sophia–Antipolis, France, e-mail: Nicholas.Ayache@inria.fr* (Ch. 3).

Stéphane Robin, *LAB PSA-Renault, Paris, France, e-mail: stephane.robin@mpsa.com* (Ch. 2).

Table of Contents

Mathematical Modeling and Numerical Simulation of the Cardiovascular System

Alfio Quarteroni [a,b], Luca Formaggia [b]

[a] *Institute of Mathematics, EPFL, Lausanne, Switzerland*

[b] *MOX, Department of Mathematics, Politecnico di Milano, Milano, Italy*

E-mail addresses: alfio.quarteroni@epfl.ch (A. Quarteroni), luca.formaggia@mate.polimi.it (L. Formaggia)

Essential Computational Modeling for the Human Body
Special Volume (N. Ayache, Guest Editor) of
HANDBOOK OF NUMERICAL ANALYSIS, VOL. XII
P.G. Ciarlet (Editor)

ISSN 1570-8659
DOI 10.1016/B978-0-444-53755-3.00001-9

Contents

CHAPTER I

1. Introduction

In these notes we will address the problem of developing models for the numerical simulation of the human circulatory system. In particular, we will focus our attention on the problem of hemodynamics in large human arteries.

Indeed, the mathematical investigation of blood flow in the human circulatory system is certainly one of the major challenges of the next years. The social and economical relevance of these studies is highlighted by the unfortunate fact that cardiovascular diseases represent the major cause of death in developed countries.

Altered flow conditions, such as separation, flow reversal, low and oscillatory shear stress areas, are now recognized by the medical research community as important factors in the development of arterial diseases. An understanding of the local hemodynamics can then have useful applications for the medical research and, in a longer term perspective, to surgical planning and therapy. The development of effective and accurate numerical simulation tools could play a crucial role in this process.

Besides their possible role in medical research, another possible use of numerical models of vascular flow is to form the basis for simulators to be used as training systems. For instance, a technique now currently used to cure a stenosis (a pathological restriction of an artery, usually due to fat deposition) is angioplasty. It consists of inflating a balloon positioned in the stenotic region by the help of a catheter. The balloon should squash the stenosis and approximately restore the original lumen area. The success of the procedure depends, among other things, on the sensitivity of the surgeon and his ability of placing the catheter in the right position. A training system which couples virtual reality techniques with the simulation of the flow field around the catheter, the balloon and the vessel walls, employing geometries extracted from real patients, could well serve as training bed for new vascular surgeons. A similar perspective could provide specific design indications concerning the realizations of surgical operations. For instance, numerical simulations could help the surgeon in understanding how the different surgical solutions may affect blood circulation and guide the selection of the most appropriate procedure for a specific patient.

In such "virtual surgery" environments, the outcome of alternative treatment plans for the individual patient can be foreseen by simulations. This numerical approach is one of the aspects of a new paradigm of the clinical practice, which is referred to as "predictive medicine" (see TAYLOR, DRANEY, KU, PARKER, STEELE, WANG and ZARINS [1999]).

Since blood flow interacts mechanically with the vessel walls, it gives rise to a rather complex fluid–structure interaction problem which requires algorithms able to correctly describe the energy transfer between the fluid (typically modeled by the Navier–Stokes equations) and the structure. This is indeed one of the main subjects of these notes, which will adopt the following steps:

(1) *Analysis of the physical problem.* We illustrate problems related to hemodynamics, focusing on those aspects which are more relevant to human physiology. This will allow us to identify the major mathematical variables useful for our investigation. This part will be covered in Section 2.

(2) *Mathematical modeling.* Starting from some basic physical principles, we will derive the partial differential equations which link the variables relevant to the problem. We will address some difficulties associated to the specific characteristics of these equations. Problems such as existence, uniqueness and data dependence of the solution will be briefly analyzed. In particular, in Section 5 we will deal with models for the fluid flow and recall the derivation of the incompressible Navier–Stokes equations starting from the basic principles of conservation of mass and momentum. In Section 15 the attention will be instead focused on the dynamics of the vessel wall structure. Some simple, yet effective, mathematical models for the vessel wall displacement will be derived and discussed.

(3) *Numerical modeling.* We present different schemes which can be employed to solve the equations that have been derived and discuss their properties. In particular, Section 10 deals with some relevant mathematical aspects related to the numerical solution of the equations governing the flow field, while Section 18 is dedicated to the coupled fluid–structure problem.

    Reduced models which make use of a one dimensional description of blood flow in arteries are often used to study the propagation of average pressure and mass flow on segments of the arterial tree. In Section 20 we present the derivation of a model of this type, together with a brief analysis of its main mathematical characteristics.

(4) *Numerical simulation.* A final section is dedicated to numerical results obtained on relevant test cases.

2. A brief description of the human vascular system

The major components of the cardiovascular system are the heart, the arteries and the veins. It is usually subdivided into two main parts: the *large circulation* system and the *small circulation* system, as shown in Fig. 2.1. The former brings oxygenated blood from the heart left ventricle to the various organs (arterial system) and then brings it back to right atrium (venous system). The latter pumps the venous blood into the pulmonary artery, where it enters the pulmonary system, get oxygenated and is finally received by the heart left atrium, ready to be sent to the large circulation system.

Fig. 2.2 shows a picture of the human heart. Its functioning is very complex and various research teams are currently trying to develop satisfactory mathematical models of its mechanics, which involves, among other things, the study of the electro-chemical activation of the muscle cells. We will not cover this aspect in these notes, where we rather concentrate on vascular flow and, in particular, flow in arteries.

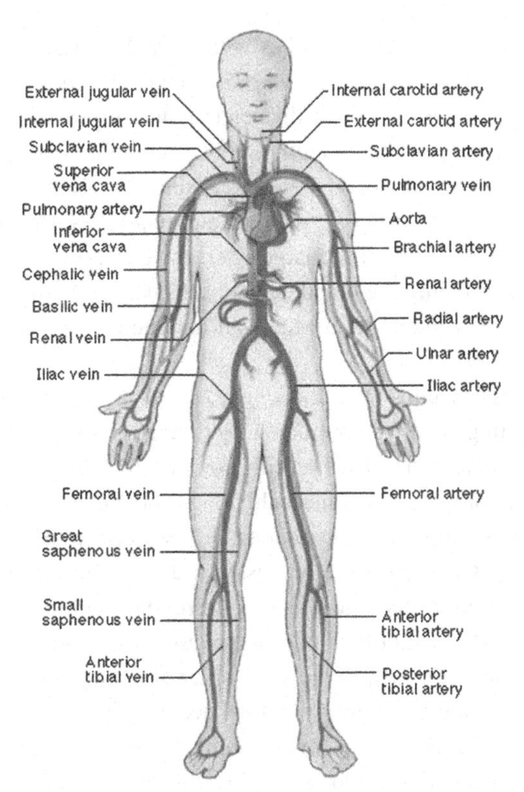

FIG. 2.1. The human circulatory system. The human cardiovascular system has the task of supplying the human organs with blood. Its correct working is obviously crucial and depends on many parameters: external temperature, muscular activity, state of health, just to mention a few. The blood pressure and flow rate then change according to the body needs.

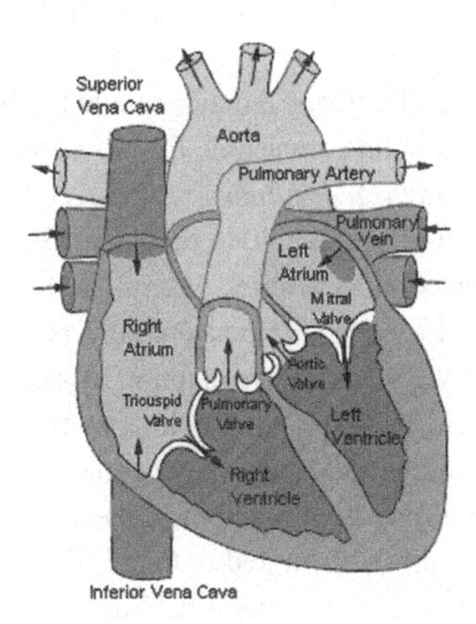

FIG. 2.2. The human heart. Courtesy of the Texas Heart®Institute.

Arteries can be regarded as hollow tubes with strongly variable diameters and can be subdivided into *large arteries*, *medium arteries* and *arterioles and capillaries*. The main role of large arteries (1–3 cm of diameter) is to carry a substantial blood flow rate from the heart to the periphery and to act as a "compliant system". They deform under blood pressure and by doing so they are capable of storing elastic energy during the systolic phase and return it during the diastolic phase. As a result the blood flow is more regular than it would be if the large arteries were rigid. We then have a *fluid–structure* interaction problem. The blood may be considered a homogeneous fluid, with "standard" behavior (Newtonian fluid), the wall may be considered elastic (or mildly visco-elastic).

The smaller arteries (0.2 mm–1 cm of diameter) are characterized by a strong branching. The vessel may in general be considered rigid (apart in the heart, where the vessel movement is mainly determined by the heart motion). Yet, the blood begins to show "non-standard" behavior typical of a shear-thinning (non-Newtonian) fluid.

The arterioles have an important muscular activity, which is aimed at regulating blood flow to the periphery. Consequently, the vessel wall mechanical characteristics may change depending on parameters such as blood pressure and others. At the smallest levels (capillaries), blood cannot be modeled anymore as a homogeneous fluid, as the dimension of the particles are now of the same order of that of the vessel. Furthermore, the effect of wall permeability on the blood flow becomes important.

The previous subdivision is not a mere taxonomy: the morphology of the vessel walls and the physical characteristics of blood change in dependence of the type of vessel.

Indeed, the blood is not a fluid but a suspension of particles in a fluid called *plasma*. Blood particles must be taken into account in the rheological model in smaller arterioles and capillaries since their size becomes comparable to that of the vessel. The most important blood particles are:

- red cells (erythrocytes), responsible for the exchange of oxygen and carbon-dioxide with the cells;
- white cells (leukocytes), which play a major role in the human immune system;
- platelets (thrombocytes), main responsible for blood coagulation.

Here, we will limit to flow in *large/medium sized vessels*. We have mentioned that the vascular system is highly complex and able to regulate itself: an excessive decrease in blood pressure will cause the smaller arteries (arterioles) to contract and the heart rate increase. On the contrary, an excessive blood pressure is counter-reacted by a relaxation of the arterioles wall (which causes a reduction of the periphery resistance to the flow) and decreasing the heart beat. Yet, it may happen that some pathological conditions develop, for example, the arterial wall may become more rigid, due to illness or excessive smoking habits, fat may accumulate in some areas causing a stenosis, that is a reduction of the vessel section as illustrated in Fig. 2.3, aneurysms may develop. The consequence of these pathologies on the blood field as well as the possible outcome of a surgical intervention may be studied by numerical tools.

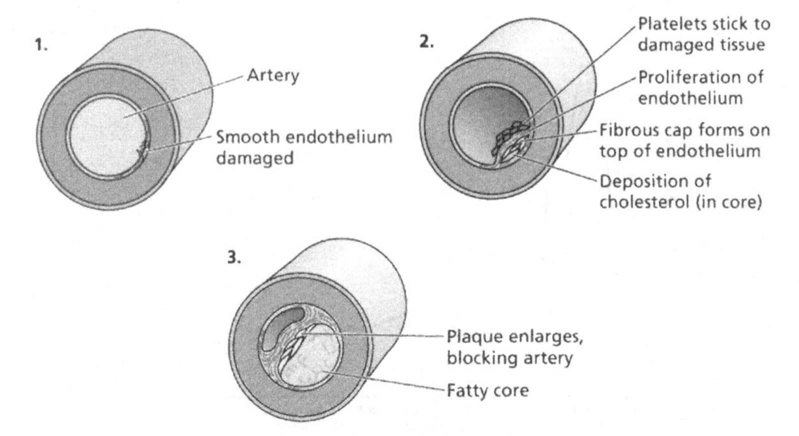

FIG. 2.3. The deposition of lipids and cholesterol in the inner wall of an artery (frequently a coronary) can cause a stenosis and eventually a dramatic reduction (or even the interruption) of blood flow. Images taken from "Life: the Science of Biology" by W.K. Purves et al., fourth edition, published by Sinauer Associates Inc. and W.H. Freeman and Company.

3. The main variables for the mathematical description of blood flow

The principal quantities which describe blood flow are the *velocity* $\mathbf{u}$ and *pressure* P. Knowing these fields allows the computation of the *stresses* to which an arterial wall is subjected due to the blood movement. Since we will treat fluid–structure interaction problems, the *displacement* of the vessel wall due to the action of the flow field is another quantity of relevance. Pressure, velocity and vessel wall displacement will be functions of time and the spatial position.

The knowledge of the *temperature* field may also be relevant in some particular context, such as the hyperthermia treatment, where some drugs are activated through an artificial localized increase in temperature. Temperature may also have a notable influence on blood properties, in particular on blood viscosity. Yet, this aspect is relevant only in the flow through very small arterioles/veins and in the capillaries, a subject which is not covered in these notes.

Another aspect of blood flow which we will not cover in these notes, is the *chemical interaction* with the vessel wall, which is relevant both for the physiology of the blood vessels and for the development of certain vascular diseases. Not mentioning the potential relevance of such investigation for the study of the propagation/absorption of pharmaceutical chemicals. Some numerical models and numerical studies for the chemical transport/diffusion process in blood and through arterial wall may be found in RAPPITSCH and PERKTOLD [1996], QUARTERONI, VENEZIANI and ZUNINO [2002].

4. Some relevant issues

Among the difficulties in the modeling of blood flow in large vessels, we mention the following ones:

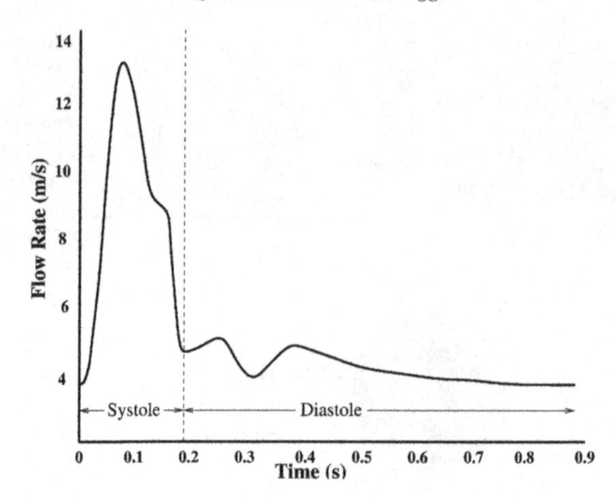

FIG. 4.1. A typical flow rate in an artery during the cardiac cycle.

- *The flow is transient.* Blood flow is obviously pulsatile. This means that one cannot neglect the time by considering a "steady state" solution, function only of the spatial position, as it is often done in many other situations (for example, the study of the flow field around an aeroplane or a car). With some approximation one may think the blood flow to be periodic in time. Yet, this is usually true only for relatively short periods, since the various human activities require to change the amount of blood sent to the various organs.

  The cardiac cycle can be subdivided into two phases. The *systole* corresponds to the instant in which the heart is pumping the blood into the arterial system. The systolic period is then characterized by the highest flow rate. The *diastole*, instead, corresponds to the instant in which the heart is filling up with the blood coming from the venous system and the aortic valve is closed. The blood flow is then at its minimum. Fig. 4.1 illustrates a typical flow rate curve on a large artery during the cardiac cycle.

  Unsteady flow is usually much more complex than its steady counterpart. For instance, if we consider a steady flow of a fluid like water inside an "infinitely long" cylindrical tube, it is possible to derive the analytical steady state solution (also called the *Poiseuille flow* solution), characterized by a parabolic velocity profile. Transient flow in the same geometrical configuration becomes much more complex. The solution may still be obtained analytically if we assume time periodicity, giving rise to the so-called *Womersley flow* (WOMERSLEY [1955]), whose expression may be found, for instance, in QUARTERONI, TUVERI and VENEZIANI [2000]. Just as an example, in Fig. 4.2 we show the velocity profile in a tube for a Poiseuille and for a Womersley flow (the latter, obviously, at a given instant) (from VENEZIANI [1998]).

- *The wall interacts mechanically with the flow field.* This aspect is relevant for relatively large vessels. In the aorta, for example, the radius may vary in a range of 5 to 10% between diastole and systole. This is quite a large displacement, which affects the flow field. The fluid–structure interaction problem is the responsible for

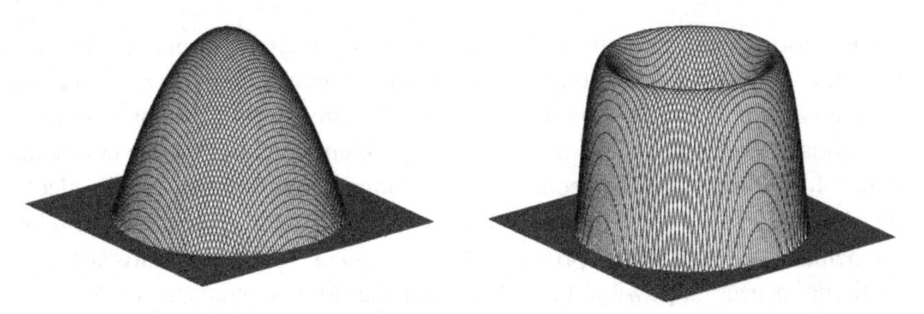

FIG. 4.2. Three-dimensional velocity profiles for a Poiseuille flow (left) and Womersley unsteady flow at a given instant (right).

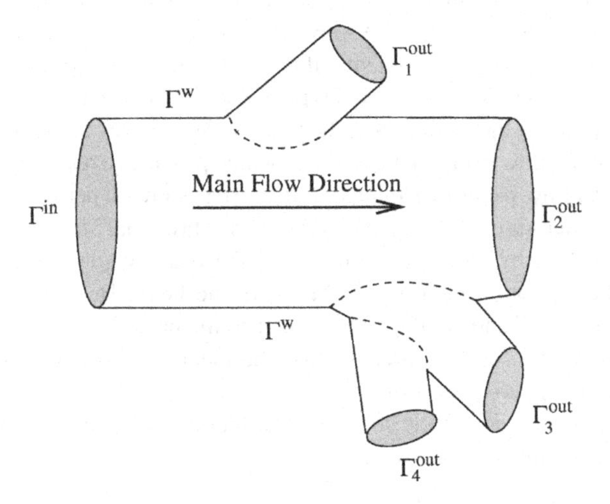

FIG. 4.3. An example of a computational domain made of a section of vascular system. We need to provide proper boundary conditions at Γ^{in}, Γ^{out} and Γ^{w}.

the propagation of pulse pressure waves. Indeed, no propagative phenomena would otherwise occur in an incompressible fluid like blood. The interaction problem is a rather complex one, since the time scales associated to the interaction phenomena are two orders of magnitude greater than those associated to the bulk flow field.

In arterioles and capillaries the movement of the wall may be considered negligible.

- *Lack of boundary data.* We are normally interested in modeling only a section of the cardiovascular system by means of partial differential equations. A proper setting of a differential problem requires to provide appropriate conditions at the domain boundary, i.e., on the sections at the ends of the region of interest. For instance, let us consider Fig. 4.3. "Standard" conditions for the inlet section Γ^{in} and the outlet sections Γ^{out}, may be derived from the analysis of the differential equations governing the fluid flow. A possible choice is to prescribe all components of the velocity on Γ^{in} and the velocity derivative along the normal direction (or the normal stress components) on Γ^{out}. Unfortunately, in practise one never has enough data for prescribing all these conditions. Normally, only "averaged" data are available (mean velocity and mean pressure), which are not sufficient for a "standard" treatment of the mathematical problem. One has thus to devise alternative formulations for the boundary conditions which, on one hand reflect the physics and exploit the available data, on the other hand, permit to formulate a mathematically well posed problem. In these notes we will not investigate this particular aspect. A possible formulation for the flow boundary conditions which is particularly suited for vascular flow problems is illustrated and analyzed in FORMAGGIA, GERBEAU, NOBILE and QUARTERONI [2002].

We have not used the terms "inflow" and "outflow" to indicate boundary conditions at Γ^{in} and Γ^{out} since they would be incorrect. Indeed, outflow would indicate the normal component of the velocity is everywhere positive (while it is negative at an inflow section). However, in vascular problems, this assumption is seldom true because the pulsating nature of blood flow might (and typically does) induce a flow reversal on portions of an artery during the cardiac beat. Indeed, the Womersley solution (WOMERSLEY [1955]) of a pulsatile flow in circular cylinders, which provides a reasonable approximation of the general flow pattern encountered in arteries, shows a periodic flow reversal.

In the medical literature, one encounters the terms "proximal" to indicate the section which is reached first by the flow exiting from the heart, while "distal" is the term associated to the sections which are farther from the heart. Here we have preferred instead the terms "inlet" and "outlet" which refer to the behavior of the mean flow rate across the section. At an inlet (outlet) section the mean flow is entering (exiting) the vascular element under consideration.

Some of the problems which the simulation of blood flow in large arteries may help in answering are summarized below.

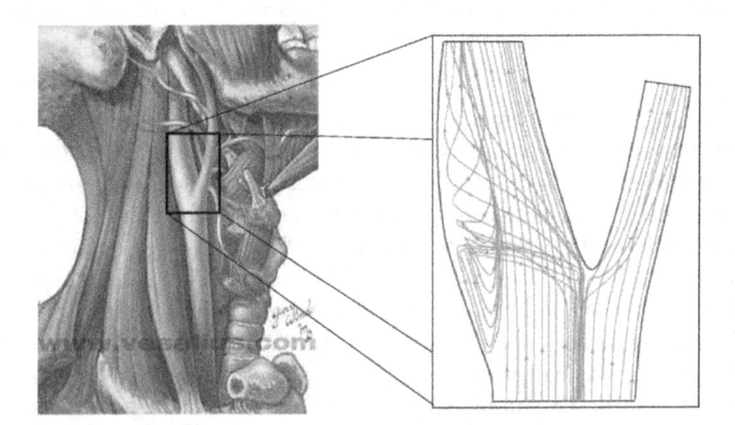

FIG. 4.4. Recirculation in the carotid bifurcation. On the left we illustrate the location of the carotid bifurcation. The image on the right shows the particle path during the diastolic period in a model of the carotid bifurcation. A strong recirculation occurs inside the carotid sinus. The image on the left is courtesy of vesalius.com.

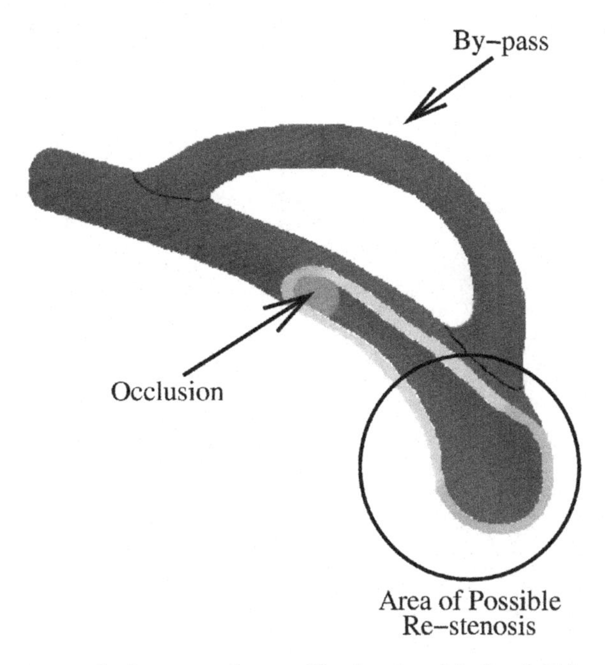

FIG. 4.5. A schematic example of a coronary by-pass. The alteration of the flow field due to the by-pass may cause the formation of a new stenosis, typically immediately downstream the by-pass.

- *Study of the physiological behavior of vessel walls.* For example, are there any characteristics of the flow field which may be related to the formation of stenoses? In particular, in some sites like the carotid bifurcation (see Fig. 4.4) it is quite usual to have a reversal of the flow during the cardiac cycle which generates a *recirculation zone*. These recirculation zones have been found to be possible sites for fat accumulation and, consequently, the appearance of stenosis. There is some evidence that one of the factors which prompt fat accumulation is linked to the oscillatory nature of the vessel wall stresses induced by the fluid in the flow reversal zone. Wall stresses are quantities very difficult to measure "in vivo" while are easily computed once the flow field is known. Numerical simulations may then help in assessing the effectiveness of such theory.
- *Study of post-surgical situations.* Is it possible to predict the flow behavior after the geometry has been modified by a surgical operation like a by-pass (see Fig. 4.5)? It has been found that the flow pattern in the by-pass region may affect the insurgence of post-surgery pathologies. Again, a zone with recirculating or stagnant fluid has negative consequences. Numerical simulations may allow to predict the post-surgery flow pattern and determine, say, the best by-pass configuration.

CHAPTER II

5. The derivation of the equations for the flow field

The flow field is governed by a set of partial differential equations in a region whose boundary changes in time. Their derivation, moving from the basic physical principles of conservation of mass and momentum, is the scope of this chapter.

6. Some nomenclature

The space $\mathbb{R}^3$ is equipped with a Cartesian coordinate system defined by the orthonormal basis $(\mathbf{e}_1, \mathbf{e}_2, \mathbf{e}_3)$, where

$$\mathbf{e}_1 = \begin{bmatrix} 1 \\ 0 \\ 0 \end{bmatrix}, \qquad \mathbf{e}_2 = \begin{bmatrix} 0 \\ 1 \\ 0 \end{bmatrix}, \qquad \mathbf{e}_3 = \begin{bmatrix} 0 \\ 0 \\ 1 \end{bmatrix}.$$

Vectors are understood as column vectors. A vector $\mathbf{f} \in \mathbb{R}^3$ may then be written as

$$\mathbf{f} = \sum_{i=1}^{3} f_i \mathbf{e}_i,$$

where f_i is the ith component of $\mathbf{f}$ with respect to the chosen basis. Vectors will be always indicated using bold letters while their components will be generally denoted by the same letter in normal typeface. Sometimes, when necessary for clarity, we will indicate the ith component of a vector $\mathbf{f}$ by $(\mathbf{f})_i$ or simply f_i. These definitions apply to vectors in $\mathbb{R}^2$ as well. With the term *domain* we will indicate an open, bounded, connected subset of $\mathbb{R}^N$, $N = 2, 3$, with orientable boundary. We will indicate with $\mathbf{n}$ the outwardly oriented unit vector normal to the boundary. We will also assume that the domain boundary be Lipschitz continuous (for instance, a piece-wise polynomial, or a C^1 curve). In Fig. 6.1 some admissible domains are shown. If a quantity f (like temperature or pressure) takes a scalar value on a domain Ω, we say that the quantity defines a *scalar field* on Ω, which we will indicate with $f : \Omega \to \mathbb{R}$. If instead a quantity $\mathbf{f}$ associates to each point in Ω a vector (as in the case of the velocity), we say that it defines a *vector field* on Ω, and we will indicate it with $\mathbf{f} : \Omega \to \mathbb{R}^3$. Finally, if a quantity $\mathbf{T}$ associates to each point in Ω a $\mathbb{R}^{N \times N}$ matrix, we will say that it defines a (second order) *tensor* field on Ω if it obeys the ordinary transformation rules for tensors (ARIS [1962]). Its components will be indicated by either $(\mathbf{T})_{ij}$, or simply T_{ij}, with $i, j = 1, \ldots, 3$.

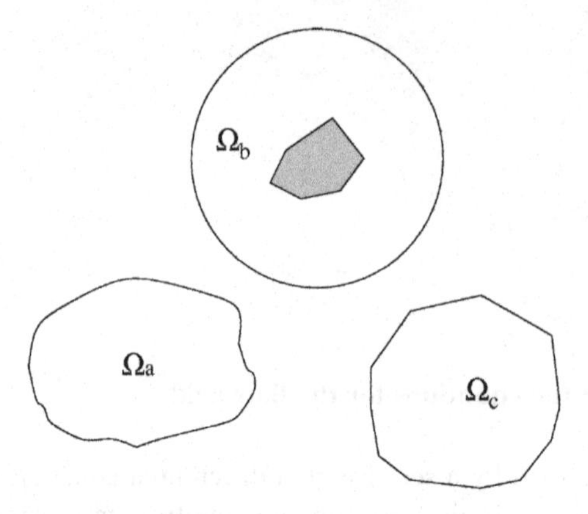

FIG. 6.1. Example of admissible domains. Ω_a has a boundary formed by piece-wise C^1 curves. Ω_b is a multi-connected domain, with a polygonal internal and a C^∞ external boundary. Finally, Ω_c has a polygonal boundary.

Given a function $f : \Omega \to \mathbb{R}$, $\mathbf{x} \to f(\mathbf{x})$, and a domain $V \subset \Omega$, we will use the shorthand notation

$$\int_V f$$

to indicate the integral

$$\int_V f(\mathbf{x})\,\mathrm{d}\mathbf{x},$$

and

$$\int_{\partial V} f$$

to indicate the surface (or line) integral

$$\int_{\partial V} f\,\mathrm{d}\sigma,$$

unless the context requires otherwise.

When referring to a physical quantity f, we will indicate with $[f]$ its measure units (in the international system). For instance, if $\mathbf{v}$ indicates a velocity, $[\mathbf{v}] = \mathrm{m/s}$, where m stands for meters and s for seconds.

7. The motion of continuous media

In order to derive the differential equations which govern the fluid motion, we need to introduce some kinematic concepts and quantities. The kinematics of a continuous medium studies the property of the motion of a medium which may be thought as continuously occupying, at each time, a portion of space. This allows the use of standard methods of analysis. We will set the derivation in $\mathbb{R}^3$, since this is the natural spatial

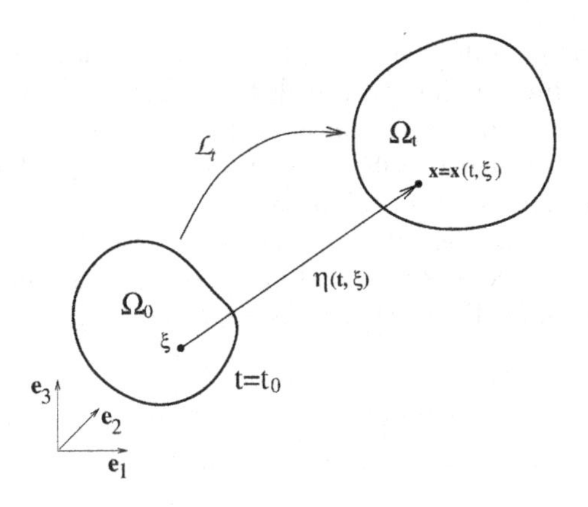

FIG. 7.1. The Lagrangian mapping.

dimension. However, the definitions and final differential equations are valid also in $\mathbb{R}^2$. Furthermore, we will assume that the motion will take place during a time interval $I = (t_0, t_1)$.

The motion itself is described by a family of mappings $\mathcal{L}_t$ which associate the position $\mathbf{x}$ of a fluid particle at time $t \in I$ to a point $\boldsymbol{\xi} \in \Omega_0$, Ω_0 being the domain occupied by the fluid at the reference initial time t_0. More precisely, we denote with Ω_t the portion of space occupied by the fluid at time t and we indicate with $\mathcal{L}_t$ the mapping

$$\mathcal{L}_t : \Omega_0 \to \Omega_t, \qquad \boldsymbol{\xi} \to \mathbf{x} = \mathbf{x}(t, \boldsymbol{\xi}) = \mathcal{L}_t(\boldsymbol{\xi}),$$

which will be denoted *Lagrangian* mapping at time t (see Fig. 7.1). We assume that $\mathcal{L}_t$ is continuous and invertible in $\overline{\Omega}_0$, with continuous inverse.

We call Ω_0 the *reference* configuration, while Ω_t is called *current* (or spatial) configuration. The position of the material particle located at the point $\mathbf{x}$ in the current configuration Ω_t is a function of time and of the position of the same material particle at the reference time.

We may thus relate the variables $(t, \mathbf{x})$ to $(t, \boldsymbol{\xi})$. The former couple is referred to as the *Eulerian* variables while the latter are called the *Lagrangian* variables.

It is worthwhile to point out that when using the Eulerian variables as independent variables, we are concentrating our attention on a position in space $\mathbf{x} \in \Omega_t$ and on the fluid particle which, at that particular time, is located at $\mathbf{x}$. When using the Lagrangian variables as independent variables (Lagrangian frame), we are instead targeting the fluid particle "labeled" $\boldsymbol{\xi}$ (that is the fluid particle which was located at position $\boldsymbol{\xi}$ at the reference time). That is, we are following the *trajectory* $T_{\boldsymbol{\xi}}$ of fluid particle $\boldsymbol{\xi} \in \Omega_0$, defined as

$$T_{\boldsymbol{\xi}} = \{(t, \mathbf{x}(t, \boldsymbol{\xi})) : t \in I\}. \tag{7.1}$$

The basic principles of mechanics are more easily formulated with reference to the moving particles, thus in the Lagrangian frame. Yet, in practice it is more convenient to work with the Eulerian variables. Therefore, we need to rewrite the equations stemming from those basic principles into the Eulerian frame. We will see later on that for the numerical approximation of the problem at hand it will be necessary to introduce yet another, intermediate, frame of reference, called *Arbitrary Lagrangian Eulerian.*

Being the mapping surjective, a quantity associated with the fluid may be described as function of either the Lagrangian or the Eulerian variables, depending on convenience. We will in general use the same symbol for the functions which describe the evolution of the same quantity in the Lagrangian and in the Eulerian frame, unless the context needs otherwise. In the latter case, we will mark with the hat symbol "^" a quantity expressed as function of the Lagrangian variables, that is, if $f : I \times \Omega_t \to \mathbb{R}$ we have the equality

$$\hat{f}(t, \boldsymbol{\xi}) = f(t, \mathbf{x}), \quad \text{with } \mathbf{x} = \mathcal{L}_t(\boldsymbol{\xi}).$$

We will often use the following alternative notation:

$$\hat{f} = f \circ \mathcal{L}_t \quad \text{or} \quad f = \hat{f} \circ \mathcal{L}_t^{-1}$$

with the understanding that the composition operator applies only to the spatial variables.

The symbol $\boldsymbol{\nabla}$ is used exclusively to indicate the gradient *with respect to the Eulerian variable* $\mathbf{x}$. When we need to indicate the gradient with respect to the Lagrangian variable $\boldsymbol{\xi}$, we will use the symbol $\boldsymbol{\nabla}_{\boldsymbol{\xi}}$, that is

$$\boldsymbol{\nabla}_{\boldsymbol{\xi}} \hat{f} = \sum_{i=1}^{3} \frac{\partial \hat{f}}{\partial \xi_i} \mathbf{e}_i .$$

The same convention applies to other spatial differential operators (divergence, Laplacian, etc.) as well.

In the following we will put $I \times \Omega_t = \{(t, \mathbf{x})\colon\ t \in I,\ x \in \Omega_t\}$ (note the little abuse of notation since technically it is not a Cartesian product).

7.1. *The velocity*

The *fluid velocity* is the major kinematic quantity of our problem. In the Lagrangian frame it is expressed by means of a vector field $\hat{\mathbf{u}} = \hat{\mathbf{u}}(t, \boldsymbol{\xi})$ defined as

$$\hat{\mathbf{u}} = \frac{\partial \mathbf{x}}{\partial t}, \quad \text{i.e.,} \quad \hat{\mathbf{u}}(t, \boldsymbol{\xi}) = \frac{\partial}{\partial t} \mathbf{x}(t, \boldsymbol{\xi}). \tag{7.2}$$

$\hat{\mathbf{u}}$ is called the *Lagrangian* velocity field (or velocity field in the Lagrangian frame), and it denotes the time derivative along the trajectory $T_{\boldsymbol{\xi}}$ of the fluid particle $\boldsymbol{\xi}$. The velocity $\mathbf{u}$ on the Eulerian frame is defined for $(t, \mathbf{x}) \in I \times \Omega_t$ as

$$\mathbf{u} = \hat{\mathbf{u}} \circ \mathcal{L}_t^{-1}, \quad \text{i.e.,} \quad \mathbf{u}(t, \mathbf{x}) = \hat{\mathbf{u}}\left(t, \mathcal{L}_t^{-1}(\mathbf{x})\right).$$

EXAMPLE 7.1. Let us consider a 2D case and the following movement law, for $t \geqslant 0$:

$$x_1 = \xi_1 e^t, \quad \xi_1 \in (-1, 1), \qquad x_2 = \xi_2, \quad \xi_2 \in (-1, 1).$$

The domain at time $t > 0$ occupies the rectangle $(-e^t, e^t) \times (-1, 1)$. The mapping is clearly invertible for all $t \geqslant 0$.

We have

$$\hat{u}_1 = \partial x_1/\partial t = \xi_1 e^t, \qquad \hat{u}_2 = \partial x_2/\partial t = 0.$$

We can immediately compute the velocity field as function of the Eulerian variable as

$$u_1 = x_1, \qquad u_2 = 0.$$

Once the velocity field and the reference configuration is known, the motion may be derived by solving the following Cauchy problem:

For any $\boldsymbol{\xi} \in \Omega_0$, find the function $\mathbf{x} = \mathbf{x}(t, \boldsymbol{\xi})$ which satisfies

$$\begin{cases} \dfrac{\partial \mathbf{x}}{\partial t}(t, \boldsymbol{\xi}) = \hat{\mathbf{u}}(t, \boldsymbol{\xi}), & \forall t \in I, \\ \mathbf{x}(t_0, \boldsymbol{\xi}) = \boldsymbol{\xi}. \end{cases}$$

7.2. The material derivative

We can relate time derivatives computed with respect to the different frames. The *material* (or *Lagrangian*) time derivative of a function f, which we will denote Df/Dt, is defined as the time derivative in the Lagrangian frame, yet expressed as function of the Eulerian variables.

That is, if $f : I \times \Omega_t \to \mathbb{R}$ and $\hat{f} = f \circ \mathcal{L}_t$,

$$\frac{Df}{Dt} : I \times \Omega_t \to \mathbb{R}, \qquad \frac{Df}{Dt}(t, \mathbf{x}) = \frac{\partial \hat{f}}{\partial t}(t, \boldsymbol{\xi}), \quad \boldsymbol{\xi} = \mathcal{L}_t^{-1}(\mathbf{x}). \tag{7.3}$$

Therefore, for any fixed $\boldsymbol{\xi} \in \Omega_0$ we may also write

$$\frac{Df}{Dt}(t, \mathbf{x}) = \frac{\mathrm{d}}{\mathrm{d}t} f(t, \mathbf{x}(t, \boldsymbol{\xi})),$$

by which we can observe that the material derivative represents the rate of variation of f along the trajectory $T_{\boldsymbol{\xi}}$.

By applying the chain-rule of derivation of composed functions, we have

$$\frac{Df}{Dt} = \frac{\partial f}{\partial t} + \mathbf{u} \cdot \nabla f. \tag{7.4}$$

Indeed,

$$\frac{Df}{Dt} = \left[\frac{\partial}{\partial t}(f \circ \mathcal{L}_t)\right] \circ \mathcal{L}_t^{-1} = \frac{\partial f}{\partial t} + \nabla f \cdot \left(\frac{\partial \mathbf{x}}{\partial t} \circ \mathcal{L}_t^{-1}\right) = \frac{\partial f}{\partial t} + \mathbf{u} \cdot \nabla f.$$

A quantity which satisfies

$$\frac{\partial f}{\partial t} = 0$$

is called *stationary*, and a motion for which

$$\frac{\partial \mathbf{u}}{\partial t} = 0$$

is said a *stationary motion*.

EXAMPLE 7.2. Let us consider again the motion of Example 7.1 and consider the function $f(x_1, x_2) = 3x_1 + x_2$ (which is independent of t). The application of relation (7.4) gives

$$\frac{Df}{Dt} = 0 + \begin{bmatrix} x_1 \\ 0 \end{bmatrix} \cdot \begin{bmatrix} 3 \\ 1 \end{bmatrix} = 3x_1.$$

On the other hand,

$$\hat{f} = 3\xi_1 e^t + \xi_2 \quad \text{and} \quad \frac{\partial \hat{f}}{\partial t} = 3\xi_1 e^t,$$

by which we deduce that,

$$\frac{\partial \hat{f}}{\partial t} \circ \mathcal{L}_t^{-1} = 3x_1.$$

This example, besides verifying relation (7.4), shows that a function $f = f(t, \mathbf{x})$ with $\partial f/\partial t = 0$ in general has $Df/Dt \neq 0$.

7.3. The acceleration

In the Lagrangian frame the acceleration is a vector field $\hat{\mathbf{a}} : I \times \Omega_0 \to \mathbb{R}^3$ defined as

$$\hat{\mathbf{a}} = \frac{\partial \hat{\mathbf{u}}}{\partial t} = \frac{\partial^2 \mathbf{x}}{\partial t^2}.$$

By recalling the definition of material derivative, we may write the acceleration in Eulerian frame as

$$\mathbf{a} = \frac{D\mathbf{u}}{Dt} = \frac{\partial \mathbf{u}}{\partial t} + (\boldsymbol{u} \cdot \nabla)\mathbf{u}. \tag{7.5}$$

Component-wise,

$$a_i = \frac{\partial u_i}{\partial t} + \sum_{j=1}^{3} u_j \frac{\partial u_i}{\partial x_j}. \tag{7.6}$$

7.4. The deformation gradient

Another kinematic quantity necessary for the derivation of the mathematical model is the *deformation gradient* $\widehat{\mathbf{F}}_t$, which is defined, for each $t \in I$, as

$$\widehat{\mathbf{F}}_t : \Omega_0 \to \mathbb{R}^{N \times N}, \quad \widehat{\mathbf{F}}_t = \nabla_m \mathcal{L}_t = \frac{\partial \mathbf{x}}{\partial \boldsymbol{\xi}}. \tag{7.7}$$

Component-wise,

$$(\widehat{\mathbf{F}}_t)_{ij} = \frac{\partial x_i}{\partial \xi_j}.$$

In particular, its determinant,

$$\widehat{J}_t = \det \widehat{\mathbf{F}}_t, \tag{7.8}$$

is called the Jacobian of the mapping $\mathcal{L}_t$. As usual, its counterpart in the Eulerian frame is indicated J_t.

It is possible to show that the time continuity and the invertibility of the Lagrangian mapping is sufficient to have, for all $t \in I$,

$$\widehat{J}_t(\boldsymbol{\xi}) > 0 \quad \forall \boldsymbol{\xi} \in \Omega_0. \tag{7.9}$$

The importance of J_t is clearly linked to the rule which transforms integrals from the current to the reference configuration. We recall the following theorem of elementary calculus (without providing its proof).

THEOREM 7.1. *Let $V_t \subset \Omega_t$ be a subdomain of Ω_t and let us consider the function $f : I \times V_t \to \mathbb{R}$. Then f is integrable on V_t if and only if $(f \circ \mathcal{L}_t) J_t$ is integrable on $V_0 = \mathcal{L}_t^{-1}(V_t)$, and*

$$\int_{V_t} f(t, \mathbf{x})\, \mathrm{d}\mathbf{x} = \int_{V_0} \hat{f}(t, \boldsymbol{\xi})\, \widehat{J}_t(\boldsymbol{\xi})\, \mathrm{d}\boldsymbol{\xi},$$

where $\hat{f}(t, \boldsymbol{\xi}) = f(t, \mathcal{L}_t(\boldsymbol{\xi}))$. In short,

$$\int_{V_t} f = \int_{V_0} \hat{f} \widehat{J}_t.$$

7.5. The Reynolds transport theorem

An interesting property of the Jacobian is that its time derivative is linked to the divergence of the velocity field.

LEMMA 7.1. *Let J_t denote the Jacobian* (7.8) *in the Eulerian frame. Then*

$$\frac{D}{Dt} J_t = J_t \operatorname{div} \mathbf{u}. \tag{7.10}$$

This relation is sometimes called Euler expansion formula.

PROOF. We have, by direct application of the chain-rule,

$$\nabla_{\boldsymbol{\xi}} \hat{\mathbf{u}} = \nabla_{\boldsymbol{\xi}} (\mathbf{u} \circ \mathcal{L}_t) = \widehat{\nabla \mathbf{u}}\, \nabla_{\boldsymbol{\xi}} \mathcal{L}_t = \widehat{\nabla \mathbf{u}}\, \widehat{\mathbf{F}}_t.$$

On the other hand, by recalling the definition of the velocity (7.2),

$$\nabla_{\boldsymbol{\xi}} \hat{\mathbf{u}} = \nabla_{\boldsymbol{\xi}} \left(\frac{\partial \mathbf{x}}{\partial t} \right) = \frac{\partial}{\partial t} \nabla_{\boldsymbol{\xi}} \mathbf{x} = \frac{\partial \widehat{\mathbf{F}}_t}{\partial t}.$$

Thus, we may write

$$\widehat{\mathbf{F}}_{t+\varepsilon} = \widehat{\mathbf{F}}_t + \varepsilon \frac{\partial \widehat{\mathbf{F}}_t}{\partial t} + \mathbf{o}(\varepsilon) = \widehat{\mathbf{F}}_t + \varepsilon \widehat{\nabla \mathbf{u}}\, \widehat{\mathbf{F}}_t + \mathbf{o}(\varepsilon) = (\mathbf{I} + \varepsilon \widehat{\nabla \mathbf{u}})\, \widehat{\mathbf{F}}_t + \mathbf{o}(\varepsilon).$$

We now exploit the well-known result that for any matrix **A**,

$$\det(\mathbf{I}+\varepsilon\mathbf{A}) = 1+\varepsilon\,\mathrm{tr}\,\mathbf{A}+o(\varepsilon),$$

where $\mathrm{tr}\,\mathbf{A} = \sum_i A_{ii}$ denotes the trace of the matrix A, to write

$$\widehat{J}_{t+\varepsilon} = \det(\widehat{\mathbf{F}}_{t+\varepsilon}) = (1+\varepsilon\,\mathrm{tr}\,\widehat{\nabla}\mathbf{u})\widehat{J}_t + o(\varepsilon) = (1+\varepsilon\,\mathrm{div}\,\hat{\mathbf{u}})\widehat{J}_t + o(\varepsilon).$$

We have used the identity $\mathrm{tr}\,\widehat{\nabla\mathbf{u}} = \mathrm{div}\,\hat{\mathbf{u}}$. Then, by applying the definition of material derivative and exploiting the continuity of the Lagrangian mapping, we may write

$$\frac{DJ_t}{Dt} = \left(\lim_{\varepsilon\to 0}\frac{\widehat{J}_{t+\varepsilon}-\widehat{J}_t}{\varepsilon}\right)\circ\mathcal{L}_t^{-1} = (\mathrm{div}\,\hat{\mathbf{u}}\,\widehat{J}_t)\circ\mathcal{L}_t^{-1} = \mathrm{div}\,\mathbf{u}\,J_t. \qquad \square$$

EXAMPLE 7.3. For the movement law given by Example 7.1, we have

$$\widehat{J}_t = \det\begin{bmatrix} e^t & 0\\ 0 & 1\end{bmatrix} = e^t$$

and $J_t = e^t$ as well. We may verify directly relation (7.10) since

$$J_t\,\mathrm{div}\,\mathbf{u} = e^t(1+0) = e^t = \frac{\mathrm{d}}{\mathrm{d}t}\widehat{J}_t = (\text{by relation (7.3)}) = \frac{D}{Dt}J_t.$$

We have now the following fundamental result.

THEOREM 7.2 (Reynolds transport theorem). *Let $V_0 \subset \Omega_0$, and $V_t \subset \Omega_t$ be its image under the mapping $\mathcal{L}_t$. Let $f : I \times \Omega_t \to \mathbb{R}$ be a continuously differentiable function with respect to both variables* **x** *and t. Then,*

$$\frac{\mathrm{d}}{\mathrm{d}t}\int_{V_t} f = \int_{V_t}\left(\frac{Df}{Dt} + f\,\mathrm{div}\,\mathbf{u}\right) = \int_{V_t}\left(\frac{\partial f}{\partial t} + \mathrm{div}\,(f\mathbf{u})\right). \tag{7.11}$$

PROOF. Thanks to Theorem 7.1 and relations (7.10) and (7.3), we have

$$\begin{aligned}\frac{\mathrm{d}}{\mathrm{d}t}\int_{V_t} f(t,\mathbf{x})\,\mathrm{d}\mathbf{x} &= \frac{\mathrm{d}}{\mathrm{d}t}\int_{V_0}\hat{f}(t,\boldsymbol{\xi})\widehat{J}_t(\boldsymbol{\xi})\,\mathrm{d}\boldsymbol{\xi} = \int_{V_0}\frac{\partial}{\partial t}\left[\hat{f}(t,\boldsymbol{\xi})\widehat{J}_t(\boldsymbol{\xi})\right]\mathrm{d}\boldsymbol{\xi}\\ &= \int_{V_0}\left[\frac{\partial\hat{f}}{\partial t}(t,\boldsymbol{\xi})\widehat{J}_t(\boldsymbol{\xi}) + \hat{f}(t,\boldsymbol{\xi})\frac{\partial}{\partial t}\widehat{J}_t(\boldsymbol{\xi})\right]\mathrm{d}\boldsymbol{\xi}.\end{aligned}$$

We now use Theorem 7.1 and the definition of material derivative (7.3) to write

$$\int_{V_0}\frac{\partial\hat{f}}{\partial t}(t,\boldsymbol{\xi})\widehat{J}_t(\boldsymbol{\xi})\,\mathrm{d}\boldsymbol{\xi} = \int_{V_t}\frac{Df}{Dt}(t,\mathbf{x})\,\mathrm{d}\mathbf{x}.$$

Furthermore, we exploit again the definition of material derivative (7.3) in order to rewrite relation (7.10) in the following equivalent form:

$$\frac{\partial}{\partial t}\widehat{J}_t(\boldsymbol{\xi}) = \widehat{J}_t(\boldsymbol{\xi})\,\mathrm{div}\,\mathbf{u}\,(t,\mathbf{x}(t,\boldsymbol{\xi})).$$

Consequently,

$$\begin{aligned}\frac{\mathrm{d}}{\mathrm{d}t}\int_{V_t} f(t,\mathbf{x})\,\mathrm{d}\mathbf{x} &= \int_{V_t} \frac{Df}{Dt}(t,\mathbf{x})\,\mathrm{d}\mathbf{x} + \int_{V_0} \hat{f}(t,\boldsymbol{\xi}) J_t\,(\mathbf{x}(t,\boldsymbol{\xi}))\ \mathrm{div}\,\mathbf{u}\,(t,\mathbf{x}(t,\boldsymbol{\xi}))\ \mathrm{d}\boldsymbol{\xi}\\ &= \int_{V_t} \frac{Df}{Dt}(t,\mathbf{x})\,\mathrm{d}\mathbf{x} + \int_{V_t} f(t,\mathbf{x})\,\mathrm{div}\,\mathbf{u}(t,\mathbf{x})\,\mathrm{d}\mathbf{x}\\ &= \int_{V_t} \frac{Df}{Dt}(t,\mathbf{x}) + f(t,\mathbf{x})\,\mathrm{div}\,\mathbf{u}(t,\mathbf{x})\,\mathrm{d}\mathbf{x}.\end{aligned}$$

The second equality in (7.11) is a consequence of (7.4). □

Relation (7.11) is given the name of *Reynolds transport formula*, or simply *transport formula* (sometimes the name *convection formula* is used as well).

By the application of the divergence theorem the previous expression becomes

$$\frac{\mathrm{d}}{\mathrm{d}t}\int_{V_t} f = \int_{V_t} \frac{\partial f}{\partial t} + \int_{\partial V_t} f\mathbf{u}\cdot\mathbf{n}.$$

8. The derivation of the basic equations of fluid mechanics

In the sequel, the symbol V_t will always be used to indicate a *material volume* at time t, i.e., V_t is the image under the Lagrangian mapping of a subdomain $V_0 \subset \Omega_0$, i.e., $V_t = \mathcal{L}_t(V_0)$ (as already done in Theorem 7.2).

8.1. *Continuity equation or mass conservation*

We assume that there exists a strictly positive, measurable function $\rho : I \times \Omega_t \to \mathbb{R}$, called *density* such that on each $V_t \subset \Omega_t$,

$$\int_{V_t} \rho = m(V_t),$$

where $m(V_t)$ is the mass of the material contained in V_t. The density ρ has dimensions $[\rho] = \mathrm{kg/m^3}$.

A fundamental principle of classical mechanics, called principle of *mass conservation*, states that mass is neither created nor destroyed during the motion. This principle translates into the following mathematical statement.

Given any material volume $V_t \subset \Omega_t$ the following equality holds:

$$\frac{\mathrm{d}}{\mathrm{d}t}\int_{V_t} \rho = 0.$$

We can apply the transport theorem, obtaining

$$\int_{V_t} \left(\frac{D\rho}{Dt} + \rho\,\mathrm{div}\,\mathbf{u}\right) = 0. \tag{8.1}$$

By assuming that the terms under the integral are continuous, the arbitrariness of V_t allows us to write the *continuity equation* in differential form

$$\frac{\partial \rho}{\partial t} + \operatorname{div} \rho \mathbf{u} = 0.$$

In these cases for which we can make the assumption that ρ is constant (like for blood flow), we obtain

$$\operatorname{div} \mathbf{u} = 0. \tag{8.2}$$

Relation (8.2), which has been derived from the continuity equation in the case of a *constant density* fluid (sometimes also called incompressible fluid), is indeed a kinematic constraint. Thanks to (7.10), relation (8.2) is equivalent to

$$\frac{D}{Dt} J_t = 0, \tag{8.3}$$

which is the *incompressibility constraint*. A flow which satisfies the incompressibility constraint is called *incompressible*. By the continuity equation, we derive the following implication:

$$\text{constant density fluid} \quad \Rightarrow \quad \text{incompressible flow},$$

whereas the converse is not true in general.

By employing the transport formula (7.11) with $f = 1$, we may note that the incompressibility constraint is equivalent to

$$\frac{\mathrm{d}}{\mathrm{d}t} \int_{V_t} \mathrm{d}\mathbf{x} = 0 \quad \forall V_t \subset \Omega_t,$$

which means that the only possible motions of an incompressible flow are those which preserve the fluid volume.

8.2. The momentum equation

Another important principle allows the derivation of an additional set of differential equations, that is the principle of *conservation of momentum*. It is an extension of the famous Newton law, "force = mass × acceleration", to a continuous medium.

REMARK 8.1. In the dimension unit specifications we will use the symbol Ne to indicate the *Newtons* (the dimension units of a force), $\mathrm{Ne} = \mathrm{kg\,m/s^2}$, instead of the more standard symbol N, since we have used the latter to indicate the number of space dimensions.

Three different types of forces may be acting on the material inside Ω_t:

- *Body forces*. These forces are proportional to the mass. They are normally represented by introducing a vector field $\mathbf{f}^b : I \times \Omega_t \to \mathbb{R}^3$, called *specific body force*, whose dimension unit, $[\mathbf{f}^b] = \mathrm{Ne/kg} = \mathrm{m/s^2}$, is that of an acceleration. The body force acting on a volume V_t is given by

$$\int_{V_t} \rho \mathbf{f}^b,$$

whose dimension unit is clearly Ne. An example is the gravity force, given by $\mathbf{f}^b = -g\mathbf{e}_3$, where $\mathbf{e}_3$ represents the vertical direction and g the gravitational acceleration.

- *Applied surface forces.* They represent that part of the forces which are imposed on the medium through its surface. We will assume that they may be represented through a vector field $\mathbf{t}^e : I \times \Gamma_t^n \to \mathbb{R}^3$, called *applied stresses*, defined on a measurable subset of the domain boundary $\Gamma_t^n \subset \partial\Omega_t$ and with dimension unit $[\mathbf{t}^e] = \text{Ne/m}^2$. The resultant force acting through the surface is then given by

$$\int_{\Gamma_t^n} \mathbf{t}^e.$$

An example of a surface stress is that caused by the friction of the air flowing over the surface of a lake.

- *Internal "continuity" forces.* These are the forces that the continuum media particles exert on each other and are responsible for maintaining material continuity during the movement. To model these forces let us recall the following principle, due to Cauchy.

8.2.1. The Cauchy principle

There exists a vector field $\mathbf{t}$, called *Cauchy stress*,

$$\mathbf{t} : I \times \Omega_t \times \mathbf{S}_1 \to \mathbb{R}^3$$

with

$$\mathbf{S}_1 = \left\{ \mathbf{n} \in \mathbb{R}^3 : \ |\mathbf{n}| = 1 \right\}$$

such that its integral on the surface of any material domain $V_t \subset \Omega_t$, given by

$$\int_{\partial V_t} \mathbf{t}(t, \mathbf{x}, \mathbf{n})\, \mathrm{d}\sigma \tag{8.4}$$

is *equivalent* to the resultant of the material continuity forces acting on V_t. In (8.4), $\mathbf{n}$ indicates the outward normal of ∂V_t.

Furthermore, we have that

$$\mathbf{t} = \mathbf{t}^e \quad \text{on } \partial V_t \cap \Gamma_t^n.$$

This principle is of fundamental importance because it states that the only dependence of the internal forces on the geometry of ∂V_t is through $\mathbf{n}$.

We may now state the following *principle of conservation of linear momentum.*

For any $t \in I$, on any sub-domain $V_t \subset \Omega_t$ *completely contained in* Ω_t, the following relation holds:

$$\frac{\mathrm{d}}{\mathrm{d}t}\int_{V_t} \rho(t, \mathbf{x})\mathbf{u}(t, \mathbf{x})\, \mathrm{d}\mathbf{x} = \int_{V_t} \rho(t, \mathbf{x})\mathbf{f}^b(t, \mathbf{x})\, \mathrm{d}\mathbf{x} + \int_{\partial V_t} \mathbf{t}(t, \mathbf{x}, \mathbf{n})\, \mathrm{d}\sigma, \tag{8.5}$$

where all terms dimension unit is Ne. Relation (8.5) expresses the property that the variation of the *linear momentum* of V_t (represented by the integral at the left-hand side) is balanced by the resultant of the internal and body forces.

With some further assumptions on the regularity of the Cauchy stresses, we are now able to relate the internal continuity forces to a *tensor field*, as follows.

THEOREM 8.1 (Cauchy stress tensor theorem). *Let us assume that* $\forall t \in I$, *the body forces* $\mathbf{f}^b$, *the density* ρ *and* $(D/Dt)\mathbf{u}$ *are all bounded functions on* Ω_t *and that the Cauchy stress vector field* $\mathbf{t}$ *is continuously differentiable with respect to the variable* $\mathbf{x}$ *for each* $\mathbf{n} \in \mathbf{S}_1$, *and continuous with respect to* $\mathbf{n}$. *Then, there exists a continuously differentiable symmetric*[1] *tensor field, called* Cauchy stress tensor,

$$\mathbf{T} : I \times \overline{\Omega_t} \to \mathbb{R}^{3\times 3}, \quad [\mathbf{T}] = \text{Ne/m}^2,$$

such that

$$\mathbf{t}(t, \mathbf{x}, \mathbf{n}) = \mathbf{T}(t, \mathbf{x}) \cdot \mathbf{n}, \quad \forall t \in I,\ \forall \mathbf{x} \in \Omega_t,\ \forall \mathbf{n} \in \mathbf{S}_1.$$

The proof is omitted. The interested reader may refer to ARIS [1962], SERRIN [1959]. Therefore, under the hypotheses of the Cauchy theorem, we have

$$\mathbf{T} \cdot \mathbf{n} = \mathbf{t}^e, \quad \text{on } \partial V_t \cap \Gamma_t^n, \tag{8.6}$$

and that the resultant of the internal forces on V_t is expressed by

$$\int_{\partial V_t} \mathbf{T} \cdot \mathbf{n}, \tag{8.7}$$

and we may rewrite the principle of conservation of linear momentum (8.5) as follows.

For all $t \in I$, on any sub-domain $V_t \subset \Omega_t$ *completely contained in* Ω_t, the following relation holds:

$$\frac{\mathrm{d}}{\mathrm{d}t} \int_{V_t} \rho \mathbf{u} = \int_{V_t} \rho \mathbf{f}^b + \int_{\partial V_t} \mathbf{T} \cdot \mathbf{n}. \tag{8.8}$$

Since ρ is constant and $\operatorname{div} \mathbf{u} = 0$, by invoking the transport formula (7.11), we obtain

$$\frac{\mathrm{d}}{\mathrm{d}t} \int_{V_t} \rho \mathbf{u} = \int_{V_t} \left(\frac{D}{Dt}(\rho \mathbf{u}) + \rho \mathbf{u} \operatorname{div} \mathbf{u} \right) = \int_{V_t} \rho \frac{D\mathbf{u}}{Dt}.$$

By using the divergence theorem and assuming that $\operatorname{div} \mathbf{T}$ is integrable, relation (8.8) becomes

$$\int_{V_t} \left[\rho \frac{D\mathbf{u}}{Dt} - \mathbf{div}\, \mathbf{T} - \rho \mathbf{f}^b \right] = 0.$$

Thanks to the arbitrariness of V_t and under the hypothesis that the terms under the integrals are continuous in space, we derive the following differential equation:

$$\rho \frac{D\mathbf{u}}{Dt} - \mathbf{div}\, \mathbf{T} = \rho \mathbf{f}^b \quad \text{in } \Omega_t. \tag{8.9}$$

[1] The symmetry of the Cauchy tensor may indeed be derived from the conservation of angular momentum.

REMARK 8.2. In deriving (8.9), we have assumed that V_t is completely contained into Ω_t. We may however extend the derivation to the case where V_t has a part of boundary in common with Γ_t^n. In that case, we should use in place of (8.8) the following:

$$\frac{\mathrm{d}}{\mathrm{d}t}\int_{V_t}\rho\mathbf{u}=\int_{\partial V_t\setminus\Gamma_t^n}\mathbf{T}\cdot\mathbf{n}+\int_{\partial V_t\cap\Gamma_t^n}\mathbf{t}^e+\int_{V_t}\rho\mathbf{f}^b. \tag{8.10}$$

Even now we would re-obtain (8.9) in view of property (8.6) of the Cauchy stress tensor, which should now be regarded as *boundary condition*.

We may note that $D\mathbf{u}/Dt$ is indeed the fluid acceleration. Referring to relation (7.5), it may be written as

$$\frac{D\mathbf{u}}{Dt}=\frac{\partial\mathbf{u}}{\partial t}+(\mathbf{u}\cdot\nabla)\mathbf{u},$$

where $(\mathbf{u}\cdot\nabla)\mathbf{u}$ is a vector whose components are

$$((\mathbf{u}\cdot\nabla)\mathbf{u})_i=\sum_{j=1}^{3}u_j\frac{\partial u_i}{\partial x_j},\quad i=1,\ldots,3.$$

For ease of notation, from now on we will omit the subscript b to indicate the body force density applied to the fluid, which will be indicated just as $\mathbf{f}$.

Relation (8.9) may finally be written as

$$\rho\frac{\partial\mathbf{u}}{\partial t}+\rho(\mathbf{u}\cdot\nabla)\mathbf{u}-\mathbf{div}\,\mathbf{T}=\rho\mathbf{f}. \tag{8.11}$$

Component-wise,

$$\rho\frac{\partial u_i}{\partial t}+\rho\sum_{j=1}^{3}u_j\frac{\partial u_i}{\partial x_j}-\sum_{j=1}^{3}\frac{\partial T_{ij}}{\partial x_j}=\rho f_i^b,\quad i=1,\ldots,3.$$

The non-linear term $\rho(\mathbf{u}\cdot\nabla)\mathbf{u}$ is called the *convective term*.

REMARK 8.3. We note the convective term may be written in the so-called *divergence form* $\mathbf{div}\,(\mathbf{u}\otimes\mathbf{u})$, where

$$(\mathbf{div}\,\mathbf{u}\otimes\mathbf{u})_i=\sum_{j=1}^{3}\frac{\partial}{\partial x_j}(u_i u_j),\quad i=1,\ldots,3.$$

Indeed, thanks to the incompressibility of the fluid,

$$(\mathbf{u}\cdot\nabla)\mathbf{u}=(\mathbf{u}\cdot\nabla)\mathbf{u}+\mathbf{u}\,\mathrm{div}\,\mathbf{u}=\mathbf{div}\,(\mathbf{u}\otimes\mathbf{u}).$$

The momentum equation in divergence form is then

$$\rho\frac{\partial\mathbf{u}}{\partial t}+\rho\,\mathbf{div}\,(\mathbf{u}\otimes\mathbf{u}-\mathbf{T})=\rho\mathbf{f}. \tag{8.12}$$

8.3. The constitutive law

In order to close the system of Eqs. (8.2) and (8.11) just derived, we need to link the Cauchy stress tensor to the kinematic quantities, and in particular, the velocity field. Such a relation, called *constitutive law*, provides a characterization of the mechanical behavior of the particular fluid under consideration.

The branch of science which studies the behavior of a moving fluid and in particular the relation between stresses and kinematic quantities is called *rheology*. We have already anticipated in the introduction that blood rheology could be complex, particularly in vessels with small size.

Here, we will assume for the fluid a *Newtonian behavior* (an approximation valid for many fluids and also for blood flow in large vessels, which is the case in our presentation). *In a Newtonian incompressible fluid*, the Cauchy stress tensor may be written as a linear function of the velocity derivatives (SERRIN [1959]), according to

$$\mathbf{T} = -P\mathbf{I} + \mu\left(\nabla\mathbf{u} + \nabla\mathbf{u}^{\mathrm{T}}\right), \tag{8.13}$$

where P is a scalar function called *pressure*, $\mathbf{I}$ is the identity matrix, μ is the *dynamic viscosity* of the fluid and is a positive quantity. The tensor

$$\mathbf{D}(\mathbf{u}) = \frac{(\nabla\mathbf{u} + \nabla\mathbf{u}^{\mathrm{T}})}{2}, \quad D_{ij} = \frac{1}{2}\left(\frac{\partial u_i}{\partial x_j} + \frac{\partial u_j}{\partial x_i}\right), \ i = 1,\ldots,3, \ j = 1,\ldots,3,$$

is called the *strain rate* tensor. Then,

$$\mathbf{T} = -P\mathbf{I} + 2\mu\mathbf{D}(\mathbf{u}).$$

The term $2\mu\mathbf{D}(\mathbf{u})$ in the definition of the Cauchy stress tensor is often referred to as *viscous stress* component of the stress tensor. We have that $[P] = \mathrm{Ne/m^2}$ and $[\mu] = \mathrm{kg/m\,s}$. The viscosity may vary with respect to time and space. For example, it may depend on the fluid temperature. The assumption of Newtonian fluid, however, implies that μ is *independent from kinematic quantities*. Simple models for non-Newtonian fluids, often used for blood flow simulations, express the viscosity as function of the strain rate, that is $\mu = \mu(\mathbf{D}(\mathbf{u}))$. The treatment of such cases is rather complex and will not be considered here, the interested reader may consult, for instance, RAJAGOPAL [1993], COKELET [1987].

We now recall that, if P is a scalar and $\boldsymbol{\Sigma}$ a vector field, then

$$\mathbf{div}\,(P\boldsymbol{\Sigma}) = \nabla P\boldsymbol{\Sigma} + P\,\mathbf{div}\,\boldsymbol{\Sigma},$$

and, therefore,

$$\mathbf{div}\,(P\mathbf{I}) = \nabla P\mathbf{I} + P\,\mathbf{div}\,\mathbf{I} = \nabla P.$$

The momentum equation may then be written as

$$\rho\frac{\partial\mathbf{u}}{\partial t} + \rho(\mathbf{u}\cdot\nabla)\mathbf{u} + \nabla P - 2\,\mathbf{div}(\mu\mathbf{D}(\mathbf{u})) = \rho\mathbf{f}.$$

Since ρ is constant, it is sometimes convenient to introduce the kinematic viscosity $\nu = \mu/\rho$, with $[\nu] = \mathrm{m}^2/\mathrm{s}$, and to write

$$\frac{\partial \mathbf{u}}{\partial t} + (\mathbf{u} \cdot \boldsymbol{\nabla})\mathbf{u} + \boldsymbol{\nabla} p - 2\,\mathbf{div}(\nu \mathbf{D}(\mathbf{u})) = \mathbf{f}, \tag{8.14}$$

where $p = P/\rho$ is a scaled pressure (with $[p] = \mathrm{m}^2/\mathrm{s}^2$).

REMARK 8.4. Under the additional hypothesis that ν is constant, the momentum equation may be further elaborated by considering that

$$\mathbf{div}\,\boldsymbol{\nabla}\mathbf{u} = \Delta\mathbf{u},$$
$$\mathbf{div}\,\boldsymbol{\nabla}\mathbf{u}^{\mathrm{T}} = \boldsymbol{\nabla}(\mathrm{div}\,\mathbf{u}) = (\text{by relation (8.2)}) = \mathbf{0}.$$

Consequently, the momentum equation for an incompressible Newtonian fluid with constant viscosity may be written in the alternative form

$$\frac{\partial \mathbf{u}}{\partial t} + (\mathbf{u} \cdot \boldsymbol{\nabla})\mathbf{u} + \boldsymbol{\nabla} p - \nu\Delta\mathbf{u} = \mathbf{f}. \tag{8.15}$$

However, for reasons that will appear clear later on (and that have to see with the different natural boundary conditions associated with the two formulations), we prefer to use the Navier–Stokes equations in the form (8.14), even when considering a constant viscosity.

9. The Navier–Stokes equations

The set of differential equations formed by the continuity equation and the momentum equations in the form derived in the previous section provides the *Navier–Stokes equations* for incompressible fluids.

They are, in particular, valid on any fixed spatial domain Ω which is for all times of interest inside the portion of space filled by the fluid, i.e., $\Omega \subset \Omega_t$. Indeed, in most cases, as with the flow around a car or an aeroplane, the flow motion is studied in a fixed domain Ω (usually called computational domain) embodying the region of interest. We will see in Section 18 that this is not possible anymore when considering the fluid–structure interaction problem arising when blood is flowing in a large artery.

Yet, before addressing this more complex situation, we will analyze the Navier–Stokes equations in a fixed domain, that is, we will consider, for any $t \in I$, the system of equations

$$\begin{aligned} &\frac{\partial \mathbf{u}}{\partial t} + (\mathbf{u} \cdot \boldsymbol{\nabla})\mathbf{u} + \boldsymbol{\nabla} p - 2\,\mathbf{div}(\nu \mathbf{D}(\mathbf{u})) = \mathbf{f}, \quad \text{in } \Omega, \\ &\mathrm{div}\,\mathbf{u} = 0, \quad \text{in } \Omega. \end{aligned} \tag{9.1}$$

Furthermore, we need to prescribe the initial status of the fluid velocity, for instance

$$\mathbf{u}(t = t_0, \mathbf{x}) = \mathbf{u}_0(\mathbf{x}), \quad \mathbf{x} \in \Omega. \tag{9.2}$$

The principal unknowns are the velocity $\mathbf{u}$ and the "scaled" pressure $p = P/\rho$.

Let us take a practical case-study, namely the blood flow in an artery, for example the carotid (ref. Fig. 2.3), which we will here consider rigid. We proceed by identifying the area of interest, which may be the carotid sinus, and a domain Ω which will contain that area and which extends into the vessels up to a certain distance. For obvious practical reasons we will need to "truncate" the domain at certain sections. Inside such domain, the Navier–Stokes equations are valid, yet in order to solve them we need to provide appropriate boundary conditions.

9.1. *Boundary conditions for the Navier–Stokes equations*

The Navier–Stokes equations must be supplemented by proper boundary conditions that allow the determination of the velocity field up to the boundary of the computational domain Ω. The more classical boundary conditions which are mathematically compatible with the Navier–Stokes equations are:

(1) *Applied stresses* (or *Neumann* boundary condition). We have already faced this condition when discussing the Cauchy principle. With the current definition for the Cauchy stresses it becomes

$$\mathbf{T}\cdot\mathbf{n} = -P\mathbf{n} + 2\mu\mathbf{D}(\mathbf{u})\cdot\mathbf{n} = \mathbf{t}^e \quad \text{on } \Gamma^n \subset \partial\Omega, \tag{9.3}$$

where Γ^n is a measurable subset (possibly empty) of the whole boundary $\partial\Omega$.

(2) *Prescribed velocity* (or *Dirichlet* boundary condition). A given velocity field is imposed on Γ^d, a measurable subset of $\partial\Omega$ (which may be empty). This means that a vector field

$$\mathbf{g} : I \times \Gamma^d \to \mathbb{R}^3$$

is prescribed and we impose that

$$\mathbf{u} = \mathbf{g} \quad \text{on } \Gamma^d.$$

Since $\operatorname{div}\mathbf{u} = 0$ in Ω, it must be noted that if $\Gamma^d = \partial\Omega$ then at any time $\mathbf{g}$ must satisfy the following compatibility condition:

$$\int_{\partial\Omega} \mathbf{g}\cdot\mathbf{n} = 0. \tag{9.4}$$

Clearly, for a proper boundary conditions specification we must have $\Gamma^n \cup \Gamma^d = \partial\Omega$.

The conditions to apply are normally driven by physical considerations. For instance, for a viscous fluid ($\mu > 0$) like the one we are considering here, physical considerations lead to impose the homogeneous Dirichlet condition $\mathbf{u} = \mathbf{0}$ at a solid fixed boundary. When dealing with an "artificial boundary", that is a boundary which truncates the space occupied by the fluid (for computational reasons) the choice of appropriate conditions is often more delicate and should in any case guarantee the well-posedness of the resulting differential problem.

For example, for the flow field inside a 2D model for the carotid artery such as the one shown in Fig. 9.1, we could impose a Dirichlet boundary condition on Γ^{in}, by prescribing a velocity field $\mathbf{g}$.

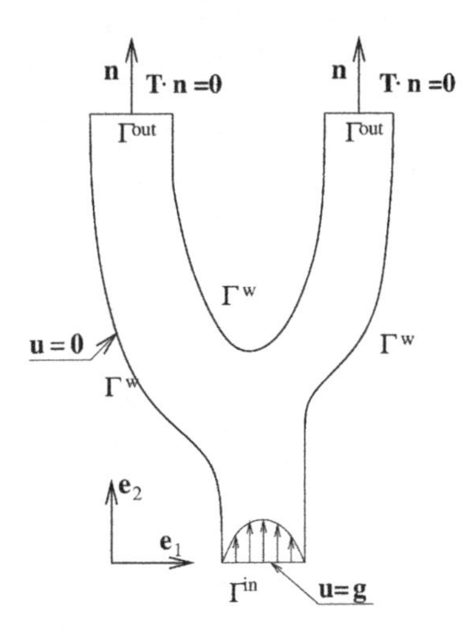

FIG. 9.1. A possible boundary subdivision for the flow in a carotid bifurcation.

On the "wall" boundary Γ^w, which is in this case assumed to be fixed, we will impose homogeneous Dirichlet conditions, that is $\mathbf{u} = \mathbf{0}$ on Γ^w. When we will consider the coupled problem between fluid and vessel wall, Γ^w will be moving, hence the homogeneous Dirichlet condition will be replaced by $\mathbf{u} = \mathbf{w}$, where $\mathbf{w}$ is the wall velocity.

At the exit Γ^{out}, we could, for instance, impose homogeneous Neumann conditions, i.e., relation (9.3) with $\mathbf{t}^e = \mathbf{0}$. For the case illustrated in Fig. 9.1 and with that choice of coordinate basis, it becomes (derivation left as exercise)

$$\mu\left(\frac{\partial u_1}{\partial x_2} + \frac{\partial u_2}{\partial x_1}\right) = 0, \qquad -P + 2\mu\frac{\partial u_2}{\partial x_2} = 0.$$

REMARK 9.1. We anticipate the fact (without providing the proof) that this choice of boundary conditions, with the hypothesis that at Γ^{out} the velocity satisfies everywhere the condition $\mathbf{u} \cdot \mathbf{n} > 0$, is sufficient to guarantee that the solution of the Navier–Stokes problem exists and is continuously dependent from the data (initial solution, boundary conditions, forcing terms), provided that the initial data and forcing term are sufficiently small.

Unfortunately, the homogeneous Neumann condition, which indeed would simulate a discharge into the open air, is rather unphysical for the case of a human vessel. As a matter of fact, it neglects completely the presence of the remaining part of the circulatory system. The difficulty in devising proper boundary conditions for this specific problem was already mentioned in Section 2 of these notes. The matter is still open and is the subject of active research. A possibility is provided by coupling the Navier–Stokes equations on the section of the arterial tree of interest with reduced models, like the one that will be presented in Section 20, which are able to represent, though in a simplified way, the presence of the remaining part of the circulatory system. Techniques of this type has been used and analyzed in FORMAGGIA, NOBILE, QUARTERONI and VENEZIANI [1999], FORMAGGIA, GERBEAU, NOBILE and QUARTERONI [2001].

CHAPTER III

10. The incompressible Navier–Stokes equations and their approximation

In this section we introduce the weak formulation of the Navier–Stokes equations for constant density (incompressible) fluids. Then, we address basic issues concerning the approximation of these equations in the context of the finite element method.

10.1. Some functional spaces

For the following discussion we need to introduce some Sobolev spaces for vector functions. We assume that the reader is already acquainted with the main definitions and results on Sobolev spaces in one dimension. A simple introduction is provided in REDDY [1998]. For a deeper insight see, for instance, BREZIS [1983].

We will indicate with $\mathbf{L}^p(\Omega)$ $(1 \leqslant p \leqslant \infty)$ the space of vector functions $\mathbf{f}: \Omega \to \mathbb{R}^N$ (with $N = 2$ or 3) whose components belong to $L^p(\Omega)$. Its norm is

$$\|\mathbf{f}\|_{\mathbf{L}^p(\Omega)} = \left(\sum_{i=1}^N \|f_i\|^p_{L^p(\Omega)}\right)^{1/p}, \quad 1 \leqslant p < \infty,$$

and

$$\|\mathbf{f}\|_{\mathbf{L}^\infty(\Omega)} = \inf\{C \in \mathbb{R}:\ |f_i| \leqslant C,\ i = 1, \dots, N, \text{ a.e. in } \Omega\},$$

where "a.e." stands for "almost everywhere". We will use the same notation for tensor fields, i.e., we will also indicate with $\mathbf{L}^p(\Omega)$ the space of tensor fields $\mathbf{T}: \Omega \to \mathbb{R}^{N\times N}$ whose components belongs to $L^p(\Omega)$. In this case

$$\|\mathbf{T}\|_{\mathbf{L}^p(\Omega)} = \left(\sum_{i=1}^N \sum_{j=1}^N \|T_{ij}\|^p_{L^p(\Omega)}\right)^{1/p}, \quad 1 \leqslant p < \infty.$$

Analogously, a vector (or a tensor) function $\mathbf{f}$ belongs to $\mathbf{H}^m(\Omega)$ if all its components belong to $H^m(\Omega)$, and we have

$$\|\mathbf{f}\|_{\mathbf{H}^m(\Omega)} = \left(\sum_{i=1}^N \|f_i\|^2_{H^m(\Omega)}\right)^{1/2},$$

while its semi-norm is

$$|\mathbf{f}|_{\mathbf{H}^m(\Omega)} = \left(\sum_{i=1}^N |f_i|^2_{H^m(\Omega)}\right)^{1/2}.$$

It is understood that, when $m = 0$,

$$\mathbf{H}^0(\Omega) \equiv \mathbf{L}^2(\Omega).$$

When equipped with the following scalar product:

$$(\mathbf{f}, \mathbf{g})_{\mathbf{H}^m(\Omega)} = \sum_{i=1}^{N} (f_i, g_i)_{H^m(\Omega)}, \quad \mathbf{f}, \mathbf{g} \in \mathbf{H}^m(\Omega),$$

the space $\mathbf{H}^m(\Omega)$ is a Hilbert space.

To ease notation, we will often use the following short-hand notation for the L^2 scalar products:

$$(\mathbf{v}, \mathbf{w}) \equiv (\mathbf{v}, \mathbf{w})_{\mathbf{L}^2(\Omega)}, \qquad (p, q) \equiv (p, q)_{L^2(\Omega)}.$$

We note that the L^2 scalar product of two tensor fields $\mathbf{T}$ and $\mathbf{G}$ belonging to $\mathbf{L}^2(\Omega)$ is defined as

$$(\mathbf{T}, \mathbf{G}) \equiv (\mathbf{T}, \mathbf{G})_{\mathbf{L}^2(\Omega)} = \int_\Omega \mathbf{T} : \mathbf{G} = \sum_{i=1}^{N} \sum_{j=1}^{N} \int_\Omega T_{ij} G_{ij}.$$

For our purposes we will usually have $m = 1$. In that case we have the equality

$$\|\mathbf{f}\|^2_{\mathbf{H}^1(\Omega)} = \|\mathbf{f}\|^2_{\mathbf{L}^2(\Omega)} + \|\nabla \mathbf{f}\|^2_{\mathbf{L}^2(\Omega)}.$$

We often utilize the space $\mathbf{H}^1_0(\Omega)$ defined as

$$\mathbf{H}^1_0(\Omega) = \left\{ \mathbf{v} \in \mathbf{H}^1(\Omega) : \ \mathbf{v}|_{\partial\Omega} = 0 \right\}.$$

We will consider *bounded* domains Ω with regular (i.e., *Lipschitz continuous*) boundary $\partial\Omega$, so that both the Sobolev embedding theorems in $\mathbb{R}^N$ and the *Green* integration formula hold. Some important results are here recalled, without providing the demonstration, which may be found in LIONS and MAGENES [1968] or BREZIS [1983].

THEOREM 10.1 (Sobolev embeddings (simplified form)). *Let Ω be a bounded domain of $\mathbb{R}^N$ with Lipschitz continuous boundary. The following properties hold:*

$$\begin{cases} \textit{If } 0 \leqslant s < \dfrac{N}{2}, & \mathbf{H}^s(\Omega) \hookrightarrow \mathbf{L}^p(\Omega), \quad p = \dfrac{2N}{N-2s}, \\ \textit{If } s = \dfrac{N}{2}, & \mathbf{H}^s(\Omega) \hookrightarrow \mathbf{L}^q(\Omega), \quad 2 \leqslant q < \infty, \\ \textit{If } s > \dfrac{N}{2}, & \mathbf{H}^s(\Omega) \hookrightarrow \left[C^0(\overline{\Omega})\right]^N, \end{cases}$$

where $A \hookrightarrow B$ means that A is included in B with continuous embedding.

THEOREM 10.2 (Green integration formula). *Let Ω be a bounded domain of $\mathbb{R}^N$ with Lipschitz continuous boundary and let $\mathbf{n}$ denote the unit outer normal along $\partial\Omega$. Let*

$u, v \in H^1(\Omega)$, then the integral

$$\int_{\partial\Omega} u v n_i$$

exists and is finite for each component n_i of **n**. *In addition we have*

$$\int_{\Omega} \frac{\partial u}{\partial x_i} v = -\int_{\Omega} u \frac{\partial v}{\partial x_i} + \int_{\partial\Omega} u v n_i, \quad i = 1, \ldots, N.$$

LEMMA 10.1 (Poincaré inequality – multidimensional case). *Let $f : \mathbb{R}^N \to \mathbb{R}$ be a function of $\mathbf{H}^1(\Omega)$, with $f = 0$ on $\Gamma \subset \partial\Omega$ of strictly positive measure. Then there exists a positive constant C_P (depending only on the domain Ω and on Γ), such that*

$$\|f\|_{L_2(\Omega)} \leqslant C_P \|\nabla f\|_{\mathbf{L}_2(\Omega)}. \tag{10.1}$$

LEMMA 10.2. *Let Ω be a bounded and connected subset of $\mathbb{R}^N$, where $N = 2$ or 3. Furthermore, let us assume that the velocity field $\mathbf{u} \in \mathbf{H}^1(\Omega)$ vanishes on $\Gamma \subset \partial\Omega$ of strictly positive measure. Then, there exists a constant $C_K > 0$ so that the following inequality holds:*

$$\int_{\Omega} \mathbf{D}(\mathbf{u}) : \mathbf{D}(\mathbf{u}) \geqslant C_K \|\nabla \mathbf{u}\|^2_{\mathbf{L}^2(\Omega)}. \tag{10.2}$$

This theorem is a consequence of the *Korn inequality*, whose precise statement may be found, for instance, in CIARLET [1988], DUVAUT and LIONS [1976].

LEMMA 10.3 (Gronwall lemma). *Let f be a non-negative function which is integrable in $I = (t_0, t_1)$ and g and ϕ be two continuous functions in I, with g non-decreasing. If*

$$\phi(t) \leqslant g(t) + \int_{t_0}^{t} f(\tau)\phi(\tau)\,\mathrm{d}\tau \quad \forall t \in I, \tag{10.3}$$

then

$$\phi(t) \leqslant g(t) \exp \int_{t_0}^{t} f(\tau)\,\mathrm{d}\tau \quad \forall t \in I. \tag{10.4}$$

11. Weak form of Navier–Stokes equations

The incompressible Navier–Stokes equations read

$$\frac{\partial \mathbf{u}}{\partial t} + (\mathbf{u} \cdot \nabla)\mathbf{u} + \nabla p - 2\,\mathbf{div}(\nu \mathbf{D}(\mathbf{u})) = \mathbf{f}, \quad \text{in } \Omega,\ t \in I, \tag{11.1a}$$

$$\operatorname{div} \mathbf{u} = 0, \quad \text{in } \Omega,\ t \in I, \tag{11.1b}$$

$$\mathbf{u} = \mathbf{u}_0, \quad \text{in } \Omega,\ t = t_0. \tag{11.1c}$$

We assume that ν is a bounded strictly positive function, precisely we assume that there exist two constants $\nu_0 > 0$ and $\nu_1 > 0$ such that $\forall t \in I$,

$$\nu_0 \leqslant \nu \leqslant \nu_1 \quad \text{almost everywhere in } \Omega.$$

We consider the case in which the system of differential equations (11.1) is equipped with the following boundary conditions:

$$\mathbf{u} = \mathbf{g} \quad \text{on } \Gamma^d,\ t \in I, \tag{11.2a}$$
$$-p\mathbf{n} + 2\nu\mathbf{D}(\mathbf{u}) \cdot \mathbf{n} = \mathbf{h} \quad \text{on } \Gamma^n,\ t \in I, \tag{11.2b}$$

We have indicated with Γ^d and Γ^n the portions of $\partial\Omega$ where Dirichlet and Neumann boundary conditions are applied, respectively. We must have $\Gamma^d \cup \Gamma^n = \partial\Omega$.

REMARK 11.1. If $\Gamma^d = \partial\Omega$ we call the problem formed by (11.1) and (11.2) a *Dirichlet problem*. We will instead use the term *Neumann problem* when $\Gamma^n = \partial\Omega$. The conditions $\mathbf{g} = \mathbf{0}$ and $\mathbf{h} = \mathbf{0}$ are called *homogeneous boundary conditions*.

In the case of a Dirichlet problem, the boundary datum has to satisfy the following compatibility relation for all $t \in I$:

$$\int_{\partial\Omega} \mathbf{g} \cdot \mathbf{n} = 0.$$

REMARK 11.2. For the problem at hand, we normally have $\mathbf{f} = \mathbf{0}$, since the only external force which one may eventually consider in blood flow is the gravity force. Even in this case, we may still adopt the Navier–Stokes equations with $\mathbf{f} = \mathbf{0}$ by replacing p with $p^*(t, \mathbf{x}) = p(t, \mathbf{x}) + gz(\mathbf{x})\mathbf{e}_z$, where g is the gravity acceleration, $\mathbf{e}_z$ the unit vector defining the vertical direction (upwardly oriented) and $z(\mathbf{x})$ the (known) quota of point $\mathbf{x}$ with respect to a reference horizontal plane. Yet, for the sake of completeness, many of the derivations of this as well as the following sections refer to the general case $\mathbf{f} \neq \mathbf{0}$.

The *weak form* of the Navier–Stokes equations is (formally) obtained by taking the scalar product of the momentum equations with a vector function $\mathbf{v}$ belonging to a functional space $\mathbf{V}$ (called *test function space*), which will be better specified later on, integrating over Ω and applying the Green integration formula. We operate similarly on the continuity equation, by multiplying it by a function $q \in Q$ and integrating. Also the space Q will be specified at a later stage.

We formally obtain

$$\left(\frac{\partial \mathbf{u}}{\partial t}, \mathbf{v}\right) + ((\mathbf{u} \cdot \nabla)\mathbf{u}, \mathbf{v}) + 2\int_{\Omega} \nu\mathbf{D}(\mathbf{u}) : \mathbf{D}(\mathbf{v}) - (p, \operatorname{div} \mathbf{v})$$
$$= (\mathbf{f}, \mathbf{v}) + \int_{\partial\Omega} \mathbf{v} \cdot (2\nu\mathbf{D}(\mathbf{u}) \cdot \mathbf{n} - p\mathbf{n}),$$
$$(\operatorname{div} \mathbf{u}, q) = 0.$$

We have exploited the identity

$$\int_{\Omega} \nu\mathbf{D}(\mathbf{u}) : \nabla\mathbf{v} = \int_{\Omega} \nu\mathbf{D}(\mathbf{u}) : \mathbf{D}(\mathbf{v}),$$

which derives from the symmetry of the tensor $\mathbf{D}(\mathbf{u})$.

The boundary term may now be split into two parts,

$$\int_{\partial\Omega} \mathbf{v}\cdot(2\nu\mathbf{D}(\mathbf{u})\cdot\mathbf{n}-p\mathbf{n}) = \int_{\Gamma^d} \mathbf{v}\cdot(2\nu\mathbf{D}(\mathbf{u})\cdot\mathbf{n}-p\mathbf{n}) + \int_{\Gamma^n} \mathbf{v}\cdot\mathbf{h}.$$

We note that the contribution from the Neumann boundary is now a *given data*, while contribution from the Dirichlet boundary can be eliminated by appropriately choosing the test space $\mathbf{V}$.

By inspection, we may recognize that all terms make sense if we choose as test function spaces

$$\mathbf{V} = \left\{\mathbf{v}\in\mathbf{H}^1(\Omega):\ \mathbf{v}|_{\Gamma^d} = \mathbf{0}\right\},$$

$$Q = \left\{q\in L^2(\Omega):\ \text{with} \int_\Omega q = 0 \text{ if } \Gamma^d = \partial\Omega\right\},$$

and if we seek, at each time t, the velocity in

$$\mathbf{V}_\mathbf{g} = \left\{\mathbf{u}\in\mathbf{H}^1(\Omega):\ \mathbf{u}|_{\Gamma^d} = \mathbf{g}\right\}$$

and the pressure in Q.

REMARK 11.3. The request that Q is formed by functions with zero mean on Ω when we treat a Dirichlet problem derives from the fact that in such a case the pressure is determined only up to a constant, as it appears in the equations only through its gradient. To compute a unique value for the pressure it is then necessary to fix the constant. This is obtained by the zero-mean constraint.

Finally, the *weak form* of the Navier–Stokes problem (11.1) and (11.2), reads:

Find, $\forall t\in I$, $\mathbf{u}(t)\in\mathbf{V}_\mathbf{g}$ *and* $p(t)\in Q$ *such that*

$$\begin{cases}\left(\dfrac{\partial\mathbf{u}}{\partial t},\mathbf{v}\right) + a(\mathbf{u},\mathbf{v}) + c(\mathbf{u},\mathbf{u},\mathbf{v}) + b(\mathbf{v},p) = (\mathbf{f},\mathbf{v}) + \int_{\Gamma^n}\mathbf{v}\cdot\mathbf{h}, & \forall\mathbf{v}\in\mathbf{V},\\ b(\mathbf{u},q) = 0, \quad \forall q\in Q,\end{cases} \tag{11.3}$$

where

$$a(\mathbf{u},\mathbf{v}) = 2\int_\Omega \nu\mathbf{D}(\mathbf{u}) : \mathbf{D}(\mathbf{v}), \tag{11.4}$$

$$c(\mathbf{w},\mathbf{u},\mathbf{v}) = \int_\Omega (\mathbf{w}\cdot\nabla)\mathbf{u}\cdot\mathbf{v}, \tag{11.5}$$

$$b(\mathbf{v},p) = -\int_\Omega p\,\mathrm{div}\,\mathbf{v}. \tag{11.6}$$

11.1. The homogeneous Dirichlet problem

In this section we will focus on the homogeneous Dirichlet problem, that is the case when $\Gamma^d = \partial\Omega$ and $\mathbf{g} = \mathbf{0}$ in (11.2a). Therefore,

$$\mathbf{V} = \mathbf{H}_0^1(\Omega), \qquad Q = L_0^2(\Omega) = \left\{q\in L^2(\Omega),\ \int_\Omega q = 0\right\} \tag{11.7}$$

and the weak form reads:

Find, $\forall t \in I$, $\mathbf{u}(t) \in \mathbf{V}$ *and* $p(t) \in Q$ *such that*

$$\begin{cases} \left(\dfrac{\partial \mathbf{u}}{\partial t}, \mathbf{v}\right) + a(\mathbf{u}, \mathbf{v}) + c(\mathbf{u}, \mathbf{u}, \mathbf{v}) + b(\mathbf{v}, p) = (\mathbf{f}, \mathbf{v}), & \forall \mathbf{v} \in \mathbf{V}, \\ b(\mathbf{u}, q) = 0, & \forall q \in Q. \end{cases} \tag{11.8}$$

LEMMA 11.1. *The forms* $a : \mathbf{V} \times \mathbf{V} \to \mathbb{R}$, $c : \mathbf{V} \times \mathbf{V} \times \mathbf{V} \to \mathbb{R}$ *and* $b : \mathbf{V} \times Q \to \mathbb{R}$ *are continuous with respect to their arguments. In addition,* $a(\cdot, \cdot)$ *is coercive, i.e.,* $\exists \alpha > 0$ *such that*

$$a(\mathbf{v}, \mathbf{v}) \geqslant \alpha \|\mathbf{v}\|^2_{\mathbf{H}^1(\Omega)}, \quad \forall \mathbf{v} \in \mathbf{V}.$$

PROOF. The continuity of the bilinear forms a and b is an immediate consequence of the Cauchy–Schwarz inequality. Indeed, $\forall \mathbf{u}, \mathbf{v} \in V$ and $\forall q \in Q$,

$$|a(\mathbf{u}, \mathbf{v})| \leqslant \nu_1 |\mathbf{u}|_{\mathbf{H}^1(\Omega)} |\mathbf{v}|_{\mathbf{H}^1(\Omega)} \leqslant \nu_1 \|\mathbf{u}\|_{\mathbf{H}^1(\Omega)} \|\mathbf{v}\|_{\mathbf{H}^1(\Omega)},$$
$$|b(\mathbf{u}, p)| \leqslant \|\operatorname{div} \mathbf{u}\|_{L^2(\Omega)} \|p\|_{L^2(\Omega)} \leqslant \|\mathbf{u}\|_{\mathbf{H}^1(\Omega)} \|p\|_{L^2(\Omega)}.$$

For the tri-linear form c we first have to note that thanks to the Sobolev embedding theorem $\mathbf{H}^1(\Omega) \hookrightarrow \mathbf{L}^6(\Omega)$ (as $N = 2, 3$) and consequently $\mathbf{H}^1(\Omega) \hookrightarrow \mathbf{L}^4(\Omega)$. Then, $\mathbf{wu} \in \mathbf{L}^2(\Omega)$, and considering the expression of $c(\cdot, \cdot, \cdot)$ component-wise, we have

$$\int_\Omega w_i \frac{\partial u_k}{\partial x_i} v_k \leqslant \|w_i v_k\|_{L^2(\Omega)} \left\| \frac{\partial u_k}{\partial x_i} \right\|_{L^2(\Omega)} \leqslant \|w_i\|_{L^4(\Omega)} \|v_k\|_{L^4(\Omega)} \left\| \frac{\partial u_k}{\partial x_i} \right\|_{L^2(\Omega)}.$$

Then

$$\begin{aligned} \int_\Omega w_i \frac{\partial u_k}{\partial x_i} v_k &\leqslant C \|w_i\|_{H^1(\Omega)} \left\| \frac{\partial u_k}{\partial x_i} \right\|_{L^2(\Omega)} \|v_k\|_{H^1(\Omega)} \\ &\leqslant C \|w_i\|_{H^1(\Omega)} |u_k|_{H^1(\Omega)} \|v_k\|_{H^1(\Omega)} \\ &\leqslant C \|w_i\|_{H^1(\Omega)} \|u_k\|_{H^1(\Omega)} \|v_k\|_{H^1(\Omega)}, \end{aligned} \tag{11.9}$$

where C is a positive constant.

It follows that, $\forall \mathbf{u}, \mathbf{v}, \mathbf{v} \in \mathbf{V}$,

$$c(\mathbf{w}, \mathbf{u}, \mathbf{v}) \leqslant C_1 \|\mathbf{w}\|_{\mathbf{H}^1(\Omega)} \|\mathbf{u}\|_{\mathbf{H}^1(\Omega)} \|\mathbf{v}\|_{\mathbf{H}^1(\Omega)},$$

by which the continuity of the tri-linear form is proved (C_1 is a positive constant).

The coercivity of the linear form a derives from inequalities (10.1) and (10.2), since

$$a(\mathbf{v}, \mathbf{v}) \geqslant 2\nu_0 \int_\Omega \mathbf{D}(\mathbf{v}) : \mathbf{D}(\mathbf{v}) \geqslant 2\nu_0 C_K |\mathbf{v}|^2_{\mathbf{H}^1(\Omega)} \geqslant \alpha \|\mathbf{v}\|^2_{\mathbf{H}^1(\Omega)}, \quad \forall \mathbf{v} \in \mathbf{V}, \tag{11.10}$$

with $\alpha = (2\nu_0 C_K)/(C_P^2 + 1)$, being C_P and C_K the constants in (10.1) and (10.2), respectively. □

We now introduce the space

$$\mathbf{V}_{\text{div}} = \{\mathbf{v} \in V : \operatorname{div} \mathbf{v} = 0 \text{ a.e. in } \Omega\}.$$

THEOREM 11.1. *If* $\mathbf{u}$ *is a solution of the weak formulation* (11.8), *then* $\mathbf{u}(t) \in \mathbf{V}_{\text{div}}$ *for all* $t \in I$ *and it satisfies*

$$\left(\frac{\partial \mathbf{u}}{\partial t}, \mathbf{v}\right) + a(\mathbf{u}, \mathbf{v}) + c(\mathbf{u}, \mathbf{u}, \mathbf{v}) = (\mathbf{f}, \mathbf{v}), \quad \forall \mathbf{v} \in \mathbf{V}_{\text{div}},\ t \in I. \tag{11.11}$$

Conversely, if, $\forall t \in I$, $\mathbf{u}(t) \in \mathbf{V}_{\text{div}}$ *is a solution of* (11.11) *and* $\partial \mathbf{u}/\partial t \in \mathbf{L}^2(\Omega)$, *then there exists a unique* $p \in Q$ *such that* $(\mathbf{u}, p)$ *satisfies* (11.8).

PROOF. The first part of the proof is trivial. If $\mathbf{u}$ satisfies (11.8) then it belongs to $\mathbf{V}_{\text{div}}$ and it satisfies (11.11), since $\mathbf{V}_{\text{div}} \subset \mathbf{V}$.

The demonstration of the inverse implication requires first to state the following result.

LEMMA 11.2. *Let* Ω *be a domain of* $\mathbb{R}^N$ *and let* $L \in \mathbf{V}'$. *Then* $L(\mathbf{v}) = 0$, $\forall \mathbf{v} \in \mathbf{V}_{\text{div}}$ *if and only if there exists a function* $p \in L^2(\Omega)$ *such that*

$$L(\mathbf{v}) = (p, \operatorname{div} \mathbf{v}), \quad \forall \mathbf{v} \in \mathbf{V}.$$

For the proof see Lemma 2.1 of GIRAULT and RAVIART [1986].

The application L defined as

$$L(\mathbf{v}) = \left(\frac{\partial \mathbf{u}}{\partial t}, \mathbf{v}\right) + a(\mathbf{u}, \mathbf{v}) + c(\mathbf{u}, \mathbf{u}, \mathbf{v}) - (\mathbf{f}, \mathbf{v}), \quad \forall \mathbf{v} \in \mathbf{V},$$

belongs to $\mathbf{V}'$, being a linear continuous functional on $\mathbf{V}$. We can therefore apply Lemma 11.2 and obtain the desired result. □

12. An energy inequality for the Navier–Stokes equations

We now prove an energy inequality for problem (11.8), by which we may assess a continuous dependence of the solution from the given data.

THEOREM 12.1 (Energy inequalities). *Let* $\mathbf{u}(t) \in \mathbf{V}_{\text{div}}$ *be a solution of* (11.8), $\forall t \in I$. *Then the following inequalities hold:*

$$\|\mathbf{u}(t)\|^2_{\mathbf{L}^2(\Omega)} + C_1 \int_0^t \|\nabla \mathbf{u}(\tau)\|^2_{\mathbf{L}^2(\Omega)}\, \mathrm{d}\tau \leqslant \left(\|\mathbf{u}_0\|^2_{\mathbf{L}^2(\Omega)} + \int_0^t \|\mathbf{f}(\tau)\|^2_{\mathbf{L}^2(\Omega)}\, \mathrm{d}\tau\right) e^t,$$

where $C_1 = 4\nu_0 C_K$, *and*

$$\|\mathbf{u}(t)\|^2_{\mathbf{L}^2(\Omega)} + C_2 \int_0^t \|\nabla \mathbf{u}(\tau)\|^2_{\mathbf{L}^2(\Omega)}\, \mathrm{d}\tau \leqslant \|\mathbf{u}_0\|^2_{\mathbf{L}^2(\Omega)} + \frac{C_P}{C_2} \int_0^t \|\mathbf{f}(\tau)\|^2_{\mathbf{L}^2(\Omega)}\, \mathrm{d}\tau,$$

where $C_2 = 2\nu_0 C_K$. *Here,* C_K *and* C_P *are the constants in the Poincaré inequality* (10.1) *and in* (10.2), *respectively.*

We first prove the following result.

LEMMA 12.1. *If* $\mathbf{u}$ *is a solution of* (11.8) *then* $c(\mathbf{u}, \mathbf{u}, \mathbf{u}) = 0$.

PROOF. It follows from the Green formula and the fact that $\mathbf{u}|_{\partial\Omega} = \mathbf{0}$. Indeed,

$$c(\mathbf{u}, \mathbf{u}, \mathbf{u}) = \int_\Omega (\mathbf{u} \cdot \nabla)\mathbf{u} \cdot \mathbf{u} = \int_\Omega \frac{1}{2}\nabla\left(|\mathbf{u}|^2\right) \cdot \mathbf{u}$$
$$= -\frac{1}{2}\int_\Omega |\mathbf{u}|^2 \operatorname{div} \mathbf{u} + \frac{1}{2}\int_{\partial\Omega} |\mathbf{u}|^2 \mathbf{u} \cdot \mathbf{n}.$$

Now, the last integral is zero since $\mathbf{u} = 0$ on $\partial\Omega$. Moreover, for the same reason

$$\int_\Omega \operatorname{div} \mathbf{u} = \int_{\partial\Omega} \mathbf{u} \cdot \mathbf{n} = 0.$$

Then, if we set

$$c = \int_\Omega |\mathbf{u}|^2,$$

we have

$$\int_\Omega |\mathbf{u}|^2 \operatorname{div} \mathbf{u} = \int_\Omega |\mathbf{u}|^2 \operatorname{div} \mathbf{u} - c\int_\Omega \operatorname{div} \mathbf{u} = \int_\Omega \left(|\mathbf{u}|^2 - c\right) \operatorname{div} \mathbf{u}$$
$$= b\left(\mathbf{u}, \left(|\mathbf{u}|^2 - c\right)\right) = 0,$$

where the last equality is obtained since $(|\mathbf{u}|^2 - c) \in Q$ and $b(\mathbf{u}, q) = 0, \ \forall q \in Q$. □

We now give the demonstration of Theorem 12.1.

PROOF. For all fixed t, take $\mathbf{v} = \mathbf{u}(t)$ in the momentum equation of (11.11). We have

$$\frac{1}{2}\frac{\mathrm{d}}{\mathrm{d}t}\|\mathbf{u}\|^2_{\mathbf{L}^2(\Omega)} + c(\mathbf{u}, \mathbf{u}, \mathbf{u}) + b(\mathbf{u}, p) + a(\mathbf{u}, \mathbf{u}) = (\mathbf{f}, \mathbf{u}). \tag{12.1}$$

Then,

$$\frac{1}{2}\frac{\mathrm{d}}{\mathrm{d}t}\|\mathbf{u}\|^2_{\mathbf{L}^2(\Omega)} + a(\mathbf{u}, \mathbf{u}) = (\mathbf{f}, \mathbf{u}).$$

Now, thanks to (10.2),

$$a(\mathbf{u}, \mathbf{u}) = 2\int_\Omega \nu\mathbf{D}(\mathbf{u}) : \mathbf{D}(\mathbf{u}) \geqslant 2\nu_0 C_K \|\nabla\mathbf{u}\|^2_{\mathbf{L}^2(\Omega)},$$

then

$$\frac{\mathrm{d}}{\mathrm{d}t}\|\mathbf{u}\|^2_{\mathbf{L}^2(\Omega)} + 4\nu_0 C_K \|\nabla\mathbf{u}\|^2_{\mathbf{L}^2(\Omega)} \leqslant 2(\mathbf{f}, \mathbf{u}) \leqslant \frac{1}{2\varepsilon}\|\mathbf{f}\|^2_{\mathbf{L}^2(\Omega)} + 2\varepsilon\|\mathbf{u}\|^2_{\mathbf{L}^2(\Omega)}, \tag{12.2}$$

for any $\varepsilon > 0$. By choosing $\varepsilon = 1/2$ and integrating between t_0 and t, we have

$$\|\mathbf{u}(t)\|^2_{\mathbf{L}^2(\Omega)} + 4\nu_0 C_K \int_{t_0}^t \|\nabla\mathbf{u}(\tau)\|^2_{\mathbf{L}^2(\Omega)}\, \mathrm{d}\tau$$
$$\leqslant \int_{t_0}^t \|\mathbf{f}(\tau)\|^2_{\mathbf{L}^2(\Omega)}\, \mathrm{d}\tau + \int_{t_0}^t \|\mathbf{u}(\tau)\|^2_{\mathbf{L}^2(\Omega)}\, \mathrm{d}\tau + \|\mathbf{u}_0\|^2_{\mathbf{L}^2(\Omega)}.$$

We apply Gronwall lemma (Lemma 10.3) by identifying

$$\|\mathbf{u}(t)\|^2_{\mathbf{L}^2(\Omega)} + 4\nu_0 C_K \int_{t_0}^{t} \|\nabla \mathbf{u}(\tau)\|^2_{\mathbf{L}^2(\Omega)}\, \mathrm{d}\tau$$

with $\phi(t)$, obtaining the first inequality.

By using instead the Poincaré inequality on the last term of (12.2), and by taking $\varepsilon = (\nu_0 C_K)/C_P^2$, we obtain

$$\frac{\mathrm{d}}{\mathrm{d}t}\|\mathbf{u}\|^2_{\mathbf{L}^2(\Omega)} + 2\nu_0 C_K \|\nabla \mathbf{u}\|^2_{\mathbf{L}^2(\Omega)} \leqslant \frac{C_P^2}{2\nu_0 C_K}\|\mathbf{f}\|^2_{\mathbf{L}^2(\Omega)}.$$

By integrating between t_0 and t, we obtain the second inequality of the theorem. □

REMARK 12.1. In the case where $\mathbf{f} = \mathbf{0}$ we may derive the simpler estimate

$$\|\mathbf{u}(t)\|^2_{\mathbf{L}^2(\Omega)} + 4\nu_0 C_K \int_{0}^{t} \|\nabla \mathbf{u}(\tau)\|^2_{\mathbf{L}^2(\Omega)}\, \mathrm{d}\tau \leqslant \|\mathbf{u}_0\|^2_{\mathbf{L}^2(\Omega)}, \quad \forall t \geqslant t_0.$$

13. The Stokes equations

The space discretization of the Navier–Stokes equations give rise to a *non-linear* set of ordinary differential equations because of the presence of the convective term. This makes both the analysis and the numerical solution more difficult. In some cases, when the fluid is highly viscous, the contribution of the non-linear convective term may be neglected. The key parameter which allow us to make that decision is the *Reynolds number Re*, which is an a-dimensional number defined as

$$Re = \frac{|\mathbf{u}|L}{\nu},$$

where L represents a length-scale for the problem at hand and $|\mathbf{u}|$ the Euclidean norm of the velocity. For the flow in a tube L is the tube diameter.

Contrary to other fluid dynamic situations, the high variation in time and space of the velocity in the vascular system does not allow to select a single representative value of the Reynolds number,[2] nevertheless in the situations where $\mathrm{Re} \ll 1$ (for instance, flow in smaller arteries or capillaries) we may say that the convective term is negligible compared to the viscous contribution and may be discarded. We have then the *Stokes* equations, which read (in the case of homogeneous Dirichlet conditions):

$$\frac{\partial \mathbf{u}}{\partial t} + \nabla p - 2\,\mathbf{div}(\nu \mathbf{D}(\mathbf{u})) = \mathbf{f}, \quad \text{in } \Omega,\ t \in I, \tag{13.1a}$$

$$\operatorname{div}\mathbf{u} = 0, \quad \text{in } \Omega,\ t \in I, \tag{13.1b}$$

$$\mathbf{u} = \mathbf{0}, \quad \text{on } \partial\Omega,\ t \in I, \tag{13.1c}$$

$$\mathbf{u} = \mathbf{u}_0, \quad \text{in } \Omega,\ t = t_0. \tag{13.1d}$$

The corresponding weak form reads:

[2]Another a-dimensional number which measures the relative importance of inertia versus viscous in oscillatory flow is the Womersley number (FUNG [1984]).

Find, $\forall t \in I$, $\mathbf{u}(t) \in \mathbf{V}$, $p(t) \in Q$, *such that*

$$\left(\frac{\partial \mathbf{u}}{\partial t}, \mathbf{v}\right) + a(\mathbf{u}, \mathbf{v}) + b(\mathbf{v}, p) = (\mathbf{f}, \mathbf{v}), \quad \forall \mathbf{v} \in \mathbf{V},$$
$$b(\mathbf{u}, q) = 0, \quad \forall q \in Q. \tag{13.2}$$

In the case of a *steady problem*, that is when we consider $\partial \mathbf{u}/\partial t = \mathbf{0}$, the solution $(\mathbf{u}, p)$ of the Stokes problem (13.2) is a saddle point for the functional

$$\mathcal{S}(\mathbf{v}, q) = \frac{1}{2}a(\mathbf{v}, \mathbf{v}) + b(\mathbf{v}, q) - (\mathbf{f}, \mathbf{v}), \quad \mathbf{v} \in \mathbf{V},\ q \in Q.$$

This means

$$\mathcal{S}(\mathbf{u}, p) = \min_{\mathbf{v} \in \mathbf{V}} \max_{q \in Q} \mathcal{S}(\mathbf{v}, q).$$

In this respect, the pressure p may be considered as a *Lagrange multiplier* associated to the incompressibility constraint.

REMARK 13.1. In those cases where $Re \gg 1$ (*high Reynolds number flows*) the flow becomes *unstable*. High frequency fluctuations in the velocity and pressure field appear, which might give rise to *turbulence*. This phenomenon is particularly complex and its numerical simulation may be extremely difficult. To make the problem amenable to numerical solution it is often necessary to adopt a *turbulence model*, which allows to give a more or less accurate description of the effect of turbulence on the main flow variables.

In normal physiological situations, the typical values of the Reynolds number reached in the cardiovascular system do not allow the formation of full scale turbulence. Some flow instabilities may occur only at the exit of the aortic valve and limited to the systolic phase. Indeed, in this region the Reynolds number may reach the value of few thousands only for the portion of the cardiac cycle corresponding the peak systolic velocity. Therefore, there is no sufficient time for a full turbulent flow to develop.

The situation is different in some pathological circumstances, e.g., in the presence of a stenotic artery. The increase of the velocity at the location of the vessel restriction may induce turbulence to develop. This fact could explain the high increase in the noise caused by the blood stream in this situation.

14. Numerical approximation of Navier–Stokes equations

In this section we give a very short account on possible numerical methods for the solution of the Navier–Stokes equations. This subject is far from being simple, and we will not make any attempt to be exhaustive. The interested reader can consult, for instance, QUARTERONI and VALLI [1994], Chapters 9, 10 and 13, and the classic books on the subject by GIRAULT and RAVIART [1986] and TEMAM [1984].

Here, we will simply mention a few methods to advance the Navier–Stokes equations from a given time-level to a new one and we will point out some of the mathematical problems that have to be faced. For the sake of simplicity we will confine ourselves to the homogeneous Dirichlet problem (11.8).

14.1. *Time advancing by finite differences*

The Navier–Stokes problem (9.1) (equivalently, its weak form (11.8)) can be advanced in time by suitable finite difference schemes.

The simulation will cover the interval $I = (0, T)$ which we subdivide into sub-intervals (time-steps) $I^k = (t^k, t^{k+1})$ with $k = 0, \ldots, N$ and where $t^{k+1} - t^k = \Delta t$ is constant. We have thus partitioned the space-time domain $I \times \Omega$ into several *time-slabs* $I^k \times \Omega$. We assume that on each slab we know the solution at $t = t^k$ and that we wish to find the solution at $t = t^{k+1}$. Clearly, for the first time slab the assumption is true since at $t = 0$ the approximate solution is obtained from the initial data. If we treat the time slabs in their natural order, as soon as the solution on the kth time slab has been found, it is made available as initial condition for the computation on the next time slab. This is a *time-advancing* procedure.

We will indicate by $(\mathbf{u}^k, p^k)$ the approximate solution at time t^k, that is

$$\left(\mathbf{u}^k, p^k\right) \approx \left(\mathbf{u}\left(t^k\right), p\left(t^k\right)\right).$$

A family of simple time-advancing schemes is obtained by using the Taylor expansion formula to write

$$\frac{\partial \mathbf{u}}{\partial t}\left(t^{k+1}\right) = \frac{\mathbf{u}\left(t^{k+1}\right) - \mathbf{u}(t^k)}{\Delta t} + \mathrm{O}(\Delta t).$$

Then, by making the first order approximation

$$\frac{\partial \mathbf{u}}{\partial t}\left(t^{k+1}\right) \approx \frac{\mathbf{u}^{k+1} - \mathbf{u}^k}{\Delta t},$$

into (9.1), we may write the following time-stepping scheme to calculate $\mathbf{u}^{k+1}$ and p^{k+1}:

$$\frac{\mathbf{u}^{k+1} - \mathbf{u}^k}{\Delta t} - 2\,\mathbf{div}\,\nu\mathbf{D}\left(\mathbf{u}^{k+1}\right) + (\mathbf{u}^* \cdot \nabla)\mathbf{u}^{**} + \nabla p^{k+1} = \mathbf{f}^{k+1}, \quad \text{in } \Omega, \tag{14.1a}$$

$$\operatorname{div} \mathbf{u}^{k+1} = 0, \quad \text{in } \Omega, \tag{14.1b}$$

$$\mathbf{u}^{k+1} = \mathbf{0}, \quad \text{on } \partial\Omega. \tag{14.1c}$$

Here, $\mathbf{f}^{k+1}$ stands for $\mathbf{f}(t^{k+1})$.

The value of $\mathbf{u}^*$ and $\mathbf{u}^{**}$ in the non-linear convective term may be taken, for instance, as follows:

$$\left(\mathbf{u}^* \cdot \nabla\right)\mathbf{u}^{**} = \begin{cases} \left(\mathbf{u}^k \cdot \nabla\right)\mathbf{u}^k, & \text{fully explicit treatment,} \\ \left(\mathbf{u}^k \cdot \nabla\right)\mathbf{u}^{k+1}, & \text{semi-implicit treatment,} \\ \left(\mathbf{u}^{k+1} \cdot \nabla\right)\mathbf{u}^{k+1}, & \text{fully implicit treatment.} \end{cases}$$

In the case of the fully implicit treatment, Eq. (14.1) gives rise to a non-linear system. The semi-implicit and fully explicit treatments, instead, perform a *linearization* of the convective term, thus eliminating the non-linearity.

Let us consider the scheme resulting from the *fully explicit treatment* of the convective term. Problem (14.1) is then rewritten as

$$\frac{1}{\Delta t}\mathbf{u}^{k+1} - 2\,\mathbf{div}\Big(\nu\mathbf{D}\left(\mathbf{u}^{k+1}\right)\Big) + \nabla p^{k+1} = \mathbf{f}^{k+1} + \frac{1}{\Delta t}\mathbf{u}^{k} - \left(\mathbf{u}^{k}\cdot\nabla\right)\mathbf{u}^{k} \quad \text{in } \Omega, \tag{14.2a}$$

$$\operatorname{div}\mathbf{u}^{k+1} = 0, \quad \text{in } \Omega, \tag{14.2b}$$

$$\mathbf{u}^{k+1} = \mathbf{0}, \quad \text{on } \partial\Omega. \tag{14.2c}$$

We will now denote $\mathbf{u}^{k+1}$ and p^{k+1} by $\mathbf{w}$ and π, respectively, and by $\mathbf{q}$ and a_0 the quantities

$$\mathbf{q} = \mathbf{f}^{k+1} + \frac{1}{\Delta t}\mathbf{u}^{k} - \left(\mathbf{u}^{k}\cdot\nabla\right)\mathbf{u}^{k}, \qquad a_0 = \frac{1}{\Delta t}. \tag{14.3}$$

Problem (14.2) may be written in the form

$$a_0\mathbf{w} - 2\,\mathbf{div}(\nu\mathbf{D}(\mathbf{w})) + \nabla\pi = \mathbf{q}, \quad \text{in } \Omega, \tag{14.4a}$$

$$\operatorname{div}\mathbf{w} = 0, \quad \text{in } \Omega, \tag{14.4b}$$

$$\mathbf{w} = \mathbf{0}, \quad \text{on } \partial\Omega, \tag{14.4c}$$

which is called the *generalized Stokes problem.*

A characteristic treatment of the time derivative would also lead at each time step to a generalized Stokes problem (see Section 14.3).

For its approximation, a Galerkin finite element procedure can be set up by considering two finite element spaces $\mathbf{V}_h$ for the velocity and Q_h for the pressure, and seeking $\mathbf{w}_h \in \mathbf{V}_h$ and $\pi_h \in Q_h$ such that

$$\begin{cases} \tilde{a}(\mathbf{w}_h, \mathbf{v}_h) + b(\mathbf{w}_h, \pi_h) = (\mathbf{q}, \mathbf{v}_h), & \forall \mathbf{v}_h \in \mathbf{V}_h, \\ b(\mathbf{w}_h, q_h) = 0, & \forall q_h \in Q_h, \end{cases} \tag{14.5}$$

where $\tilde{a}(\mathbf{w}, \mathbf{v}) = a_0(\mathbf{w}, \mathbf{v}) + a(\mathbf{w}, \mathbf{v})$.

The algebraic form of problem (14.5) is derived by denoting with

$$\{\boldsymbol{\varphi}_i,\ i = 1, \ldots, N_{\mathbf{V}_h}\}, \qquad \{\psi_i,\ i = 1, \ldots, N_{Q_h}\}$$

the bases of $\mathbf{V}_h$ and Q_h, respectively. Here $N_{\mathbf{V}_h} = \dim(\mathbf{V}_h)$ and $N_{Q_h} = \dim(Q_h)$. Then, by setting

$$\mathbf{w}_h(\mathbf{x}) = \sum_{i=1}^{N_{\mathbf{V}_h}} w_i \boldsymbol{\varphi}_i(\mathbf{x}), \qquad p_h(\mathbf{x}) = \sum_{i=1}^{N_{Q_h}} \pi_i \psi_i(\mathbf{x}), \tag{14.6}$$

we obtain the following system from (14.5):

$$\begin{pmatrix} C & D^{\mathrm{T}} \\ D & 0 \end{pmatrix}\begin{pmatrix} \mathbf{W} \\ \boldsymbol{\Pi} \end{pmatrix} = \begin{pmatrix} \mathbf{F}_s \\ \mathbf{0} \end{pmatrix}, \tag{14.7}$$

where $\mathbf{W}$, $\boldsymbol{\Pi}$ and $\mathbf{F}_s$ denote three vectors defined respectively as

$$(\mathbf{W})_i = w_i, \qquad (\boldsymbol{\Pi})_i = \pi_i, \qquad (\mathbf{F}_s)_i = (\mathbf{q}, \boldsymbol{\varphi}_i),$$

while C, K and D are matrices whose components are defined as

$$(C)_{ij} = \tilde{a}(\boldsymbol{\varphi}_j, \boldsymbol{\varphi}_i), \qquad (D)_{ij} = b(\boldsymbol{\varphi}_j, \psi_i).$$

The global matrix

$$A = \begin{pmatrix} C & D^{\mathrm{T}} \\ D & 0 \end{pmatrix} \tag{14.8}$$

is a square matrix with dimension $(N_{\mathbf{V}_h} + N_{Q_h}) \times (N_{\mathbf{V}_h} + N_{Q_h})$.

In the case of a finite element approximation, p_i represents the pressure at the ith mesh node. The interpretation of w_i is made more complex by the fact that the velocity is a vector function, while w_i is a scalar. Let us assume that we are considering a three-dimensional problem and let the basis for $\mathbf{V}_h$ be chosen by grouping the vector functions $\boldsymbol{\varphi}_i$ into 3 families, as follows:

$$\boldsymbol{\varphi}_i^1 = \begin{bmatrix} \varphi_i \\ 0 \\ 0 \end{bmatrix}, \qquad \boldsymbol{\varphi}_i^2 = \begin{bmatrix} 0 \\ \varphi_i \\ 0 \end{bmatrix}, \qquad \boldsymbol{\varphi}_i^3 = \begin{bmatrix} 0 \\ 0 \\ \varphi_i \end{bmatrix}.$$

Finally, let $M_{\mathbf{V}_h} = N_{\mathbf{V}_h}/3$. Then, we may rewrite the first expansion in (14.6) as

$$\mathbf{w}_h(\mathbf{x}) = \sum_{i=1}^{M_{\mathbf{V}_h}} \sum_{j=1}^{3} w_i^j \boldsymbol{\varphi}_i^j(\mathbf{x}),$$

where w_i^j here represents the jth component of $\mathbf{w}$ at the ith mesh node.

LEMMA 14.1. *If* $\ker D^{\mathrm{T}} = \mathbf{0}$, *then matrix A is non-singular.*

PROOF. We first prove the non-singularity of C. For any $\mathbf{W} \in \mathbb{R}^{N_{\mathbf{V}_h}}$, $\mathbf{W} \neq \mathbf{0}$,

$$\mathbf{W}^{\mathrm{T}} C \mathbf{W} = \sum_{i=1}^{N_{\mathbf{V}_h}} \sum_{j=1}^{N_{\mathbf{V}_h}} w_i w_j C_{ij} = \tilde{a}(\mathbf{w}, \mathbf{w}) > 0,$$

where $\mathbf{w} = \sum_{i=1}^{N_{\mathbf{V}_h}} w_i \boldsymbol{\varphi}_i$. Consequently, C is positive-definite, and thus non-singular. From (14.7) we have

$$\mathbf{W} = C^{-1}\left(\mathbf{F}_s - D^{\mathrm{T}} \boldsymbol{\Pi}\right), \qquad D\mathbf{W} = \mathbf{0}.$$

Then we may formally compute the discrete pressure terms by

$$-\left(DC^{-1}D^{\mathrm{T}}\right)\boldsymbol{\Pi} = -DC^{-1}\mathbf{F}_s.$$

Proving that A is non-singular thus reduces to show that the matrix

$$S = DC^{-1}D^{\mathrm{T}}$$

is non-singular. If we take any $\mathbf{q} \in \mathbb{R}^{N_{Q_h}}$ with $|\mathbf{q}| \neq 0$ we have by hypothesis that $D^{\mathrm{T}}\mathbf{q} \neq 0$. Then

$$\mathbf{q}^{\mathrm{T}} S \mathbf{q} = \left(D^{\mathrm{T}}\mathbf{q}\right)^{\mathrm{T}} C^{-1} D^{\mathrm{T}} \mathbf{q} \neq 0,$$

since C^{-1} is symmetric positive definite. Thus, matrix S (which is clearly symmetric) has all eigenvalues different from zero and, consequently, is non-singular. This concludes the proof. □

The scheme we have presented, with an explicit treatment of just the convective term, is only one of the many possible ways of producing a time discretization of the Navier–Stokes equations. Another choice is to resort to a fully implicit scheme.

14.2. Fully implicit schemes

By employing in (14.1) a full implicit treatment of the convective part, we would obtain a non-linear system of the following type:

$$\begin{pmatrix} E(\mathbf{W}) & D^{\mathrm{T}} \\ D & 0 \end{pmatrix} \begin{pmatrix} \mathbf{W} \\ \boldsymbol{\Pi} \end{pmatrix} = \begin{pmatrix} \mathbf{F}_s \\ \mathbf{0} \end{pmatrix}, \tag{14.9}$$

where now the matrix E is a function of the unknown velocity,

$$(E(\mathbf{W}))_{ij} = \tilde{a}(\varphi_i, \varphi_j) + c\left(\mathbf{u}^{k+1}, \varphi_j, \varphi_i\right) = C_{ij} + \sum_{m=1}^{N_{\mathbf{V}_h}} c(\varphi_m, \varphi_j, \varphi_i) W_m.$$

A possible way to solve it is to resort to *Newton's method*:
Given $\binom{\mathbf{W}^0}{\boldsymbol{\Pi}^0}$, solve for $l = 0, \ldots,$

$$\begin{aligned} &\begin{pmatrix} \dfrac{\partial E}{\partial \mathbf{W}}(\mathbf{W}^l) \cdot \mathbf{W}^l + E\left(\mathbf{W}^l\right) & D^{\mathrm{T}} \\ D & 0 \end{pmatrix} \begin{pmatrix} \mathbf{W}^{l+1} - \mathbf{W}^l \\ \boldsymbol{\Pi}^{l+1} - \boldsymbol{\Pi}^l \end{pmatrix} \\ &\quad = \begin{pmatrix} \mathbf{F}_s \\ \mathbf{0} \end{pmatrix} - \begin{pmatrix} E\left(\mathbf{W}^l\right) & D^{\mathrm{T}} \\ D & 0 \end{pmatrix} \begin{pmatrix} \mathbf{W}^l \\ \boldsymbol{\Pi}^l \end{pmatrix}, \end{aligned} \tag{14.10}$$

until a suitable convergence criterion is met.

The solution of a non-linear system is now reduced to a series of solutions of linear systems. Going back to the Navier–Stokes equations, we may note that a full implicit scheme would require to solve *at each time step* a series of linear systems of form (14.10), that resembles the Stokes problem. The resulting numerical scheme is thus in general very *computationally intensive*.

14.3. Semi-Lagrangian schemes

An alternative way to treat the non-linear term in the Navier–Stokes equations is obtained by performing an operator splitting that separates the effect of the convective term. The technique exploits the fact that the convective term is indeed the material derivative of $\mathbf{u}$,

$$\frac{\partial \mathbf{u}}{\partial t} + \mathbf{u} \cdot \nabla \mathbf{u} = \frac{D\mathbf{u}}{Dt},$$

that is the derivative along the particle trajectories T_ξ (also called characteristic lines) defined in Section 7.

On each time-slab I^k we then have that

$$\int_{t^k}^{t^{k+1}} \frac{D\mathbf{u}(\tau,\mathbf{x})}{D\tau}\,\mathrm{d}\tau = \mathbf{u}\left(t^{k+1},\mathbf{x}\right) - \mathbf{u}\left(t^{k+1},\mathbf{x}^*\right) \approx \mathbf{u}^{k+1}(\mathbf{x}) - \mathbf{u}^k(\mathbf{x}^*), \qquad (14.11)$$

where $\mathbf{x}^*$ is position at $t = t^k$ of the fluid particle located in $\mathbf{x}$ at $t = t^{k+1}$, i.e., $\mathbf{x}^* = \mathbf{y_x}(1)$ where $\mathbf{y_x}(s)$ is the solution of

$$\begin{cases} \dfrac{\mathrm{d}\mathbf{y_x}(s)}{\mathrm{d}s} = -\mathbf{u}\left(t^{k+1} - s\Delta t, \mathbf{y_x}(s)\right), \\ \mathbf{y_x}(0) = \mathbf{x}. \end{cases} \qquad (14.12)$$

The point $\mathbf{x}^*$ is often denoted as the "foot" of the characteristic line $\mathbf{y_x}$.

This interpretation leads to the *semi-Lagrangian* schemes, so called because we treat the convective operator in the Lagrangian frame. For instance, a backward Euler semi-Lagrangian scheme will lead at each time step I^k a generalized Stokes problem like (14.4), where now

$$\mathbf{q}(\mathbf{x}) = \mathbf{f}^{k+1}(\mathbf{x}) + \frac{1}{\Delta t}\mathbf{u}^k(\mathbf{x}^*),$$

that may then be treated by a Galerkin finite element procedure as described in Section 14.1.

Clearly, system (14.12) has to be approximated as well. A first-order approximation leads to

$$\mathbf{x}^* = \mathbf{x} - \mathbf{u}^k(\mathbf{x})\Delta t.$$

This explicit treatment will eventually entail a stability condition which depends on the fluid velocity. Higher-order schemes may be devised as well, see, for instance, BOUKIR, MADAY, MÉTIVET and RAZAFINDRAKOTO [1997].

The major drawback of semi-Lagrangian schemes is the computation of the approximation of $\mathbf{u}^k(\mathbf{x}^*)$. In a finite element context it requires to locate the mesh element where the foot of the characteristic passing through each mesh point lies (or each quadrature point if a quadrature rule is used to compute the space integrals). An efficient implementation calls for the use of special data structures. Furthermore, a proper treatment is needed when $\mathbf{x}^*$ falls outside the computational domain. In that case the boundary conditions have to be properly taken into account.

14.4. Projection methods

We now follow another route for the solution of the incompressible Navier–Stokes equations which does not lead to a Stokes problem but to a series of simpler systems of partial differential equations. We start from the Navier–Stokes equations already discretized in time and we will consider again a single time step, that is

$$\frac{\mathbf{u}^{k+1} - \mathbf{u}^k}{\Delta t} + \left(\mathbf{u}^k \cdot \nabla\right)\mathbf{u}^{k+1} - 2\,\mathbf{div}\left(\nu\mathbf{D}\left(\mathbf{u}^{k+1}\right)\right) + \nabla p^{k+1} = \mathbf{f}^{k+1}, \quad \text{in } \Omega, \qquad (14.13)$$

plus (14.1b) and (14.1c). Here, for the sake of simplicity (and without any loss of generality) we have chosen a semi-implicit treatment of the convective term. We wish now to split the system in order to consider the effects of the velocity and the pressure terms separately. We define an intermediate velocity $\tilde{\mathbf{u}}$, obtained by solving the momentum equation where the pressure contribution has been dropped, precisely

$$\frac{\tilde{\mathbf{u}}-\mathbf{u}^k}{\Delta t}+\left(\mathbf{u}^k\cdot\nabla\right)\tilde{\mathbf{u}}-2\,\mathbf{div}\big(\nu\mathbf{D}(\tilde{\mathbf{u}})\big)=\mathbf{f}^{k+1},\quad \text{in } \Omega, \tag{14.14a}$$

$$\tilde{\mathbf{u}}=\mathbf{0},\quad \text{on } \partial\Omega. \tag{14.14b}$$

We may recognize that (14.14a) is now a problem on the velocity only, which could be re-interpreted as the time discretization of a parabolic differential equation of the following type:

$$\frac{\partial\tilde{\mathbf{u}}}{\partial t}+(\mathbf{w}\cdot\nabla)\tilde{\mathbf{u}}-2\,\mathbf{div}\big(\nu\mathbf{D}(\tilde{\mathbf{u}})\big)=\mathbf{f},$$

with $\mathbf{w}$ a given vector field. At this stage, we cannot impose the incompressibility condition because we would obtain an over-constrained system.

We then consider the contribution given by the pressure term and the incompressibility constraint, that is

$$\frac{\mathbf{u}^{k+1}-\tilde{\mathbf{u}}}{\Delta t}+\nabla p^{k+1}=\mathbf{0},\quad \text{in } \Omega, \tag{14.15a}$$

$$\operatorname{div}\mathbf{u}^{k+1}=0,\quad \text{in } \Omega. \tag{14.15b}$$

System (14.15) depends on both the velocity and pressure, yet we may derive an *equation only for the pressure* by taking (formally) the divergence of (14.15a) and exploiting the incompressibility constraint (14.15b). That is,

$$\begin{aligned}0=\operatorname{div}\left(\frac{\mathbf{u}^{k+1}-\tilde{\mathbf{u}}}{\Delta t}+\nabla p^{k+1}\right)&=-\frac{1}{\Delta t}\operatorname{div}\tilde{\mathbf{u}}+\operatorname{div}\nabla p^{k+1}\\&=-\frac{1}{\Delta t}\operatorname{div}\tilde{\mathbf{u}}+\Delta p^{k+1},\end{aligned}$$

by which we obtain a *Poisson equation* for the pressure in the form

$$\Delta p^{k+1}=\frac{1}{\Delta t}\operatorname{div}\tilde{\mathbf{u}},\quad \text{in } \Omega. \tag{14.16}$$

Eq. (14.16) must be supplemented by boundary conditions, which are *not directly available from the original problem* (14.13). For that, we need to resort to the following theorem, also known as *Ladhyzhenskaja theorem.*

THEOREM 14.1 (Helmholtz decomposition principle). *Let Ω be a domain of $\mathbb{R}^N$ with smooth boundary. Any vector function $\mathbf{v}\in\mathbf{L}^2(\Omega)$ (with $N=2,3$) can be uniquely represented as $\mathbf{v}=\mathbf{w}+\nabla\psi$ with $\mathbf{w}\in\mathbf{H}_{\text{div}}(\Omega)$, where*

$$\mathbf{H}_{\text{div}}(\Omega)=\Big\{\mathbf{w}:\ \mathbf{w}\in\mathbf{L}^2(\Omega),\ \operatorname{div}\mathbf{w}=0,\ \textit{a.e. } \mathbf{w}\cdot\mathbf{n}=0 \textit{ on } \partial\Omega\Big\},$$

and $\psi\in H^1(\Omega)$.

The proof is rather technical and is here omitted. An outline, valid for the case $\mathbf{v} \in \mathbf{H}^1(\Omega)$, is given in CHORIN and MARSDEN [1990]. A more general demonstration is found in TEMAM [1984], Theorems 1 and 5.

If we now consider the expression

$$\tilde{\mathbf{u}} = \mathbf{u}^{k+1} + \nabla\left(\Delta t p^{k+1}\right), \tag{14.17}$$

derived from (14.15a), we may identify $\tilde{\mathbf{u}}$ with $\mathbf{v}$ and $(\Delta t p^{k+1})$ with ψ in the Helmholtz decomposition principle. Then, the natural space for $\mathbf{u}^{k+1}$ is $\mathbf{H}_{\text{div}}(\Omega)$, by which we should impose

$$\mathbf{u}^{k+1} \cdot \mathbf{n} = 0, \quad \text{on } \partial\Omega. \tag{14.18}$$

Unfortunately, (14.18) is still a condition on the velocity, while we are looking for a boundary condition for the pressure. The latter is found by considering the normal component of (14.17) on the boundary,

$$\tilde{\mathbf{u}} \cdot \mathbf{n} = \mathbf{u}^{k+1} \cdot \mathbf{n} + \Delta t \nabla p^{k+1} \cdot \mathbf{n}, \quad \text{on } \partial\Omega,$$

and noting that on $\partial\Omega$ we have $\tilde{\mathbf{u}} \cdot \mathbf{n} = 0$, because of (14.14b), and $\mathbf{u}^{k+1} \cdot \mathbf{n} = 0$. Then,

$$\nabla p^{k+1} \cdot \mathbf{n} = \frac{\partial p^{k+1}}{\partial n} = 0, \quad \text{on } \partial\Omega,$$

which is a *homogeneous Neumann boundary condition* for the Poisson problem (14.16).

The *projection method* here presented for the solution of the Navier–Stokes equations consists then in solving at each time-step a sequence of simpler problems, listed in the following:

(1) *Advection–diffusion problem for the velocity* $\tilde{\mathbf{u}}$. Solve problem (14.14a)–(14.14b).

(2) *Poisson problem for the pressure*

$$\Delta p^{k+1} = \frac{1}{\Delta t} \operatorname{div} \tilde{\mathbf{u}}, \quad \text{in } \Omega, \tag{14.19a}$$

$$\frac{\partial}{\partial n} p^{k+1} = 0, \quad \text{on } \partial\Omega. \tag{14.19b}$$

(3) *Computation of* $\mathbf{u}^{k+1}$ (this is an explicit step)

$$\mathbf{u}^{k+1} = \tilde{\mathbf{u}} - \Delta t \nabla p^{k+1}. \tag{14.20}$$

For an analysis of projection methods as well as the set-up of higher order schemes the reader may consult PROHL [1997] and GUERMOND [1999]. We point out that projection schemes may also be used in conjunction with the semi-Lagrangian treatment of the convective term (ACHDOU and GUERMOND [2000]).

14.5. Algebraic factorization methods

An alternative way of reducing the computational cost of the solution of the full Navier–Stokes problem is to operate at algebraic level. We will consider the generalized Stokes problem in its algebraic form (14.7). This is the typical system that arises at each time step of a time advancing scheme for the solution of the Navier–Stokes by a finite element method, when the convective term is treated explicitly. In this case, the matrix C has the form

$$C = \frac{M}{\Delta t} + K + B,$$

where M is the mass matrix, K the stiffness matrix and B the matrix arising from the explicit treatment of the convective term.

The matrix D derives from the discretization of the divergence term, while D^{T} represents a discrete gradient operator. We may formally solve for $\mathbf{W}$,

$$\mathbf{W} = C^{-1}\left(\mathbf{F}_s - D^{\mathrm{T}}\boldsymbol{\Pi}\right), \tag{14.21}$$

and by substituting into (14.7), we have

$$DC^{-1}D^{\mathrm{T}}\boldsymbol{\Pi} = DC^{-1}\mathbf{F}_s. \tag{14.22}$$

The matrix $DC^{-1}D^{\mathrm{T}}$ is called *Stokes pressure matrix* and is somehow akin to a discrete Laplace operator. Having obtained $\boldsymbol{\Pi}$ from (14.22), we can then compute the velocity by solving (14.21).

However, the inversion of C is in general prohibitive in terms of memory and computational cost (indeed, C is sparse, but C^{-1} is not).

A way to simplify the computation can be found by recognizing that steps (14.22) and (14.21) may be derived from the following LU factorization of the global matrix A:

$$A = \begin{pmatrix} C & D^{\mathrm{T}} \\ D & 0 \end{pmatrix} = \begin{pmatrix} C & 0 \\ D & -DC^{-1}D^{\mathrm{T}} \end{pmatrix} \begin{pmatrix} I_{\mathbf{W}} & C^{-1}D^{\mathrm{T}} \\ D & I_{\Pi} \end{pmatrix} = LU, \tag{14.23}$$

where $I_{\mathbf{W}}$ and I_{Π} indicate the identity matrices of dimension equal to the number of velocity and pressure degrees of freedom, respectively. We then consider the LU solution

$$\begin{cases} C\widetilde{\mathbf{W}} = \mathbf{F}_s, \\ D\widetilde{\mathbf{W}} - DC^{-1}D^{\mathrm{T}}\widetilde{\boldsymbol{\Pi}} = 0, \end{cases} \qquad \begin{cases} \mathbf{W} + C^{-1}D^{\mathrm{T}}\boldsymbol{\Pi} = \widetilde{\mathbf{W}}, \\ \boldsymbol{\Pi} = \widetilde{\boldsymbol{\Pi}}, \end{cases}$$

where $\widetilde{\mathbf{W}}$ and $\widetilde{\boldsymbol{\Pi}}$ are intermediate velocities and pressures.

The scheme may be written in the following alternative form:

$$\textit{Intermediate velocity} \quad C\widetilde{\mathbf{W}} = \mathbf{F}_s, \tag{14.24a}$$

$$\textit{Pressure computation} \quad -DC^{-1}D^{\mathrm{T}}\boldsymbol{\Pi} = -D\widetilde{\mathbf{W}}, \tag{14.24b}$$

$$\textit{Velocity update} \quad \mathbf{W} = \widetilde{\mathbf{W}} - C^{-1}D^{\mathrm{T}}\boldsymbol{\Pi}. \tag{14.24c}$$

The key to reduce complexity is to replace C by a matrix simpler to invert, which, however, is "similar" to C, in a sense that we will make precise. This technique is called *inexact factorization*. In practise, we replace A in (14.23) by an approximation A^* obtained by replacing in the LU factorization the matrix C^{-1} by convenient approximations, which we indicate by H_1 and H_2, that is

$$A^* = L^*U^* = \begin{pmatrix} C & 0 \\ D & -DH_1D^{\mathrm{T}} \end{pmatrix} \begin{pmatrix} I_{\mathbf{W}} & H_2D^{\mathrm{T}} \\ D & I_{\Pi} \end{pmatrix}$$
$$= \begin{pmatrix} C & CH_2D^{\mathrm{T}} \\ D & D(H_2 - H_1)D^{\mathrm{T}} \end{pmatrix}. \tag{14.25}$$

If we choose $H_2 = H_1$, the discrete continuity equation is unaltered, that means that the approximated system still guarantees *mass conservation* at discrete level. If $H_2 = C^{-1}$, the discrete momentum equations are unaltered, and the resulting scheme satisfies the discrete conservation of momentum. In particular, we can consider the two special cases

$$H_1 = H_2 = H \quad \Rightarrow \quad A^* = \begin{pmatrix} C & CHD^{\mathrm{T}} \\ D & 0 \end{pmatrix},$$
$$H_1 = C^{-1} \neq H_2 \quad \Rightarrow \quad A^* = \begin{pmatrix} C & D^{\mathrm{T}} \\ D & Q \end{pmatrix}, \quad Q = D\left(H_1 - C^{-1}\right)D^{\mathrm{T}}.$$

14.5.1. The algebraic Chorin–Temam scheme

We note that

$$C = \frac{M}{\Delta t} + K + B = \frac{1}{\Delta t}\left(M + \Delta t(K + B)\right) = \frac{1}{\Delta t}M\left(I_{\mathbf{W}} + \Delta t M^{-1}(K + B)\right).$$

We recall the well-known Neumann expansion formula (MEYER [2000])

$$(I + \varepsilon A)^{-1} = \sum_{j=0}^{\infty}(-1)^j(\varepsilon A)^j,$$

which converges for any matrix A and any positive number ε small enough to guarantee that the spectral radius of εA is strictly less than one. We can apply this formula to C^{-1} to get

$$C^{-1} = \Delta t\left(I_{\mathbf{W}} + \Delta t M^{-1}(K + B)\right)^{-1} M^{-1}$$
$$= \Delta t \sum_{j=0}^{\infty}(-1)^j\left[\Delta t M^{-1}(K + B)\right]^j M^{-1}$$
$$= \Delta t\left(I_{\mathbf{W}} - \Delta t M^{-1}S + \cdots\right)M^{-1}, \tag{14.26}$$

where we have put $S = K + B$.

A way to find a suitable approximation is to replace C^{-1} with a few terms of the series. The simplest choice considers just a first order approximation, which corresponds to put into (14.25)

$$H_1 = H_2 = H = \Delta t M^{-1}. \tag{14.27}$$

Consequently,

$$A^* = A_{CT} = \begin{pmatrix} C & \Delta t C M^{-1}D^{\mathrm{T}} \\ D & 0 \end{pmatrix} = \begin{pmatrix} C & D^{\mathrm{T}} + \Delta t S M^{-1}D^{\mathrm{T}} \\ D & 0 \end{pmatrix}. \tag{14.28}$$

The scheme obtained by applying the corresponding LU decomposition reads:

$$\text{Intermediate velocity} \quad C\widetilde{\mathbf{W}} = \mathbf{F}_s, \tag{14.29a}$$

$$\text{Pressure computation} \quad -\Delta t D M^{-1} D^{\mathrm{T}} \boldsymbol{\Pi} = -D\widetilde{\mathbf{W}}, \tag{14.29b}$$

$$\text{Velocity update} \quad \mathbf{W} = \widetilde{\mathbf{W}} - \Delta t M^{-1} D^{\mathrm{T}} \boldsymbol{\Pi}. \tag{14.29c}$$

This algorithm is known as *algebraic Chorin–Temam scheme*. Comparing with the standard projection method, we may note that the algebraic scheme replaces in the pressure computation step (14.29b) the Laplace operator of the Poisson problem (14.9) with a "discrete Laplacian" $DM^{-1}D^{\mathrm{T}}$, which incorporates the boundary condition of the original problem. No additional boundary condition is required for the pressure, contrary to the standard (differential type) scheme.

REMARK 14.1. The finite element mass matrix M is sparse and with the same graph structure as C. Therefore, it may seem that there is little gain in the computational efficiency with respect to the original factorization (14.24). However, the matrix M may be approximated by a diagonal matrix called *lumped mass matrix* (QUARTERONI and VALLI [1994]), whose inversion is now trivial.

REMARK 14.2. It is possible to write the algebraic Chorin–Temam scheme in incremental form, as it has been done for its differential counterpart.

14.5.2. The Yosida scheme

If we make the special choice

$$H_1 = \Delta t M^{-1}, \qquad H_2 = C^{-1}, \tag{14.30}$$

we have

$$A^* = A_Y = \begin{pmatrix} C & D^{\mathrm{T}} \\ D & Q \end{pmatrix} \quad \text{with } Q = -D\left(\Delta t M^{-1} - C^{-1}\right) D^{\mathrm{T}}. \tag{14.31}$$

The corresponding scheme reads

$$\text{Intermediate velocity} \quad C\widetilde{\mathbf{W}} = \mathbf{F}_s, \tag{14.32a}$$

$$\text{Pressure computation} \quad -\Delta t D M^{-1} D^{\mathrm{T}} \boldsymbol{\Pi} = -D\widetilde{\mathbf{W}}, \tag{14.32b}$$

$$\text{Velocity update} \quad \mathbf{W} = \widetilde{\mathbf{W}} - \Delta t C^{-1} D^{\mathrm{T}} \boldsymbol{\Pi}. \tag{14.32c}$$

The last step (14.32c) is more expensive than its counterpart (14.29c) in the Chorin–Temam scheme, since now we need to invert the full matrix C. An analysis of this method is found in QUARTERONI, SALERI and VENEZIANI [1999].

REMARK 14.3. If we consider the Stokes problem, we have $C = (\Delta t)^{-1}M + K$ and consequently the matrix $Q = -D(\Delta t M^{-1} - C^{-1})D^{\mathrm{T}}$ in (14.31) may be written as

$$Q = -\Delta t D\left[I_{\mathbf{W}} - (I_{\mathbf{W}} + \Delta t K)^{-1}\right] D^{\mathrm{T}} = -(\Delta t)^2 D Y D^{\mathrm{T}},$$

where

$$Y = \frac{1}{\Delta t}\left[I_{\mathrm{W}} - (I_{\mathrm{W}} + \Delta t K)^{-1}\right],$$

may be regarded as the *Yosida* regularization of K, which is the discretization of the Laplace operator. That is Q may be interpreted as the discretization of the differential operator

$$(\Delta t)^2 \operatorname{div}(\mathcal{Y}_{\Delta t}\nabla),$$

where $\mathcal{Y}_{\Delta t}$ is the Yosida operator (BREZIS [1983]).

REMARK 14.4. An incremental form may be found as follows. If $\boldsymbol{\Pi}^n$ represents the known value of the pressure degrees of freedom from the previous time step, we have

Intermediate velocity	$C\widetilde{\mathbf{W}} = \mathbf{F}_s - D^{\mathrm{T}}\boldsymbol{\Pi}^n,$
Pressure increment	$-\Delta t D M^{-1} D^{\mathrm{T}}\left(\boldsymbol{\Pi} - \boldsymbol{\Pi}^n\right) = -D\widetilde{\mathbf{W}},$
Velocity update	$\mathbf{W} = \widetilde{\mathbf{W}} - \Delta t C^{-1} D^{\mathrm{T}}\left(\boldsymbol{\Pi} - \boldsymbol{\Pi}^n\right).$

More details on algebraic fractional step methods may be found in PEROT [1993] and QUARTERONI, SALERI and VENEZIANI [2000].

A major advantage of the algebraic factorization schemes with respect to projection methods is that they do not require to devise special boundary conditions for the pressure problem, a task which is not always trivial.

All the techniques here presented may be extended to moving domains using the procedure that will be illustrated in Section 18. In a moving domain context the various matrices of the final algebraic system have to be recomputed at each times step to reflect the change of domain geometry. As a consequence, a fully implicit approach is even less computationally attractive, and factorization schemes (at differential or algebraic level), possibly with a semi-Lagrangian treatment of the convective term, are normally preferred.

In the context of hemodynamics, algebraic factorization schemes are particularly attractive because of their flexibility with respect to the application of boundary conditions. In particular, they can easily accommodate defective boundary conditions (FORMAGGIA, GERBEAU, NOBILE and QUARTERONI [2002]).

CHAPTER IV

15. Mathematical modeling of the vessel wall

The vascular wall has a very complex nature and devising an accurate model for its mechanical behavior is rather difficult. Its structure is indeed formed by many layers with different mechanical characteristics (FUNG [1993], HOLZAPFEL, GASSER and OGDEN [2000]) (see Fig. 15.1). Moreover, experimental results obtained by specimens are only partially significant. Indeed, the vascular wall is a living tissue with the presence of muscular cells which contribute to its mechanical behavior. It may then be expected that the dead tissue used in the laboratory will have different mechanical characteristics than the living one. Moreover, the arterial mechanics depend also on the type of the surrounding tissues, an aspect almost impossible to reproduce in a laboratory. We are then facing a problem whose complexity is enormous. It is the role of mathematical modeling to find reasonable simplifying assumptions by which major physical characteristics remain present, yet the problem becomes amenable to numerical analysis and computational solution.

The set up of a general mathematical model of the mechanics of a solid continuum may follow the same general route that we have indicated for fluid mechanics. In particular, it is possible to identify again a Cauchy stress tensor $\mathbf{T}$. The major difference between solids and fluids is in the constitutive relation which links $\mathbf{T}$ to kinematics

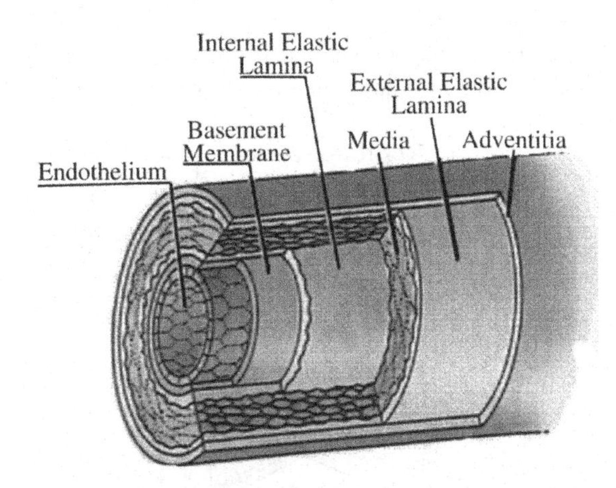

FIG. 15.1. The vessel wall is formed by many layers made of tissues with different mechanical characteristics. Image taken from "Life: the Science of Biology" by W.K. Purves et al., fourth edition, published by Sinauer Associates Inc. and W.H. Freeman and Company.

quantities. We have seen in Section 5 that for a fluid such a kinematic quantity is the velocity gradient or, more precisely, the strain rate **D**. For a solid, the Cauchy stress tensor is instead a function of the *deformation gradient*, which we have already defined in (7.7). That is, the *constitutive law* for a solid may be written as

$$\mathbf{T} = \mathbf{T}(\mathbf{F}_t).$$

If we assume that both the deformation gradient and the displacements are small, under the hypothesis of *linear elasticity* and homogeneous material it is possible to derive relatively simple relations for **T**. For sake of space, we will not pursue that matter here. The interested reader may consult, for instance, the book by L.A. Segel (SEGEL [1987], Chapter 4), or, for a more extensive treatment, the book by P.G. Ciarlet (CIARLET [1988]).

Another possible situation is the one that involves a constitutive law of the form

$$\mathbf{T} = \mathbf{T}(\mathbf{D}, \mathbf{F}_t), \tag{15.1}$$

which describes the mechanical behavior of a material with characteristics intermediate to those of a liquid and a solid. In such case, the continuum is said to be *viscoelastic*. An example of such behavior is given by certain plastics or by liquid suspensions. In particular, also blood exhibits a viscoelastic nature, particularly when flowing in small vessels, e.g. in arterioles and capillaries. Indeed, in that case the presence of suspended particles and their interaction during the motion strongly affect the blood mechanical behavior. Again, we will not cover this topic here. The book by Y.C. Fung (FUNG [1993]) may be used by the reader interested on the peculiar aspects of the mechanics of living tissues.

The geometry of a section of an artery where no branching is present may be described by using a curvilinear cylindrical coordinate system (r, θ, z) with the corresponding base unit vectors $\mathbf{e}_r$, $\mathbf{e}_\theta$ and $\mathbf{e}_z$, where $\mathbf{e}_z$ is aligned with the axis of the artery, as shown in Fig. 15.2.

Clearly, the vessel structure may be studied using full three-dimensional models, which may also account for its multilayer nature. However, it is common practice to

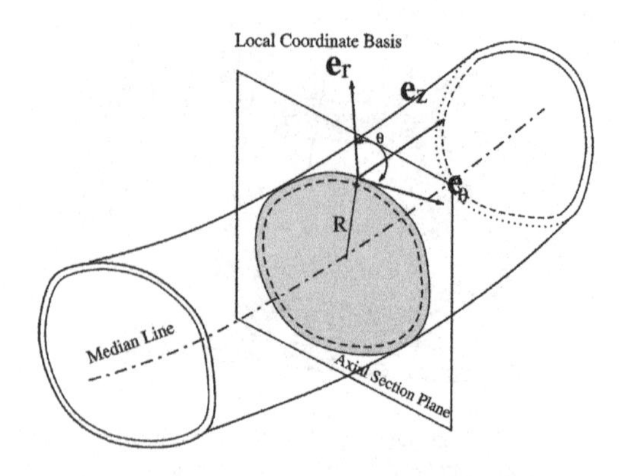

FIG. 15.2. A model of a "realistic" section of an artery with the principal geometrical parameters.

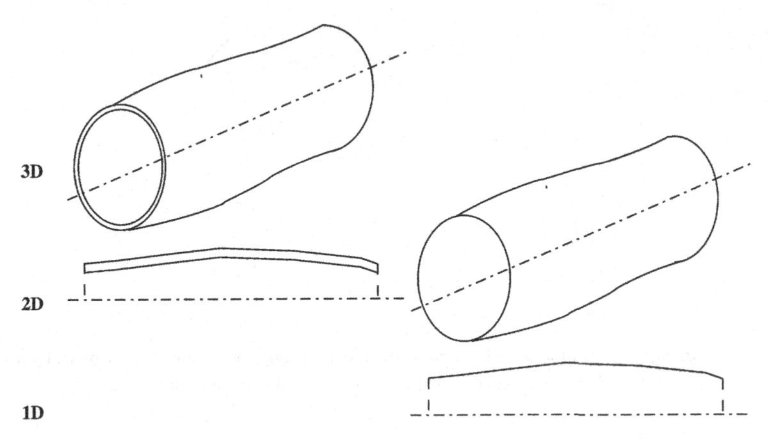

FIG. 15.3. Different models for arterial wall mechanics.

resort to simplified 2D or even 1D mechanical models in order to reduce the overall computational complexity when the final aim is to study the coupled fluid–structure problem. In Fig. 15.3 we sketch some of the approximations normally made. A 2D model may be obtained by either resorting to a shell-type description or considering longitudinal Sections ($\theta = \text{const.}$) of the vessels. In the first case we exploit the fact that the effective wall thickness is relatively small to reduce the whole structure to a surface. A rigorous mathematical derivation (for the linear case) may be found in CIARLET [1998]. In the second case we neglect the variations of the stresses in the circumferential direction. In this way we are able to eliminate all terms containing derivatives with respect to θ in the equations and we may consider each plane $\theta = \text{const.}$ independently. The resulting displacement field will depend only parametrically on θ. If, in addition, we assume that the problem has an axial symmetry (which implies the further assumption of a straight axis) the dependence on θ is completely neglected. In this case, also the fluid would be described by a 2D axi-symmetric model.

The simplest models, called 1D models, are derived by making the same assumption on the wall thickness made for the shell model, yet starting from a 2D model. The structure will then be represented by a line on a generic longitudinal section, as shown in the last picture of Fig. 15.3.

Even with all these simplifying assumptions an accurate model of the vessel wall mechanics is rather complex. Therefore, in these notes we will only present the simplest models, whose derivation is now detailed.

16. Derivation of 1D models of vessel wall mechanics

We are going to introduce a hierarchy of 1D models for the vessel structure of variable complexity. We first present the assumptions common to all models.

The relatively small thickness of the vessel wall allows us to use as basis model a shell model, where the vessel wall geometry is fully described by its median surface, see Fig. 16.1.

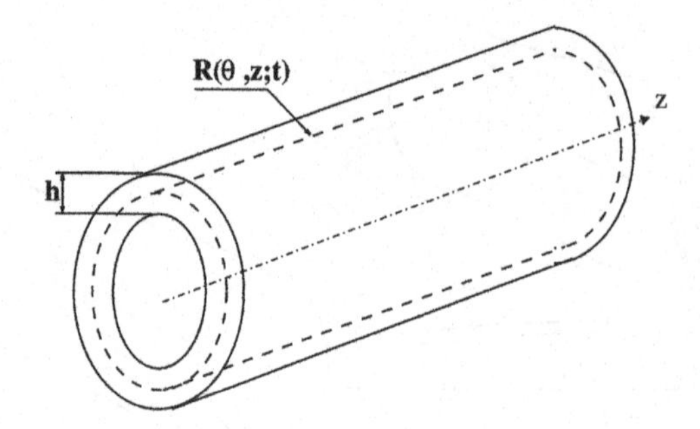

FIG. 16.1. A cylindrical model of the vessel geometry. The latter is approximated, at any time t, by a surface $r = R(\theta, z; t)$, which is outlined with dashed lines in figure.

We take as *reference configuration* Γ_0^w the one assumed by the vessel at rest when filled with fluid with zero velocity and whose pressure is equal to the pressure P_{ext} exerted by the tissues external to the vessel. Although in principle P_{ext} can change along the vessel (for instance, because of the effect of gravity), for the sake of simplicity (and without any loss of generality) we will consider only the case where P_{ext} is constant.

The cylindrical-like aspect of sections of the arterial system allows us to derive simplified mathematical models for the movement of the arterial wall assuming a straight cylindrical geometry. We thus assume that the reference configuration Γ_0^w be a cylindrical surface with radius R_0 (a regular strictly positive function of z), i.e.,

$$\Gamma_0^w = \{(r, \theta, z):\ r = R_0(z),\ \theta \in [0, 2\pi),\ z \in [0, L]\},$$

where L indicates the length of the arterial element under consideration. In our cylindrical coordinate system (r, θ, z), the z coordinate is aligned along the vessel axes and a plane $z = \bar{z}$ (= constant) defines an *axial section*.

We assume that the displacement vector $\boldsymbol{\eta}$ has only a radial component, that is

$$\boldsymbol{\eta} = \eta \mathbf{e}_r = (R - R_0)\mathbf{e}_r, \tag{16.1}$$

where $R = R(\theta, z; t)$ is the function that provides, at each t, the radial coordinate $r = R(\theta, z; t)$ of the wall surface. The *current configuration* Γ_t^w at time t of the vessel surface is then given by

$$\Gamma_t^w = \{(r, \theta, z):\ r = R(\theta, z; t),\ \theta \in [0, 2\pi),\ z \in [0, L]\}.$$

As a consequence, the length of the vessel does not change with time. We will indicate with $\mathbf{n}$ the outwardly oriented unit normal to the surface Γ_t^w at a given point. In Fig. 16.2 we sketch the reference and current configuration for the model of the section of an artery.

Another important assumption is that of *plain stresses*. We neglect the stress components along the normal direction $\mathbf{n}$, i.e., we assume that the stresses lie on the vessel surface.

We itemise here the main assumptions:

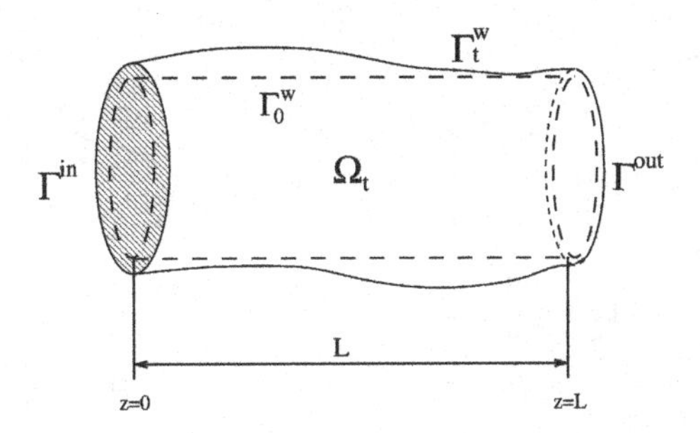

FIG. 16.2. The reference configuration Γ_0^w used for the derivation of our models is that of a circular cylinder. Γ_t^w indicates the current configuration at a given time t, while Ω_t is the domain occupied by the fluid.

(A1) *Small thickness and plain stresses.* The vessel wall thickness h is sufficiently small to allow a shell-type representation of the vessel geometry. In addition, we will also suppose that it is constant in the reference configuration. The vessel structure is subjected to plain stresses.

(A2) *Cylindrical reference geometry and radial displacements.* The reference vessel configuration is described by a circular cylindrical surface with straight axes.[3] The displacements are only in the radial direction.

(A3) *Small deformation gradients.* We assume that the deformation gradients are small, so that the structure basically behaves like a linear elastic solid and $\partial R/\partial\theta$ and $\partial R/\partial z$ remain uniformly bounded during motion.

(A4) *Incompressibility.* The vessel wall tissue is incompressible, i.e., it maintains its volume during the motion. This is a reasonable assumption since biological tissues are indeed nearly incompressible.

The models that we are going to illustrate could be derived from the general laws of solid mechanics. Yet, this is not the route we will follow, preferring to describe them in a more direct way, while trying to give some insight on the physical meaning of the various terms that we are about to introduce.

16.1. Forces acting on the vessel wall

Let us consider the vessel configuration at a given time t and a generic point on the vessel surface of coordinates $\theta = \bar{\theta}$, $z = \bar{z}$ and $r = R(\bar{\theta}, \bar{z}; t)$, with $\bar{z} \in (0, L)$ and $\theta \in (0, 2\pi)$. In the following derivation, if not otherwise indicated, all quantities are computed at location $(R(\bar{\theta}, \bar{z}; t), \bar{\theta}, \bar{z})$ and at time t.

We will indicate with $d\sigma$ the measure of the following elemental surface:

$$dS = \left\{(r, \theta, z):\ r = R(\theta, z; t),\ \theta \in \left[\bar{\theta} - \frac{d\theta}{2}, \bar{\theta} + \frac{d\theta}{2}\right],\ z \in \left[\bar{z} - \frac{dz}{2}, \bar{z} + \frac{dz}{2}\right]\right\}.$$

[3]This assumption may be partially dispensed with, by assuming that the reference configuration is "close" to that of a circular cylinder. The model here derived may be supposed valid also in that situation.

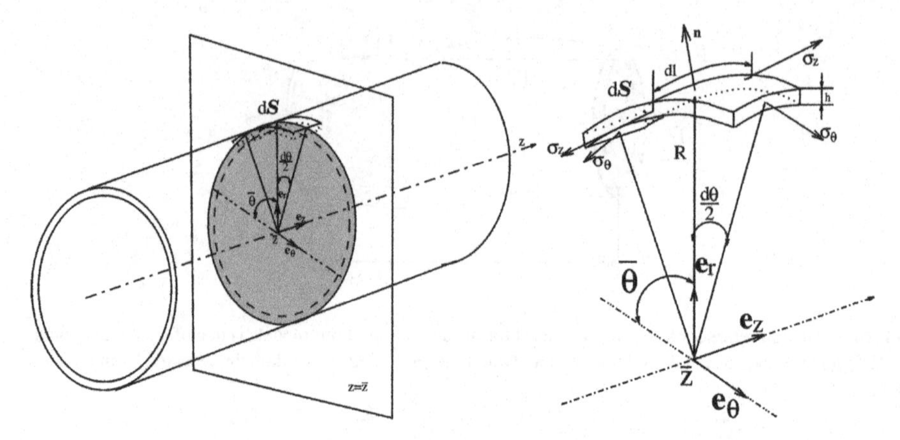

FIG. 16.3. A cylindrical model of the vessel geometry (left) and the infinitesimal portion of vessel wall used for the derivation of the equations (right).

In Fig. 16.3 we have also indicated the two main stresses, the circumferential stress and the longitudinal stress $\boldsymbol{\sigma}_\theta$ and $\boldsymbol{\sigma}_z$, which represent the internal forces acting on the portion under consideration.

We may derive the following expression for $\mathbf{n}$ and $\mathrm{d}\sigma$:

$$\mathbf{n} = (R_0 g)^{-1}\left(R\mathbf{e}_r - \frac{\partial R}{\partial \theta}\mathbf{e}_\theta - R\frac{\partial R}{\partial z}\mathbf{e}_z\right), \tag{16.2}$$

$$\mathrm{d}\sigma = g R_0\,\mathrm{d}\theta\,\mathrm{d}z = g\,\mathrm{d}\sigma_0, \tag{16.3}$$

where

$$g = \frac{R}{R_0}\sqrt{1 + \left(\frac{1}{R}\frac{\partial R}{\partial \theta}\right)^2 + \left(\frac{\partial R}{\partial z}\right)^2},$$

and $\sigma_0 = R_0\,\mathrm{d}\theta\,\mathrm{d}z$ is the measure of the image of $\mathrm{d}S$ in the reference configuration Γ_0^w. In particular, we have

$$\mathbf{n}\cdot\mathbf{e}_r = \frac{R}{R_0}g^{-1} \tag{16.4}$$

and

$$\mathbf{n}\cdot\mathbf{e}_r\,\mathrm{d}\sigma = R\,\mathrm{d}\theta\,\mathrm{d}z. \tag{16.5}$$

The linear dimension of the elemental surface $\mathrm{d}S$ along the longitudinal direction has been indicated with $\mathrm{d}l$. It can be easily verified that

$$\mathrm{d}l = \sqrt{1 + \left(\frac{\partial R}{\partial z}\right)^2}\,\mathrm{d}z. \tag{16.6}$$

Let us now consider the external forces acting through the elemental surface $\mathrm{d}S$.

- *Forces from the surrounding tissues.* As the tissue surrounding the vessel interacts with the vessel wall structure by exerting a constant pressure P_{ext}, the resulting force acting on $\mathrm{d}S$ is simply given by

$$\mathbf{f}_{\text{tissue}} = -P_{\text{ext}}\mathbf{n}\,\mathrm{d}\sigma + \mathrm{o}(\mathrm{d}\sigma). \tag{16.7}$$

- *Forces from the fluid.* The forces the fluid exerts on the vessel wall are represented by the Cauchy stresses on the wall. Then, if we indicate with $\mathbf{T}_f$ the Cauchy stress tensor for the fluid, we have

$$\mathbf{f}_{\text{fluid}} = -\mathbf{T}_f \cdot \mathbf{n}\,\mathrm{d}\sigma + \mathrm{o}(\mathrm{d}\sigma) = P\mathbf{n}\,\mathrm{d}\sigma - 2\mu\mathbf{D}(\mathbf{u}) \cdot \mathbf{n}\,\mathrm{d}\sigma + \mathrm{o}(\mathrm{d}\sigma). \tag{16.8}$$

16.2. The independent ring model

The independent ring model is expressed by a differential equation for the time evolution of η, for each z and θ. For the derivation of this model, we will make some additional assumptions:

(IR-1) *Dominance of circumferential stresses* $\boldsymbol{\sigma}_\theta$. The stresses $\boldsymbol{\sigma}_z$ acting along longitudinal direction are negligible with respect to $\boldsymbol{\sigma}_\theta$ and are thus neglected when writing the momentum equation.

(IR-2) *Cylindrical configuration.* The vessel remains a circular cylinder during motion, i.e., $\partial R/\partial\theta = 0$. This hypothesis may be partially dispensed with, by allowing small circumferential variations of the radius, yet we will neglect $\partial R/\partial\theta$ in our model.

(IR-3) *Linear elastic behavior*. Together with hypotheses (IR-1) and (IR-2) it allows us to write that the circumferential stress is proportional to the relative circumferential elongation, i.e.,

$$\sigma_\theta = \frac{E}{1-\xi^2}\frac{\eta}{R_0}, \tag{16.9}$$

where ξ is the *Poisson ratio* (which may be taken equal to 0.5 thanks to the hypothesis (A4)) and E is the Young modulus.[4]

We will write the balance of momentum along the radial direction by analyzing the system of forces acting on $\mathrm{d}S$. We have already examined the external forces, we need now to look in more details at the effect of the internal forces, which, by assumption, are only due to the circumferential stress $\boldsymbol{\sigma}_\theta$.

We may note in Fig. 16.4 that the two vectors

$$\mathbf{e}_\theta\left(\bar{\theta} + \frac{\mathrm{d}\theta}{2}\right) \quad \text{and} \quad \mathbf{e}_\theta\left(\bar{\theta} - \frac{\mathrm{d}\theta}{2}\right)$$

form with $\mathbf{e}_r$ an angle of $\pi/2 + \mathrm{d}\theta/2$ and $-(\pi/2 + \mathrm{d}\theta/2)$, respectively. The component of the resultant of the internal forces on the radial direction is then

$$f_{\text{int}} = \left(\sigma_\theta\mathbf{e}_\theta\left(\bar{\theta} + \frac{\mathrm{d}\theta}{2}\right) + \sigma_\theta\mathbf{e}_\theta\left(\bar{\theta} - \frac{\mathrm{d}\theta}{2}\right)\right) \cdot \mathbf{e}_r h\,\mathrm{d}l$$

[4] The presence of the term $1 - \xi^2$ is due to the assumption of planar stresses. Some authors (like FUNG [1984]) consider that the hypothesis of mono-axial stresses is more realistic for the problem at hand. In that case one has to omit the term $1 - \xi^2$ from the stress–strain relation and write simply $\sigma_\theta = E\eta R_0^{-1}$.

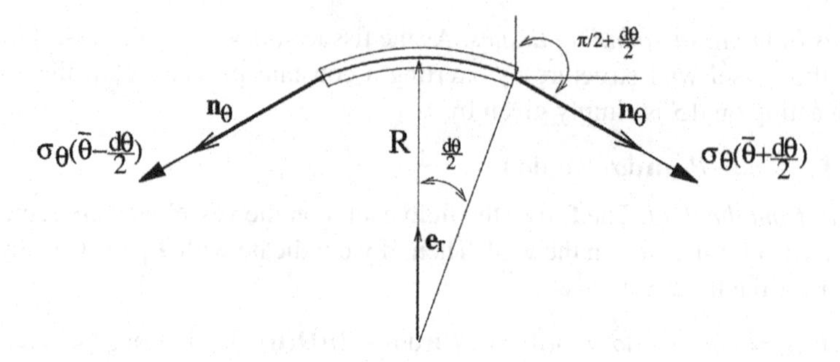

FIG. 16.4. Computation of the angle between σ_θ and the radial direction $\mathbf{e}_r$.

$$= -2\sigma_\theta \sin\frac{\mathrm{d}\theta}{2} h\,\mathrm{d}l = -\sigma_\theta h\,\mathrm{d}\theta\,\mathrm{d}l + \mathrm{o}(\mathrm{d}\theta\,\mathrm{d}l). \tag{16.10}$$

Owing to the incompressibility assumption (A4), the volume in the current configuration is unchanged with respect to that in the reference configuration, i.e.,

$$hR\,\mathrm{d}\theta\,\mathrm{d}l = h_0 R_0\,\mathrm{d}\theta\,\mathrm{d}z.$$

Then, being $\mathrm{o}(\mathrm{d}l) = \mathrm{o}(\mathrm{d}z)$, we may write (16.10) as

$$f_{\mathrm{int}} = -\frac{\sigma_\theta}{R} h_0 R_0 + \mathrm{o}(\mathrm{d}\theta\,\mathrm{d}z) = \frac{Eh_0}{1-\xi^2}\frac{\eta}{R}\,\mathrm{d}\theta\,\mathrm{d}z + \mathrm{o}(\mathrm{d}\theta\,\mathrm{d}z).$$

Finally, the mass of the portion of vessel wall under consideration is

$$\mathrm{mass} = \rho_w hR\,\mathrm{d}\theta\,\mathrm{d}l = \rho_w h_0 R_0\,\mathrm{d}\theta\,\mathrm{d}z,$$

where ρ_w is the density of the vessel tissue, whereas the acceleration along the radial direction is given by

$$\frac{\partial^2 R}{\partial t^2} = \frac{\partial^2 \eta}{\partial t^2}.$$

By balancing the resultant of the internal and external forces, provided in (16.7) and (16.8), with the inertia term, we have

$$\rho_w h_0 R_0 \frac{\partial^2 \eta}{\partial t^2}\,\mathrm{d}\theta\,\mathrm{d}z + \frac{Eh_0}{1-\xi^2}\frac{\eta}{R}\,\mathrm{d}\theta\,\mathrm{d}z$$
$$= -(2\mu\mathbf{D}(\mathbf{u})\cdot\mathbf{n})\cdot\mathbf{e}_r\,\mathrm{d}\sigma + (P - P_{\mathrm{ext}})\mathbf{n}\cdot\mathbf{e}_r\,\mathrm{d}\sigma + \mathrm{o}(\mathrm{d}\theta\,\mathrm{d}z). \tag{16.11}$$

By dividing either side by $\mathrm{d}\theta\,\mathrm{d}z$ and passing to the limit for $\mathrm{d}\theta \to 0$ and $\mathrm{d}z \to 0$, and recalling that $\mathrm{d}\sigma = gR_0\,\mathrm{d}\theta\,\mathrm{d}z = R(\mathbf{n}\cdot\mathbf{e}_r)^{-1}\,\mathrm{d}\theta\,\mathrm{d}z$, thanks to (16.2) and (16.4), we obtain

$$\rho_w h_0 R_0 \frac{\partial^2 \eta}{\partial t^2} + \frac{Eh_0}{1-\xi^2}\frac{\eta}{R} = -(2\mu\mathbf{D}(\mathbf{u})\cdot\mathbf{n})\cdot\mathbf{e}_r gR_0 + (P - P_{\mathrm{ext}})R.$$

Since the derivation has been made by considering an arbitrary plane $\theta = \bar{\theta}$ and time t, we may finally obtain the *independent ring model*

$$\frac{\partial^2 \eta}{\partial t^2} + b\eta = H, \quad \text{in } \Gamma_0^w,\ t \in I, \tag{16.12}$$

where

$$b = \frac{E}{\rho_w(1-\xi^2)R_0^2}, \tag{16.13}$$

is a positive coefficient linked to the wall mechanical properties, while

$$\begin{aligned} H &= \frac{1}{\rho_w h_0}\left[\frac{R}{R_0}(P - P_{\text{ext}}) - 2g\mu\,(\mathbf{D}(\mathbf{u})\cdot\mathbf{n})\cdot\mathbf{e}_r\right] \\ &= \frac{\rho}{\rho_w h_0}\left[\frac{R}{R_0}(p - p_{\text{ext}}) - 2g\nu\,(\mathbf{D}(\mathbf{u})\cdot\mathbf{n})\cdot\mathbf{e}_r\right], \end{aligned} \tag{16.14}$$

is the *forcing term* which accounts for the action of external forces.

REMARK 16.1. Often, the term R/R_0 in the right-hand side of (16.14) is neglected as well as the contribution to the forcing term due to the fluid viscous stresses. In this case, we have just

$$H = \frac{P - P_{\text{ext}}}{\rho_w h_0} \tag{16.15}$$

and the forcing term does not depend anymore on the current geometrical configuration.

By neglecting the acceleration term in (16.12), we obtain the following *algebraic model*, which is often found in the medical and bioengineering literature:

$$b\eta = H, \quad \text{in } \Gamma_0^w,\ t \in I, \tag{16.16}$$

according to which the wall displacement is proportional to the normal component of the applied external stresses.

REMARK 16.2. One may account for the *viscoelastic* nature of the vessel wall structure even in this simple model by adding to the constitutive relation (16.9) a term proportional to the displacement velocity, as in a simple Voigt–Kelvin model (FUNG [1993]), that is by writing

$$\sigma_\theta = \frac{E}{1-\xi^2}\frac{\eta}{R_0} + \frac{\gamma}{R_0}\frac{\partial \eta}{\partial t},$$

where γ (whose unit is $[\gamma] = \text{kg/m s}$) is a positive constant damping parameter.

Then, the resulting differential equation would read:

$$\frac{\partial^2\eta}{\partial t^2} + \frac{\gamma}{R_0^2\rho_w h_0}\frac{\partial \eta}{\partial t} + b\eta = H, \quad \text{in } \Gamma_0^w,\ t \in I. \tag{16.17}$$

We may note that the term $\frac{1}{R_0}\frac{\partial\eta}{\partial t}$ plays the role of the strain rate $\mathbf{D}$ into the general relation for viscoelastic materials (15.1).

Models (16.12), (16.16) and (16.17) are all apt to provide a solution η for every possible value of θ. In principle, since no differentiation with respect to θ is present in the model, nothing would prevent us to get significant variations of η with θ (or even a discontinuity), which would contradict assumption (IR-2). This potential drawback could be eliminated by enriching the models with further terms involving derivatives along θ, as in the case of models derived from shell theory (CIARLET [2000]). On the other hand, a more heuristic and less rigorous argument can be put forward moving from (16.16). Since b is relatively large, smooth variations of the forcing term H with respect to θ are damped to tiny one on η. This observation may be extended also to models (16.12) and (16.17) in view of the fact that for the problems at hand the term $b\eta$ dominates the other terms on the left-hand side. Similar considerations apply to the model that we will introduce in the next subsection.

16.3. The generalized string model

A more complete model (QUARTERONI, TUVERI and VENEZIANI [2000]) considers also the effects of the longitudinal stresses $\boldsymbol{\sigma}_z$. Experimental and physiological analysis (FUNG [1993]) show that vessel walls are in a "pre-stressed" state. In particular, when an artery is extracted from a body tends to "shrink", i.e., to reduce its length. This fact implies that arteries in the human body are normally subjected to a longitudinal tension.

At the base of the generalized string model is the assumption that this longitudinal tension is indeed the dominant component of the longitudinal stresses.

More precisely, let us refer to Fig. 16.5; we replace assumption (IR-1) by the following:

(GS-1) The longitudinal stress $\boldsymbol{\sigma}_z$ is not negligible and, in particular,

$$\boldsymbol{\sigma}_z = \pm\sigma_z\boldsymbol{\tau}, \tag{16.18}$$

where $\boldsymbol{\tau}$ is the unitary vector tangent to the curve

$$r = R(\bar{\theta}, z; t), \tag{16.19}$$

and its modulus σ_z is constant. Moreover, we assume that it is a *traction stress* (that is with a versus equal to that of the normal to the surface on which it applies).

We also maintain assumption (IR-2) of the independent ring model. When considering the forces acting on $\mathrm{d}S$, we have now a further term, namely (referring again to Fig. 16.5)

$$\begin{aligned}\mathbf{f}_z &= [\boldsymbol{\sigma}_z(\bar{z}+\mathrm{d}z/2) + \boldsymbol{\sigma}_z(\bar{z}-\mathrm{d}z/2)]\,hR\,\mathrm{d}\theta \\ &= \sigma_z\frac{\boldsymbol{\tau}(\bar{z}+\mathrm{d}z/2) - \boldsymbol{\tau}(\bar{z}-\mathrm{d}z/2)}{\mathrm{d}l}\,\mathrm{d}l\,hR\,\mathrm{d}\theta = \sigma_z\frac{\mathrm{d}\boldsymbol{\tau}}{\mathrm{d}l}R_0h_0\,\mathrm{d}l\,\mathrm{d}\theta + \mathbf{o}(\mathrm{d}z\,\mathrm{d}\theta).\end{aligned}$$

We now exploit the Frenet–Serret formulae to write

$$\frac{\mathrm{d}\boldsymbol{\tau}}{\mathrm{d}l} = \kappa\mathbf{n},$$

where κ is the curvature of the line $r = R(\bar{\theta}, z; t)$, whose expression is

$$\kappa = \frac{\partial^2 R}{\partial z^2}\left[1 + \left(\frac{\partial R}{\partial z}\right)^2\right]^{-3/2}. \tag{16.20}$$

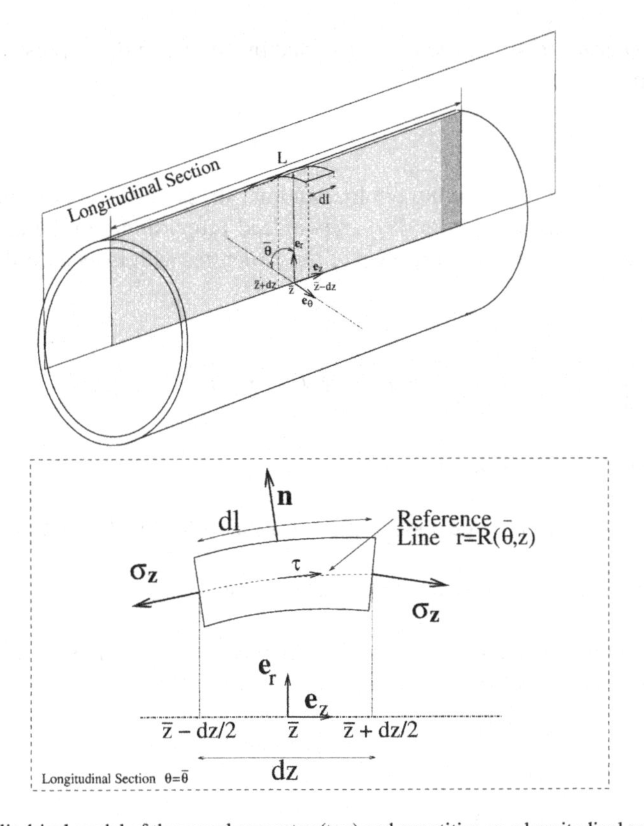

FIG. 16.5. A cylindrical model of the vessel geometry (top) and quantities on a longitudinal section (bottom).

By recalling (16.6) and (16.4), we obtain

$$\mathbf{f}_z \cdot \mathbf{e}_r = \sigma_z \frac{\partial^2 R}{\partial z^2} \left[1 + \left(\frac{\partial R}{\partial z} \right)^2 \right]^{-3/2} R_0 h_0 \, \mathrm{d}z \, \mathrm{d}\theta + \mathrm{o}(\mathrm{d}z \, \mathrm{d}\theta).$$

We eliminate the geometric non-linearity in the model by neglecting the term $(\partial R/\partial z)^2$. Furthermore, we replace $\partial^2 R/\partial z^2$ by $\partial^2 \eta/\partial z^2$.[5]

By proceeding like in the previous section, we may modify the independent ring model into the following differential equation:

$$\frac{\partial^2 \eta}{\partial t^2} - a \frac{\partial^2 \eta}{\partial z^2} + b\eta = H, \quad \text{in } \Gamma_0^w, \ t \in I, \tag{16.21}$$

where

$$a = \frac{\sigma_z}{\rho_w h_0}.$$

[5]This last equality is clearly true whenever R_0 is varying linearly with z.

The final *generalized string* model is obtained by adding to the expression for $\boldsymbol{\sigma}_z$ in (16.18) a term

$$c\frac{\partial}{\partial t}\frac{\partial \eta}{\partial z}, \quad c>0,$$

which is a viscoelastic term linking the longitudinal stress to the rate of rotation of the structure. For small displacements, $\partial\eta/\partial z$ is indeed proportional to the angle of rotation around the circumferential direction of the structure, with respect to the reference configuration.

The result is

$$\frac{\partial^2\eta}{\partial t^2} - a\frac{\partial^2\eta}{\partial z^2} + b\eta - c\frac{\partial^3\eta}{\partial t\partial z^2} = H, \quad \text{in } \Gamma_0^w,\ t\in I. \tag{16.22}$$

17. Analysis of vessel wall models

In the following we will provide some a-priori estimates for the differential models just proposed.

We recall Poincaré inequality for the one-dimensional case.

LEMMA 17.1 (Poincaré inequality – one-dimensional case). *Let $f \in H^1(a,b)$ with $f(a)=0$. Then there exists a positive constant C_p such that*

$$\|f\|_{L^2(0,L)} \leqslant C_p \left\|\frac{\mathrm{d}f}{\mathrm{d}x}\right\|_{L^2(0,L)}. \tag{17.1}$$

PROOF. For all $x\in[a,b]$ we have,

$$f(x) = f(a) + \int_a^x \frac{\mathrm{d}f}{\mathrm{d}x}(\tau)\,\mathrm{d}\tau = \int_a^x \frac{\mathrm{d}f}{\mathrm{d}x}(\tau)\,\mathrm{d}\tau.$$

Then,

$$\begin{aligned}\int_a^b f^2(s)\,\mathrm{d}s &= \int_a^b \left(\int_a^s \frac{\mathrm{d}f}{\mathrm{d}x}(\tau)\,\mathrm{d}\tau\right)^2 \mathrm{d}s\\ &\leqslant \int_a^b \left(\left(\int_a^s 1^2\,\mathrm{d}\tau\right)^{1/2} \left\{\int_a^s \left[\frac{\mathrm{d}f}{\mathrm{d}x}(\tau)\right]^2 \mathrm{d}\tau\right\}^{1/2}\right)^2 \mathrm{d}s\\ &\qquad \text{(by Cauchy–Schwarz inequality)}\\ &\leqslant \int_a^b (b-a)\left\|\frac{\mathrm{d}f}{\mathrm{d}x}\right\|^2_{L^2(a,b)} \mathrm{d}s = (b-a)^2 \left\|\frac{\mathrm{d}f}{\mathrm{d}x}\right\|^2_{L^2(a,b)},\end{aligned}$$

by which inequality (17.1) is proved by taking $C_p=(b-a)$. The same inequality holds if $f(b)=0$. □

Thanks to the fact that no derivatives with respect to the variable θ are present in the equations, we may carry out some further analysis of the structure models illustrated so far by considering the equations for a fixed value of θ and z.

We will consider Eq. (16.12) and address then the following problem:

$$\frac{\partial^2 \eta}{\partial t^2} + b\eta = H, \quad \text{in } \Gamma_0^w,\ t \in I, \tag{17.2}$$

with the following initial values for the displacement and its time rate:

$$\eta = \eta_0, \qquad \frac{\partial \eta}{\partial t} = \eta_1, \quad \text{in } \Gamma_0^w,\ t = t_0. \tag{17.3}$$

We also introduce the space $L^2(I; L^2(\Gamma_0^w))$ of functions $f : \Gamma_0^w \times I \to \mathbb{R}$ that are square integrable in Γ_0^w for almost every (a.e.) $t \in I$ and such that

$$\int_{t_0}^{t_1} \|f(\tau)\|^2_{L^2(\Gamma_0^w)}\, \mathrm{d}\tau < \infty.$$

LEMMA 17.2. *If $H \in L^2(I; L^2(\Gamma_0^w))$, the following inequality holds for a.e. $t \in I$:*

$$\begin{aligned} &\left\| \frac{\partial \eta}{\partial t}(t) \right\|^2_{L^2(\Gamma_0^w)} + b\, \|\eta(t)\|^2_{L^2(\Gamma_0^w)} \\ &\quad \leqslant \left(\|\eta_1\|^2_{L^2(\Gamma_0^w)} + b\|\eta_0\|^2_{L^2(\Gamma_0^w)} + \int_{t_0}^{t} \|H(\tau)\|^2_{L^2(\Gamma_0^w)}\, \mathrm{d}\tau \right) e^{(t-t_0)}. \end{aligned} \tag{17.4}$$

PROOF. It can be obtained by multiplying (17.2) by $\partial\eta/\partial t$ and applying Gronwall lemma (Lemma 10.3). □

Relation (17.4) asserts that the sum of the total kinetic and elastic potential energy associated to Eq. (17.2) is bounded, at each time t, by a quantity which depends only on the initial condition and the forcing term.

Let us consider the generalized string model (16.22) with the following initial and boundary conditions:

$$\eta = \eta_0, \qquad \frac{\partial \eta}{\partial t} = \eta_1 \quad \text{in } \Gamma_0^w,\ t = t_0, \tag{17.5a}$$

$$\eta|_{z=0} = \alpha, \qquad \eta|_{z=L} = \beta, \quad t \in I. \tag{17.5b}$$

Let us define the following energy function:

$$e_s(t) = \frac{1}{2}\left(\left\| \frac{\partial \eta}{\partial t}(t) \right\|^2_{L^2(\Gamma_0^w)} + a \left\| \frac{\partial \eta}{\partial z}(t) \right\|^2_{L^2(\Gamma_0^w)} + b\, \|\eta(t)\|^2_{L^2(\Gamma_0^w)} \right). \tag{17.6}$$

LEMMA 17.3. *If $H \in L^2(I; L^2(\Gamma_0^w))$ and $\alpha = \beta = 0$, the following inequality holds for a.e. $t \in I$:*

$$e_s(t) + \frac{c}{2} \int_{t_0}^{t} \left\| \frac{\partial^2 \eta}{\partial t \partial z}(\tau) \right\|^2_{L^2(\Gamma_0^w)} \mathrm{d}\tau \leqslant e_s(0) + k \int_{t_0}^{t_1} \|H(\tau)\|^2_{L^2(\Gamma_0^w)}\, \mathrm{d}\tau, \tag{17.7}$$

where $k = C_p^2/(2c)$ and C_p is the Poincaré constant.

PROOF. We use the short-hand notations $\dot{\eta}$ and $\ddot{\eta}$ for the time derivatives of η. We first multiply the generalized string equation (16.22) by $\dot{\eta}$ and integrate w.r. to z:

$$\begin{aligned}
&\int_0^L \dot{\eta}\ddot{\eta} - a\int_0^L \dot{\eta}\frac{\partial^2\eta}{\partial z^2} - c\int_0^L \dot{\eta}\frac{\partial^3\eta}{\partial t\partial z^2} + b\int_0^L \dot{\eta}\eta \\
&\quad = \frac{1}{2}\frac{\mathrm{d}}{\mathrm{d}t}\int_0^L \dot{\eta}^2 + a\int_0^L \frac{\partial^2\eta}{\partial t\partial z}\frac{\partial\eta}{\partial z} - a\left[\dot{\eta}\frac{\partial\eta}{\partial z}\right]_0^L + c\int_0^L \left(\frac{\partial^2\eta}{\partial t\partial z}\right)^2 \\
&\qquad - c\left[\frac{\partial^2\eta}{\partial t\partial z}\dot{\eta}\right]_0^L + \frac{b}{2}\frac{\mathrm{d}}{\mathrm{d}t}\int_0^L \eta^2 = \int_0^L \dot{\eta}H.
\end{aligned} \tag{17.8}$$

By exploiting the homogeneous boundary conditions and the fact that

$$\frac{\partial^2\eta}{\partial t\partial z}\frac{\partial\eta}{\partial z} = \frac{1}{2}\frac{\partial}{\partial t}\left(\frac{\partial\eta}{\partial z}\right)^2,$$

we have

$$\frac{1}{2}\frac{\mathrm{d}}{\mathrm{d}t}\int_0^L \dot{\eta}^2 + \frac{a}{2}\frac{\mathrm{d}}{\mathrm{d}t}\int_0^L \frac{\partial\eta}{\partial z}^2 + c\int_0^L \left(\frac{\partial^2\eta}{\partial t\partial z}\right)^2 + \frac{b}{2}\frac{\mathrm{d}}{\mathrm{d}t}\int_0^L \eta^2 = \int_0^L \dot{\eta}H.$$

Thanks to the hypothesis of axial symmetry, we have

$$\frac{\mathrm{d}e_s}{\mathrm{d}t} + c\left\|\frac{\partial^2\eta}{\partial t\partial z}\right\|^2_{L^2(\Gamma_0^w)} = \int_{\Gamma_0^w} \dot{\eta}H. \tag{17.9}$$

The application the Cauchy–Schwarz, Young and Poincaré inequalities to the right-hand side gives

$$\begin{aligned}
\frac{\mathrm{d}e_s}{\mathrm{d}t} + c\left\|\frac{\partial^2\eta}{\partial t\partial z}\right\|^2_{L^2(\Gamma_0^w)} &\leqslant \frac{1}{4\varepsilon}\|H\|^2_{L^2(\Gamma_0^w)} + \varepsilon\|\dot{\eta}\|^2_{L^2(\Gamma_0^w)} \\
&\leqslant \frac{1}{4\varepsilon}\|H\|^2_{L^2(\Gamma_0^w)} + C_p^2\varepsilon\left\|\frac{\partial^2\eta}{\partial t\partial z}\right\|^2_{L^2(\Gamma_0^w)}
\end{aligned}$$

for any positive ε. If we choose ε such that $C_p^2\varepsilon = c/2$ and integrate in time between t_0 and t, we finally obtain the desired result. □

CHAPTER V

18. The coupled fluid structure problem

In this part we will treat the situation arising when the flow in a vessel interacts mechanically with the wall structure. This aspect is particularly relevant for blood flow in large arteries, where the vessel wall radius may vary up to 10% because of the forces exerted by the flowing blood stream.

We will first illustrate a framework for the Navier–Stokes equations in a moving domain which is particularly convenient for the analysis and for the set up of numerical solution methods.

18.1. The Arbitrary Lagrangian Eulerian (ALE) formulation of the Navier–Stokes equation

In Section 9 we have introduced the Navier–Stokes equations in a fixed domain Ω, according to the *Eulerian* approach where the independent spatial variables are the coordinates of a fixed Eulerian system. We now consider the case where the domain is moving. In practical situations, such as the flow inside a portion of a compliant artery, we have to compute the flow solution in a *computational domain* Ω_t varying with time.

The boundary of Ω_t may in general be subdivided into two parts. The first part coincides with the physical fluid boundary, i.e., the vessel wall. In the example of Fig. 18.1, this part is represented by Γ_t^w, which is moving under the effect of the flow field. The other part of $\partial\Omega_t$ corresponds to "fictitious boundaries" (also called artificial boundaries) which delimit the region of interest. They are necessary because solving the fluid equation on the whole portion of space occupied by the fluid under study is in general

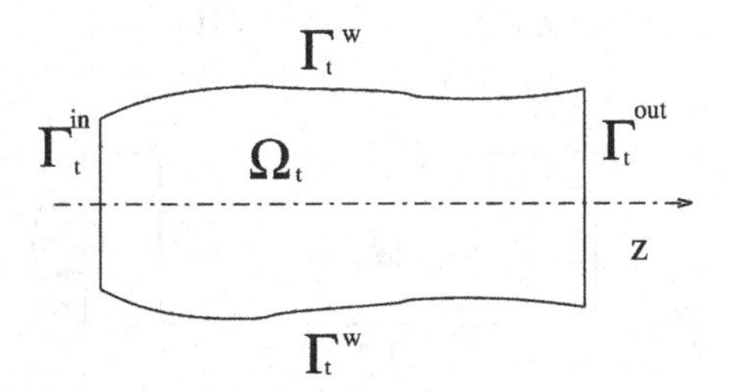

FIG. 18.1. The longitudinal section of a model of an artery. The vessel wall Γ_t^w is moving. The location along the z axis of Γ_t^{in} and Γ_t^{out} are fixed.

impractical, if not impossible. In our case, that would mean solving the whole circulatory system!

In the example of Fig. 18.1, the artificial boundaries are the inlet and outlet boundaries, there indicated by Γ_t^{in} and Γ_t^{out}, respectively. The location of these boundaries is fixed a priori. More precisely, Γ_t^{in} and Γ_t^{out} may change with time because of the displacement of Γ_t^w, however they remain planar and their position along the vessel axis is fixed.

Clearly in this case the Eulerian approach becomes impractical.

A possible alternative would be to use the *Lagrangian approach*. Here, we identify the computational domain on a reference configuration Ω_0 and the corresponding domain in the current configuration, which we indicate with $\Omega_{\mathcal{L}_t}$, will be provided by the Lagrangian mapping (which has been introduced in Section 7), i.e.,

$$\Omega_{\mathcal{L}_t} = \mathcal{L}_t(\Omega_0), \quad t \in I. \tag{18.1}$$

Fig. 18.2 illustrates the situation for the flow inside an artery whose wall is moving. Since the fluid velocity at the wall is equal to the wall velocity, the Lagrangian mapping effectively maps Γ_0^w to the correct wall position Γ_t^w at each time t. However, the "fictitious" boundaries Γ_0^{in} and Γ_0^{out} in the reference configuration will now be transported along the fluid trajectories, into $\Gamma_{\mathcal{L}_t}^{\text{in}}$ and $\Gamma_{\mathcal{L}_t}^{\text{out}}$. This is clearly not acceptable, particularly if one wants to study the problem for a relatively large time interval. Indeed, the domain rapidly becomes highly distorted.

The ideal situation would then be that indicated in Fig. 18.2(b). Even if the wall is moving, one would like to keep the inlet and outlet boundaries at the same spatial location along the vessel axis.

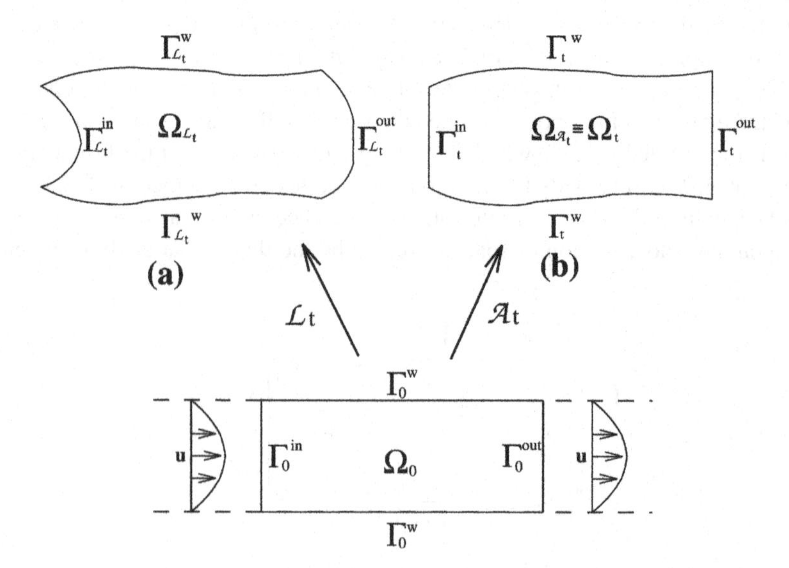

FIG. 18.2. Comparison between the Lagrangian and the ALE approach. The reference computational domain Ω_0 is mapped by (a) the Lagrangian mapping $\mathcal{L}_t$ and by (b) the Arbitrary Lagrangian Eulerian mapping.

With that purpose, we introduce the *Arbitrary Lagrangian Eulerian* (ALE) mapping

$$\mathcal{A}_t : \Omega_0 \to \Omega_{\mathcal{A}_t}, \qquad \mathbf{Y} \to \mathbf{y}(t,\mathbf{Y}) = \mathcal{A}_t(\mathbf{Y}), \tag{18.2}$$

which provides the spatial coordinates $(t, \mathbf{y})$ in terms of the so-called *ALE coordinates* $(t, \mathbf{Y})$, with the basic requirement that $\mathcal{A}_t$ retrieves, at each time $t \in I$, the desired computational domain, i.e.,

$$\Omega_{\mathcal{A}_t} \equiv \mathcal{A}_t(\Omega_0) = \Omega_t, \quad \forall t \in I.$$

The ALE mapping should be continuous and bijective in $\overline{\Omega_0}$. Once given, we may define the *domain* velocity field as

$$\widetilde{\mathbf{w}}(t,\mathbf{Y}) = \frac{\partial}{\partial t}\mathbf{y}(t,\mathbf{Y}), \tag{18.3}$$

which in the spatial coordinates is expressed as

$$\mathbf{w} = \widetilde{\mathbf{w}} \circ \mathcal{A}_t^{-1}, \quad \text{i.e.,} \quad \mathbf{w}(t,\mathbf{y}) = \widetilde{\mathbf{w}}\left(t, \mathcal{A}_t^{-1}(\mathbf{y})\right). \tag{18.4}$$

Similarly to what has been done for the Lagrangian mapping in Section 7 we use the convention of indicating by $\tilde{f}$ the composition of a function f with the ALE mapping, i.e., $\tilde{f} = f \circ \mathcal{A}_t$.

We define the ALE trajectory $T_\mathbf{Y}$ for every $\mathbf{Y} \in \Omega_0$ as

$$T_\mathbf{Y} = \{(t, \mathbf{y}(t,\mathbf{Y})) : \ t \in I\} \tag{18.5}$$

and the ALE derivative of a function f, which we denote by $(D^{\mathcal{A}}/Dt\, f)$, as the time derivative along a trajectory $T_\mathbf{Y}$, that is if

$$f : I \times \Omega_t \to \mathbb{R},$$

then

$$\frac{D^{\mathcal{A}}}{Dt} f : I \times \Omega_t \to \mathbb{R}, \qquad \frac{D^{\mathcal{A}}}{Dt} f(t,\mathbf{y}) = \frac{\partial \tilde{f}}{\partial t}(t,\mathbf{Y}), \qquad \mathbf{Y} = \mathcal{A}_t^{-1}(\mathbf{y}). \tag{18.6}$$

Similarly to what already obtained for the Lagrangian mapping (relation (7.4)), we have

$$\frac{D^{\mathcal{A}}}{Dt} f = \frac{\partial f}{\partial t} + \mathbf{w} \cdot \nabla f, \tag{18.7}$$

where now the gradient is made with respect to the $\mathbf{y}$-coordinates.

The Jacobian of the ALE mapping $J^{\mathcal{A}_t}$, defined as

$$J^{\mathcal{A}_t} = \det\left(\frac{\partial \mathbf{y}}{\partial \mathbf{Y}}\right), \tag{18.8}$$

is, for all $t \in I$, a positive quantity because the ALE mapping is surjective and at time t_0 is equal to the identity mapping. It satisfies the following relation:

$$\frac{D^{\mathcal{A}}}{Dt} J^{\mathcal{A}_t} = J^{\mathcal{A}_t} \operatorname{div} \mathbf{w}. \tag{18.9}$$

Again in a way all analogous to what seen for the Lagrangian mapping, we may derive the following result.

THEOREM 18.1 (ALE transport theorem). *Let $V_0 \subset \Omega_0$, and let $V^{\mathcal{A}_t} \subset \Omega_t$ be its image under the mapping $\mathcal{A}_t$. Furthermore, let $f : I \times \Omega_t \to \mathbb{R}$ be continuously differentiable with respect to both variables. Then*

$$\frac{\mathrm{d}}{\mathrm{d}t}\int_{V^{\mathcal{A}_t}} f = \int_{V^{\mathcal{A}_t}} \left(\frac{D^{\mathcal{A}}}{Dt} f + f \operatorname{div} \mathbf{w}\right) = \int_{V^{\mathcal{A}_t}} \left(\frac{\partial f}{\partial t} + \operatorname{div}(f\mathbf{w})\right)$$
$$= \int_{V^{\mathcal{A}_t}} \frac{\partial f}{\partial t} + \int_{\partial V^{\mathcal{A}_t}} f \mathbf{w} \cdot \mathbf{n}. \tag{18.10}$$

The proof is similar to that of Theorem 7.2 and is omitted.

The Navier–Stokes equations (9.1) are clearly valid on Ω_t, yet it may be convenient to recast them in order to put into evidence the ALE time derivative. We obtain, by a straightforward application of (18.7) to (9.1),

$$\begin{aligned} &\frac{D^{\mathcal{A}}}{Dt}\mathbf{u} + [(\mathbf{u} - \mathbf{w}) \cdot \nabla]\mathbf{u} + \nabla p - 2\,\mathbf{div}(\nu \mathbf{D}(\mathbf{u})) = \mathbf{f}, \\ &\operatorname{div} \mathbf{u} = 0, \end{aligned} \tag{18.11}$$

in Ω_t and for all $t \in I$.

18.2. Coupling with the structure model

We now study the properties of the coupled fluid–structure problem, using for the structure the generalized string model (16.22). Referring to Fig. 16.2, we recall that Γ_t^w is the current configuration of the vessel structure, while Γ_0^w is the reference configuration in which the structure equation is written. We also recall that we take $\mathbf{n}$ always to be the outwardly vector normal to the fluid domain boundary.

We will then address the following problem:

For all $t \in I$, find $\mathbf{u}$, p, η such that

$$\begin{aligned} &\frac{D^{\mathcal{A}}}{Dt}\mathbf{u} + [(\mathbf{u} - \mathbf{w}) \cdot \nabla]\mathbf{u} + \nabla p - 2\,\mathbf{div}(\nu \mathbf{D}(\mathbf{u})) = \mathbf{f}, \\ &\operatorname{div} \mathbf{u} = 0, \quad \text{in } \Omega_t, \end{aligned} \tag{18.12}$$

and

$$\frac{\partial^2 \eta}{\partial t^2} - a\frac{\partial^2 \eta}{\partial z^2} + b\eta - c\frac{\partial^3 \eta}{\partial t \partial^2 z} = H, \quad \text{in } \Gamma_0^w \tag{18.13}$$

with the following initial conditions for $t = t_0$:

$$\mathbf{u} = \mathbf{u}_0, \qquad \mathbf{x} \in \Omega_0, \tag{18.14a}$$
$$\eta = \eta_0, \quad \dot{\eta} = \eta_1, \quad \text{in } \Gamma_0^w, \tag{18.14b}$$

boundary conditions for $t \in I$,

$$[2\nu\mathbf{D}(\mathbf{u}) - (p - P_{\text{ext}})\mathbf{I}] \cdot \mathbf{n} = \mathbf{0}, \quad \text{on } \Gamma_t^{\text{out}}, \tag{18.15a}$$

$$\mathbf{u} = \mathbf{g}, \quad \text{on } \Gamma_t^{\text{in}}, \tag{18.15a}$$

$$\eta|_{z=0} = \alpha, \qquad \eta|_{z=L} = \beta, \tag{18.15b}$$

and the interface condition

$$\tilde{\mathbf{u}} = \mathbf{u} \circ \mathcal{A}_t = \frac{\partial \eta}{\partial t}\mathbf{e}_r, \quad \text{on } \Gamma_0^w,\ t \in I. \tag{18.16}$$

Another interface condition is implicitly provided by the fact that the forcing term H is function of the fluid variables (see (16.14)).

Here, $\mathbf{u}_0$, $\mathbf{g}$, α and β are given functions, H is the forcing term (16.14) and $\mathcal{A}_t$ is an ALE mapping such that $\mathcal{A}_t^{-1}(\partial\Omega_t) = \Gamma^{\text{in}} \cup \Gamma^{\text{out}} \cup \Gamma_0^w$. We have used the ALE form for the Navier–Stokes equations since it is best suited in view of the numerical solution, as it will be detailed in the next section.

We may then recognize the sources of the coupling between the fluid and the structure models, which are twofold (in view of a possible iterative solution strategy):

- fluid $\to$ structure. The fluid solution provides the value of H, which is function of the fluid stresses at the wall.
- structure $\to$ fluid. The movement of the vessel wall changes the geometry on which the fluid equations must be solved. In addition, the proper boundary conditions for the fluid velocity in correspondence to vessel wall are not anymore homogeneous Dirichlet conditions, but they impose the equality between the fluid and the structure velocity. They express the fact that the fluid particle in correspondence of the vessel wall should move at the same velocity as the wall.

Note that we have made some changes with respect to the nomenclature used in (9.1) to indicate that the domain is now moving. We rewrite the expression of the forcing term H, given in (16.14), by noting that while the fluid velocity and pressure are written in the current configuration, H lives in the reference configuration for the vessel wall Γ_0^w. Therefore, following the nomenclature introduced in the previous subsection, we write

$$H = \frac{\rho}{\rho_w h_0}\left[(\tilde{p} - p_0)\frac{R}{R_0} - 2g\tilde{\nu}\left(\widetilde{\mathbf{D}(\mathbf{u}) \cdot \mathbf{n}}\right) \cdot \mathbf{e}_r\right]. \tag{18.17}$$

18.3. An energy inequality for the coupled problem

In this section we will obtain an a-priori inequality for the coupled fluid–structure problem just presented. We will consider only the case of homogeneous boundary conditions, that is

$$\mathbf{g} = \mathbf{0}, \qquad \alpha = \beta = 0,$$

for the coupled problem (18.12)–(18.16).

LEMMA 18.1. *The coupled problem* (18.12)–(18.16) *with* $\mathbf{g} = \mathbf{0}$ *and* $\alpha = \beta = 0$ *satisfies the following energy equality for all* $t \in I$*:*

$$\frac{\mathrm{d}}{\mathrm{d}t}\left[\frac{\omega}{2}\,\|\mathbf{u}(t)\|_{\mathbf{L}^2(\Omega_t)} + e_s(t)\right] + 2\omega\int_{\Omega_t} \nu\mathbf{D}(\mathbf{u}) : \mathbf{D}(\mathbf{u}) + c\left\|\frac{\partial^2\eta}{\partial z\partial t}\right\|^2_{L^2(\Gamma_0^w)}$$

$$+\frac{\omega}{2}\int_{\Gamma_t^{\text{out}}}|\mathbf{u}|^2\mathbf{u}\cdot\mathbf{n}=\omega\int_{\Omega_t}\mathbf{f}\cdot\mathbf{u}, \tag{18.18}$$

where e_s was defined in (17.6) *and*

$$\omega=\frac{\rho}{\rho_w h_0}. \tag{18.19}$$

Moreover, if we assume that the net kinetic energy flux is non-negative on the outlet section, i.e.,

$$\int_{\Gamma_t^{\text{out}}}|\mathbf{u}|^2\mathbf{u}\cdot\mathbf{n}\geqslant 0\quad\forall t\in I, \tag{18.20}$$

we obtain the a-priori *energy estimate*

$$\begin{aligned}&\frac{\omega}{2}\|\mathbf{u}(t)\|_{\mathbf{L}^2(\Omega_t)}+e_s(t)+C_K\omega\nu_0\int_{t_0}^t\|\nabla\mathbf{u}(\tau)\|^2_{\mathbf{L}^2(\Omega_\tau)}\,\mathrm{d}\tau\\&\quad+c\int_{t_0}^t\left\|\frac{\partial^2\eta}{\partial z\partial t}(\tau)\right\|^2_{L^2(\Gamma_w^0)}\mathrm{d}\tau\\&\leqslant\frac{\omega}{2}\|\mathbf{u}_0\|_{\mathbf{L}^2(\Omega_t)}+e_s(t_0)+\frac{\omega C_P^2}{4\nu C_K}\int_{t_0}^t\|\mathbf{f}(\tau)\|^2_{\mathbf{L}^2(\Omega_\tau)}\,\mathrm{d}\tau,\quad t\in I.\end{aligned} \tag{18.21}$$

PROOF. We recall expression (17.9) and we recast the right-hand side on the current configuration Γ_t^w. By exploiting (16.3) and (16.4), we have

$$\begin{aligned}\int_{\Gamma_0^w}H\frac{\partial\eta}{\partial t}\,\mathrm{d}\sigma_0&=\frac{\rho}{\rho_w h_0}\int_{\Gamma_0^w}\left[\frac{R}{R_0}(\tilde p-p_{\text{ext}})-2g\nu\left(\widetilde{\mathbf{D}(\mathbf{u})\cdot\mathbf{n}}\right)\cdot\mathbf{e}_r\right]\frac{\partial\eta}{\partial t}\,\mathrm{d}\sigma_0\\&=\omega\int_{\Gamma_0^w}\left[(\tilde p-p_{\text{ext}})\mathbf{n}\cdot\mathbf{e}_r-2\nu\left(\widetilde{\mathbf{D}(\mathbf{u})\cdot\mathbf{n}}\right)\cdot\mathbf{e}_r\right]\frac{\partial\eta}{\partial t}g\,\mathrm{d}\sigma_0\\&=\omega\int_{\Gamma_0^w}\left[(\tilde p-p_{\text{ext}})\mathbf{n}-2\nu\left(\widetilde{\mathbf{D}(\mathbf{u})\cdot\mathbf{n}}\right)\right]\cdot\tilde{\mathbf{u}}g\,\mathrm{d}\sigma_0\\&=\omega\int_{\Gamma_t^w}[(p-p_{\text{ext}})\mathbf{n}-2\nu\,(\mathbf{D}(\mathbf{u})\cdot\mathbf{n})]\cdot\mathbf{u}\,\mathrm{d}\sigma,\end{aligned}$$

where we have used the interface conditions (18.16). Then,

$$\frac{1}{2}\frac{\mathrm{d}e_s}{\mathrm{d}t}+c\left\|\frac{\partial^2\eta}{\partial z\partial t}\right\|^2_{L^2(\Gamma_0^w)}=\omega\int_{\Gamma_t^w}[(p-p_{\text{ext}})\mathbf{n}-2\nu\,(\mathbf{D}(\mathbf{u})\cdot\mathbf{n})]\cdot\mathbf{u}\,\mathrm{d}\sigma. \tag{18.22}$$

As for the fluid equations, we follow the same route of Theorem 12.1. In particular, we begin by multiplying (18.12) by $\mathbf{u}$ and integrating over Ω_t, obtaining

$$\int_{\Omega_t}\mathbf{u}\cdot\frac{D^{\mathcal{A}}}{Dt}\mathbf{u}+\int_{\Omega_t}\mathbf{u}\cdot[(\mathbf{u}-\mathbf{w})\cdot\nabla]\mathbf{u}+\int_{\Omega_t}\mathbf{u}\cdot(\nabla p-2\nu\,\mathbf{div}\,\mathbf{D}(\mathbf{u}))=(\mathbf{f},\mathbf{u}). \tag{18.23}$$

We now analyze each term in turn. By exploiting the ALE transport theorem (18.10), we may derive that

$$\begin{aligned}\int_{\Omega_t} \mathbf{u} \cdot \frac{D^{\mathcal{A}}}{Dt}\mathbf{u} &= \int_{\Omega_0} J_t \tilde{\mathbf{u}} \cdot \frac{\partial \tilde{\mathbf{u}}}{\partial t} = \frac{1}{2}\int_{\Omega_0} J_t \frac{\partial |\tilde{\mathbf{u}}|^2}{\partial t} \\ &= \frac{1}{2}\int_{\Omega_t} \frac{D^{\mathcal{A}}}{Dt}|\mathbf{u}|^2 = \frac{1}{2}\frac{\mathrm{d}}{\mathrm{d}t}\int_{\Omega_t} |\mathbf{u}|^2 - \frac{1}{2}\int_{\Omega_t} |\mathbf{u}|^2 \operatorname{div} \mathbf{w}. \end{aligned} \tag{18.24}$$

The convective term gives

$$\begin{aligned}&\int_{\Omega_t} \mathbf{u} \cdot [(\mathbf{u} - \mathbf{w}) \cdot \nabla]\mathbf{u} \\ &= -\frac{1}{2}\int_{\Omega_t} |\mathbf{u}|^2 \operatorname{div} \mathbf{u} + \frac{1}{2}\int_{\Omega_t} |\mathbf{u}|^2 \operatorname{div} \mathbf{w} + \frac{1}{2}\int_{\partial\Omega_t} |\mathbf{u}|^2 (\mathbf{u} - \mathbf{w}) \cdot \mathbf{n} \\ &= \frac{1}{2}\int_{\Omega_t} |\mathbf{u}|^2 \operatorname{div} \mathbf{w} + \frac{1}{2}\int_{\Gamma_t^{\mathrm{out}}} |\mathbf{u}|^2 \mathbf{u} \cdot \mathbf{n}, \end{aligned} \tag{18.25}$$

since $\operatorname{div} \mathbf{u} = 0$ in Ω_t while $\mathbf{w} = \mathbf{u}$ on Γ_t^w and $\mathbf{w} = \mathbf{0}$ on $\partial\Omega_t \setminus \Gamma_t^w$.

The other terms provide

$$\begin{aligned}\int_{\Omega_t} \mathbf{u} \cdot \nabla p &= (\text{since } p_{\mathrm{ext}} = \text{const.}) \int_{\Omega_t} \mathbf{u} \cdot \nabla (p - p_{\mathrm{ext}}) \\ &= -\int_{\Omega_t} (p - p_{\mathrm{ext}}) \operatorname{div} \mathbf{u} + \int_{\partial\Omega_t} (p - p_{\mathrm{ext}})\mathbf{u} \cdot \mathbf{n} \\ &= \int_{\Gamma_t^{\mathrm{out}}} (p - p_{\mathrm{ext}})\mathbf{u} \cdot \mathbf{n} + \int_{\Gamma_t^w} (p - p_{\mathrm{ext}})\mathbf{u} \cdot \mathbf{n} \end{aligned} \tag{18.26}$$

and

$$\begin{aligned}\int_{\Omega_t} \nu \mathbf{u} \cdot \operatorname{\mathbf{div}} \mathbf{D}(\mathbf{u}) &= -\int_{\Omega_t} \nu \nabla \mathbf{u} : \mathbf{D}(\mathbf{u}) + \int_{\partial\Omega_t} \nu \mathbf{u} \cdot \mathbf{D}(\mathbf{u}) \cdot \mathbf{n} \\ &= -\int_{\Omega_t} \nu \mathbf{D}(\mathbf{u}) : \mathbf{D}(\mathbf{u}) + \int_{\partial\Omega_t} \nu \mathbf{u} \cdot \mathbf{D}(\mathbf{u}) \cdot \mathbf{n} \\ &= -\int_{\Omega_t} \nu \mathbf{D}(\mathbf{u}) : \mathbf{D}(\mathbf{u}) + \int_{\Gamma_t^{\mathrm{out}}} \nu (\mathbf{D}(\mathbf{u}) \cdot \mathbf{n}) \cdot \mathbf{u} \\ &\quad + \int_{\Gamma_t^w} \nu (\mathbf{D}(\mathbf{u}) \cdot \mathbf{n}) \cdot \mathbf{u}, \end{aligned} \tag{18.27}$$

where we have exploited again the symmetry of $\mathbf{D}(\mathbf{u})$.

Using the results obtained in (18.24)–(18.27) into (18.23), rearranging the terms and recalling the boundary condition (18.15a), we can write

$$\begin{aligned}&\frac{1}{2}\frac{\mathrm{d}}{\mathrm{d}t}\|\mathbf{u}\|^2_{\mathbf{L}^2(\Omega_t)} + 2\int_{\Omega_t} \nu \mathbf{D}(\mathbf{u}) : \mathbf{D}(\mathbf{u}) + \frac{1}{2}\int_{\Gamma_t^{\mathrm{out}}} |\mathbf{u}|^2 \mathbf{u} \cdot \mathbf{n} \\ &\quad + \int_{\Gamma_t^w} [(p - p_{\mathrm{ext}})\mathbf{n} - 2\nu \mathbf{D}(\mathbf{u}) \cdot \mathbf{n}] \cdot \mathbf{u} = \int_{\Omega_t} \mathbf{f} \cdot \mathbf{u}. \end{aligned}$$

We now recall expression (18.22) and recognize the equivalence of the integrals over Γ_t^w, which express the exchange of power (rate of energy) between fluid and structure. We multiply then the last equality by ω and add it to (18.22), obtaining (18.18).

Using (18.20), (10.2) and the fact that $\nu \geqslant \nu_0 > 0$,

$$\frac{\mathrm{d}}{\mathrm{d}t}\left[\frac{\omega}{2}\|\mathbf{u}(t)\|_{\mathbf{L}^2(\Omega_t)} + e_s(t)\right] + 2C_K\omega\nu_0\|\nabla\mathbf{u}\|^2_{\mathbf{L}^2(\Omega_t)} + c\left\|\frac{\partial^2\eta}{\partial z\partial t}\right\|^2_{L^2(\Gamma_0^w)}$$

$$\leqslant \frac{\mathrm{d}}{\mathrm{d}t}\left[\frac{\omega}{2}\|\mathbf{u}(t)\|_{\mathbf{L}^2(\Omega_t)} + e_s(t)\right] + 2\omega\int_{\Omega_t}\nu\mathbf{D}(\mathbf{u}) : \mathbf{D}(\mathbf{u}) + c\left\|\frac{\partial^2\eta}{\partial z\partial t}\right\|^2_{L^2(\Gamma_0^w)}$$

$$\leqslant \omega\int_{\Omega_t}\mathbf{f}\cdot\mathbf{u} \leqslant \frac{\omega}{4\varepsilon}\|\mathbf{f}\|^2_{\mathbf{L}^2(\Omega_t)} + \omega\varepsilon\|\mathbf{u}\|^2_{\mathbf{L}^2(\Omega_t)} \leqslant \frac{\omega}{4\varepsilon}\|\mathbf{f}\|^2_{\mathbf{L}^2(\Omega_t)} + C_P^2\omega\varepsilon\|\nabla\mathbf{u}\|^2_{\mathbf{L}^2(\Omega_t)},$$

for any positive ε. To derive the last inequality we have applied the Poincaré inequality (10.1).

The desired result is then obtained by taking $\varepsilon = (\nu_0 C_K)/C_P^2$ and integrating in time between t_0 and t. □

This last result shows that the energy associated to the coupled problem is bounded, at any time, by quantities which depend only on the initial condition and the applied volume forces. Moreover, since in blood flow simulation we neglect the volume force term $\mathbf{f}$ in the Navier–Stokes equations, estimate (18.21) simplifies into

$$\frac{\omega}{2}\|\mathbf{u}(t)\|_{\mathbf{L}^2(\Omega_t)} + e_s(t) + 2C_K\omega\nu_0\int_{t_0}^{t}\|\nabla\mathbf{u}(\tau)\|^2_{\mathbf{L}^2(\Omega_\tau)}\,\mathrm{d}\tau$$

$$+ c\int_{t_0}^{t}\left\|\frac{\partial^2\eta}{\partial z\partial t}(\tau)\right\|^2_{L^2(\Gamma_w^0)}\mathrm{d}\tau$$

$$\leqslant \frac{\omega}{2}\|\mathbf{u}_0\|_{\mathbf{L}^2(\Omega_t)} + e_s(t_0), \quad \forall t \in I.$$

REMARK 18.1. We may note that the non-linear convective term in the Navier–Stokes equations is crucial to obtain the stability result, because it generates a boundary term which compensates that coming from the treatment of the velocity time derivative. These two contributions are indeed only present in the case of a moving boundary.

REMARK 18.2. Should we replace the boundary condition (18.15a) by

$$2\nu\mathbf{D}(\mathbf{u})\cdot\mathbf{n} - \left(p - p_{\text{ext}} + \frac{1}{2}|\mathbf{u}|^2\right)\mathbf{n} = \mathbf{0} \quad \text{on } \Gamma_t^{\text{out}},\ t \in I, \tag{18.28}$$

we would obtain the stability results without the restrictions on the outlet velocity (18.20).

Let us note that the above boundary condition amounts to imposing a zero value for the *total stress* at the outflow surface.

REMARK 18.3. Under slightly different assumptions, that is periodic boundary conditions in space and the presence of a further dissipative term proportional to $\partial^4\eta/\partial z^4$ in the generalized string model, BEIRÃO DA VEIGA [2004] has recently proven an existence result of strong solutions to the coupled fluid–structure problem. The well posedness of fluid–structure interaction solutions in more general settings is still a largely open problem. A review of recent theoretical results may be found in GRANDMONT and MADAY [2000].

The hypothesis (18.20) is obviously satisfied if Γ_t^{out} is indeed an *outflow section*, i.e., $\mathbf{u} \cdot \mathbf{n} \geqslant 0$ for all $\mathbf{x} \in \Gamma_t^{\text{out}}$. As already pointed out, this is seldom true for vascular flow, particularly in large arteries.

We may observe that the "viscoelastic term" $-c(\partial^3\eta/(\partial t\partial^2 z))$ in (16.22) allows to obtain the appropriate regularity of the velocity field $\mathbf{u}$ on the boundary (see NOBILE [2001]).

In the derivation of the energy inequality (18.21), we have considered homogeneous boundary conditions both for the fluid and the structure. However, the conditions $\eta = 0$ at $z = 0$ and $z = L$, which correspond to hold the wall fixed at the two ends, are not realistic in the context of blood flow. Since the model (16.22) for the structure is of propagative type, the first order absorbing boundary conditions

$$\frac{\partial \eta}{\partial t} - \sqrt{a}\frac{\partial \eta}{\partial z} = 0 \quad \text{at } z = 0, \tag{18.29}$$

$$\frac{\partial \eta}{\partial t} + \sqrt{a}\frac{\partial \eta}{\partial z} = 0 \quad \text{at } z = L \tag{18.30}$$

look more suited to the problem at hand. An inequality of the type (18.21) could still be proven. Indeed, the boundary term which appears in (17.8) would now read

$$\begin{aligned} -\left[a\frac{\partial \eta}{\partial z}\frac{\partial \eta}{\partial t} + c\frac{\partial^2 \eta}{\partial z \partial t}\frac{\partial \eta}{\partial t}\right]_{z=0}^{z=L} &= \sqrt{a}\left[\left(\frac{\partial \eta}{\partial t}\bigg|_{z=0}\right)^2 + \left(\frac{\partial \eta}{\partial t}\bigg|_{z=L}\right)^2\right] \\ &\quad + \frac{c}{2}\sqrt{a}\frac{\mathrm{d}}{\mathrm{d}t}\left[\left(\frac{\partial \eta}{\partial t}\bigg|_{z=0}\right)^2 + \left(\frac{\partial \eta}{\partial t}\bigg|_{z=L}\right)^2\right]. \end{aligned}$$

This term, integrated in time, would eventually appear on the left-hand side of inequality (18.21). We may note, however, that we obtain both for $z = 0$ and $z = L$ the following expression:

$$\sqrt{a}\int_{t_0}^{t}\left(\frac{\partial \eta}{\partial t}(\tau)\right)^2 \mathrm{d}\tau + \frac{c}{2}\sqrt{\frac{1}{a}}\left(\frac{\partial \eta}{\partial t}(t)\right)^2 = \frac{c}{2}\sqrt{\frac{1}{a}}\left(\frac{\partial \eta}{\partial t}(t_0)\right)^2. \tag{18.31}$$

This additional term is positive and depends only on initial conditions.

Yet, conditions (18.29) and (18.30) are not compatible with the homogeneous Dirichlet boundary conditions for the fluid; indeed, if $\eta|_{z=0} \neq 0$ and $\mathbf{u} = \mathbf{0}$ on Γ_t^{in}, the trace of $\mathbf{u}$ on the boundary is discontinuous and thus not compatible with the regularity required on the solution of (18.12) (see, e.g., QUARTERONI and VALLI [1994]).

A possible remedy consists of changing the condition $\mathbf{u} = \mathbf{0}$ on Γ_t^{in} into

$$\mathbf{u} \cdot \mathbf{e}_z = g_z \circ \mathcal{A}_t^{-1}, \qquad (\mathbf{T} \cdot \mathbf{n}) \times \mathbf{e}_z = \mathbf{0}$$

on Γ_t^{in}, where g_z is a given function defined on Γ_0^{in}, with $g_z = 0$ on $\partial\Gamma_0^{\text{in}}$. Here $\mathbf{T}$ is the stress tensor defined in (8.13). An energy inequality for the coupled problem can be derived also in this case with standard calculations, taking a suitable harmonic extension $\tilde{g}_z$ of the non-homogeneous data g_z. The calculations are here omitted for the sake of brevity.

19. An iterative algorithm to solve the coupled fluid–structure problem

In this section we outline an algorithm that at each time-level allows the decoupling of the sub-problem related to the fluid from that related to the vessel wall. As usual, t^k, $k = 0, 1, \dots$ denotes the kth discrete time level; $\Delta t > 0$ is the time-step, while v^k is the approximation of the function (scalar or vector) v at time t^k.

The numerical solution of the fluid–structure interaction problem (18.12), (18.13) will be carried out by constructing a suitable finite element approximation of each sub-problem. In particular, for the fluid we need to devise a finite element formulation suitable for moving domains (or, more precisely, moving grids). In this respect, the ALE formulation will provide an appropriate framework.

To better illustrate the situation, we refer to Fig. 19.1 where we have drawn a 2D fluid–structure interaction problem. The fluid domain is Ω_t and the movement of its upper boundary Γ_t^w is governed by a generalized string model. This geometry could be derived from an axisymmetric model of the flow inside a cylindrical vessel. However, in this case we should employ the Navier–Stokes equations in axisymmetric coordinates. Since this example is only for the purpose of illustrating a possible set-up for a coupled fluid–structure algorithm, for the sake of simplicity we consider here a two-dimensional fluid–structure problem governed by Eqs. (18.12), (18.13), with interface conditions (18.16), initial and boundary conditions (18.14) and the additional condition

$$\mathbf{u}|_{\Gamma^0} = \mathbf{0}, \quad t \in I.$$

The algorithm here presented may be readily extended to three-dimensional problems.

The structure on Γ_0^w will be discretized by means of a finite element triangulation $\mathcal{T}_h^s$, like the one we illustrate in Fig. 19.2. We have considered the space S_h of piece-wise linear continuous (P1) finite elements functions to represent the approximate vessel wall displacement η_h. In the same figure we show the position at time t of the discretized vessel wall boundary $\Gamma_{t,h}^w$, corresponding to a given value of the discrete displacement field $\eta_h \in S_h$. Consequently, the fluid domain will be represented at every time by a polygon, which we indicate by $\Omega_{t,h}$. Its triangulation $\mathcal{T}_{t,h}^f$ will be constructed as the

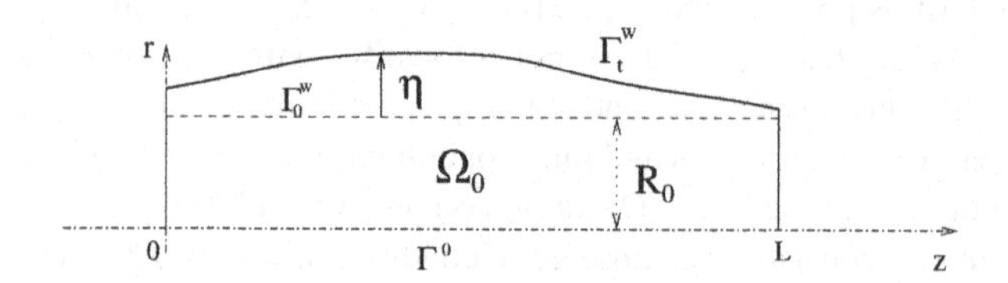

FIG. 19.1. A simple fluid–structure interaction problem.

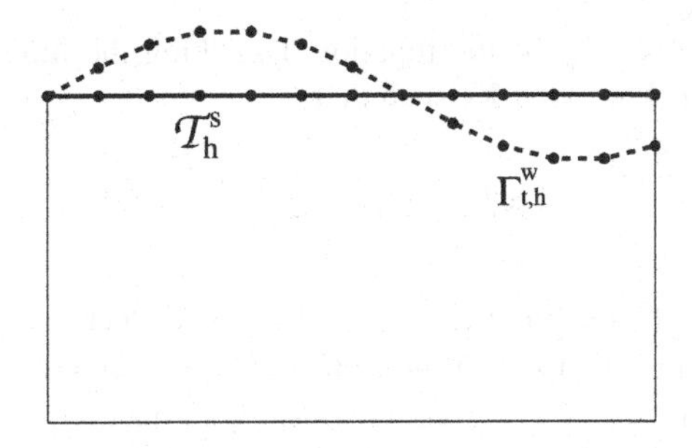

FIG. 19.2. Position of the discretized vessel wall corresponding to a possible value of η_h.

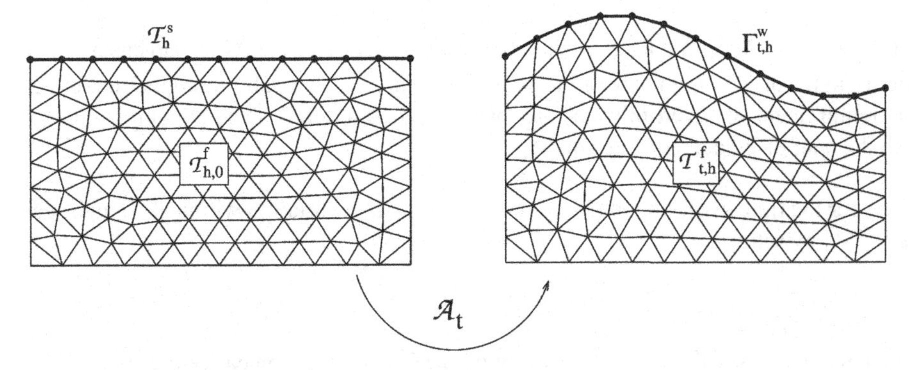

FIG. 19.3. The triangulation used for the fluid problem at each time t is the image through a map $\mathcal{A}_t$ of a mesh constructed on Ω_0.

image by an appropriate ALE mapping $\mathcal{A}_t$ of a triangulation $\mathcal{T}_{0,h}^f$ of Ω_0, as shown in Fig. 19.3. Correspondingly, $\Omega_{t,h} = \mathcal{A}_t\Omega_{0,h}$, where $\Omega_{0,h}$ is the approximation of Ω_0 induced by the triangulation $\mathcal{T}_{t,h}^f$ (clearly, if Ω_0 has a polygonal boundary we have $\Omega_{0,h} = \Omega_0$.) The trace of $\mathcal{T}_{0,h}^f$ on Γ_0^w will coincide with the "triangulation" $\mathcal{T}_h^s$ of the vessel wall, thus we consider *geometrically conforming* finite elements between the fluid and the structure. The possibility of using a geometrically non-conforming finite element representation has been investigated in GRANDMONT and MADAY [1998].

We then have to face the following problem. Suppose that we know at $t = t^{k+1}$ a discrete displacement field η_h^{k+1} and thus the corresponding position of the domain boundary $\partial\Omega_{t^{k+1},h}$. How to build a map $\mathcal{A}_{t^{k+1}}$ such that $\mathcal{A}_{t^{k+1}}(\mathcal{T}_{0,h}^f)$ is an acceptable finite element mesh for the fluid domain? This task is in general not simple. However, if we can assume that $\Omega_{t,h}$ is convex for all t and that the displacements are relatively small, the technique known as *harmonic extension* may well serve the purpose. Let $\mathbf{X}_h$ be the P1 finite element vector space associated to $\mathcal{T}_{0,h}^f$, while

$$\mathbf{X}_h^0 = \{\mathbf{w}_h \in \mathbf{X}_h:\ \mathbf{w}_h|_{\partial\Omega_{0,h}} = \mathbf{0}\}$$

and let $\mathbf{g}_h : \partial\Omega_{0,h} \to \partial\Omega_{t^{k+1},h}$ be the function describing the fluid domain boundary. We build the map by seeking $\mathbf{y}_h \in \mathbf{X}_h$ such that

$$\int_{\Omega_0} \nabla\mathbf{y}_h : \nabla\mathbf{z}_h = 0 \quad \forall\mathbf{z}_h \in \mathbf{X}_h^0, \qquad \mathbf{y}_h = \mathbf{g}_h, \quad \text{on } \partial\Omega_{0,h}, \tag{19.1}$$

and then setting $\mathcal{A}_{t^{k+1}}(\mathbf{Y}) = \mathbf{y}_h(\mathbf{Y}),\ \forall\mathbf{Y} \in \Omega_{0,h}$. This technique has indeed been adopted for the mesh in Fig. 19.3. From a practical point of view, the value of $\mathbf{y}_h$ in correspondence to the nodes of $\mathcal{T}_{0,h}^f$ gives the position of the corresponding node in $\mathcal{T}_{t,h}^f$ at time t^{k+1}. A more general discussion on the construction of the ALE mapping may be found in FORMAGGIA and NOBILE [1999], NOBILE [2001] as well as in GASTALDI [2001].

REMARK 19.1. Adopting P1 elements for the construction of the ALE map ensures that the triangles of $\mathcal{T}_{h,0}^f$ are mapped into triangles, thus $\mathcal{T}_{h,t}^f$ is a valid triangulation, under the requirement of invertibility of the map (which is assured if the domain is convex and the wall displacements are small).

As for the time evolution, we may adopt a linear time variation within each time slab $[t^k, t^{k+1}]$ by setting

$$\mathcal{A}_t = \frac{t - t^k}{\Delta t}\mathcal{A}_{t^{k+1}} - \frac{t - t^{k+1}}{\Delta t}\mathcal{A}_{t^k}, \quad t \in \left[t^k, t^{k+1}\right].$$

Then, the corresponding domain velocity $\mathbf{w}_h$ will be constant on each time slab.

We are now in the position of describing a possible finite element scheme for both the structure and the fluid problem, to be adopted in the sub-structuring algorithm. We first give more details on the adopted finite element discretization.

19.1. The discretization of the structure

For the structure we consider a *mid-point* scheme. We introduce the additional variable $\dot\eta^k$ which is the approximation of the displacement velocity at time t^k.

The time advancing scheme reads:

$\forall k \geqslant 0$ find η^{k+1} and $\dot\eta^{k+1}$ that satisfy the following system:

$$\frac{\eta^{k+1} - \eta^k}{\Delta t} = \frac{\dot\eta^k + \dot\eta^{k+1}}{2}, \tag{19.2a}$$

$$\frac{\dot\eta^{k+1} - \dot\eta^k}{\Delta t} - a\frac{\partial^2}{\partial z^2}\frac{\eta^k + \eta^{k+1}}{2} + b\frac{\eta^{k+1} + \eta^k}{2} - c\frac{\partial^2}{\partial z^2}\frac{\dot\eta^{k+1} + \dot\eta^k}{2} = H^{k+1/2}, \tag{19.2b}$$

with

$$\eta^{k+1}|_{z=0} = \alpha\left(t^{k+1}\right), \qquad \eta^{k+1}|_{z=L} = \beta\left(t^{k+1}\right), \tag{19.3a}$$

and

$$\dot{\eta}^{k+1}|_{z=0} = \frac{\partial}{\partial t}\alpha\left(t^{k+1}\right), \quad \dot{\eta}^{k+1}|_{z=L} = \frac{\partial}{\partial t}\beta\left(t^{k+1}\right), \tag{19.3b}$$

while the value of η^0 *and* $\dot{\eta}^0$ *are given by the initial conditions.*

Here, $H^{k+1/2}$ is a suitable approximation of H at time $t^k + \frac{1}{2}\Delta t$ which in the context of a sub-structuring iteration for the coupled problem is a known quantity and whose calculation from the Navier–Stokes data will be made precise later.

System (19.2) is then discretized in space by taking $\eta_h^k \in S_h$ and $\dot{\eta}_h^k \in S_h$. We set $S_h^0 = \{s_h \in S_h\colon\ s_h(0) = 0,\ s_h(L) = 0\}$ and the finite element problem reads:

For all $k \geqslant 0$ *find* $\eta_h^{k+1} \in S_h$ *and* $\dot{\eta}_h^{k+1} \in S_h$ *that satisfy the following system*:

$$\left(2\eta_h^{k+1} - \Delta t\dot{\eta}_h^{k+1}, s_h\right) = \left(2\eta_h^k + \Delta t\dot{\eta}_h^k, s_h\right), \tag{19.4a}$$

$$\begin{aligned} &\left(\frac{1}{\Delta t}\dot{\eta}_h^{k+1} + \frac{b}{2}\eta_h^{k+1}, s_h\right) + \frac{1}{2}\left(a\frac{\partial \eta_h^{k+1}}{\partial z} + c\frac{\partial \dot{\eta}_h^{k+1}}{\partial z}, \frac{\partial s_h}{\partial z}\right) \\ &\quad = \left(H^{k+1/2}, s_h\right) + \left(\frac{1}{\Delta t}\dot{\eta}_h^k + \frac{b}{2}\eta_h^k, s_h\right) - \frac{1}{2}\left(a\frac{\partial \eta_h^k}{\partial z} + c\frac{\partial \dot{\eta}_h^k}{\partial z}, \frac{\partial s_h}{\partial z}\right), \end{aligned} \tag{19.4b}$$

$\forall s_h \in S_h^0$, *together with the boundary conditions*

$$\eta_h^{k+1}|_{z=0} = \alpha\left(t^{k+1}\right), \quad \eta_h^{k+1}|_{z=L} = \beta\left(t^{k+1}\right), \tag{19.5a}$$

$$\dot{\eta}_h^{k+1}|_{z=0} = \frac{\partial}{\partial t}\alpha\left(t^{k+1}\right), \quad \dot{\eta}_h^{k+1}|_{z=L} = \frac{\partial}{\partial t}\beta\left(t^{k+1}\right), \tag{19.5b}$$

and the initial conditions

$$\eta_h^0 = \pi_{S_h}\eta^0, \qquad \eta_h^0 = \pi_{S_h}\eta^0,$$

being π^{S_h} *the standard interpolation operator upon* S_h.

19.2. The discretization of the fluid problem

In the frame of our splitting scheme the velocity field at Γ_t^w as well as the current domain configuration are provided by the calculation of η_h; they can thus be considered as given data. We consider the following finite element spaces. $\widetilde{Q}_h$ is the space of continuous piece-wise linear finite elements, while $\widetilde{\mathbf{V}}_h$ is that of vector functions whose components are in the space $\widetilde{V}_h$ of continuous piece-wise quadratic (or P1-isoP2) finite elements. Both refer to the triangulation $\mathcal{T}_{0,h}^f$ of Ω_0. For a precise definition of these finite element spaces the reader may refer to QUARTERONI and VALLI [1994] or BREZZI and FORTIN [1991].

We will also need to define

$$\widetilde{\mathbf{V}}_h^0 = \{\tilde{\mathbf{v}}_h \in \widetilde{\mathbf{V}}_h\colon\ \tilde{\mathbf{v}}_h|_{\partial\Omega_0 \setminus \Gamma_0^{\text{in}}} = \mathbf{0}\}$$

and the space $\widetilde{V}_h^{\Gamma_0^w}$ formed by function in Γ_0^w which are the trace of a function in $\widetilde{V}_h$.

The corresponding spaces on the current configuration will be given by

$$Q_{h,t} = \{q_h\colon\ q_h \circ \mathcal{A}_t \in \widetilde{Q}_h\}, \qquad \mathbf{V}_{h,t} = \{\mathbf{v}_h\colon\ \mathbf{v}_h \circ \mathcal{A}_t \in \widetilde{\mathbf{V}}_h\},$$

and analogously for $\mathbf{V}^0_{h,t}$.

REMARK 19.2. The functions belonging to $Q_{h,t}$ and $\mathbf{V}_{h,t}$ depend also on time through the ALE mapping. A thorough presentation of finite element spaces in an ALE framework is contained in FORMAGGIA and NOBILE [1999] and NOBILE [2001].

We will employ an implicit Euler time advancing scheme with a semi-explicit treatment of the convective term. Let us assume that the solution $(\mathbf{u}_h^k, p_h^k)$ at time step t^k is known, as well as the domain configuration $\Omega_{t^{k+1},h}$ at time t^{k+1} (and thus the corresponding ALE map).

The numerical solution at t^{k+1} can be computed as follows:

Find $\mathbf{u}_h^{k+1} \in \mathbf{V}_{h,t^{k+1}}$ *and* $p_h^{k+1} \in Q_{h,t^{k+1}}$ *such that*

$$\begin{aligned}
&\frac{1}{\Delta t}\left(\mathbf{u}^{k+1}, \tilde{\mathbf{v}}_h\right)_{k+1} - c_{k+1/2}\left(\mathbf{w}^{k+1/2}, \mathbf{u}^{k+1}, \tilde{\mathbf{v}}_h\right) + c_{k+1}\left(\mathbf{u}^k, \mathbf{u}^{k+1}, \tilde{\mathbf{v}}_h\right)\\
&\qquad + d_{k+1/2}\left(\mathbf{w}^{k+1/2}, \mathbf{u}^{k+1}, \tilde{\mathbf{v}}_h\right) + b_{k+1}\left(\tilde{\mathbf{v}}_h, p^{k+1}\right) + a_{k+1}\left(\mathbf{u}^{k+1}, \tilde{\mathbf{v}}_h\right)\\
&\qquad = \left(\mathbf{f}^{k+1}, \tilde{\mathbf{v}}_h\right)_{k+1} + \frac{1}{\Delta t}\left(\mathbf{u}^k, \tilde{\mathbf{v}}_h\right)_k, \quad \forall \tilde{\mathbf{v}}_h \in \widetilde{\mathbf{V}}_h^0 \qquad (19.6a)\\
&b_{k+1}\left(\mathbf{u}^{k+1}, \tilde{q}_h\right) = 0, \quad \forall \tilde{q}_h \in \widetilde{Q}_h, \qquad (19.6b)
\end{aligned}$$

and

$$\mathbf{u}_h^{k+1} = \mathbf{g}_h^{k+1}, \quad \text{on } \Gamma^{\text{in}}_{t^{k+1}}, \qquad (19.7a)$$

$$\mathbf{u}_h^{k+1} = \left(\Pi_h^{\Gamma_0^w} \dot{\eta}_h^{k+1}\right) \circ \mathcal{A}^{-1}_{t^{k+1}} \mathbf{e}_r, \quad \text{on } \Gamma^w_{t^{k+1}}. \qquad (19.7b)$$

We have defined

$$(\mathbf{w}, \tilde{\mathbf{v}})_k = \int_{\Omega_{t^k}} \mathbf{w} \cdot \left(\tilde{\mathbf{v}} \circ \mathcal{A}^{-1}_{t^k}\right),$$

$$c_k(\mathbf{w}, \mathbf{z}, \tilde{\mathbf{v}}) = \int_{\Omega_{t^k}} ((\mathbf{w} \cdot \nabla)\mathbf{z}) \cdot \left(\tilde{\mathbf{v}} \circ \mathcal{A}^{-1}_{t^k}\right),$$

$$d_k(\mathbf{w}, \mathbf{z}, \tilde{\mathbf{v}}) = \int_{\Omega_{t^k}} (\operatorname{div} \mathbf{w})\mathbf{z} \cdot \left(\tilde{\mathbf{v}} \circ \mathcal{A}^{-1}_{t^k}\right),$$

$$b_k(\mathbf{w}, \tilde{q}) = \int_{\Omega_{t^k}} \operatorname{div}\mathbf{w} \left(\tilde{q} \circ \mathcal{A}^{-1}_{t^k}\right), \qquad b_k(\widetilde{\mathbf{w}}, q) = \int_{\Omega_{t^k}} \operatorname{div}\left(\widetilde{\mathbf{w}} \circ \mathcal{A}^{-1}_{t^k}\right) q,$$

$$a_k(\mathbf{w}, \tilde{\mathbf{v}}) = \int_{\Omega_{t^k}} 2\nu \mathbf{D}(\mathbf{w}) : \mathbf{D}\left(\tilde{\mathbf{v}} \circ \mathcal{A}^{-1}_{t^k}\right).$$

The function $\mathbf{g}_h^{k+1}$ is the finite element interpolant of the boundary data $\mathbf{g}(t^{k+1})$ on the space of restrictions of $\mathbf{V}_{h,t^{k+1}}$ on $\Gamma^{\text{in}}_{t^{k+1}}$. Moreover, $\Pi_h^{\Gamma_0^w} : S_h \to \widetilde{V}_h^{\Gamma_0^w}$ is the interpolation operator required to project the discrete vessel velocity computed by the structure solver on the trace space of discrete fluid velocity on the vessel wall. Since we are using geometrically conforming finite elements, this operator is quite simple to build up.

It is understood that when the approximation of $\mathbf{u}$ and $\mathbf{w}$ in (19.6) are not evaluated at the same time as the integral, they need to be mapped on the correct domain by means of the ALE transformation.

REMARK 19.3. The term involving the domain velocity $\mathbf{w}$ has been computed on the intermediate geometry $\Omega_{t^{k+1/2}}$ in order to satisfy the so-called Geometry Conservation Law (GCL) (GUILLARD and FARHAT [2000]). A discussion on the significance of the GCL for the problem at hand may be found in NOBILE [2001].

19.3. Recovering the forcing term for the vessel wall

We need now to compute the forcing term $H^{k+1/2}$ in (19.4) as the residual of the discrete momentum equation (19.6b) for time step t^{k+1}. Let us define

$$\begin{aligned} R_h^{k+1}(\tilde{\mathbf{v}}_h) &= \left(\mathbf{f}^{k+1}, \tilde{\mathbf{v}}_h\right)_{k+1} + \frac{1}{\Delta t}\left(\mathbf{u}^k, \tilde{\mathbf{v}}_h\right)_k - \frac{1}{\Delta t}\left(\mathbf{u}^{k+1}, \tilde{\mathbf{v}}_h\right)_{k+1} \\ &\quad + c_{k+1/2}\left(\mathbf{w}^{k+1/2}, \mathbf{u}^{k+1}, \tilde{\mathbf{v}}_h\right) - c_{k+1}\left(\mathbf{u}^k, \mathbf{u}^{k+1}, \tilde{\mathbf{v}}_h\right) \\ &\quad - d_{k+1/2}\left(\mathbf{w}^{k+1/2}, \mathbf{u}^{k+1}, \tilde{\mathbf{v}}_h\right) - b_{k+1}\left(\tilde{\mathbf{v}}_h, p^{k+1}\right) - a_{k+1}\left(\mathbf{u}^{k+1}, \tilde{\mathbf{v}}_h\right), \quad \forall \tilde{\mathbf{v}}_h \in \widetilde{\mathbf{V}}_h. \end{aligned}$$

Note that $R_h^{k+1}(\tilde{\mathbf{v}}_h) = 0$, for all $\tilde{\mathbf{v}}_h \in \widetilde{\mathbf{V}}_h^0$. We define the following operator:

$$\mathcal{S}_h : S_h \to \widetilde{\mathbf{V}}_h, \qquad \mathcal{S}_h s_h = \left[\mathcal{R}_h\left(\Pi_h^{\Gamma_0^w} s_h\right)\right]\mathbf{e}_r,$$

where $\mathcal{R}_h : \widetilde{V}_h^{\Gamma_0^w} \to \widetilde{V}_h$ is a finite element extension operator such that

$$(\mathcal{R}_h v_h)|_{\Gamma_0^w} = v_h, \quad \forall v_h \in \widetilde{V}_h^{\Gamma_0^w},$$

for instance the one obtained by extending by zero at all internal nodes (see QUARTERONI and VALLI [1999]). We then take

$$\left(H^{k+1/2}, s_h\right) = \frac{\omega}{2}\left[R_h^{k+1}(\mathcal{S}_h s_h) + R_h^k(\mathcal{S}_h s_h)\right]. \tag{19.8}$$

19.4. The algorithm

We are now in the position of describing an iterative algorithm for the solution of the coupled problem. As usual, we assume to have all quantities available at $t = t^k$, $k \geqslant 0$, provided either by previous calculations or by the initial data and we wish to advance to the new time step t^{k+1}. For ease of notation we here omit the subscript h, with the understanding that we are referring exclusively to finite element quantities.

The algorithm requires to choose a *tolerance* $\tau > 0$, which is used to test the convergence of the procedure, and a *relaxation parameter* $0 < \theta \leqslant 1$. In the following, the subscript $j \geqslant 0$ denotes the sub-iteration counter.

The algorithm reads:

(A1) Extrapolate the vessel wall structure displacement and velocity:

$$\eta_{(0)}^{k+1} = \eta^k + \Delta t \dot{\eta}^k, \qquad \dot{\eta}_{(0)}^{k+1} = \dot{\eta}^k.$$

(A2) Set $j = 0$.

(A2.1) By using $\eta_{(j)}^{k+1}$, compute the new grid for the fluid domain Ω_t and the ALE map by solving (19.1).

(A2.2) Solve the Navier–Stokes problem (19.6) to compute $\mathbf{u}_{(j+1)}^{k+1}$ and $p_{(j+1)}^{k+1}$, using as velocity on the wall boundary the one calculated from $\dot{\eta}_{(j)}^{k+1}$.

(A2.3) Solve (19.4) to compute η_*^{k+1} and $\dot{\eta}_*^{k+1}$ using as forcing term the one recovered from $\mathbf{u}_{(j+1)}^{k+1}$ and $p_{(j+1)}^{k+1}$ using (19.8).

(A2.4) Unless $\|\eta_*^{k+1} - \eta_{(j)}^{k+1}\|_{L^2(\Gamma_0^w)} + \|\dot{\eta}_*^{k+1} - \dot{\eta}_{(j)}^{k+1}\|_{L^2(\Gamma_0^w)} \leqslant \tau$, set

$$\eta_{(j+1)}^{k+1} = \theta \eta_{(j)}^{k+1} + (1-\theta)\eta_*^{k+1},$$
$$\dot{\eta}_{(j+1)}^{k+1} = \theta \dot{\eta}_{(j)}^{k+1} + (1-\theta)\dot{\eta}_*^{k+1},$$

and $j \leftarrow j+1$. Then return to step (A2.1).

(A3) Set

$$\eta^{k+1} = \eta_*^{k+1}, \qquad \dot{\eta}^{k+1} = \dot{\eta}_*^{k+1}.$$
$$\mathbf{u}^{k+1} = \mathbf{u}_{(j+1)}^{k+1}, \qquad p^{k+1} = p_{(j+1)}^{k+1}.$$

If the algorithm converges, $\lim_{j\to\infty} \mathbf{u}_{(j)}^{k+1} = \mathbf{u}^{k+1}$ and $\lim_{j\to\infty} \eta_{(j)}^{k+1} = \eta^{k+1}$, where $\mathbf{u}^{k+1}$ and η^{k+1} are the solution at time step t^{k+1} of the coupled problem.

The algorithm entails, at each sub-iteration, the computation of the generalized string equation (19.4)–(19.5), the Navier–Stokes equations and the solution of two Laplace equations (19.1), one for every displacement component.

Improvements on the computational efficiency of the coupled procedure just described may be obtained either by employing standard acceleration techniques like Aitken extrapolation, or by using an altogether different approach to the non-linear problem (like Newton–Krylov techniques or multilevel schemes). The matter is still the subject of current active research investigations.

More explicit schemes for the fluid–structure interaction problem, known as "serial staggered" procedures, have been successfully applied to aeroelastic 2D and 3D problems (FARHAT, LESOINNE and MAMAN [1995], FARHAT and LESOINNE [2000], PIPERNO and FARHAT [2001]). However, it has been found that in the case of an incompressible fluid they become unstable when the density of the structure mass is comparable to that of the fluid (LE TALLEC and MOURO [2001]), which is unfortunately our situation. An analysis of decoupling technique for unsteady fluid structure interaction, carried out on a simplified, yet representative, one-dimensional model may be found in GRANDMONT, GUIMET and MADAY [2001].

CHAPTER VI

20. One-dimensional models of blood flow in arteries

In this section we introduce a simple 1D model to describe the flow motion in arteries and its interaction with the wall displacement. In the absence of branching, a short section of an artery may be considered as a cylindrical compliant tube. As before we denote by $I = (t_0, t_1)$ the time interval of interest and by Ω_t the spatial domain which is supposed to be a circular cylinder filled with blood. The reason why one-dimensional models for blood flow may be attractive is that full 3D investigations are quite computationally expensive. Yet, in many situations we might desire to have just information of the evolution of averaged quantities along the arterial tree, such as mass flux and average pressure. In this context simplified models are able to provide an reasonable answer in short times.

As already done in Section 15, we will employ cylindrical coordinates and indicate with $\mathbf{e}_r$, $\mathbf{e}_\theta$ and $\mathbf{e}_z$ the radial, circumferential and axial unit vectors, respectively, and with (r, θ, z) the corresponding coordinates system. The vessel extends from $z = 0$ to $z = L$ and the vessel length L is constant with time.

The basic model is deduced by making the following assumptions, some of which are analogous to the ones made in Section 16:

(A1) *Axial symmetry*. All quantities are independent from the angular coordinate θ. As a consequence, every axial section $z =$ const. remains circular during the wall motion. The tube radius R is a function of z and t.

(A2) *Radial displacements*. The wall displaces along the radial direction solely, thus at each point on the tube surface we may write $\boldsymbol{\eta} = \eta \mathbf{e}_r$, where $\eta = R - R_0$ is the displacement with respect to the reference radius R_0.

(A3) *Constant pressure*. We assume that the pressure P is constant on each section, so that it depends only on z and t.

(A4) *No body forces*. We neglect body forces (the inclusion of the gravity force, if needed, is straightforward); thus we put $\mathbf{f} = \mathbf{0}$ in the momentum equation (11.1a).

(A5) *Dominance of axial velocity*. The velocity components orthogonal to the z axis are negligible compared to the component along z. The latter is indicated by u_z and its expression in cylindrical coordinates reads

$$u_z(t, r, z) = \bar{u}(t, z) s\left(\frac{r}{R(t, z)}\right), \tag{20.1}$$

where $\bar{u}$ is the *mean velocity* on each axial section and $s : \mathbb{R} \to \mathbb{R}$ is a *velocity profile*.[6]

A generic axial section will be indicated by $\mathcal{S} = \mathcal{S}(t, z)$. Its measure A is given by

$$A(t, z) = \text{meas } (\mathcal{S}(t, z)) = \pi R^2(t, z) = \pi \, (R_0(z) + \eta(t, z))^2. \tag{20.2}$$

The mean velocity $\bar{u}$ is then given by

$$\bar{u} = A^{-1} \int_{\mathcal{S}} u_z \, \mathrm{d}\sigma,$$

and from (20.1) and the definition of $\bar{u}$ it follows that

$$\int_0^1 s(y) y \, \mathrm{d}y = \frac{1}{2}.$$

We will indicate with α the *momentum-flux correction coefficient* (sometimes called Coriolis coefficient), defined as

$$\alpha = \frac{\int_{\mathcal{S}} u_z^2 \, \mathrm{d}\sigma}{A\bar{u}^2} = \frac{\int_{\mathcal{S}} s^2 \, \mathrm{d}\sigma}{A}, \tag{20.3}$$

where the dependence of the various quantities on the spatial and time coordinates is understood. It is immediate to verify that $\alpha \geqslant 1$. In general, this coefficient will vary in time and space, yet in our model it is taken constant as a consequence of (20.1).

One possible choice for the profile law is the parabolic profile $s(y) = 2(1 - y^2)$, which corresponds to the Poiseuille solution characteristic of steady flows in circular tubes. In this case we have $\alpha = 4/3$. However, for blood flow in arteries it has been found that the velocity profile is, on average, rather flat. Indeed, a profile law often used for blood flow in arteries (see, for instance, SMITH, PULLAN and HUNTER [2003]) is a power law of the type $s(y) = \gamma^{-1}(\gamma + 2)(1 - y^\gamma)$, with typically $\gamma = 9$ (the value $\gamma = 2$ gives again the parabolic profile). Correspondingly, we have $\alpha = 1.1$. Furthermore, we will see that the choice $\alpha = 1$, which indicates a completely flat velocity profile, would lead to a certain simplification in our analysis.

The mean flux Q, defined as

$$Q = \int_{\mathcal{S}} u_z \, \mathrm{d}\sigma = A\bar{u},$$

is one of the main variables of our problem, together with A and the pressure P.

20.1. The derivation of the model

There are (at least) three ways of deriving our model. The first one moves from the incompressible Navier–Stokes equations with constant viscosity and performs an asymptotic analysis by assuming that the ratio R_0/L is small, thus discarding the higher order terms with respect to R_0/L (see BARNARD, HUNT, TIMLAKE and VARLEY [1966]).

[6] The fact that the velocity profile does not vary is in contrast with experimental observations and numerical results carried out with full scale models. However, it is a necessary assumption for the derivation of the reduced model. One may then think s as being a profile representative of an average flow configuration.

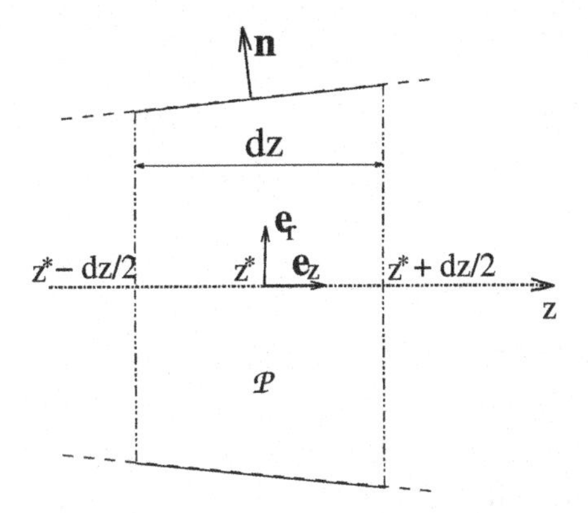

FIG. 20.1. A longitudinal section (θ = const.) of the tube and the portion between $z = z^* - \mathrm{d}z/2$ and $z = z^* + \mathrm{d}z/2$ used for the derivation of the 1D reduced model.

The second approach derives the model directly from the basic conservation laws written in integral form. The third approach consists of integrating the Navier–Stokes equations on a generic section $\mathcal{S}$.

We will indicate with Γ_t^w the wall boundary of Ω_t, which now reads

$$\Gamma_t^w = \{(r,\theta,z)\colon\ r = R(z,t),\ \theta \in [0,2\pi),\ z \in (0,L)\}$$

while $\mathbf{n}$ is the outwardly oriented normal to $\partial\Omega_t$. Under the previous assumption, the momentum and continuity equations along z are:

$$\frac{\partial u_z}{\partial t} + \operatorname{div}(u_z\mathbf{u}) + \frac{1}{\rho}\frac{\partial P}{\partial z} - \nu\Delta u_z = 0, \quad z \in (0,L),\ t \in I, \tag{20.4a}$$

$$\operatorname{div}\mathbf{u} = 0, \quad z \in (0,L),\ t \in I, \tag{20.4b}$$

and on the tube wall we have

$$\mathbf{u} = \dot{\boldsymbol{\eta}}, \quad \text{on } \Gamma_t^w,\ t \in I.$$

We have written the convective term in divergence form, like in (8.12), because it simplifies the further derivation.

To ease notation, in this section we will omit to explicitly indicate the time dependence, with the understanding that all variables are considered at time t. Let us consider the portion $\mathcal{P}$ of Ω_t, sketched in Fig. 20.1, comprised between $z = z^* - \mathrm{d}z/2$ and $z = z^* + \mathrm{d}z/2$, with $z^* \in (0,L)$ and $\mathrm{d}z > 0$ small enough so that $z^* + \mathrm{d}z/2 < L$ and $z^* - \mathrm{d}z/2 > 0$. The part of $\partial\mathcal{P}$ laying on the tube wall is indicated by $\Gamma_{\mathcal{P}}^w$. The reduced model is derived by integrating (20.4b) and (20.4a) on $\mathcal{P}$ and passing to the limit as $\mathrm{d}z \to 0$, assuming that all quantities are smooth enough.

We will first illustrate a result derived from the application of the ALE transport theorem (Theorem 18.1) to $\mathcal{P}$.

LEMMA 20.1. *Let $f : \Omega_t \times I \to \mathbb{R}$ be an axisymmetric function, i.e., $\partial f/\partial\theta = 0$. Let us indicate by f_w the value of f on the wall boundary and by $\bar{f}$ its mean value on each* *axial section, defined by*

$$\bar{f} = A^{-1} \int_{\mathcal{S}} f \, \mathrm{d}\sigma.$$

We have the following relation:

$$\frac{\partial}{\partial t}(A\bar{f}) = A\overline{\frac{\partial f}{\partial t}} + 2\pi R\dot{\eta} f_w. \tag{20.5}$$

In particular, taking $f = 1$ yields

$$\frac{\partial A}{\partial t} = 2\pi R\dot{\eta}. \tag{20.6}$$

PROOF. The application of (18.10) to $\mathcal{P}$ gives

$$\frac{\mathrm{d}}{\mathrm{d}t}\int_{\mathcal{P}} f = \int_{\mathcal{P}} \frac{\partial f}{\partial t} + \int_{\partial\mathcal{P}} f\mathbf{g}\cdot\mathbf{n}, \tag{20.7}$$

where $\mathbf{g}$ denotes the velocity of the boundary of $\mathcal{P}$, i.e.,

$$\mathbf{g} = \begin{cases} \dot{\boldsymbol{\eta}} & \text{on } \Gamma^w_{\mathcal{P}}, \\ \mathbf{0} & \text{on } \partial\mathcal{P} \setminus \Gamma^w_{\mathcal{P}}. \end{cases} \tag{20.8}$$

Then, by applying the mean-value theorem to both sides of (20.7), we have

$$\frac{\mathrm{d}}{\mathrm{d}t}\left[A(z^*)\bar{f}(z^*)\,\mathrm{d}z + \mathrm{o}(\mathrm{d}z)\right] = A\overline{\frac{\partial f}{\partial t}} + \mathrm{o}(\mathrm{d}z) + \int_{\Gamma^w_{\mathcal{P}}} f\dot{\eta}\mathbf{e}_r\cdot\mathbf{n}.$$

We recall relation (16.5), already used in the derivation of the models for the wall structure dynamics, to write

$$\int_{\Gamma^w_{\mathcal{P}}} f\dot{\eta}\mathbf{e}_r\cdot\mathbf{n}$$
$$= \int_0^{2\pi}\int_{z^*-\mathrm{d}z/2}^{z^*+\mathrm{d}z/2} f\dot{\eta}R\,\mathrm{d}z\,\mathrm{d}\theta = \left[2\pi\dot{\eta}(z^*)R(z^*)f_w(z^*)\,\mathrm{d}z + \mathrm{o}(\mathrm{d}z)\right]. \tag{20.9}$$

By substituting into (20.7), dividing by dz and passing to the limit as d$z \to 0$, we obtain the desired result. □

We are now ready to derive our reduced model. We start first from the continuity equation. Using the divergence theorem, we obtain

$$\begin{aligned} 0 = \int_{\mathcal{P}} \operatorname{div}\mathbf{u} &= -\int_{\mathcal{S}^-} u_z + \int_{\mathcal{S}^+} u_z + \int_{\Gamma^w_{\mathcal{P}}} \mathbf{u}\cdot\mathbf{n} \\ &= -\int_{\mathcal{S}^-} u_z + \int_{\mathcal{S}^+} u_z + \int_{\Gamma^w_{\mathcal{P}}} \dot{\boldsymbol{\eta}}\cdot\mathbf{n}. \end{aligned} \tag{20.10}$$

We have exploited (20.8) and the fact that $\mathbf{n} = -\mathbf{e}_z$ on $\mathcal{S}^-$ while $\mathbf{n} = \mathbf{e}_z$ on $\mathcal{S}^+$. Now, since $\dot{\boldsymbol{\eta}} = \dot{\eta}\mathbf{e}_r$, we deduce

$$\int_{\Gamma^w_{\mathcal{P}}} \dot{\boldsymbol{\eta}}\cdot\mathbf{n} = \left[2\dot{\eta}\pi R(z^*)\,\mathrm{d}z + \mathrm{o}(\mathrm{d}z)\right] = (\text{by (20.6)}) = \frac{\partial}{\partial t}A(z^*)\,\mathrm{d}z + \mathrm{o}(\mathrm{d}z).$$

By substituting into (20.10), using the definition of Q, and passing to the limit as $\mathrm{d}z \to 0$, we finally obtain

$$\frac{\partial A}{\partial t} + \frac{\partial Q}{\partial z} = 0,$$

which is the reduced form of the continuity equation.

We will now consider all terms in the momentum equation in turn. Again, we will integrate them over $\mathcal{P}$ and consider the limit as $\mathrm{d}z$ tends to zero,

$$\int_{\mathcal{P}} \frac{\partial u_z}{\partial t} = \frac{\mathrm{d}}{\mathrm{d}t}\int_{\mathcal{P}} u_z - \int_{\partial\mathcal{P}} u_z \mathbf{g}\cdot\mathbf{n} = \frac{\mathrm{d}}{\mathrm{d}t}\int_{\mathcal{P}} u_z.$$

In order to eliminate the boundary integral, we have exploited the fact that $u_z = 0$ on $\Gamma^w_{\mathcal{P}}$ and $\mathbf{g} = \mathbf{0}$ on $\mathcal{S}^-$ and $\mathcal{S}^+$. We may then write

$$\int_{\mathcal{P}} \frac{\partial u_z}{\partial t} = \frac{\partial}{\partial t}\left[A(z^*)\bar{u}(z^*)\,\mathrm{d}z + \mathrm{o}(\mathrm{d}z)\right] = \frac{\partial Q}{\partial t}(z^*)\,\mathrm{d}z + \mathrm{o}(\mathrm{d}z).$$

Moreover, we have

$$\begin{aligned}
\int_{\mathcal{P}} \mathrm{div}\,(u_z\mathbf{u}) &= \int_{\partial\mathcal{P}} u_z\mathbf{u}\cdot\mathbf{n} = -\int_{\mathcal{S}^-} u_z^2 + \int_{\mathcal{S}^+} u_z^2 + \int_{\Gamma^w_{\mathcal{P}}} u_z\mathbf{g}\cdot\mathbf{n} \\
&= \alpha\left[A\left(z^* + \frac{\mathrm{d}z}{2}\right)\bar{u}^2\left(z^* + \frac{\mathrm{d}z}{2}\right) - A\left(z^* - \frac{\mathrm{d}z}{2}\right)\bar{u}^2\left(z^* - \frac{\mathrm{d}z}{2}\right)\right] \\
&= \frac{\partial \alpha A\bar{u}^2}{\partial z}(z^*)\,\mathrm{d}z + \mathrm{o}(\mathrm{d}z).
\end{aligned}$$

Again, we have exploited the condition $u_z = 0$ on $\Gamma^w_{\mathcal{P}}$.

Since the pressure is assumed to be constant on each section, we obtain

$$\begin{aligned}
\int_{\mathcal{P}} \frac{\partial P}{\partial z} &= -\int_{\mathcal{S}^-} P + \int_{\mathcal{S}^+} P + \int_{\Gamma^w_{\mathcal{P}}} P n_z \\
&= A\left(z^* + \frac{\mathrm{d}z}{2}\right)P\left(z^* + \frac{\mathrm{d}z}{2}\right) - A\left(z^* - \frac{\mathrm{d}z}{2}\right)P\left(z^* - \frac{\mathrm{d}z}{2}\right) \\
&\quad + \int_{\Gamma^w_{\mathcal{P}}} P n_z. \qquad (20.11)
\end{aligned}$$

Since

$$\int_{\partial\mathcal{P}} n_z = 0,$$

we may write that

$$\begin{aligned}
\int_{\Gamma^w_{\mathcal{P}}} P n_z &= P(z^*)\int_{\Gamma^w_{\mathcal{P}}} n_z + \mathrm{o}(\mathrm{d}z) = -P(z^*)\int_{\partial\mathcal{P}\setminus\Gamma^w_{\mathcal{P}}} n_z + \mathrm{o}(\mathrm{d}z) \\
&= -P(z^*)\left(A\left(z^* + \frac{\mathrm{d}z}{2}\right) - A\left(z^* - \frac{\mathrm{d}z}{2}\right)\right) + \mathrm{o}(\mathrm{d}z).
\end{aligned}$$

By substituting the last result into (20.11), we have

$$\int_{\mathcal{P}} \frac{\partial P}{\partial z} = A\left(z^* + \frac{\mathrm{d}z}{2}\right) P\left(z^* + \frac{\mathrm{d}z}{2}\right) - A\left(z^* - \frac{\mathrm{d}z}{2}\right) P\left(z^* - \frac{\mathrm{d}z}{2}\right)$$
$$- P(z^*)\left[A\left(z^* + \frac{\mathrm{d}z}{2}\right) - A\left(z^* - \frac{\mathrm{d}z}{2}\right)\right] + \mathrm{o}(\mathrm{d}z)$$
$$= \frac{\partial(AP)}{\partial z}(z^*)\,\mathrm{d}z - P(z^*)\frac{\partial A}{\partial z}(z^*)\,\mathrm{d}z + \mathrm{o}(\mathrm{d}z) = A\frac{\partial P}{\partial z}(z^*)\,\mathrm{d}z + \mathrm{o}(\mathrm{d}z).$$

We finally consider the viscous term,

$$\int_{\mathcal{P}} \Delta u_z = \int_{\partial\mathcal{P}} \nabla u_z \cdot \mathbf{n} = -\int_{\mathcal{S}^-} \frac{\partial u_z}{\partial z} + \int_{\mathcal{S}^+} \frac{\partial u_z}{\partial z} + \int_{\Gamma_{\mathcal{P}}^w} \nabla u_z \cdot \mathbf{n}.$$

We neglect $\partial u_z/\partial z$ by assuming that its variation along z is small compared to the other terms. Moreover, we split $\mathbf{n}$ into two vector components, the radial component $\mathbf{n}_r = n_r \mathbf{e}_r$ and $\mathbf{n}_z = \mathbf{n} - \mathbf{n}_r$. Owing to the cylindrical geometry, $\mathbf{n}$ has no component along the circumferential coordinate and, consequently, $\mathbf{n}_z$ is indeed oriented along z. We may thus write

$$\int_{\mathcal{P}} \Delta u_z = \int_{\Gamma_{\mathcal{P}}^w} (\nabla u_z \cdot \mathbf{n}_z + \nabla u_z \cdot \mathbf{e}_r n_r)\,\mathrm{d}\sigma.$$

Again, we neglect the term $\nabla u_z \cdot \mathbf{n}_z$, which is proportional to $\partial u_z/\partial z$. We recall relation (20.1) to write

$$\int_{\mathcal{P}} \Delta u_z = \int_{\Gamma_{\mathcal{P}}^w} n_r \nabla u_z \cdot \mathbf{e}_r\,\mathrm{d}\sigma = \int_{\Gamma_{\mathcal{P}}^w} \bar{u} R^{-1} s'(1)\mathbf{n} \cdot \mathbf{e}_r\,\mathrm{d}\sigma = 2\pi \int_{z^*-\mathrm{d}z/2}^{z^*+\mathrm{d}z/2} \bar{u} s'(1)\,\mathrm{d}z,$$

where we have used the relation $n_r\,\mathrm{d}\sigma = 2\pi R\,\mathrm{d}z$ and indicated by s' the first derivative of s.

Then,

$$\int_{\mathcal{P}} \Delta u_z \approx 2\pi \bar{u}(z^*) s'(1)\,\mathrm{d}z.$$

By substituting all results into (20.4a), dividing all terms by $\mathrm{d}z$ and passing to the limit as $\mathrm{d}z \to 0$, we may finally write the momentum equation of our one-dimensional model as follows:

$$\frac{\partial Q}{\partial t} + \frac{\partial(\alpha A \bar{u}^2)}{\partial z} + \frac{A}{\rho}\frac{\partial P}{\partial z} + K_r \bar{u} = 0,$$

where

$$K_r = -2\pi \nu s'(1)$$

is a *friction parameter*, which depends on the type of profile chosen, i.e., on the choice of the function s in (20.1). For a profile law given by $s(y) = \gamma^{-1}(\gamma + 2)(1 - y^\gamma)$, we have $K_r = 2\pi\nu(\gamma + 2)$. In particular, for a parabolic profile $K_r = 8\pi\nu$, while for $\gamma = 9$ we obtain $K_r = 22\pi\nu$.

To conclude, the final system of equations reads

$$\frac{\partial A}{\partial t}+\frac{\partial Q}{\partial z}=0, \quad z\in(0,L),\ t\in I, \tag{20.12a}$$

$$\frac{\partial Q}{\partial t}+\alpha\frac{\partial}{\partial z}\left(\frac{Q^2}{A}\right)+\frac{A}{\rho}\frac{\partial P}{\partial z}+K_r\left(\frac{Q}{A}\right)=0, \quad z\in(0,L),\ t\in I, \tag{20.12b}$$

where the unknowns are A, Q and P and α is here taken constant.

20.2. *Accounting for the vessel wall displacement*

In order to close system (20.12), we provide a relation for the pressure. A possibility is to resort to an algebraic relation linking pressure to the wall deformation and consequently to the vessel section A.

More generally, we may assume that the pressure satisfies a relation like

$$P(t,z)-P_{\text{ext}}=\psi\left(A(t,z);A_0(z),\boldsymbol{\beta}(z)\right), \tag{20.13}$$

where we have outlined that the pressure will in general depend also on $A_0=\pi R_0^2$ and on a set of coefficients $\boldsymbol{\beta}=(\beta_0,\beta_1,\dots,\beta_p)$, related to physical and mechanical properties, that are, in general, *given* functions of z. Here P_{ext} indicates, as in Section 15, the external pressure. We require that ψ be (at least) a C^1 function of all its arguments and be defined for all $A>0$ and $A_0>0$, while the range of variation of $\boldsymbol{\beta}$ will depend by the particular mechanical model chosen for the vessel wall.

Furthermore, we require that for all allowable values of A, A_0 and $\boldsymbol{\beta}$,

$$\frac{\partial\psi}{\partial A}>0 \quad \text{and} \quad \psi(A_0;A_0,\boldsymbol{\beta})=0. \tag{20.14}$$

By exploiting the linear elastic law provided in (16.16), with the additional simplifying assumption (16.15), and using the fact that

$$\eta=\left(\sqrt{A}-\sqrt{A_0}\right)/\sqrt{\pi}, \tag{20.15}$$

we can obtain the following expression for ψ:

$$\psi(A;A_0,\beta_0)=\beta_0\frac{\sqrt{A}-\sqrt{A_0}}{A_0}. \tag{20.16}$$

We have identified $\boldsymbol{\beta}$ with the single parameter

$$\beta_0=\frac{\sqrt{\pi}h_0E}{1-\xi^2}.$$

The latter depends on z only in those cases where the Young modulus E or the vessels thickness h_0 are not constant.

For ease of notation, the dependence of A, A_0 and $\boldsymbol{\beta}$ from their arguments will be understood. It is immediate to verify that all the requirements in (20.14) are indeed satisfied.

Another commonly used expression for the pressure-area relationship is given by HAYASHI, HANDA, NAGASAWA and OKUMURA [1980], SMITH, PULLAN and HUNTER [2003]:

$$\psi(A; A_0, \boldsymbol{\beta}) = \beta_0 \left[\left(\frac{A}{A_0} \right)^{\beta_1} - 1 \right].$$

In this case, $\boldsymbol{\beta} = (\beta_0, \beta_1)$, where $\beta_0 > 0$ is an elastic coefficient while $\beta_1 > 0$ is normally obtained by fitting the stress-strain response curves obtained by experiments.

Another alternative formulation (LANGEWOUTERS, WESSELING and GOEDHARD [1984]) is

$$\psi(A; A_0, \boldsymbol{\beta}) = \beta_0 \tan \left[\pi \left(\frac{A - A_0}{2A_0} \right) \right],$$

where again the coefficients vector $\boldsymbol{\beta}$ reduces to the single coefficient β_0.

In the following, whenever not strictly necessary we will omit to indicate the dependence of the various quantities on A_0 and $\boldsymbol{\beta}$, which is however always understood.

20.3. The final model

By exploiting relation, (20.12), we may eliminate the pressure P from the momentum equation. To that purpose we will indicate by $c_1 = c_1(A; A_0, \boldsymbol{\beta})$ the following quantity:

$$c_1 = \sqrt{\frac{A}{\rho} \frac{\partial \psi}{\partial A}}, \tag{20.17}$$

which has the dimension of a velocity and, as we will see later on, is related to the speed of propagation of simple waves along the tube.

By simple manipulations (20.12) may be written in *quasi-linear* form as follows:

$$\frac{\partial}{\partial t} U + \mathbf{H}(U) \frac{\partial U}{\partial z} + B(U) = \mathbf{0}, \quad z \in (0, L),\ t \in I \tag{20.18}$$

where,

$$U = \begin{bmatrix} A \\ Q \end{bmatrix},$$

$$\mathbf{H}(U) = \begin{bmatrix} 0 & 1 \\ \frac{A}{\rho} \frac{\partial \psi}{\partial A} - \alpha \bar{u}^2 & 2\alpha \bar{u} \end{bmatrix} = \begin{bmatrix} 0 & 1 \\ c_1^2 - \alpha \left(\frac{Q}{A} \right)^2 & 2\alpha \frac{Q}{A} \end{bmatrix}, \tag{20.19}$$

and

$$B(U) = \begin{bmatrix} 0 \\ K_R \left(\frac{Q}{A} \right) + \frac{A}{\rho} \frac{\partial \psi}{\partial A_0} \frac{\mathrm{d}A_0}{\mathrm{d}z} + \frac{A}{\rho} \frac{\partial \psi}{\partial \boldsymbol{\beta}} \frac{\mathrm{d}\boldsymbol{\beta}}{\mathrm{d}z} \end{bmatrix}.$$

Clearly, if A_0 and $\boldsymbol{\beta}$ are constant the expression for B becomes simpler. A *conservation form* for (20.18) may be found as well and reads

$$\frac{\partial U}{\partial t} + \frac{\partial}{\partial z} [F(U)] + S(U) = 0, \tag{20.20}$$

where

$$F(U)=\begin{bmatrix} Q \\ \alpha\dfrac{Q^2}{A}+C_1 \end{bmatrix}$$

is the vector of fluxes,

$$S(U)=B(U)-\begin{bmatrix} 0 \\ \dfrac{\partial C_1}{\partial A_0}\dfrac{\mathrm{d}A_0}{\mathrm{d}z}+\dfrac{\partial C_1}{\partial \boldsymbol{\beta}}\dfrac{\mathrm{d}\boldsymbol{\beta}}{\mathrm{d}z}, \end{bmatrix}$$

and C_1 is a primitive of c_1^2 with respect to A, given by

$$C_1(A;A_0,\boldsymbol{\beta})=\int_{A_0}^{A} c_1^2(\tau;A_0,\boldsymbol{\beta})\,\mathrm{d}\tau.$$

Again, if A_0 and $\boldsymbol{\beta}$ are constant, the source term S simplifies and becomes $S=B$. System (20.20) allows to identify the vector U as the *conservation variables* of our problem.

REMARK 20.1. In the case we use relation (20.16), we have

$$c_1=\sqrt{\frac{\beta_0}{2\rho A_0}A^{1/4}},\qquad C_1=\frac{\beta_0}{3\rho A_0}A^{3/2}. \tag{20.21}$$

LEMMA 20.2. *If $A\geqslant 0$, the matrix* $\mathbf{H}$ *possesses two real eigenvalues. Furthermore, if $A>0$ the two eigenvalues are distinct and* (20.18) *is a strictly hyperbolic system of partial differential equations.*

PROOF. By straightforward computations, we have the following expression for the eigenvalues of $\mathbf{H}$:

$$\lambda_{1,2}=\alpha\bar{u}\pm c_\alpha, \tag{20.22}$$

where

$$c_\alpha=\sqrt{c_1^2+\bar{u}^2\alpha(\alpha-1)}.$$

Since $\alpha\geqslant 1$, c_α is a real number. If $c_\alpha>0$ the two eigenvalues are distinct. A sufficient condition to have $c_\alpha>0$ is $c_1>0$ and, thanks to the definition of c_1 and (20.14), this is always true if $A>0$. If $\alpha=1$, this condition is also necessary.

The existence of a complete set of (right and left) eigenvectors is an immediate consequence of $\mathbf{H}$ having distinct eigenvalues. □

REMARK 20.2. System (20.12) shares many analogies with the 1D compressible Euler equations, after identifying the section area A with the density. The equivalence is not complete since the term $\partial P/\partial z$ in the Euler equations is here replaced by $A\partial P/\partial z$.

20.3.1. Characteristics analysis

Let $(\mathbf{l}_1, \mathbf{l}_2)$ and $(\mathbf{r}_1, \mathbf{r}_2)$ be two couples of left and right eigenvectors of the matrix $\mathbf{H}$ in (20.19), respectively. The matrices $\mathbf{L}$, $\mathbf{R}$ and $\boldsymbol{\Lambda}$ are defined as

$$\mathbf{L} = \begin{bmatrix} \mathbf{l}_1^{\mathrm{T}} \\ \mathbf{l}_2^{\mathrm{T}} \end{bmatrix}, \qquad \mathbf{R} = \begin{bmatrix} \mathbf{r}_1 & \mathbf{r}_2 \end{bmatrix}, \qquad \boldsymbol{\Lambda} = \operatorname{diag}(\lambda_1, \lambda_2) = \begin{bmatrix} \lambda_1 & 0 \\ 0 & \lambda_2 \end{bmatrix}. \tag{20.23}$$

Since right and left eigenvectors are mutually orthogonal, without loss of generality we choose them so that $\mathbf{LR} = \mathbf{I}$. Matrix $\mathbf{H}$ may then be decomposed as

$$\mathbf{H} = \mathbf{R}\boldsymbol{\Lambda}\mathbf{L}, \tag{20.24}$$

and system (20.18) written in the equivalent form

$$\mathbf{L}\frac{\partial U}{\partial t} + \boldsymbol{\Lambda}\mathbf{L}\frac{\partial U}{\partial z} + \mathbf{L}B(U) = \mathbf{0}, \quad z \in (0, L),\ t \in I. \tag{20.25}$$

If there exist two quantities W_1 and W_2 which satisfy

$$\frac{\partial W_1}{\partial U} = \mathbf{l}_1, \qquad \frac{\partial W_2}{\partial U} = \mathbf{l}_2, \tag{20.26}$$

we will call them *characteristic variables* of our hyperbolic system. We point out that in the case where the coefficients A_0 and $\boldsymbol{\beta}$ are not constant, W_1 and W_2 are not autonomous functions of U.

By setting $W = [W_1, W_2]^{\mathrm{T}}$, system (20.25) may be elaborated into

$$\frac{\partial W}{\partial t} + \boldsymbol{\Lambda}\frac{\partial W}{\partial z} + G = \mathbf{0}, \quad z \in (0, L),\ t \in I, \tag{20.27}$$

where

$$G = \mathbf{L}B - \frac{\partial W}{\partial A_0}\frac{\mathrm{d}A_0}{\mathrm{d}z} - \frac{\partial W}{\partial \boldsymbol{\beta}}\frac{\mathrm{d}\boldsymbol{\beta}}{\mathrm{d}z}. \tag{20.28}$$

We note that the extra terms on the right-hand side are a consequence of the fact that the characteristic variables depend parametrically on the coefficient A_0 and β, which may by a function of z, and thus

$$\frac{\partial W}{\partial z} = \frac{\partial W}{\partial U}\frac{\partial U}{\partial z} + \frac{\partial W}{\partial A_0}\frac{\mathrm{d}A_0}{\mathrm{d}z} + \frac{\partial W}{\partial \boldsymbol{\beta}}\frac{\mathrm{d}\boldsymbol{\beta}}{\mathrm{d}z} = \mathbf{L}\frac{\partial U}{\partial z} + \frac{\partial W}{\partial A_0}\frac{\mathrm{d}A_0}{\mathrm{d}z} + \frac{\partial W}{\partial \boldsymbol{\beta}}\frac{\mathrm{d}\boldsymbol{\beta}}{\mathrm{d}z}.$$

In the case where $B = 0$ and the coefficients A_0 and $\boldsymbol{\beta}$ are constant, (20.27) takes the simpler form

$$\frac{\partial W}{\partial t} + \boldsymbol{\Lambda}\frac{\partial W}{\partial z} = \mathbf{0}, \quad z \in (0, L),\ t \in I, \tag{20.29}$$

which component-wise reads

$$\frac{\partial W_i}{\partial t} + \lambda_i\frac{\partial W_i}{\partial z} = 0, \quad z \in (0, L),\ t \in I,\ i = 1, 2. \tag{20.30}$$

REMARK 20.3. From definition (20.26) and the fact that the left and right eigenvectors $\mathbf{l}_i$ and $\mathbf{r}_i$ are mutually orthogonal it follows that

$$\frac{\partial W_1}{\partial U}(U) \cdot \mathbf{r}_2(U) = 0,$$

thus W_1 is a 2-*Riemann invariant* of our hyperbolic system (GODLEWSKI and RAVIART [1996]). Analogously, one may show that W_2 is a 1-Riemann invariant.

From (20.30) we have that W_1 and W_2 are constant along the two *characteristic curves* in the (z, t) plane described by the differential equations

$$\frac{dz}{dt} = \lambda_1 \quad \text{and} \quad \frac{dz}{dt} = \lambda_2,$$

respectively. In the more general case (20.27) we may easily show that W_1 and W_2 satisfy a coupled system of ordinary differential equations.

The expression for the left eigenvectors $\mathbf{l}_1$ and $\mathbf{l}_2$ is given by

$$\mathbf{l}_1 = \zeta \begin{bmatrix} c_\alpha - \alpha\bar{u} \\ 1 \end{bmatrix}, \qquad \mathbf{l}_2 = \zeta \begin{bmatrix} -c_\alpha - \alpha\bar{u} \\ 1 \end{bmatrix},$$

where $\zeta = \zeta(A, \bar{u})$ is any arbitrary smooth function of its arguments with $\zeta > 0$. Here we have expressed $\mathbf{l}_1$ and $\mathbf{l}_2$ as functions of $(A, \bar{u})$ instead of (A, Q) as is more convenient for the next developments. Thus, relations (20.26) become

$$\frac{\partial W_1}{\partial A} = \zeta\,[c_\alpha - \bar{u}(\alpha - 1)], \qquad \frac{\partial W_1}{\partial \bar{u}} = \zeta A, \tag{20.31a}$$

$$\frac{\partial W_2}{\partial A} = \zeta\,[-c_\alpha - \bar{u}(\alpha - 1)], \qquad \frac{\partial W_2}{\partial \bar{u}} = \zeta A. \tag{20.31b}$$

For a hyperbolic system of two equations it is always possible to find the characteristic variables (or, equivalently, the Riemann invariants) locally, that is in a sufficiently small neighborhood of any point U (GODLEWSKI and RAVIART [1996], LAX [1973]), yet the existence of global characteristic variables is not in general guaranteed. However, in the special case $\alpha = 1$, (20.31) takes the much simpler form

$$\frac{\partial W_1}{\partial A} = \zeta c_1, \qquad \frac{\partial W_1}{\partial \bar{u}} = \zeta A,$$
$$\frac{\partial W_2}{\partial A} = -\zeta c_1, \qquad \frac{\partial W_2}{\partial \bar{u}} = \zeta A.$$

Let us show that a set of global characteristic variables for our problem does exist in this case. We remind that the characteristic variable W_1 exists if and only if

$$\frac{\partial^2 W_1}{\partial A \partial \bar{u}} = \frac{\partial^2 W_1}{\partial \bar{u} \partial A},$$

for all allowable values of A and $\bar{u}$. Since now c_1 does not depend on $\bar{u}$, the above condition yields

$$c_1 \frac{\partial \zeta}{\partial \bar{u}} = \zeta + A \frac{\partial \zeta}{\partial A}.$$

In order to satisfy this relation, it is sufficient to take $\zeta = \zeta(A)$ such that $\zeta = -A(\partial\zeta/\partial A)$. A possible instance is $\zeta = A^{-1}$. The resulting differential form is

$$\partial W_1 = \frac{c_1}{A}\partial A + \partial\bar{u},$$

and by proceeding in the same way for W_2, we have

$$\partial W_2 = -\frac{c_1}{A}\partial A + \partial\bar{u},$$

To integrate it in the $(A, \bar{u})$ plane, we need to fix the value at a reference state, for instance $W_1 = W_2 = 0$ for $(A, \bar{u}) = (A_0, 0)$. We finally obtain

$$W_1 = \bar{u} + \int_{A_0}^{A} \frac{c_1(\tau)}{\tau}\,\mathrm{d}\tau, \qquad W_2 = \bar{u} - \int_{A_0}^{A} \frac{c_1(\tau)}{\tau}\,\mathrm{d}\tau. \tag{20.32}$$

REMARK 20.4. If we adopt relation (20.16) and use the expression for c_1 given in (20.21), after simple computations we have

$$W_1 = \bar{u} + 4(c_1 - c_{1,0}), \qquad W_2 = \bar{u} - 4(c_1 - c_{1,0}), \tag{20.33}$$

where $c_{1,0}$ is the value of c_1 corresponding to the reference vessel area A_0.

Under physiological conditions, typical values of the flow velocity and mechanical characteristics of the vessel wall are such that $c_\alpha \gg \alpha\bar{u}$. Consequently, $\lambda_1 > 0$ and $\lambda_2 < 0$, i.e., the flow is sub-critical everywhere. Furthermore, the flow is smooth. Discontinuities, which would normally appear when treating a non-linear hyperbolic system, do not have indeed the time to form in our context because of the pulsatility of the boundary conditions. It may be shown (CANIC and KIM [2003]) that, for the typical values of the mechanical and geometric parameters in physiological conditions and the typical vessel lengths in the arterial tree, the solution of our hyperbolic system remains smooth, in accordance to what happens in the actual physical problem (which is however dissipative, a feature which has been neglected in our one-dimensional model).

In the light of the previous considerations, from now on we will always assume sub-critical regime and smooth solutions.

20.3.2. Boundary conditions

System (20.12) must be supplemented by proper boundary conditions. The number of conditions to apply at each end equals the number of characteristics entering the domain through that boundary. Since we are only considering sub-critical flows, we need to impose exactly one boundary condition at both $z = 0$ and $z = L$.

An important class of boundary conditions, called non-reflecting or 'absorbing', are those that allow the simple wave associated to the outgoing characteristic to exit the computational domain with no reflections. Following THOMPSON [1987], HEDSTROM [1979], non-reflecting boundary conditions for one-dimensional systems of non-linear hyperbolic equation like (20.20) may be written as

$$\mathbf{l}_1\left(\frac{\partial U}{\partial t} + S(U)\right) = 0 \quad \text{at } z = 0, \qquad \mathbf{l}_2\left(\frac{\partial U}{\partial t} + S(U)\right) = 0 \quad \text{at } z = L,$$

for all $t \in I$. When $S = 0$ these conditions are equivalent to impose a constant value (typically set to zero) to the incoming characteristic variable. When $S \neq 0$ they take into account the "natural variation" of the characteristic variables due to the presence of the source term. A boundary condition of this type is quite convenient at the outlet section.

At the inlet instead one usually desires to impose values of pressure or mass flux derived from measurements or other means. Let us suppose that $z = 0$ is an inlet section (the following discussion may be readily extended to the boundary $z = L$). Whenever an explicit formulation of the characteristic variables is available, the boundary condition may be expressed directly in terms of the entering characteristic variable W_1, i.e., for all $t \in I$,

$$W_1(t) = g_1(t) \quad \text{at } z = 0, \tag{20.34}$$

g_1 being a given function. However, seldom one has directly g_1 at disposal, as the available boundary data is normally given in terms of physical variables. Let us suppose that we know the time variation of both pressure and mass flux at that boundary (for instance, taken from measurements). We may derive the corresponding value of g_1 using directly the definition of the characteristic variable W_1. If $P_m = P_m(t)$ and $Q_m = Q_m(t)$ are the measured average pressure and mass flux at $z = 0$ for $t \in I$ and $W_1(A, Q)$ indicates the characteristic variable W_1 as function of A and Q, we may pose

$$g_1(t) = W_1\left(\psi^{-1}(P_m(t) - P_{\text{ext}}), Q_m(t)\right), \quad t \in I,$$

in (20.34). This means that P_m and Q_m are not imposed exactly at $z = 0$ (this would not be possible since our system accounts for only one boundary condition at each end of the computational domain), yet we require that at all times t the value of A and Q at $z = 0$ lies on the curve in the (A, Q) plane defined by

$$W_1(A, Q) - W_1\left(\psi^{-1}(P_m(t) - P_{\text{ext}}), Q_m(t)\right) = 0.$$

If instead one has at disposal the time history $q(t)$ of a just one physical variable $\phi = \phi(A, Q)$, the boundary condition

$$\phi(A(t), Q(t)) = q(t), \quad \forall t \in I, \quad \text{at } z = 0,$$

is admissible under certain restrictions (QUARTERONI and VALLI [1994]), which in our case reduce to exclude the case where ϕ may be expressed as function of only W_2. In particular, it may be found that for the problem at hand the imposition of either average pressure or mass flux are both admissible.

REMARK 20.5. If the integration of (20.26) is not feasible (as, for instance, in the case $\alpha \neq 1$), one may resort to the *pseudo-characteristic* variables (QUARTERONI and VALLI [1994]), $Z = [Z_1, Z_2]^{\mathrm{T}}$, defined by linearizing (20.26) around an appropriately chosen reference state. One obtains

$$Z = \overline{Z} + \mathbf{L}(\overline{U})(U - \overline{U}), \tag{20.35}$$

where $\overline{U}$ is the chosen reference state and $\overline{Z}$ the corresponding value for $\mathbf{Z}$. One may then use the pseudo-characteristic variables instead of W, by imposing

$$Z_1(t) = g_1(t) \quad \text{at } z = 0, \qquad Z_2(t) = 0 \quad \text{at } z = L.$$

In the context of a time advancing scheme for the numerical solution of (20.20) the pseudo-characteristics are normally computed linearizing around the solution computed at the previous time step.

REMARK 20.6. When considering the numerical discretization, we need in general to provide an additional equation at each end point in order to close the resulting algebraic system. Typically, this extra relation is provided by the so-called *compatibility conditions* (QUARTERONI and VALLI [1994]), which read as follows:

$$\mathbf{l}_2^{\mathrm{T}}\left(\frac{\partial}{\partial t}U+\mathbf{H}\frac{\partial U}{\partial z}+B\right)=0,\quad z=0,\ t\in I, \tag{20.36a}$$

$$\mathbf{l}_1^{\mathrm{T}}\left(\frac{\partial}{\partial t}U+\mathbf{H}\frac{\partial U}{\partial z}+B\right)=0,\quad z=L,\ t\in I. \tag{20.36b}$$

20.3.3. Energy conservation for the 1D model

Most of the results presented in this section are taken from FORMAGGIA, GERBEAU, NOBILE and QUARTERONI [2001] and CANIC and KIM [2003].

LEMMA 20.3. *Let us consider the hyperbolic problem* (20.18) *and assume that the initial and boundary conditions are such that* $\forall z \in (0, L)$,

$$A(0,z)>0 \quad \textit{and} \quad A(t,0)>0,\quad A(t,L)>0,\quad \forall t\in I,$$

and that the solution U *is smooth for all* $(t,z)\in I\times(0,L)$. *Then* $A(t,z)>0$ *for all* $(t,z)\in I\times(0,L)$.

PROOF. Let us suppose that we have $A(t^*,z^*)=0$ at a generic point $(t^*,z^*)\in I\times(0,L)$. From the definition of λ_1 and λ_2 the line $l=\{(t,z):\ z=z_u(t)\}$ satisfying

$$\frac{\mathrm{d}z_u}{\mathrm{d}t}(t)=\bar{u}\,(t,z_u(t))$$

and ending at the point (t^*,z^*), lies between the two characteristic curves passing through the same point. Therefore, it completely lies inside the domain of dependence of (t^*,z^*) and either intersects the segment $z\in(0,L)$ at $t=0$ or one of the two semi-lines $z=0$ or $z=L$ at $t\geqslant 0$. We indicate this intersection point by $(\bar{t},\bar{z})$. The corresponding value of A, call it $\overline{A}$, is positive by hypothesis. From the continuity equation, A satisfies along the line l the following ordinary differential equation:

$$\frac{\mathrm{d}A}{\mathrm{d}t}=-A\frac{\partial\bar{u}}{\partial z},$$

where here the $\mathrm{d}A/\mathrm{d}t$ indicates the directional derivative along l. Therefore,

$$A(t^*,z^*)=\overline{A}\int_{\bar{t}}^{t^*}\frac{\partial\bar{u}}{\partial z}(\tau,z_u(\tau))\,\mathrm{d}\tau>0,$$

in contradiction with the hypothesis. Therefore, we must have $A>0$. □

Here we derive now an *a priori* estimate for the solution of system (20.18) under the hypotheses of $\alpha = 1$, sub-critical smooth flow, and $A > 0$. We will consider the following initial and boundary conditions:

$$\begin{aligned} &\text{initial conditions} \quad && A(0,z) = A^0(z), \quad Q(0,z) = Q^0(z), \quad z \in (0,L) && (20.37)\\ &\text{boundary conditions} \quad && W_1(t,0) = g_1(t), \quad t \in I, && (20.38)\\ & && W_2(t,L) = g_2(t), \quad t \in I. && \end{aligned}$$

Let the quantity e be defined as

$$e = \frac{\rho}{2} A\bar{u}^2 + \Psi, \tag{20.39}$$

where $\Psi = \Psi(A)$ is given by

$$\Psi(A) = \int_{A_0}^{A} \psi(\zeta)\,\mathrm{d}\zeta. \tag{20.40}$$

Here and in what follows we omit to indicate the dependence of ψ on A_0 and $\boldsymbol{\beta}$, since it is not relevant to obtain the desired result, which can be however extended also to the general case where the coefficients A_0 and $\boldsymbol{\beta}$ depend on z.

An energy of the 1D model is given by

$$\mathcal{E}(t) = \int_0^L e(t,z)\,\mathrm{d}z, \quad t \in I. \tag{20.41}$$

Indeed, owing to the assumptions we have made on ψ in (20.14), we may observe that ψ attains a minimum at $A = A_0$, since

$$\Psi(A_0) = \Psi'(A_0) = 0 \quad \text{and} \quad \Psi''(A) > 0, \quad \forall A > 0.$$

It follows that $\Psi(A) \geqslant 0,\ \forall A > 0$. Consequently, $\mathcal{E}(t)$ is a positive function for all Q and $A > 0$ and, moreover,

$$\mathcal{E}(t) = 0 \quad \text{iff} \quad (A(t,z), Q(t,z)) = (A_0, 0), \quad \forall z \in (0,L).$$

The following lemma holds.

LEMMA 20.4. *In the special case $\alpha = 1$, system (20.12), supplied with an algebraic pressure-area relationship of the form (20.13) and under conditions (20.14), satisfies the following conservation property, $\forall t \in I$:*

$$\mathcal{E}(t) + \rho K_R \int_{t_0}^{t}\int_0^L \bar{u}^2\,\mathrm{d}z\,\mathrm{d}\tau + \int_{t_0}^{t} Q(P_{\mathrm{tot}} - P_{\mathrm{ext}})\Big|_0^L\,\mathrm{d}\tau = \mathcal{E}(0), \tag{20.42}$$

where $\mathcal{E}(0)$ depends only on the initial data A^0 and Q^0, while $P_{\mathrm{tot}} = P + \frac{1}{2}\rho\bar{u}^2$ is the fluid total pressure.

PROOF. Let us multiply the second equation of (20.12) by $\bar{u}$ and integrate over $(0,L)$. We will analyze separately the four terms that are obtained.

- First term:

$$I_1 = \int_0^L \frac{\partial (A\bar{u})}{\partial t}\bar{u}\,\mathrm{d}z = \frac{1}{2}\int_0^L A\frac{\partial \bar{u}^2}{\partial t}\,\mathrm{d}z + \int_0^L \frac{\partial A}{\partial t}\bar{u}^2\,\mathrm{d}z$$

$$= \frac{1}{2}\frac{\mathrm{d}}{\mathrm{d}t}\int_0^L A\bar{u}^2\,\mathrm{d}z + \frac{1}{2}\int_0^L \bar{u}^2\frac{\partial A}{\partial t}\,\mathrm{d}z. \tag{20.43}$$

- Second term:

$$I_2 = \alpha\int_0^L \frac{\partial (A\bar{u}^2)}{\partial z}\bar{u}\,\mathrm{d}z = \alpha\left[\int_0^L \frac{\partial (A\bar{u})}{\partial z}\bar{u}^2\,\mathrm{d}z + \int_0^L A\bar{u}^2\frac{\partial \bar{u}}{\partial z}\,\mathrm{d}z\right]$$

$$= \alpha\left[\frac{1}{2}\int_0^L \frac{\partial (A\bar{u})}{\partial z}\bar{u}^2\,\mathrm{d}z + \frac{1}{2}\int_0^L \frac{\partial A}{\partial z}\bar{u}^3\,\mathrm{d}z + \frac{3}{2}\int_0^L A\bar{u}^2\frac{\partial \bar{u}}{\partial z}\,\mathrm{d}z\right]$$

$$= \alpha\left[\frac{1}{2}\int_0^L \frac{\partial Q}{\partial z}\bar{u}^2\,\mathrm{d}z + \frac{1}{2}\int_0^L \frac{\partial (A\bar{u}^3)}{\partial z}\,\mathrm{d}z\right]. \tag{20.44}$$

Now, using the continuity equation, we obtain

$$I_2 = \frac{\alpha}{2}\left[-\int_0^L \frac{\partial A}{\partial t}\bar{u}^2\,\mathrm{d}z + \left(A\bar{u}^3\right)\Big|_0^L\right]. \tag{20.45}$$

- Third term:

$$I_3 = \int_0^L \frac{A}{\rho}\frac{\partial P}{\partial z}\bar{u}\,\mathrm{d}z = \frac{1}{\rho}\int_0^L A\frac{\partial}{\partial z}(P - P_{\mathrm{ext}})\bar{u}\,\mathrm{d}z$$

$$= \frac{1}{\rho}\left[-\int_0^L \frac{\partial Q}{\partial z}\psi(A)\,\mathrm{d}z + (P - P_{\mathrm{ext}})Q|_0^L\right]. \tag{20.46}$$

Again, using the first of (20.12), we have

$$I_3 = \frac{1}{\rho}\left[\int_0^L \frac{\partial A}{\partial t}\psi(A)\,\mathrm{d}z + (P - P_{\mathrm{ext}})Q\Big|_0^L\right]$$

$$= \frac{1}{\rho}\left[\frac{\mathrm{d}}{\mathrm{d}t}\int_0^L \Psi(A)\,\mathrm{d}z + (P - P_{\mathrm{ext}})Q\Big|_0^L\right].$$

- Fourth term:

$$I_4 = \int_0^L K_r\frac{Q}{A}\bar{u}\,\mathrm{d}z = K_r\int_0^L \bar{u}^2\,\mathrm{d}z. \tag{20.47}$$

By summing the four terms and multiplying by ρ, we obtain the following equality when $\alpha = 1$:

$$\frac{1}{2}\rho\frac{\mathrm{d}}{\mathrm{d}t}\int_0^L A\bar{u}^2\,\mathrm{d}z + \frac{\mathrm{d}}{\mathrm{d}t}\int_0^L \Psi(A)\,\mathrm{d}z + \rho K_r\int_0^L \bar{u}^2\,\mathrm{d}z + Q(P_{\mathrm{tot}} - P_{\mathrm{ext}})\Big|_0^L = 0. \tag{20.48}$$

Integrating Eq. (20.48) in time between t_0 and t leads to the desired result. □

In order to draw an energy inequality from (20.42), we need to investigate the sign of the last term on the left-hand side. With this aim, let us first analyze the homogeneous case $g_1 = g_2 = 0$.

We will rewrite the boundary term in (20.42) as a function of A, $\psi(A)$ and c_1 (which, in its turn, depends on A, see (20.17)).

If $g_1 = g_2 = 0$ in (20.38), then

$$\text{at } z=0, \quad W_1 = \bar{u} + \int_{A_0}^{A} \frac{c_1(\zeta)}{\zeta}\,\mathrm{d}\zeta = 0 \quad \Longrightarrow \quad \bar{u}(t,0) = -\int_{A_0}^{A} \frac{c_1(\zeta)}{\zeta}\,\mathrm{d}\zeta,$$

$$\text{at } z=L, \quad W_2 = \bar{u} - \int_{A_0}^{A} \frac{c_1(\zeta)}{\zeta}\,\mathrm{d}\zeta = 0 \quad \Longrightarrow \quad \bar{u}(t,L) = \int_{A_0}^{A} \frac{c_1(\zeta)}{\zeta}\,\mathrm{d}\zeta$$

and thus

$$Q(P_{\mathrm{tot}} - P_{\mathrm{ext}})\Big|_0^L = F(A(t,0)) + F(A(t,L)), \tag{20.49}$$

where

$$F(A) = A \int_0^{A} \frac{c_1(\zeta)}{\zeta}\,\mathrm{d}\zeta \left[\psi(A) + \frac{1}{2}\rho \left(\int_{A_0}^{A} \frac{c_1(\zeta)}{\zeta}\,\mathrm{d}\zeta\right)^2\right]. \tag{20.50}$$

From our assumption of sub-critical flow we have $|\bar{u}| < c_1$ which implies that at $z = 0$ and $z = L$ we have

$$\left|\int_{A_0}^{A} \frac{c_1(\zeta)}{\zeta}\,\mathrm{d}\zeta\right| < c_1(A). \tag{20.51}$$

We are now in the position to conclude with the following result.

LEMMA 20.5. *If the function pressure-area relationship $P = \psi(A)$ is such that $F(A) > 0$ whenever (20.51) is satisfied, then inequality*

$$\mathcal{E}(t) + \rho K_r \int_{t_0}^{t} \int_0^L \bar{u}^2\,\mathrm{d}z\,\mathrm{d}\tau \leqslant \mathcal{E}(0) \tag{20.52}$$

holds for system (20.12), provided homogeneous conditions on the characteristic variables, $W_1 = 0$ and $W_2 = 0$, are imposed at $z = 0$ and $z = L$, respectively.

PROOF. It is an immediate consequence of (20.42), (20.49) and (20.50). □

By straightforward computations, one may verify that the pressure–area relationship given in (20.16) satisfies the hypotheses of Lemma 20.5 (see FORMAGGIA, GERBEAU, NOBILE and QUARTERONI [2001]). Therefore, in that case the 1D model satisfies the energy inequality (20.52).

Under relation (20.16), we can prove a more general energy estimate, valid also in the case of non homogeneous boundary conditions. We state the following result.

LEMMA 20.6. *If the pressure–area relationship is given by (20.16), and the boundary data satisfy*

$$g_1(t) > -4c_{1,0}(0) \quad \textit{and} \quad g_2(t) < 4c_{1,0}(L), \quad \forall t \in I, \tag{20.53}$$

where

$$c_{1,0}(z) = \sqrt{\frac{\beta_0(z)}{\rho}} A_0(z)^{-1/4}$$

is the value of c_1 at the reference vessel area, then there exists a positive quantity $G(t)$ which continuously depends on the boundary data $g_1(t)$ and $g_2(t)$, as well as on the values of the coefficients A_0 and $\boldsymbol{\beta}$, at $z = 0$ and $z = L$, such that, for all $t \in I$,

$$\mathcal{E}(t) + \rho K_r \int_{t_0}^{t} \int_0^L \bar{u}^2 \, dz \, d\tau \leqslant \mathcal{E}(0) + \int_0^t G(t) \, dt. \tag{20.54}$$

PROOF. We will consider only the case where $g_1 \neq 0$ and $g_2 = 0$, since the most general case may be derived in a similar fashion. We recall that relationship (20.16) together with the assumption of sub-critical flow, complies with the conditions stated for $F(A)$ in Lemma 20.5. Then from (20.48) we obtain the following inequality:

$$\frac{d}{dt}\mathcal{E} + \rho K_r \int_0^L \bar{u}^2 \, dz \leqslant Q(P_{\text{tot}} - P_{\text{ext}})\Big|_{z=0}$$

$$\leqslant \left(A|\bar{u}| \left|\psi(A)\right| + \frac{1}{2}\rho A |\bar{u}|^3 \right)\Big|_{z=0}. \tag{20.55}$$

At $z = 0$, we have from (20.33) that

$$\bar{u} + 4(c_1 - c_{1,0}) = g_1.$$

On the other hand, the condition $\lambda_1 = \bar{u} + c_1 > 0$ gives

$$c_1 < \frac{1}{3}(g_1 + 4c_{1,0}). \tag{20.56}$$

Since c_1 is a non-negative quantity, we must necessarily have $g_1 > -4c_{1,0}$. We now note that from (20.16) and the definition of $c_{1,0}$ we may write

$$\psi(A) = 2\rho \left(c_1^2(A) - c_{1,0}^2 \right),$$

which together with (20.56) and the fact that $c_{1,0}$ is a positive function, allows us to state that, at $z = 0$,

$$\psi(A) \leqslant \frac{2\rho}{9} \left(g_1^2 + 15c_{1,0}^2 + 8g_1 c_{1,0} \right) \equiv f_1(g_1), \tag{20.57}$$

where f_1 is a positive continuous function depending parametrically on the values of A_0 and β_0 at $z = 0$. Furthermore, condition $|\bar{u}| < c_1$ together with inequality (20.56) imply that

$$|\bar{u}| \leqslant f_2(g_1),$$

being f_2 another positive and continuous function. Finally, from the definition of ψ and $c_{1,0}$ we have

$$A = \frac{A_0^2}{\beta_0^2}\left(\psi(A) + \sqrt{A_0}\right)^2 \leqslant \frac{2A_0^2}{\beta_0^2}\left(\psi^2(A) + A_0\right) \leqslant \frac{2A_0^2}{\beta_0^2}\left(f_1^2(g1) + A_0\right) \equiv f_3(g_1),$$

where we have exploited (20.57). By combining all previous inequalities, we deduce that the right-hand side in (20.55) may be bounded by a positive and continuous function of the boundary data g_1 that depend parametrically on the value of A_0 and β_0 at $z = 0$. By repeating a similar argument for the boundary conditions at $z = L$, we then obtain the desired stability inequality. □

20.3.4. Weak form

We consider the hyperbolic system (20.20) with initial condition $U = U_0$, at $t = t_0$, and appropriate boundary conditions at $z = 0$ and $z = L$. We indicate by $C_0^1((0, L) \times [t_0, t_1))$ the set of functions which are the restriction to $(0, L) \times [t_0, t_1)$ of C^1 functions with compact support in $(0, L) \times (-\infty, t_1)$. We will assume that U_0 is a bounded measurable function in $(0, L)$.

A function $U \in [L^\infty((0, L) \times [t_0, t_1))]^2$ is a weak solution of the equation in conservation form (20.20) if for all $\boldsymbol{\phi} \in [C_0^1((0, L) \times [t_0, t_1))]^2$ we have

$$\int_{t_0}^{t^1}\int_0^L \left(U \cdot \frac{\partial \boldsymbol{\phi}}{\partial t} + F(U) \cdot \frac{\partial \boldsymbol{\phi}}{\partial z} - S(U) \cdot \boldsymbol{\phi}\right) \mathrm{d}z\, \mathrm{d}t + \int_0^L U_0 \cdot \boldsymbol{\phi}|_{t=0} = 0. \quad (20.58)$$

Moreover, we will require that U complies given boundary conditions.

A solution of (20.58) is called a weak solution of our hyperbolic system. Clearly, "classical" smooth solutions of (20.20) are also weak solutions. Conversely, it may be shown that a smooth weak solution, i.e., belonging to $[C^1((0, L) \times [t_0, t_1))]^2$, is also solution of (20.20) in a classical sense. However, the weak form accommodates also for less regular U. In particular, weak solutions of our hyperbolic problem may be discontinuous. The weak form is furthermore the basis of a class of numerical schemes, in particular, the finite element method, as already seen for the Navier–Stokes equations.

REMARK 20.7. The conservation formulation (20.20) accounts also for mechanical properties which vary smoothly along z. However, there are some fundamental difficulties in extending it to the case of discontinuous mechanical characteristics (e.g., discontinuous $\boldsymbol{\beta}$). On the other hand, this situation has a certain practical relevance, for instance in stented arteries or in the presence of a vascular prosthesis. A stent is a metal meshed wire structure inserted into a stenotic artery (typically a coronary) by angioplasty, in order to restore the original lumen dimension. Vascular prostheses are used to treat degenerative pathologies, such as aneurysms, or when angioplasty is not possible.

A possibility (FORMAGGIA, NOBILE and QUARTERONI [2002]) is to model the sharp variation of the Young modulus at the interface between the artery and the prosthesis

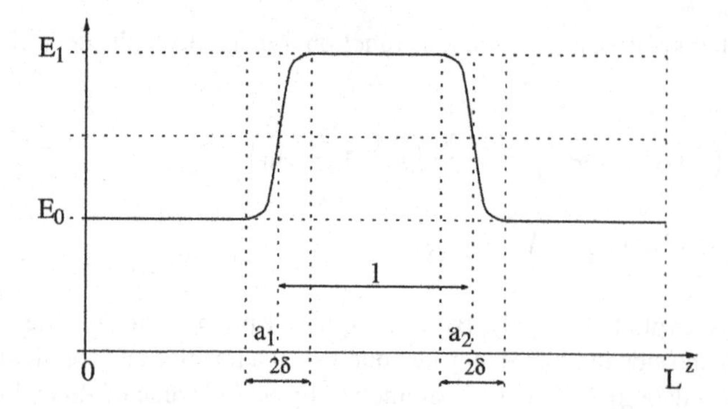

FIG. 20.2. The sharp variation of the Young modulus E from the value E_0 to the value E_1, due to the presence of a prosthesis, is modeled by a smooth function. One may argue what would happen when the parameter δ in figure tends to zero.

by a regular function. Fig 20.2 illustrates a possible description of the change in the Young modulus due to the presence of a prosthesis. One may argue what would happen when the parameter δ in figure tends to zero. Numerical experiments have shown that the solution remains bounded although it becomes discontinuous at the location of the discontinuity in the Young modulus. This fact has been recently investigated in CANIC [2002] where an expression for the jump of mass flow and area across the discontinuity is derived by computing a particular limit of weak solutions of a regularized problem. More details are found in the cited reference.

20.3.5. An entropy function for the 1D model

Let us consider the hyperbolic system written in quasi-linear form (20.18). A pair of functions $e : \mathbb{R}^2 \to \mathbb{R}$ and $F_e : \mathbb{R}^2 \to \mathbb{R}$ is called *entropy pair* for the system if e is a convex function of U (called *entropy*) and the following condition is satisfied:

$$\left(\frac{\mathrm{d}e}{\mathrm{d}U}\right)^{\mathrm{T}} \mathbf{H}(U) = \frac{\partial F_e}{\partial U} \tag{20.59}$$

for all admissible values of U.

F_e is the *entropy flux* associated to the entropy e. If the hyperbolic system admits an entropy pair then the entropy function satisfies a conservation law of the form

$$\frac{\partial e}{\partial t} + \frac{\partial F_e}{\partial z} + B_e(U) = 0,$$

where

$$B_e(U) = \frac{\mathrm{d}e}{\mathrm{d}U} \cdot B(U) - \frac{\partial F_e}{\partial A_0}\frac{\mathrm{d}A_0}{\mathrm{d}z} - \frac{\partial F_e}{\partial \boldsymbol{\beta}}\frac{\mathrm{d}\boldsymbol{\beta}}{\mathrm{d}z}$$

is a source term. The last two terms in the previous expression account for the possible dependence of the coefficients A_0 and $\boldsymbol{\beta}$ on z.

The existence of an entropy pair is of a certain importance when studying the weak solution of the hyperbolic problem and, in particular, discontinuous solutions (more details in LAX [1973] and GODLEWSKI and RAVIART [1996]). Although we have here considered only smooth solutions, the identification of an entropy for our problem is important to set the basis for the extension of the model to more general situations.

In the case $\alpha = 1$,

$$e = \frac{1}{2}\rho A\bar{u}^2 + \Psi(A) = \frac{1}{2}\rho\frac{Q^2}{A} + \Psi(A)$$

is indeed an entropy for the problem at hand, with associated flux

$$F_e = Q\left(\psi(A) + \frac{1}{2}\rho\bar{u}^2\right) = Q(P_{\text{tot}} - P_{\text{ext}}).$$

Indeed, we have

$$\frac{\partial e}{\partial U} = \begin{bmatrix} -\frac{\rho\bar{u}^2}{2} + \psi(A) \\ \rho\bar{u} \end{bmatrix}, \qquad \frac{\partial F_e}{\partial U} = \begin{bmatrix} Q\frac{\partial\psi}{\partial A}(A) - \rho\bar{u}^3 \\ \psi(A) + \frac{3}{2}\rho\bar{u}^2 \end{bmatrix}$$

and we may directly verify condition (20.59) by recalling (20.19). Furthermore, $B_e = \rho K_r\bar{u}^2$ and the entropy balance equation thus read

$$\frac{\partial}{\partial t}\left(\frac{1}{2}\rho A\bar{u}^2 + \Psi(A)\right) + \frac{\partial}{\partial z}\left[Q\left(\psi(A) + \frac{1}{2}\rho\bar{u}^2\right)\right] + \rho K_r\bar{u}^2 = 0. \tag{20.60}$$

It is valid for any smooth solution of our hyperbolic model. Furthermore, the following lemma ensures the convexity of e.

LEMMA 20.7. *The entropy*

$$e(A, Q) = \frac{\rho}{2}\frac{Q^2}{A} + \Psi(A)$$

is convex for all $A > 0$.

PROOF. By a straightforward calculation one finds that the Hessian of e is given by

$$H_e = \begin{bmatrix} \dfrac{\partial^2 e}{\partial A^2} & \dfrac{\partial^2 e}{\partial A\partial Q} \\ \dfrac{\partial^2 e}{\partial A\partial Q} & \dfrac{\partial^2 e}{\partial Q^2} \end{bmatrix} = \frac{\rho}{A}\begin{bmatrix} \bar{u}^2 + c_1^2 & -\bar{u} \\ -\bar{u} & 1 \end{bmatrix}.$$

Its eigenvalues are

$$\lambda_{1,2}(H_e) = \rho\frac{c_1^2 + \bar{u}^2 + 1 \pm \sqrt{(c_1^2 + \bar{u}^2 + 1)^2 - 4c_1^2}}{2A}.$$

The condition for the discriminant to be positive is

$$4c_1^2 \leqslant \left(c_1^2 + \bar{u}^2 + 1\right)^2.$$

Since $c_1 > 0$ whenever $A > 0$, this inequality is equivalent to impose that

$$c_1^2 + \bar{u}^2 + 1 - 2c_1 = (c_1 - 1)^2 + \bar{u}^2 \geqslant 0,$$

which it is always true. Therefore, the two eigenvalues are strictly positive for all $A > 0$. This completes our proof. □

20.4. More complex wall laws that account for inertia and viscoelasticity

The algebraic relation (20.13) assumes that the wall is instantaneously in equilibrium with the pressure forces acting on it. Indeed, this approach correspond to the independent ring model introduced in Section 15.

At the price of some approximations it is possible to maintain the simple structure of a two-equations system while introducing effects, such as the inertia, which depend on the time derivative of the wall displacement.

We will consider as starting point relation (16.17) where we account for the inertia term and we model the viscoelastic property of the wall by adding a term proportional to the displacement rate, while we will still use the approximation (16.15) for the forcing term. We may thus write

$$P - P_{\text{ext}} = \gamma_0 \frac{\partial^2 \eta}{\partial t^2} + \gamma_1 \frac{\partial \eta}{\partial t} + \psi(A; A_0, \boldsymbol{\beta}), \tag{20.61}$$

where $\gamma_0 = \rho_w h_0$, $\gamma_1 = \gamma / R_0^2$ and the last term is the elastic response, modeled is the same way as done before. Here γ is the same viscoelasticity coefficient of (16.17) and η is the wall displacement, linked to A by (20.2).

In the following, we indicate by $\dot{A}$ and $\ddot{A}$ the first and second time derivative of A. We will substitute the following identities:

$$\frac{\partial \eta}{\partial t} = \frac{1}{2\sqrt{\pi A}} \dot{A}, \qquad \frac{\partial^2 \eta}{\partial t^2} = \pi^{-1/2} \left(\frac{1}{2\sqrt{A}} \ddot{A} - \frac{1}{4\sqrt{A^3}} \dot{A}^2 \right),$$

that are derived from (20.2), into (20.61) to obtain a relation that links the pressure also to the time derivatives of A, which we write in all generality as

$$P - P_{\text{ext}} = \tilde{\psi}(A, \dot{A}, \ddot{A}; A_0) + \psi(A; A_0, \boldsymbol{\beta}),$$

where $\tilde{\psi}$ is a non-linear function which derives from the treatment of the terms containing the time derivative of η. Since it may be assumed that the contribution to the pressure is in fact dominated by the term ψ, we will simplify this relationship by linearizing $\tilde{\psi}$ around the state $A = A_0$, $\dot{A} = \ddot{A} = 0$. By doing that, after some simple algebraic manipulations, one finds

$$P - P_{\text{ext}} = \frac{\gamma_0}{2\sqrt{\pi A_0}} \ddot{A} + \frac{\gamma_1}{2\sqrt{\pi A_0}} \dot{A} + \psi(A; A_0, \boldsymbol{\beta}). \tag{20.62}$$

Replacing this expression for the pressure in the momentum equation requires to compute the term

$$\frac{A}{\rho} \frac{\partial P}{\partial z} = \frac{\gamma_0 A}{2\rho\sqrt{\pi A_0}} \frac{\partial^3 A}{\partial z \partial t^2} + \frac{\gamma_1 A}{2\rho\sqrt{\pi A_0}} \frac{\partial^2 A}{\partial z \partial t} + \frac{A}{\rho} \frac{\partial \psi}{\partial z}.$$

The last term in this equality may be treated as previously, while the first two terms may be further elaborated by exploiting the continuity equation. Indeed, we have

$$\frac{\partial^2 A}{\partial z \partial t} = -\frac{\partial^2 Q}{\partial z^2}, \qquad \frac{\partial^3 A}{\partial z \partial t^2} = -\frac{\partial^3 Q}{\partial t \partial z^2}.$$

Therefore, the momentum equation with the additional terms deriving from inertia and viscoelastic forces becomes

$$\frac{\partial Q}{\partial t} + \frac{\partial F_2}{\partial z} - \frac{\gamma_0 A}{2\rho\sqrt{\pi A_0}} \frac{\partial^3 Q}{\partial t \partial z^2} - \frac{\gamma_1 A}{2\rho\sqrt{\pi A_0}} \frac{\partial^2 Q}{\partial z^2} + S_2 = 0, \tag{20.63}$$

where with F_2 and S_2 we have indicated the second component of F and S, respectively.

REMARK 20.8. This analysis puts into evidence that the wall inertia introduces a *dispersive* term into the momentum equation, while the viscoelasticity has a *diffusion* effect.

20.5. Some further extensions

More general one-dimensional models may be derived by accounting for vessel curvature. This may be accomplished by enriching the description of the velocity field on each vessel section to allow asymmetries of the velocity profile to develop.

Another enhancement of the model is to account for vessel branching. By employing domain decomposition techniques, each branch is simulated by a separate one-dimensional model and interface conditions are used to account for the appropriate "transfer" of mass and momentum across the branching point. All these aspects are not covered in these notes. They are subject of current research and preliminary results may be found in FORMAGGIA, LAMPONI and QUARTERONI [2003].

Beside providing valuable information about average pressure and mass flux along an arterial segment, a one-dimensional model of blood flow may be used in the context of a multiscale/multimodel description of the cardiovascular system. In the multiscale framework, models of different level of complexity of the various cardiovascular elements are coupled together with the objective of simulating the whole cardiovascular system. Only the elements of major interest for the problem under study will be simulated at the highest level of detail (e.g., by employing a three-dimensional fluid–structure interaction model), while reduced models are adopted in the remaining parts. This technique allows us to account (at least partially) for the complex feedback mechanisms of the complete cardiovascular system, while keeping the overall computational costs at a reasonable level. More details on this technique may be found in FORMAGGIA, NOBILE, QUARTERONI and VENEZIANI [1999], FORMAGGIA, GERBEAU, NOBILE and QUARTERONI [2001], QUARTERONI, RAGNI and VENEZIANI [2001] while in PIETRABISSA, QUARTERONI, DUBINI, VENEZIANI, MIGLIAVACCA and RAGNI [2000] a first example on the use of this multiscale approach for a realistic clinical application is presented.

CHAPTER VII

21. Some numerical results

We provide some numerical results to illustrate applications of the techniques discussed in the previous Sections. The aim here is to show the potential of the numerical modeling to reproduce realistic flow fields relevant for medical investigations. Many of the results here presented are substantially taken from previous works of the authors, in particular from QUARTERONI, TUVERI and VENEZIANI [2000], FORMAGGIA, GERBEAU, NOBILE and QUARTERONI [2001] and FORMAGGIA, NOBILE and QUARTERONI [2002]. More details and other examples may be found in the cited references.

21.1. Compliant pipe

Here we consider two examples of a fluid–structure interaction problem like the one presented in Section 19, namely a 2D and a 3D computation of a pressure wave in a compliant tube.

In the 2D case, we have considered a rectangular domain of height 1 cm and length $L = 6$ cm. The fluid is initially at rest and an over pressure of 15 mmHg ($2 \cdot 10^4$ dynes/cm^2) has been imposed at the inlet for 0.005 seconds. The viscosity of the fluid is equal to 0.035 poise, its density is 1 g/cm^3, the Young modulus of the structure is equal to $0.75 \cdot 10^6$ dynes/cm^2, its Poisson coefficient is 0.5, its density is 1.1 g/cm^3 and its thickness is 0.1 cm.

In the 3D case, our computation has been made on a cylindrical domain of radius $R_0 = 0.5$ cm and length $L = 5$ cm, with the following physical parameters: fluid viscosity: 0.03 poise, fluid density: 1 g/cm^3, Young modulus of the structure: $3 \cdot 10^6$ dynes/cm^2, Poisson coefficient: 0.3 and structure density: 1.2 g/cm^3. Again, an over-pressure of 10 mmHg ($1.3332 \cdot 10^4$ dynes/cm^2) is imposed at the inlet for 0.005 seconds.

The fluid equations are solved using the ALE approach, with a piece-wise linear finite element space discretization. More precisely, for the 2D case the pressure is piece-wise linear on triangular elements and the velocity is linear over each of the four sub-triangles obtained by joining the midpoints of the edges of each pressure triangle (this is the so called P1isoP2–P1 discretization). We have employed the Yosida technique illustrated in Section 14.5.2. For the 3D case we have used a stabilized scheme (HUGHES, FRANCA and BALESTRA [1986]) and piece-wise linear elements for both velocity and pressure.

For the 2D case, the equation for the structure displacement (18.13) has been solved using a P^1 finite element space discretization, with nodes coincident with the ones of the pressure discretization. In the 3D case, we have used a shell-type formulation (SIMO and FOX [1989], SIMO, FOX and RIFAI [1989]) to describe the dynamics of the wall structure. In both cases, the coupling scheme adopts a sub-iterations strategy of the type illustrated in Section 19.

In order to reduce spurious wave reflections at the outlet, we have coupled the fluid–structure interaction problem with a one-dimensional system of the type described in Section 20. For more details on this technique see FORMAGGIA, GERBEAU, NOBILE and QUARTERONI [2001], as well as FORMAGGIA, GERBEAU, NOBILE and QUARTERONI [2002].

Figs. 21.1 and Figs. 21.2 show the fluid pressure and the domain deformation in the 2D and the 3D case, respectively. For the sake of clarity, the displacements shown in Fig. 21.2 are magnified by a factor 10.

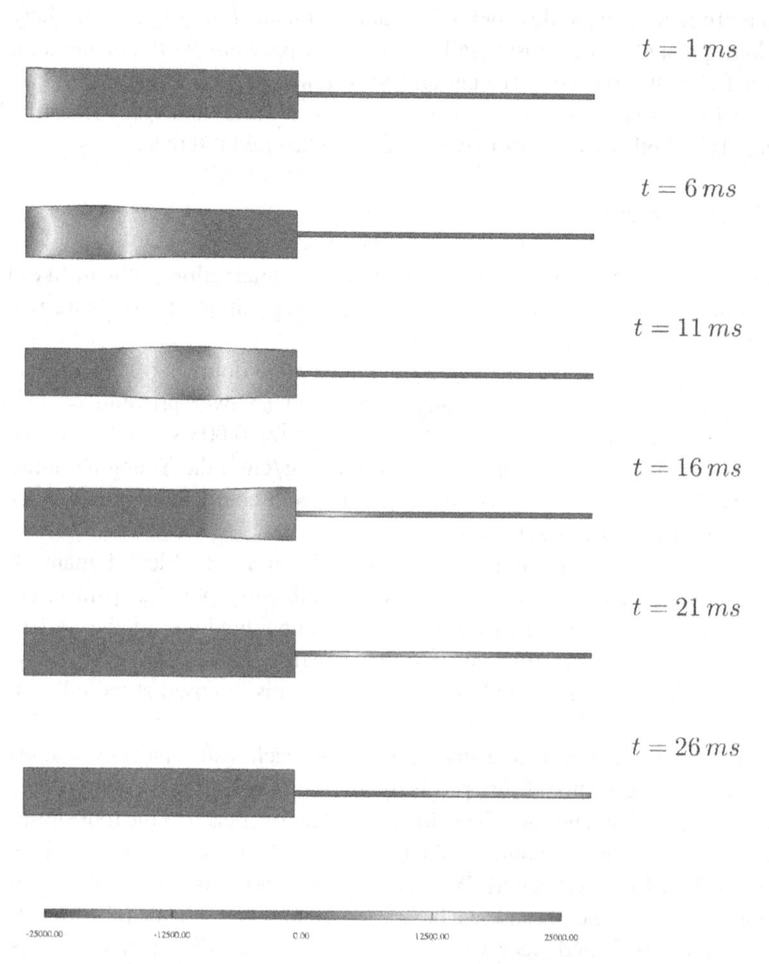

FIG. 21.1. Pressure pulse entering at the *inflow*. A non-reflecting boundary condition at the outlet has been obtained by the coupling with a 1D hyperbolic model. Solutions every 5 ms.

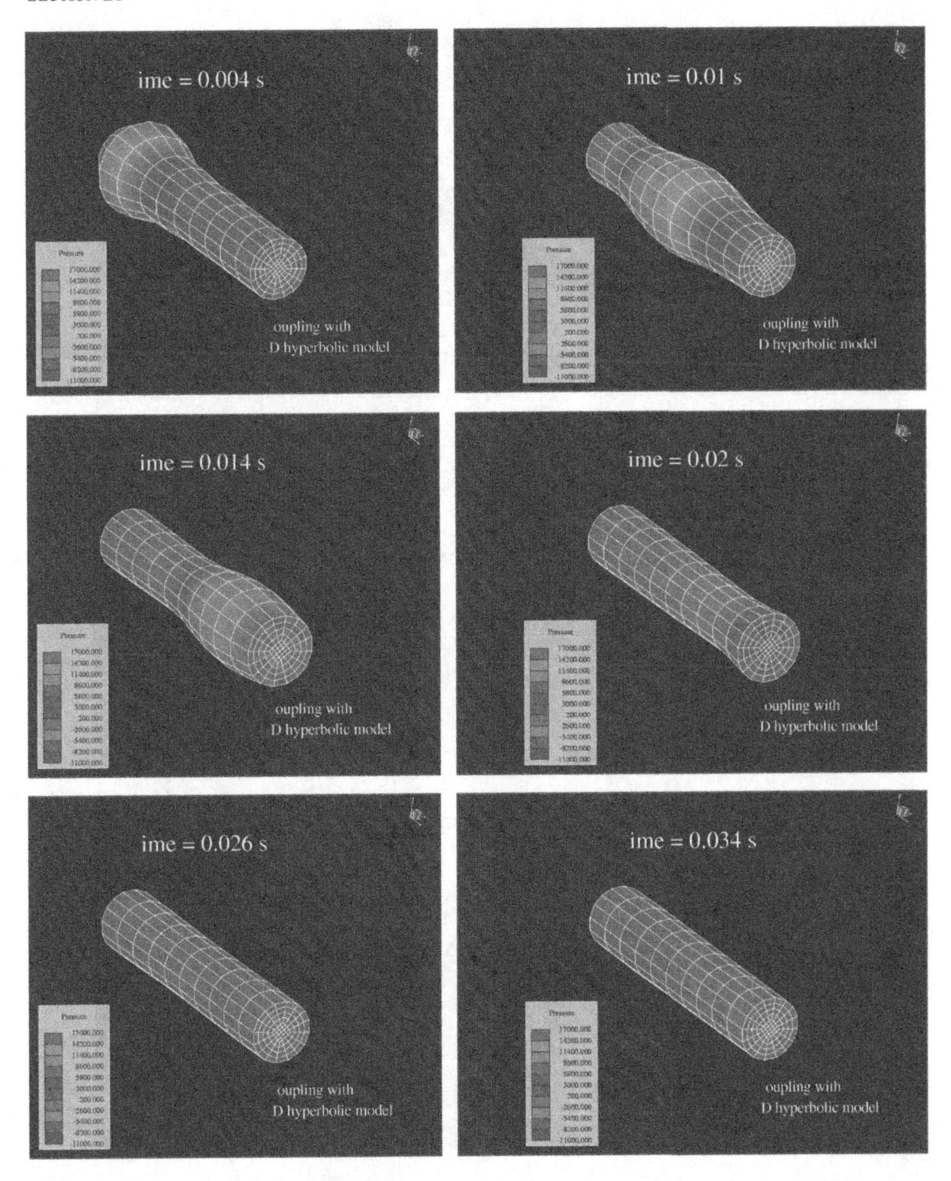

FIG. 21.2. A pressure pulse traveling in a 3D compliant vessel. The displacement of the structure has been magnified by a factor 10. A non-reflecting boundary condition at the outlet has been obtained by the coupling with a 1D hyperbolic model (not shown in the picture).

21.2. *Anastomosis models*

Anastomosis is the a surgical operation by which the functionality of a blocked artery (typically a coronary) is restored thanks to by-pass. The flow condition when the blood

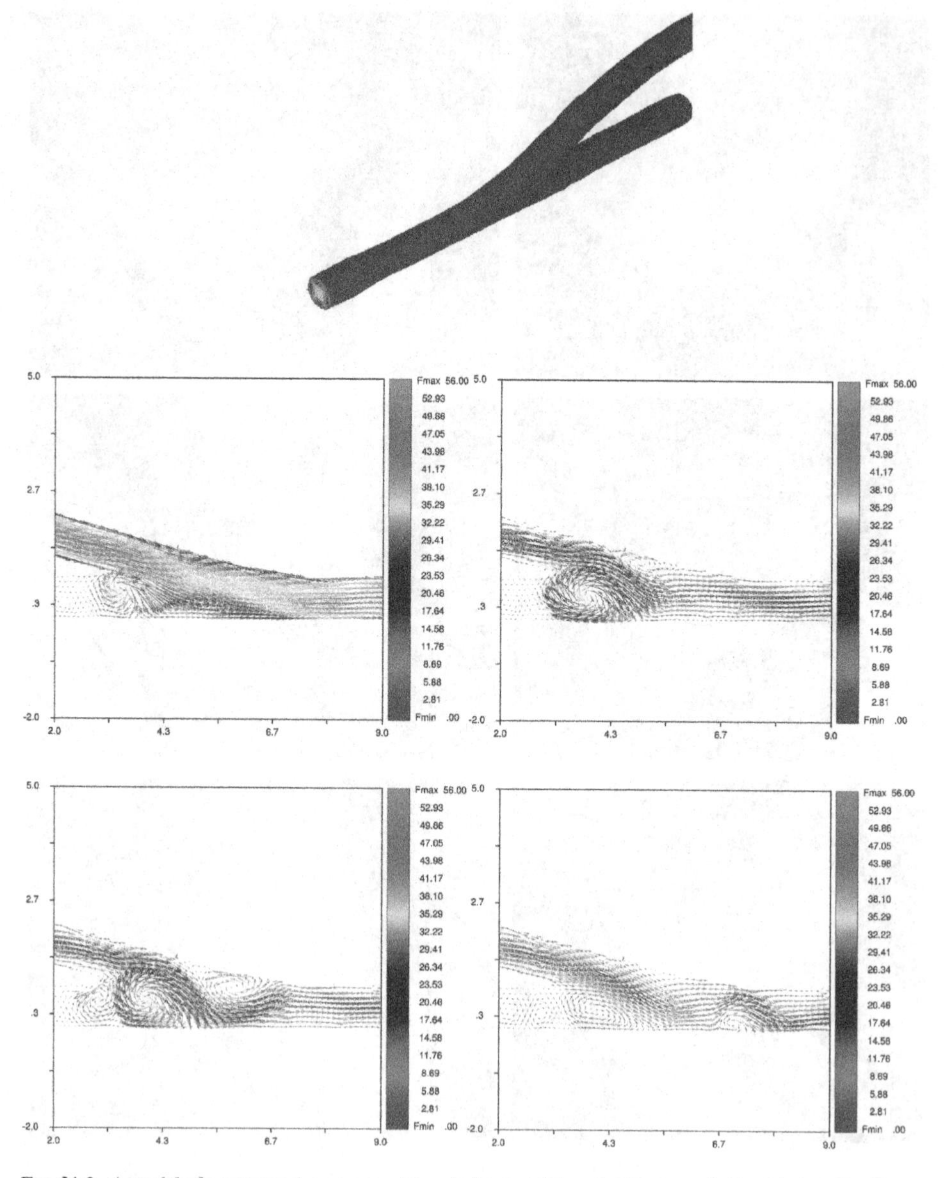

FIG. 21.3. A model of a coronary by-pass anastomosis (top) and the velocity vector field on the median plane at four different instants of the heart beat. Flow at systole (top, left), initial deceleration phase (top, right), beginning of diastole (bottom, left) and end of diastole (bottom, right). The recirculation regions upstream and downstream of the junction are evident.

in the by-pass re-joins the main artery may be critical. If we have a large recirculation area, the higher latency time of blood particles there may favor plaque growing and cause a new blockage further downstream.

The simulations here presented aim at highlight the problem. We illustrate the flow in the median plane of a 3D model of an anastomosis.[7] The junction angle is 15 degrees. The diameter of the occluded branch (below) is 1 cm, and the one of the by-pass (above) is 0.96 cm. The simulations have been carried out setting the dynamic viscosity $\mu = 0.04\ \mathrm{g\,cm^{-1}\,s^{-1}}$ and the density $\rho = 1\ \mathrm{g\,cm^{-3}}$. In this simulation the vessel wall has been assumed fixed and the boundary conditions prescribe null velocity on the walls and on the upstream section of the stenotic branch (100% stenosis), while a parabolic velocity profile has been prescribed at the inlet section with a peak velocity of $56\ \mathrm{cm\,s^{-1}}$, corresponding to a flow rate of $1320\ \mathrm{ml\,min^{-1}}$. On the downstream section a Neumann-type condition has been assigned.

Fig. 21.3 clearly illustrates the appearance and the evolution of the flow recirculation zones during the different phases of the heart beat.

21.3. Pressure wave modification caused by a prosthesis

Here we present a numerical simulation obtained using the one-dimensional model (20.12) to investigate the effect of a prosthesis in an artery, in particular with respect to the alteration of the pressure wave pattern. To that purpose we have considered the portion of an artery of length L and a prosthesis of length l (see Fig. 21.4) and a Young modulus varying as already illustrated in Fig. 20.2.

In order to assess the effect of the changes in vessel wall elastic characteristic on the pressure pattern, we have devised several numerical experiments. Two types of pressure input have been imposed at $z = 0$, namely an impulse input, that is a single sine wave with a small time period and a single sine wave with a more realistic time period (see Fig. 21.5). The impulse has been used to better highlight the reflections induced by the vascular prosthesis.

The part that simulates the presence of the prosthesis or stent of length L is comprised between coordinates a_1 and a_2. The corresponding Young's modulus has been taken as a multiple of the basis Young's modulus E_0 associated to the physiological tissue.

Three locations along the vessel have been identified and indicated by the letters D (distal), M (medium) and P (proximal). They will be taken as monitoring point

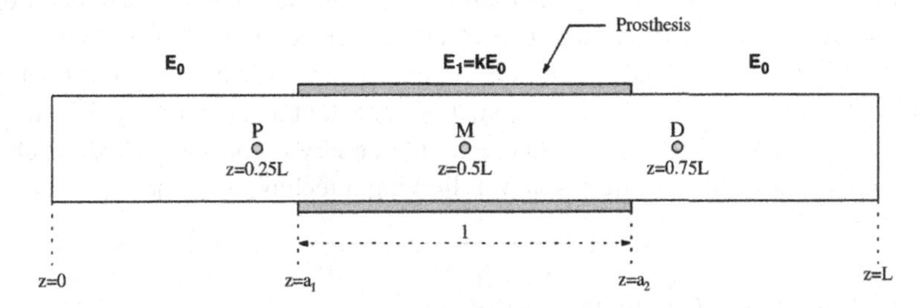

FIG. 21.4. The layout of our numerical experiment. The points P, M and D are used as 'monitoring stations' to assess the modifications on the pressure wave caused by the prosthesis.

[7]The model geometry has been provided by the Vascular Surgery Skejby Sygheus of the Aahrus University Hospital in Denmark.

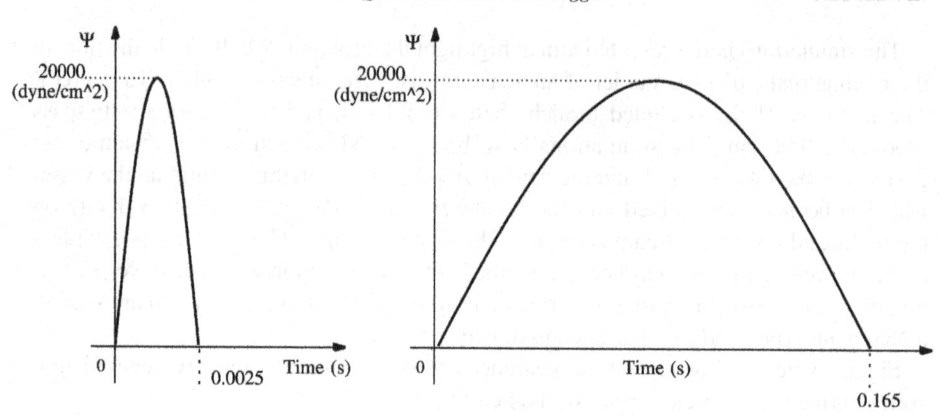

FIG. 21.5. The two types of pressure input profiles used in the numerical experiments: an impulse (left) and a more realistic sine wave (right).

TABLE 21.1
Data used in the numerical experiments

	Parameters	Value
Fluid	Input pressure amplitude	20×10^3 dyne/cm^2
	Viscosity, ν	0.035 poise
	Density, ρ	1 g/cm^3
Structure	Young's modulus, E_0	3×10^6 dyne/cm^2
	Wall thickness, h	0.05 cm
	Reference radius, R_0	0.5 cm

for the pressure variation. Different prosthesis length L have been considered; in all cases points P and D are located outside the region occupied by the prosthesis. Table 21.1 indicates the basic data which have been used in all numerical experiments. In this numerical experiment we have considered the conservation form (20.20) setting the friction term K_r to zero. The numerical scheme adopted is a second order Taylor–Galerkin (DONEA, GIULIANI, LAVAL and QUARTAPELLE [1984]). A time step $\Delta t = 2 \times 10^{-6}$ s and the initial values $A = A_0$ and $Q = 0$ have been used throughout.

At the outlet boundary $z = L$ we have kept W_2 constant and equal to its initial value (non-reflecting boundary condition). At the inlet boundary we have imposed the chosen pressure input in an approximate fashion, following a technique of the type illustrated in Section 20.3.2.

21.3.1. *Case of an impulsive pressure wave*

In Fig. 21.6 we show the results obtained for the case of a pressure impulse. We compare the results obtained with uniform Young modulus E_0 and the corresponding solution when $E_1 = 100E_0$, $l = 5$ cm and the transition zone between healthy artery and prosthesis is $\delta = 0.5$ cm. We have taken $L = 15$ cm and a non-uniform mesh of 105

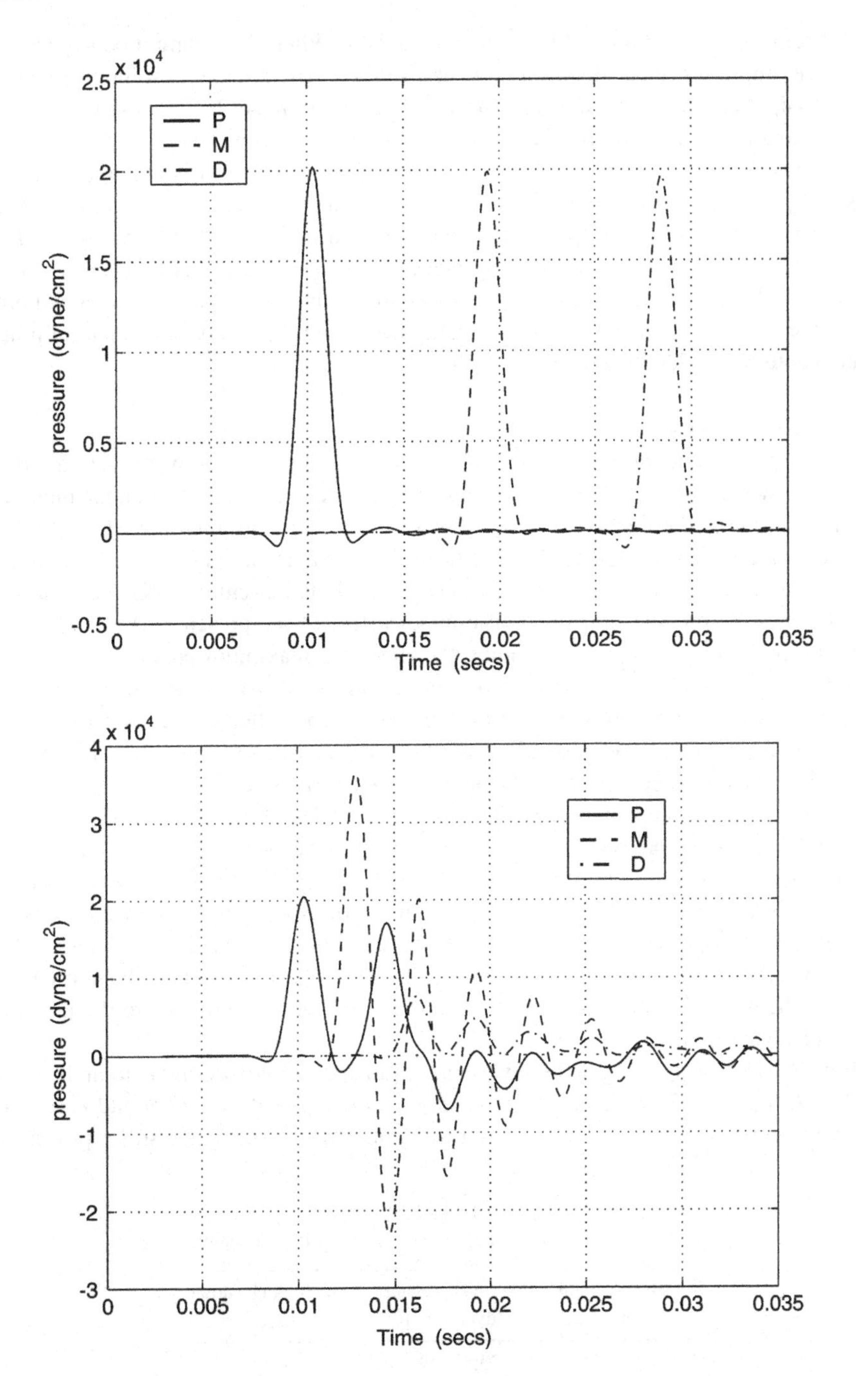

FIG. 21.6. Pressure history at points P, M and D of Fig. 21.4, for an impulsive input pressure, in the case of constant (upper) and variable (lower) E.

finite elements, refined around the points a_1 and a_2. When the Young modulus is uniform, the impulse travels along the tube undisturbed. The numerical solution shows a little dissipation and dispersion due to the numerical scheme. In the case of variable E the situation changes dramatically. Indeed, as soon as the wave enters the region at higher Young's modulus it gets partially reflected (the reflection is registered by the positive pressure value at point P and $t \approx 0.015$ s) and it accelerates. Another reflection occurs at the exit of the 'prosthesis', when E returns to its reference value E_0. The point M indeed registers an oscillatory pressure which corresponds to the waves that are reflected back and forth between the two ends of the prosthesis. The wave at point D is much weaker, because part of the energy has been reflected back and part of it has been 'captured' inside the prosthesis itself.

21.3.2. Case of a sine wave

Now, we present the case of the pressure input given by the sine wave with a larger period shown in Fig. 21.5, which describes a situation closer to reality than the impulse. We present again the results for both cases of a constant and a variable E. All other problem data have been left unchanged from the previous simulation. Now, the interaction among the reflected waves is more complex and eventually results in a less oscillatory solution (see Fig. 21.7). The major effect of the presence of the stent is a pressure increase at the proximal point P, where the maximum pressure is approximately 2500 dynes/cm^2 higher than in the constant case. At a closer inspection one may note that the interaction between the incoming and reflected waves shows up in discontinuities in the slope, particularly for the pressure history at point P. In addition, the wave is clearly accelerated inside the region where E is larger.

In Table 21.2 we show the effect of a change in the length of the prosthesis by comparing the maximum pressure value recorded for a prosthesis of 4, 14 and 24 cm, respectively. The values shown are the maximal values in the whole vessel, over one period. Here, we have taken $L = 60$ cm, $\delta = 1$ cm, a mesh of 240 elements and we have positioned in the three cases the prosthesis in the middle of the model. The maximum value is always reached at a point upstream the prosthesis. In the table we give the normalized distance between the upstream prosthesis section and of the point where the pressure attains its maximum.

Finally, we have investigated the variation of the pressure pattern due to an increase of $k = E/E_0$. Fig. 21.8 shows the result corresponding to $L = 20$ cm and $\delta = 1$ cm and various values for k. The numerical result confirms the fact that a stiffer prosthesis

TABLE 21.2
Maximum pressure value for prosthesis of different length

Prosthesis length (cm)	Maximal pressure (dyne/cm^2)	Maximum location z_{max}/l
4	23.5×10^3	0.16
14	27.8×10^3	0.11
24	30.0×10^3	0.09

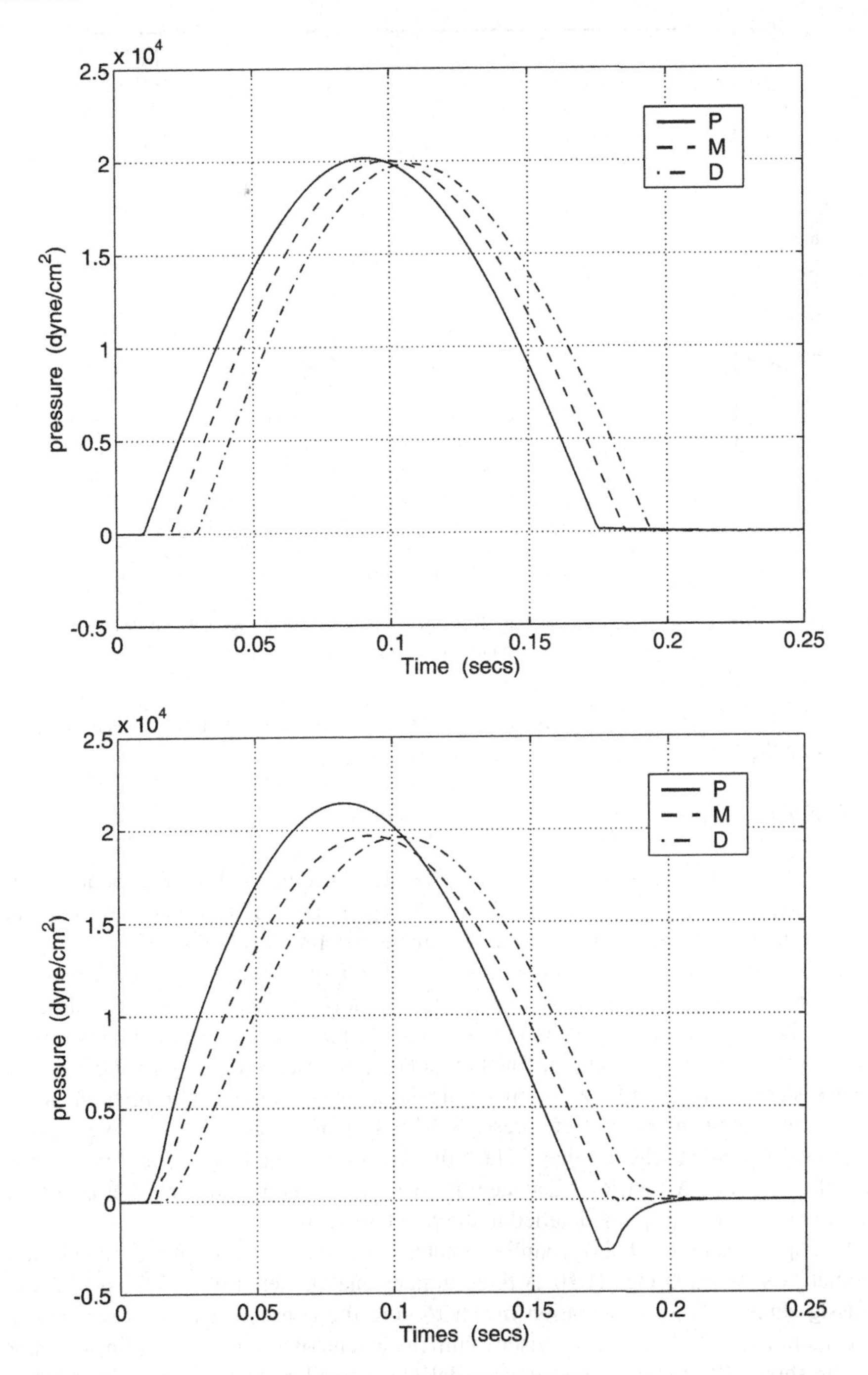

FIG. 21.7. Pressure history at points P, M and D of Fig. 21.4, for a sine wave input pressure, in the case of constant (upper) and variable (lower) E.

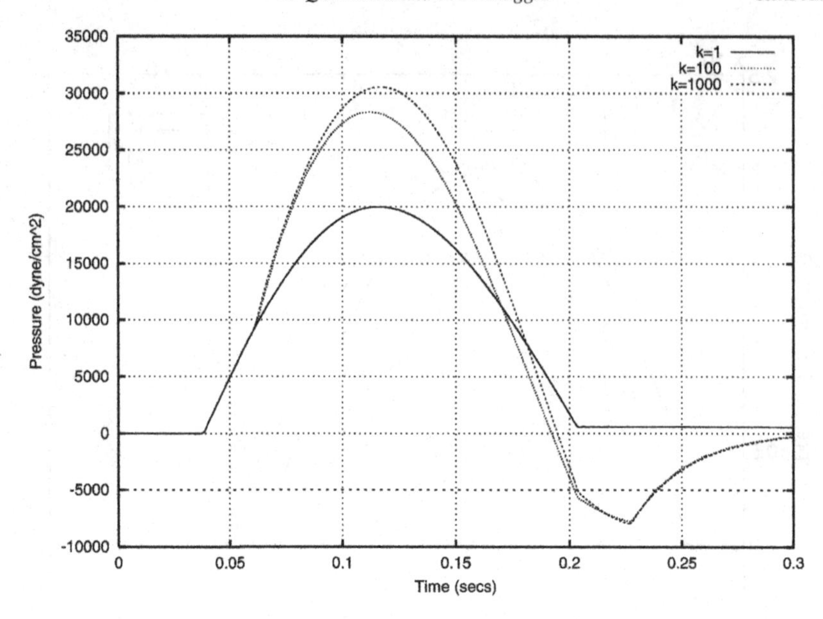

FIG. 21.8. Pressure history at point P of Fig. 21.4, for a sine wave input pressure and different Young's moduli $E = kE_0$.

causes a higher excess pressure in the proximal region, a fact that may have negative effects on the heart.

21.4. Some examples of the geometrical multiscale approach

We end this section by giving some examples of the geometrical multiscale approach, where models of different geometrical complexity are coupled together to provide the simulation of the global cardiovascular system, at different level of detail.

Fig. 21.9 shows an example of the simulation of a by-pass, with the interplay between three-dimensional, one-dimensional and lumped parameters models. A detailed description of the flow in the by-pass is obtained by solving the fluid–structure interaction problem (here using a two-dimensional model). The presence of the global cardiovascular system is provided by a system of algebraic and ordinary differential equations (ODE) for average mass flow and pressure. This system is here illustrated by means of an electrical analog, where voltage plays the role of average pressure and the current that of mass flow. A transition between the two models is provided by the use of the one-dimensional description detailed in the previous section.

A simpler example of this coupling strategy, yet on a realistic three-dimensional geometry, is shown in Fig. 21.10. A three-dimensional model of the modified Blalock–Taussig shunt a surgical operation meant to cure the consequences of a severe cardiac malformation, has been devised with the intent of finding the optimal design for the shunt. The three-dimensional model (on a fixed geometry) has been coupled

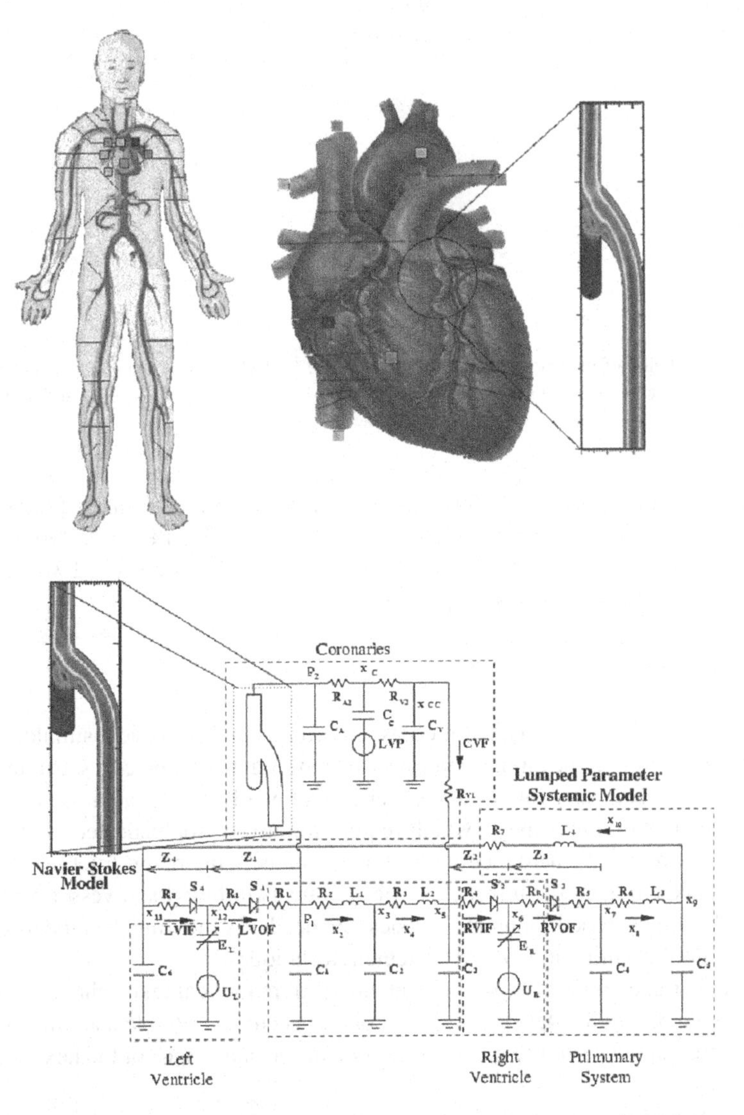

FIG. 21.9. On top we show a global model of the circulatory system where a coronary by-pass is being simulated by a Navier–Stokes fluid–structure interaction model. The rest of the circulatory system is described by means of a lumped parameter model, based on the solution of a system of ODEs, is here represented by an electrical circuit analog in the bottom part of the figure.

with the systemic lumped parameter model, which provides the boundary conditions for the Navier–Stokes equations at the inlet and outlet sections. Thanks to this multi-scale approach it has been possible to compute velocity profiles and flow patterns which are closer to reality than those obtained by using more standard boundary conditions.

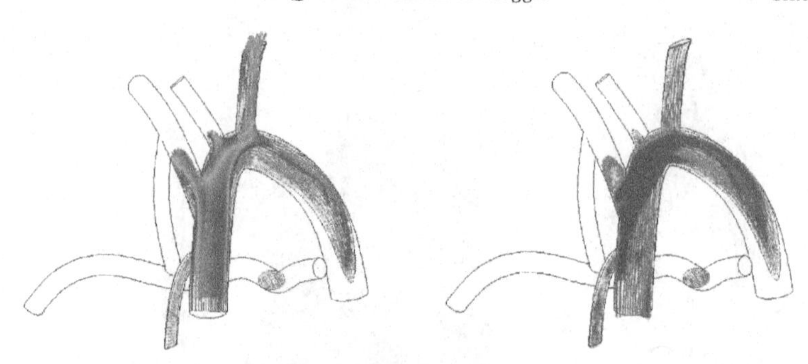

FIG. 21.10. Simulation of the hemodynamics in the modified Blalock–Taussig shunt obtained using a geometrical multiscale approach. Velocity field in the ascending aorta at two different times during the cardiac cycle.

An analysis of the technique is found in QUARTERONI and VENEZIANI [2003], while more details on this and other test cases may be found in PIETRABISSA, QUARTERONI, DUBINI, VENEZIANI, MIGLIAVACCA and RAGNI [2000], MIGLIAVACCA, LAGANÁ, PENNATI, DE LEVAL, BOVE and DUBINI [2004].

22. Conclusions

The development of mathematical models, algorithms and numerical simulation tools for the investigation of the human cardiovascular system has received a great impulse in the last years. These notes intended to cover just a few of the relevant issues. There are however other important aspects which require the use of sophisticated mathematical and numerical tools. We here mention just a few, namely the reconstruction of geometries from medical data; the transport of biochemicals in blood and vessel wall tissue; the heart dynamics; blood rheology. Besides, the need of validating the models calls for development of accurate in-vivo measurement techniques.

The number and complexity of the mathematical, numerical and technological problems involved makes the development of tools for accurate, reliable and efficient simulations of the human cardiovascular system one of the challenges of the next decades.

Acknowledgement

The authors thank Prof. Alessandro Veneziani and Dr. Fabio Nobile for their valuable contributions during the preparation of these notes and for having provided most of the numerical results here presented. We thank Dr. G. Dubini and Dr. F. Migliavacca for the availability of the numerical results for the modified Blalock–Taussig shunt.

Our research activity on the mathematical modeling of the cardiovascular system has been partially supported by grants from various research agencies, which we gratefully acknowledge. In particular, grants 21-54139.98, 21-59230.99 and 20-61862.00 from the Swiss National Science Foundation, the project of Politecnico di Milano "LSC-Multiscale Computing in Biofluiddynamics", the project "Agenzia-2000" by the

Italian CNR, titled "Modeling the fluid structure interaction in the arterial system", and a research contract "Cofin-2000" by the Italian Ministry of Education (MURST) titled "Scientific Computing: Innovative Models and Numerical Methods". Finally, the authors acknowledge the support by the European Union through the Research Training and Network project "HaeMOdel", contract number HPRN-CT-2002-002670.

References

ACHDOU, Y., GUERMOND, J.L. (2000). Convergence analysis of a finite element projection/Lagrange–Galerkin method for the incompressible Navier–Stokes equations. *SIAM J. Numer. Anal.* **37** (3), 799–826.

ARIS, R. (1962). *Vectors, Tensors and the Basic Equations of Fluid Mechanics* (Prentice Hall, New York).

BARNARD, A.C.L., HUNT, W.A., TIMLAKE, W.P., VARLEY, E. (1966). A theory of fluid flow in compliant tubes. *Biophys. J.* **6**, 717–724.

BEIRÃO DA VEIGA, H. (2004). On the existence of strong solutions to a coupled fluid–structure evolution problem. *J. Math. Fluid Mech.* **6**, 21–52.

BOUKIR, K., MADAY, Y., MÉTIVET, B., RAZAFINDRAKOTO, E. (1997). A high-order characteristics/finite element method for the incompressible Navier–Stokes equations. *Internat. J. Numer. Methods Fluids* **25** (12), 1421–1454.

BREZIS, H. (1983). *Analyse Fonctionnelle* (Masson, Paris).

BREZZI, F., FORTIN, M. (1991). *Mixed and Hybrid Finite Elements*, Springer Ser. Comput. Math. **5** (Springer-Verlag, Berlin).

CANIC, S. (2002). Blood flow through compliant vessels after endovascular repair: wall deformations induced by the discontinuous wall properties. *Comput. Visual. Sci.* **4** (3), 147–155.

CANIC, S., KIM, E. (2003). Mathematical analysis of the quasilinear effects in a hyperbolic model of blood flow through compliant axi-symmetric vessels. *Math. Methods Appl. Sci.* **26** (14), 1161–1186.

CHORIN, A.J., MARSDEN, J.E. (1990). *A Mathematical Introduction to Fluid Mechanics*, third ed., Text. Appl. Math. **4** (Springer-Verlag, New York).

CIARLET, P.G. (1988). *Mathematical Elasticity, Volume I: Three-Dimensional Elasticity*, Stud. Math. Appl. **20** (North-Holland, Amsterdam).

CIARLET, P.G. (1998). Introduction to Linear Shell Theory (Gauthier–Villars, Paris).

CIARLET, P.G. (2000). *Mathematical Elasticity, Volume III: Theory of Shells* (North-Holland, Amsterdam).

COKELET, G.R. (1987). The rheology and tube flow of blood. In: Skalak, R., Chen, S. (eds.), *Handbook of Bioengineering* (McGraw–Hill, New York).

DONEA, J., GIULIANI, S., LAVAL, H., QUARTAPELLE, L. (1984). Time-accurate solutions of advection–diffusion problems by finite elements. *Comput. Methods Appl. Mech. Engrg.* **45**, 123–145.

DUVAUT, G., LIONS, J.-L. (1976). *Inequalities in Mechanics and Physics* (Springer-Verlag, Berlin).

FARHAT, C., LESOINNE, M. (2000). Two efficient staggered algorithms for the serial and parallel solution of three-dimensional nonlinear transient aeroelastic problems. *Comput. Methods Appl. Mech. Engrg.* **182**, 499–515.

FARHAT, C., LESOINNE, M., MAMAN, N. (1995). Mixed explicit/implicit time integration of coupled aeroelastic problems: three-field formulation, geometry conservation and distributed solution. *Internat. J. Numer. Methods Fluids* **21**, 807–835.

FORMAGGIA, L., GERBEAU, J.-F., NOBILE, F., QUARTERONI, A. (2001). On the coupling of 3D and 1D Navier–Stokes equations for flow problems in compliant vessels. *Comput. Methods Appl. Mech. Engrg.* **191**, 561–582.

FORMAGGIA, L., GERBEAU, J.-F., NOBILE, F., QUARTERONI, A. (2002). Numerical treatment of defective boundary conditions for Navier–Stokes equations. *SIAM J. Numer. Anal.* **40** (1), 376–401.

FORMAGGIA, L., LAMPONI, D., QUARTERONI, A. (2003). One-dimensional models for blood flow in arteries. *J. Engrg. Math.* **47**, 251–276.

FORMAGGIA, L., NOBILE, F. (1999). A stability analysis for the Arbitrary Lagrangian Eulerian formulation with finite elements. *East–West J. Numer. Math.* **7**, 105–131.

FORMAGGIA, L., NOBILE, F., QUARTERONI, A. (2002). A one-dimensional model for blood flow: application to vascular prosthesis. In: Babuska, I., Miyoshi, T., Ciarlet, P.G. (eds.), *Mathematical Modeling and Numerical Simulation in Continuum Mechanics*. In: Lect. Notes Comput. Sci. Eng. **19** (Springer-Verlag. Berlin), pp. 137–153.

FORMAGGIA, L., NOBILE, F., QUARTERONI, A., VENEZIANI, A. (1999). Multiscale modelling of the circulatory system: a preliminary analysis. *Comput. Visual. Sci.* **2**, 75–83.

FUNG, Y.C. (1984). *Biodynamics: Circulation* (Springer-Verlag, New York).

FUNG, Y.C. (1993). *Biomechanics: Mechanical Properties of Living Tissues* (Springer-Verlag, New York).

GASTALDI, L. (2001). A priori error estimates for the arbitrary Lagrangian Eulerian formulation with finite elements. *East–West J. Numer. Math.* **9** (2), 123–156.

GIRAULT, V., RAVIART, P.-A. (1986). *Finite Element Methods for Navier–Stokes Equations, Theory and Algorithms*, Springer Ser. Comput. Math. **5** (Springer-Verlag, Berlin).

GODLEWSKI, E., RAVIART, P.-A. (1996). *Numerical Approximation of Hyperbolic Systems of Conservation Laws*, Appl. Math. Sci. **118** (Springer-Verlag, New York).

GRANDMONT, C., GUIMET, V., MADAY, Y. (2001). Numerical analysis of some decoupling techniques for the approximation of the unsteady fluid structure interaction. *Math. Models Methods Appl. Sci.* **11** (8), 1349–1377.

GRANDMONT, C., MADAY, Y. (1998). Nonconforming grids for the simulation of fluid–structure interaction. In: *Domain Decomposition Methods 10*, Boulder, CO, 1997. In: Contemp. Math. **218** (Amer. Math. Soc., Providence, RI), pp. 262–270.

GRANDMONT, C., MADAY, Y. (2000). Fluid structure interaction: a theoretical point of view. In: Dervieux. A. (ed.), Revue européenne des éléments finis **9** (Hermes Science), pp. 633–653.

GUERMOND, J.-L. (1999). Un résultat de convergence d'ordre deux en temps pour l'approximation des équations de Navier–Stokes par une technique de projection incrémentale. *M2AN Math. Model. Numer. Anal.* **33** (1), 169–189.

GUILLARD, H., FARHAT, C. (2000). On the significance of the geometric conservation law for flow computations on moving meshes. *Comput. Methods Appl. Mech. Engrg.* **190** (11–12), 1467–1482.

HAYASHI, K., HANDA, K., NAGASAWA, S., OKUMURA, A. (1980). Stiffness and elastic behaviour of human intracranial and extracranial arteries. *J. Biomech.* **13**, 175–184.

HEDSTROM, G.W. (1979). Nonreflecting boundary conditions for nonlinear hyperbolic systems. *J. Comput. Phys.* **30**, 222–237.

HOLZAPFEL, G.A., GASSER, T.C., OGDEN, R.W. (2000). A new constitutive framework for arterial wall mechanics and a comparative study of material models. *J. Elasticity* **61**, 1–48.

HUGHES, T.J., FRANCA, L.P., BALESTRA, M. (1986). A new finite element formulation for computational fluid dynamics: V. Circumventing the Babuska Brezzi condition: a stable Petrov–Galerkin formulation of the Stokes problem accommodating equal-order interpolation. *Comput. Methods Appl. Mech. Engrg.* **59**, 85–99.

LANGEWOUTERS, G.L., WESSELING, K.H., GOEDHARD, W.J.A. (1984). The elastic properties of 45 human thoracic and 20 abdominal aortas *in vitro* and the parameters of a new model. *J. Biomech.* **17**, 425–435.

LAX, P.D. (1973). *Hyperbolic Systems of Conservation Laws and the Mathematical Theory of Shock Waves*, Ser. Appl. Math. **11** (SIAM, Philadelphia, PA).

LE TALLEC, P., MOURO, J. (2001). Fluid structure interaction with large structural displacements. *Comput. Methods Appl. Mech. Engrg.* **190**, 3039–3067.

LIONS, J.L., MAGENES, E. (1968). *Problèmes aux Limites non Homogènes et Applications, 1* (Dunod, Paris).

MEYER, C.D. (2000). *Matrix Analysis and Applied Linear Algebra* (SIAM, Philadelphia, PA).

MIGLIAVACCA, F., LAGANÁ, K., PENNATI, G., DE LEVAL, M., BOVE, E., DUBINI, G. (2004). Global mathematical modeling of the norwood circulation: a multiscale approach for the study of pulmonary and coronary perfusions. *Cardiology in the Young*. In press.

NOBILE, F. (2001). Numerical approximation of fluid–structure interaction problems with application to hemodynamics. PhD thesis, École Polytechnique Fédérale de Lausanne (EPFL), thesis N. 2458.

PEROT, B. (1993). An analysis of the fractional step method. *J. Comput. Phys.* **108**, 51–58.

PIETRABISSA, R., QUARTERONI, A., DUBINI, G., VENEZIANI, A., MIGLIAVACCA, F., RAGNI, S. (2000). From the global cardiovascular hemodynamics down to the local blood motion: preliminary applications of a multiscale approach. In: Oñate, E., et al. (eds.), *ECCOMAS 2000*, Barcelona.

PIPERNO, S., FARHAT, C. (2001). Partitione procedures for the transient solution of coupled aeroelastic problems. Part ii: energy transfer and three-dimensional applications. *Comput. Methods Appl. Mech. Engrg.* **190**, 3147–3170.

PROHL, A. (1997). *Projection and Quasi-compressibility Methods for Solving the Incompressible Navier–Stokes Equations* (Teubner, Stuttgart).

QUARTERONI, A., RAGNI, S., VENEZIANI, A. (2001). Coupling between lumped and distributed models for blood problems. *Comput. Visual. Sci.* **4**, 111–124.

QUARTERONI, A., SALERI, F., VENEZIANI, A. (1999). Analysis of the Yosida method for the incompressible Navier–Stokes equations. *J. Math. Pure Appl.* **78**, 473–503.

QUARTERONI, A., SALERI, F., VENEZIANI, A. (2000). Factorization methods for the numerical approximation of the incompressible Navier–Stokes equations. *Comput. Methods Appl. Mech. Engrg.* **188**, 505–526.

QUARTERONI, A., TUVERI, M., VENEZIANI, A. (2000). Computational vascular fluid dynamics: Problems, models and methods. *Comput. Visual. Sci.* **2**, 163–197.

QUARTERONI, A., VALLI, A. (1994). *Numerical Approximation of Partial Differential Equations* (Springer-Verlag, Berlin).

QUARTERONI, A., VALLI, A. (1999). *Domain Decomposition Methods for Partial Differential Equations* (Oxford Univ. Press, New York).

QUARTERONI, A., VENEZIANI, A. (2003). Analysis of a geometrical multiscale model based on the coupling of ODE's and PDE's for blood flow simulations. *Multiscale Model. Simul.* **1** (2), 173–195.

QUARTERONI, A., VENEZIANI, A., ZUNINO, P. (2002). Mathematical and numerical modelling of solute dynamics in blood flow and arterial walls. *SIAM J. Numer. Anal.* **39** (5), 1488–1511.

RAJAGOPAL, K.R. (1993). Mechanics of non-Newtonian fluids. In: Galdi, G., Necas, J. (eds.), *Recent Developments in Theoretical Fluid Mechanics*. In: Pitman Res. Notes Math. Ser. **291** (Longman, Harlow).

RAPPITSCH, G., PERKTOLD, K. (1996). Pulsatile albumin transport in large arteries: a numerical simulation study. *ASME J. Biomech. Eng.* **118**, 511–519.

REDDY, B.D. (1998). *Introductory Functional Analysis. With Applications to Boundary Value Problems and Finite Elements* (Springer-Verlag, New York).

SEGEL, L.A. (1987). *Mathematics Applied to Continuum Mechanics* (Dover, New York).

SERRIN, J. (1959). Mathematical principles of classical fluid mechanics. In: Flugge, S., Truesdell, C. (eds.), *Handbuch der Physik, VIII/1* (Springer-Verlag, Berlin).

SIMO, J.C., FOX, D.D. (1989). On a stress resultant geometrically exact shell model, Part I: formulation and optimal parametrization. *Comput. Methods Appl. Mech. Engrg.* **72**, 267–304.

SIMO, J.C., FOX, D.D., RIFAI, M.S. (1989). On a stress resultant geometrically exact shell model, Part II: the linear theory; computational aspects. *Comput. Methods Appl. Mech. Engrg.* **73**, 53–92.

SMITH, N., PULLAN, A., HUNTER, P. (2003). An anatomically based model of coronary blood flow and myocardial mechanics. *SIAM J. Appl. Math.* **62** (3), 990–1018.

TAYLOR, C.A., DRANEY, M.T., KU, J.P., PARKER, D., STEELE, B.N., WANG, K., ZARINS, C.K. (1999). Predictive medicine: Computational techniques in therapeutic decision-making. *Comput. Aide. Surgery* **4** (5), 231–247.

TEMAM, R. (1984). *Navier–Stokes Equations, Theory and Numerical Analysis*, second ed. (North-Holland. Amsterdam).

THOMPSON, K.W. (1987). Time dependent boundary conditions for hyperbolic systems. *J. Comput. Phys.* **68**, 1–24.

VENEZIANI, A. (1998). Mathematical and numerical modelling of blood flow problems, PhD thesis, Politecnico di Milano, Italy.

WOMERSLEY, J.R. (1955). Method for the calculation of velocity, rate of flow and viscous drag in arteries when the pressure gradient is known. *J. Physiol.* **127**, 553–563.

Human Models for Crash and Impact Simulation

Eberhard Haug

ESI Software S.A., 99, rue des Solets, BP 80112,
94513 Rungis Cedex, France
E-mail: eha@esi-group.com
URL: http://www.esi-group.com

Hyung-Yun Choi

Hong-Ik University, Seoul, South Korea
E-mail: hychoi@hongik.ac.kr

Stéphane Robin

LAB PSA-Renault, Paris, France
E-mail: stephane.robin@mpsa.com

Muriel Beaugonin

ESI Software S.A., Paris, France
E-mail: mbe@esi-group.com

Essential Computational Modeling for the Human Body
Special Volume (N. Ayache, Guest Editor) of
HANDBOOK OF NUMERICAL ANALYSIS, VOL. XII
P.G. Ciarlet (Editor)

ISSN 1570-8659
DOI 10.1016/B978-0-444-53755-3.00002-0

Contents

Preface

This article deals with the application of computational impact biomechanics to the consequences of real world passenger car accidents on human occupants, using computer models in numerical simulations with industrial crash codes. The corresponding developments are illustrated on the subject of safety simulations of human passenger car occupants. With some adaptations, the developed models apply equally well to the simulation of pedestrian accidents and to the design for occupant safety of motorbikes, trucks, railway vehicles, airborne vehicles, seagoing vessels and more.

The human models elaborated in this article belong to the class of finite element models. They can be adapted, specialized and packaged for other industrial applications in human ergonomics and comfort analysis and design, in situations where humans operate at their work place, as military combatants, or in sports and leisure activities and more. In the medical field, biomechanical human models can serve as a basis for the simulation and design of orthopedic prostheses, for bone fracture planning, physical rehabilitation analysis, the simulation of blood flow, artificial blood vessels, artificial heart valves, bypass operations, and heart muscle activity, virtual organ surgery, etc.

There exists indeed a large overlap, and a pressing urge and opportunity for creating a synergy of very diverse disciplines, which all deal with the simulation of the biomechanical response of the human body.

Most considerations of this article are related to the application of modern crash codes, which discretize space with the finite element method and which apply the explicit time integration scheme of the dynamic equations of motion to discrete numerical models. The reader is assumed to be familiar with the associated basic theory, needed for the use of such codes.

The article is structured as follows.

Chapter I provides an introduction on the interest, need and difficulties of using human models in occupant safety design and analysis. It contains a short overview on mechanical dummies, often used for the design of occupant safety of transport vehicles, and it summarizes some so far existing biomechanical human computer models.

Chapter II discusses "MB (multi-body)" or "HARB (Human Articulated Rigid Body)" or "ATB (Articulated Total Body)" models. These simplest human models consist in rigid body segments, joined at the locations of their skeletal articulations, which can provide gross overall kinematic responses of the human body to static and dynamic load scenarios. For more detailed investigations, they can serve as a basis for modular plug-in of more elaborate and deformable segment models, for making zooms on the detailed response of various body parts. The chapter closes with applications to occupant safety of HARB models, including the fifth percentile female and a six year old child model.

Chapter III discusses deformable human models. In a first section, the results of the first European HUMOS (Human Models for Safety) project (1999–2001) are summarized. The HUMOS-1 project was funded by the European Commission in the Industrial and Materials Technologies (IMT) program (Brite–EuRam III). In this project the geometry of a near fiftieth percentile human cadaver geometry was acquired in a passenger car driving posture and human models were derived from the anthropometrical, biomaterial and validation database, compiled and generated within this project.

In a second section, a systematic presentation of the generation of human models and sub-models is given and illustrated on the example of a deformable fiftieth percentile human model (H-Model). This section first outlines the HARB version of the model and then the deformable sub-models of the head, skull and brain, the neck and cervical spine, the torso with the rib cage, thoracic and abdominal organs, the upper extremity with the shoulder and arm complex, the lower extremity with the knee, thigh and hip complex and the ankle-foot complex. For each deformable sub-model the relevant anatomy, the main injury mechanisms, the basic model structure, its calibration and the basic validations of the models are outlined. A validation of an abdomen model is discussed in the first section. A final section outlines the emerging deformable models of the fifth percentile female.

Appendix A gives an overview on the basic theory of explicit solution and on contact treatment. Appendix B contains data on biomaterials. Appendix C outlines the Hill type muscle models. Appendix D discusses the numerical entities of air- and bio-bags, used to simulate protective airbags and hollow organs. Appendix E provides an insight into the management of the interaction of parts and organs in biomechanical simulation of the human body.

It is clear that this article can only be an incomplete outline over the fast growing, vast and stimulating subject of biomechanical (and biomedical) modeling techniques of the human body. The presented models and methodologies will undoubtedly be upgraded by the time this article is printed. The interested reader is therefore encouraged to keep a close watch on the corresponding web sites and the open literature.

CHAPTER I

Introduction

1. On the interest, need and difficulties of using human models in virtual passenger car crash tests

1.1. Crash design

Crash tests. In car design, standardized "legal laboratory crash tests" are made in order to assess the protective and life saving performance of the car body and its built-in passive occupant safety devices, such as airbags, protective paddings and seat belts. Conventionally, the response of car occupants under accidental conditions, as in frontal crashes, lateral side impacts, rollover accidents, etc., is studied, using re-usable "mechanical occupant surrogate devices", often called "mechanical dummies" or "legal crash dummies". According to existing regulations, passenger transport vehicles must be designed to pass standard crash tests safely to obtain legal certification for selling them to customers. The achieved safety levels are assessed through the mechanical responses of the used dummy devices, as recorded by instruments in physical crash tests carried out in crash laboratories. These recordings are correlated heuristically with human injury. Safe crash design methodologies have their widest use in passenger car design, but apply to all road, water and airborne passenger transport vehicles and working devices. Recent efforts towards "legal virtual testing" try to establish regulatory frameworks that can be used to replace laboratory tests for the purpose of legal certification of vehicles with simulation. While desirable for working with dummies, such process will be mandatory for working with human models since no real world tests can be made to back up the simulations.

Crash simulation. In numerical passenger car crash simulations, numerical models of the car structure, the passive safety restraints (seat belts, airbags, cushions) and the dummy devices are made, the latter as simpler multi-body models, or as more elaborate deformable finite element dummy models. Care must be taken that the passive safety devices are modeled with enough detail, so that their deployment, deformation and energy absorption capacities are well represented in the simulation of a car crash. The numerical dummy models are placed inside the numerical models of the vehicle structure, and their performance under an imposed crash scenario is evaluated. Models of passive safety devices, such as airbags, seat belts, knee bolsters, etc., will be designed and optimized to improve the car safety or crashworthiness performance with respect to the used dummy models. The safety of the car for human occupants is assessed through the simulated response output of the virtual dummy devices, which are modeled and "instrumented" to behave like the real world mechanical dummy devices. If human models were used instead of models of mechanical occupant surrogates, or dummies, a more direct access to human injury could be provided.

Crash codes overview. Numerical crash simulations are performed with specialized crash codes, which were conceived during the eighties of the last century (Pam–Crash, Radioss, Dyna3D), following an urgent need for economy, safety and speed of passenger car design. This need was expressed by the world's passenger car manufacturers. Since the standard safety regulations in all countries became more and more strict, the conventional methods to hand-make ever lighter new car prototype structures and to crash test them became increasingly uneconomical, time consuming and unsafe. The only answer to satisfy the pressure for crashworthiness, safety, quick time to market and economy of design lay in the emerging methodologies of virtual prototyping and design, using high performance computing. This is why several commercial crash codes have emerged, all based in essence on the dynamic explicit finite element method of structural analysis, which uses the proven finite element method for discretizing space, and the explicit direct integration scheme of the non-linear equations of motion to discretize time. One early account of the practical application of a commercial crash code is given by HAUG and ULRICH [1989].

The numerical models treated by these codes started with the car body-in-white (mostly steel structures), modeled with thin shells and contacts. Soon increasingly trade specific models of passive safety devices (airbags, seat belts, knee bolsters, etc.), modeled with cables, bars, joints, membranes, shells and solids followed. Within a few years, models of mechanical dummies, impact barriers and crash obstacles appeared. Today numerical models of human occupants are under active development, with worldwide active support of national agencies for traffic safety. Like always in numerical simulation, a trade-off between computational efficiency, robustness of execution and accuracy must be found. It is therefore legitimate to create numerical models of the human body at different levels of discretization, where the less discretized models execute faster to provide more approximate answers in early design stages, and the more elaborate models cost more computer time and resources, but provide more information and yield more accurate results for the final design.

The correct simulation of contact events or collisions is one of the most crucial features of crash codes. Collisions can occur between the structure of interest and objects in its environment, such as contact between a car and a rigid wall, car-to-car contact, or contact of an occupant with an airbag or seat belt. Contact can also occur between different parts of a crashed structure, such as between the engine and the car body, tire-to-wheel case, roof-to-steering wheel, occupant arm to occupant chest. Finally, self-contact can occur within a single car body component after buckling and wrinkling of its constituent thin sheet metal parts. The correct and efficient treatment of collision events is therefore of great importance, and crash codes have been conceived giving great attention to contact algorithms. Early accounts on the conception of such algorithms are found in HUGHES, TAYLOR, SACKMAN, CURNIER and KANOKNUKULCHAI [1976], HALLQUIST, GOUDREAU and BENSON [1985], and others.

1.2. Occupant safety design

Occupant surrogates. In real world crash tests, it is common practice to use mechanical dummies as surrogates for the human vehicle occupants. Mechanical dummies are instrumented biofidelic occupant surrogate devices, made of metallic, rubber, foam and plastic materials, that are widely used by car makers in real vehicle crash tests. The impact of car accidents on human occupants is inferred from the impact performance of the used mechanical dummies, expressed in terms of standard response measurements, such as head accelerations, chest deflections, femur loads, etc. These measurements can be correlated with human injury via so-called injury criteria. The latter give rough insight into the real injuries a human occupant might experience in each studied crash scenario.

The consistent use of dummies in crash tests is not ideal, because even the best crash dummies can only approximate the behavior of real humans in a crash. Humans undergo wider trajectories inside a vehicle than dummies. Therefore ever more advanced dummies are needed to provide more representative injury data. Nevertheless, dummies and dummy models enabled car manufacturers to very significantly increase the passive safety performances of their products. Most of the current safety devices were indeed developed with the well-known Hybrid III frontal dummy, or with the EuroSID 1 side impact dummy. Since humans cannot be used in real world crash tests, dummies are the only workable alternative, and their use is mandatory. Crash dummies are under continuous improvement, and next generation mechanical occupant surrogate devices are under development (e.g., the THOR dummy developed by NHTSA), often with the help of numerical simulations using human models.

Human subjects. The direct use of humans in the everyday safety design of transport vehicles is excluded due to obvious ethical and practical constraints. Some exceptional uses of live and dead humans (cadavers or PMHS for post mortem human subjects) for research purposes and for indirect studies of the response of the human body in crash situations are listed next. All tests involving human volunteers and human cadavers are subject to very rigorous screening procedures by competent ethics committees in all countries. Adult persons can dedicate their bodies in case of decease by an act of will to science. Children can not grasp such an act, and their parents can not, in general, decide for their children. Child cadavers can therefore hardly be used for destructive tests. Exceptions may exist in using body scan images of children when the parents give their consent. Modern non-destructive bone density measurements and similar existing or emerging techniques can be used to circumnavigate this ethical dilemma.

Human volunteers. Human volunteers can only exceptionally serve in experimental impact tests. One well known historical contribution were the human tests carried by US Air Force Colonel John P. Stapp, who studied from 1946 to 1958 the effects of deceleration on both humans and animals at the Edwards Air Force Base in California and at the Holloman Air Force Base in New Mexico. Stapp exposed belt restrained volunteers, including himself, to decelerations of up to 40 g, using rocket driven sleds. Since 1955 by now yearly Stapp Car Crash Conferences take place (46th by 2002). One recent example involving human volunteers is given by low energy rear and front impact crash tests, where the principal effect of neck "whiplash" motions is studied in purely research oriented projects under medically controlled conditions. ONO [1999] studies the relationship between localized spine deformation and cervical vertebral motions for low speed rear impacts using human volunteers. In such exceptional test setups, human volunteers are subjected to sub-injury rear or front impact equivalent acceleration levels. In particular, these studies employ X-ray cine-radiography, accelerometer recordings and electro-myographic recordings on the neck response in very low speed rear-end car impacts. In such recordings the activation level of the neck muscles can be monitored via their variable electrical characteristics. The resulting data are used to develop models to evaluate neck injuries caused by higher-speed rear-end impacts, and to improve the accuracy of conventional crash-test dummies.

Accidentological studies and accident reconstructions. Such studies can give insight into crash events after a real life accident has occurred. These investigations can determine what might have happened to the involved human occupants. Accidentological studies provide data about the ways the accidents occurred, the involved vehicles, vehicle trajectories and collision with obstacles, and data about the injuries and the medical consequences for the human occupants. Accident reconstruction studies often re-enact reported crashes in the laboratory, or use numerical simulation of the reported crashes. In such re-enactions and simulations, mechanical dummies and their models can be employed. In numerical simulations of the re-enacted crashes, the use of human models is of particular interest, since the regulations which prescribe the mandatory use of mechanical occupant surrogates in certification tests do not apply. Car companies re-enact reported crashes in order to better understand the causes of injury and to improve the car design.

Human cadaver tests. Tests with human cadavers (PMHS tests) can be carried out at the exceptional research level in experimental impact studies. Most cadaver tests study the basic biomechanical mechanisms that lead to injuries of the human body (e.g., SCHMIDT, KALLIERIS, BARZ, MATTERN, SCHULZ and SCHÜLER [1978]). In no case can cadaver tests be used in everyday car design. Only principal injury mechanisms can be deduced from cadaver tests, and each cadaver tends to be different. Average human response "corridors" can be derived from test campaigns which may involve many different cadavers, each subjected to the same test. In the past, the design of mechanical dummies was based largely on the knowledge derived from specific series of different types of cadaver tests. For example, KALLIERIS and SCHMIDT [1990] describe the neck response and injury assessment using cadavers and the US-SID side impact dummy for far-side lateral impacts of rear seat occupants with inboard-anchored shoulder belts.

Cadaver test results can produce valuable information for the construction of human numerical models, rather than to be of direct value in everyday car design. For example, cadaver test studies on the human skull and the mechanisms of brain injury can clarify the relationship between different types of impacts and the nature and extent of injury. Tests on the brain and the skull are carried out in order to improve two- and three-dimensional models of the head for computer simulations, to understand the mechanisms through which injuries develop in the brain and skull. Neck tests improve the knowledge about human neck injury tolerance and mechanisms. Pendulum impact tests on the thorax and pelvis shed light on the response of the skeleton and organs in frontal and side impacts. Impact tests on the abdomen can give insight in the action of lap belts on the visceral organs. Upper extremity impact tests yield information about aggression from side impact airbags. Cadaver test research into leg injuries typically involves examining intrusion of the fire wall of passenger cars into the occupant compartment, the sitting position and kinematics of the occupant, the effectiveness of knee bolsters, the position of the pedals, and the anatomical nature of these injuries.

Animal tests. Tests which involve life or dead animals are subject to ethics committee constraints, as are tests involving human cadavers or human volunteers. In order to discern the different behavior of body segments, organs and biomaterials of the live organism, animal tests have be performed on live anesthetized animals. Again, such tests can not serve as a basis for everyday car design, but are sometimes carried out in purely research type projects where the use of humans is excluded. For example, some brain injury mechanisms were studied in the past on primates by ONO, KIKUCHI, NAKAMURA, KOBAYASHI and NAKAMURA [1980]. Pigs were also used to study the consequences of chest impacts by KROELL, ALLEN, WARNER and PERL [1986].

Humans in crash tests. While humans cannot replace mechanical dummies in real world crash tests, this is the case in virtual crash simulations. It is therefore of great potential advantage to build human models, and to use them in accident simulations. By combining crash analysis and biomechanical analysis, it is possible to advance the understanding of how injuries occur. This is the most important step towards creating safer automobiles and safer roads. As a by-product, human models can be used for improving the design of mechanical occupant surrogate devices.

1.3. Injury and trauma

Humans vs. dummies. Human models represent “bone, soft tissue, flesh and organs” instead of “steel, rubber, plastic materials and foam”, as it is the case with dummy models. Injury in the sense of biological damage does not exist in todays mechanical dummies, because dummy devices are designed for multiple re-use without repair. The danger of injury to humans is deduced indirectly from the instrument responses of the mechanical dummies (or their models), as obtained during a real world (or simulated) crash test.

Human bone, soft tissue, flesh and organ injury prediction is the primary goal of impact biomechanics. If injuries can be predicted directly and reliably, then cars can be designed safer. In impact biomechanics, two classes of human parts and organs may be distinguished from a purely structural point of view: first the ones who have an identifiable structural function, and second the ones who have not. Skeletal bones, for example, are "structural" elements in the sense that they must carry the body weight and mass, and their material resembles conventional engineering materials. The brain, on the other hand, has hardly any structural function, and it resembles a tofu-like material with a maze of reinforcing small and tiny blood vessels, not unlike a soft "composite" material. The structural response of skeletal bone can be modeled more readily with standard engineering procedures than the structural response of the brain, and the injury to skeletal bone can be inferred easily as fracture from its structural response, while neuronal brain injury is not easily derived from the structural response of the brain material.

Injury prediction. Injury of human parts, before any healing takes place, can either be defined as instantaneously irreversible mechanical damage, for example damaged articulations, broken bones, aorta rupture or soft tissue and organ laceration, or, as a reduction of the physiological functioning, for example of the neurological functions of the brain, sometimes without much visible physical damage.

Bone fracture, on the one hand, is largely characterized by the mechanical levels of stress, strain and rate of strain in the skeletal bones, as calculated readily from accurate mechanical models in the simulation of an impact event. Long bones (femur, ribs, humerus, etc.), short bones (calcaneous, wrist bones, etc.) and flat bones (skull, pelvis, scapula) can often be modeled using standard brittle material models for the harder cortical bone, and standard collapsible foam material models for the softer, spongy, trabecular or cancellous bone. Ligaments and tendons, and sometimes passive muscles, skin, etc., can be modeled fairly well using standard non-linear rubber-like hyper-visco-elastic materials.

Internal organs have physiological functions. Their structural attachment inside the body cavities is given by mutual sliding contact, by in and outgoing vessels, by ligaments and by sliding contact with the body cavity walls. Their structural response to impact is harder to calculate and the calculated mechanical response fields are hard to correlate with their physiological functioning or injury.

The heart can act like a structural vessel, for example, when it is compressed and shifted in a chest impact. Gross shifts may cause strain and rupture of the aorta, an event which can be modeled with advanced solid–fluid interaction simulation techniques. The mechanical simulation of this process requires a detailed model of the heart, the aorta walls, and of the way the heart and aorta are anchored inside the chest. The blood should then be modeled as a fluid medium.

The other internal organs are either solid (liver, spleen, kidneys, etc.), hollow (stomach, intestines, bladder, etc.), or spongeous (lungs). The solid organs respond with their bulk matter to mechanical aggression in crash events. For example, the liver might be lacerated by the action of a lap belt. However, the tender liver parenchyma is invested by tough-walled vessels which render the material heterogeneous and anisotropic. The hollow organs should be modeled as hollow cavities, with an adequate model of their contents, which might interact with the organ walls during a mechanical aggression.

For the brain, the mechanical stress and strain fields and their histories, once calculated, must yet be linked to neurological damage. After impact, the neurons are still there, but they may have ceased to function properly because they became disconnected

TABLE 1.1
AIS injury scale

AIS code	Description	General injury	Thorax injury (example)
0	No injury	–	–
1	Minor	Abrasions, sprains, cuts, bruises	–
2	Moderate	Extended abrasions and bruises; extended soft tissue wounds; mild brain concussions without loss of consciousness	Single rib fracture
3	Serious (not life threatening)	Open wounds with injuries of vessels and nerves; skull fractures; brain concussions with loss of consciousness (5–10 minutes)	2–3 rib fractures sternum fracture
4	Severe (life threatening; probability of survival)	Severe bleeding; multiple fractures with organ damage; brain concussion with neurological signs; amputations	>4 rib fractures 2–3 rib fracture with hemo/ pneumothorax
5	Critical (survival is uncertain)	Rupture of organs; severe skull and brain trauma; epidural and subdural hematoma; unconsciousness over 24 hours	>4 rib fractures with hemo/ pneumothorax
6	Maximum (treatment not possible; virtually unsurvivable)	Aorta rupture; collapse of thoracic cage; brain stem laceration; annular fracture of base of skull; separation of the trunk; destruction of the skull	Aorta laceration

at certain strain levels, or because these cells were asphyxiated from the pressure generated by hematomas, which may prevent proper blood supply to uninjured parts of the brain. While this may or may not create visible mechanical "material" damage, it will cause reduction or total loss of the brain functions, hence injury.

The definition of biological and medical injury to the internal organs and its correlation with mechanical output fields as obtained from impact biomechanics models remains an open field for intensive research.

Injury scales. Criteria for injury potential were proposed by GADD [1961], GADD [1966]. The most often used injury scale for impact accidents is the Abbreviated Injury Scale (AIS). Table 1.1 contains AIS scores and some associated injuries.

References on injury and trauma. Detailed descriptions and further extensive bibliographies of injury and trauma of the skull and facial bone, the brain, the head, the cervical spine, the thorax, the abdomen, the thoraco-lumbar spine and pelvis and the extremities can be found in the book by NAHUM and MELVIN (eds.) [1993] *Accidental Injury – Biomechanics and Prevention.*

In this book first general aspects related to impact biomechanics are discussed in the following chapters: Chapter 1: The Application of Biomechanics to the Understanding of Injury and Healing (FUNG [1993b]); Chapter 2: Instrumentation in Experimental

Design (HARDY [1993]); Chapter 3: The Use of Public Crash Data in Biomechanical Research (COMPTON [1993]); Chapter 4: Anthropometric Test Devices (MERTZ [1993]); Chapter 5: Radiologic Analysis of Trauma (PATHRIA and RESNIK [1993]); Chapter 6: A Review of Mathematical Occupant Simulation Models (PRASAD and CHOU [1993]); Chapter 7: Development of Crash Injury Protection in Civil Aviation (CHANDLER [1993]); Chapter 8: Occupant Restraint Systems (EPPINGER [1993]); Chapter 9: Biomechanics of Bone (GOLDSTEIN, FRANKENBURG and KUHN [1993]); Chapter 10: Biomechanics of Soft Tissues (HAUT [1993]).

It is recommended to read these chapters for obtaining a good background for the following chapters, which are devoted to the trauma and injury of the individual body segments: ALLSOP [1993] Skull and Facial Bone Trauma: Experimental aspects (Chapter 11); MELVIN, LIGHTHALL and UENO [1993] Brain Injury Biomechanics (Chapter 12); NEWMAN [1993] Biomechanics of Head Trauma: Head Protection (Chapter 13); MCELHANEY and MYERS [1993] Biomechanical Aspects of Cervical Trauma (Chapter 14); CAVANAUGH [1993] The Biomechanics of Thoracic Trauma (Chapter 15); ROUHANA [1993] Biomechanics of Abdominal Trauma (Chapter 16); KING [1993] Injury to the Thoraco–Lumbar Spine and Pelvis (Chapter 17); LEVINE [1993] Injury to the Extremities (Chapter 18).

These chapters provide a broad overview and many references on injury and trauma of the human body parts, and most of the brief discussions of injury and trauma in this article are based on this book. The book further contains chapters on child passenger protection (Chapter 19), isolated tissue and cellular biomechanics (Chapter 20) and on vehicle interactions with pedestrians (Chapter 21).

1.4. Human models

Models of mechanical dummies simulate their metallic, rubber and plastic parts. Human models simulate the response of bone, flesh, muscles, and hollow and solid organs humans are made of. While humans cannot replace mechanical dummies in real world crash tests, numerical models of humans can readily replace numerical models of mechanical dummies in virtual crash simulations.

Generic and specific models. Depending on the application, human models can be conceived either as "generic" or as "specific" models.

"Generic" models describe the geometry and the physical properties of average size members of the population. They are needed for industrial design, whenever objects are designed for the "average" human user. The average size of the human body can be expressed in statistical "percentiles" of a given population, where the "*n*th height percentile" means that "*n*" percent of the population is smaller in height. For example, 40% of a population is smaller than its 40th height percentile specimen, while 60% is taller. Each average height and weight percentile specimen can still have different relative size distribution of its body segments, as well as different biomechanical properties. The variations around an average percentile specimen of a population are called its "stochastic variants".

"Specific" models describe the geometry and physical properties of given human subjects. They are needed, for example, for virtual surgery, where the surgeon wants to deal with the precise bone or organ of a given patient. In the case of generic models, the acquisition process of the geometry of the body may take time, whereas, for practical reasons, the time needed for the establishment of specific models must be short. Therefore, slow mechanical slicing techniques on cadavers can serve for the data acquisition of generic models, while fast X-ray and scanning techniques on patients are required to construct specific models.

Generic human models should be comprehensive in the sense that all body sizes, genders, ages, races and body morph-types are covered. To achieve this goal, great amounts of anthropometrical and biomechanical data must be acquired and collected in databases, including for children. In fact, human computer modeling and simulation created new demands for data which were not needed or collected before, and novel physical experiments are required. Concerning model validation, modern practice of simulation tends to reverse the role of physical experiments, or laboratory tests, which tend to back up model calibration and validation, rather than to yield primary results, now obtained by the simulations.

Scaling, morphing, aging. Generic models of any type and size should be made available in data bases and through mathematical scaling, morphing and aging techniques, which can generate any given percentile human model and its stochastic variants, with long or short trunks and extremities, thin or fat, older or younger, male or female. Experimental results of standardized validation test cases, together with simulation accuracy norms, must also be provided in such data bases, allowing the human modeler to judge the performance and the quality of the models under controlled conditions.

As almost none of these new requirements are met fully today, there is plenty of room for human model development work. In this article, only the fiftieth percentile "average" male human models for passenger car occupant safety analysis and design are discussed in detail. Female and child models are discussed more briefly, since they are less advanced and their modeling techniques resemble the average male models.

Omnidirectionality. Unlike the well-known existing families of mechanical passenger car occupant surrogates ("dummies"), which are widely used in standard real crash tests by the world's car manufacturers for distinct frontal, side and rear impacts, human models should not be specialized to certain types of crash. They should rather be modeled as "omni-directional" objects, to the image of their real counterparts, i.e., respond equally well for all conceivable types of crash scenarios, impact directions and locations.

1.5. Biomaterials

Biomaterials are "exotic" as compared to most conventional structural materials. A good starting point for their analysis is nevertheless the existing library of material models, offered in modern dynamic structural analysis codes, or crash codes. The theory of the available standard material descriptions can be found in the handbooks of the commercial crash codes and need not to be discussed here in their mathematical detail. Ongoing work will adapt and refine these existing models as new knowledge on the mechanical behavior of biomaterials emerges. A condensation of the abundant literature on biomaterials is added below in brief discussions on fundamental works on biomaterials.

Basic literature. The older book by Yamada (YAMADA [1970]) contains global information about basic material properties of most biological tissues, such as the average Young's modulus and the fracture strength of the tested parts and organs, which permits to get first rough ideas about the mechanical properties of human tissues and organs. The editor of Yamada's book, F.G. Evans, states in his 1969 preface:

> "*... It is a unique book in several respects. First, it contains more data on strength of more tissues from more individuals of different ages than any other study of which I am aware. Second, all of the material used in the study was fresh and unembalmed. Third, the tests were made with standard testing machines of known accuracy or with machines that, after consultation with the manufacturer, had been specifically modified for testing biological materials. Fourth, all of the human material was obtained from one ethnic group. Thus the strength characteristics and other mechanical properties of organs and tissues from Japanese can be easily compared with those from other racial groups. Fifth, data were included on the strength characteristics of organs and tissues from other mammals as well as birds, reptiles, amphibians, and fish.*"

These remarks clearly express not only the durable value of this introductory book, but contain the fundamental specifications for the structure and contents of a comprehensive data base of biological materials. Among the tests that were carried out in order to characterize the strengths of the materials were tests in tension, compression, bending, impact bending, impact snapping, torsion, expansion, bursting, tearing, cleavage, shearing, extraction, occlusion, abrasion, crushing and hardness. The book by Yamada next contains an impressive array of basic information about the mechanical properties for humans and animals of the loco-motor organs and tissues (bone, cartilage, ligaments, muscle and tendons); the circulatory organs and tissues (heart, arteries, veins and red blood cells); the respiratory and digestive organs and tissues (larynx, trachea, lungs, teeth, masticatory muscles, esophagus, stomach, small intestine, large intestine, liver and gall bladder); the uro-genital organs and tissues (kidney, ureter, urinary bladder, uterus, vagina, amnion membrane and umbilical cord); the nervous system, integument, sense organs and tissues (nerves, dura mater, skin, panniculus adiposus (fat), hair, nails, horn sheath, cornea and sclera (eye), auricle and tympanic membrane (ear)). The mechanical properties of certain organs such as the brain, the tongue, the spleen are missing, however. Then the book compares the mechanical properties of human organs and tissues according to their strength and with respect to other materials from industry and nature. Finally, varations (scatter), age effects and aging rates are discussed.

More recently, VIANO [1986] describes the biological structures, material properties and failure characteristics of bone, articular cartilage, ligament and tendon. In his article, the load-deformation of biological tissues is presented with particular reference to the microstructure of the material.

> *"Although many of the tissues have been characterized as linear, elastic and isotropic materials, they actually have a more complicated response to load, which includes stiffening with increasing strain, inelastic yield and strain rate sensitivity. Failure of compact and cancellous bone depends on the rate, type and direction of the loading. Soft biological tissues are visco-elastic and exhibit a higher load tolerance with an increasing rate of loading."*

Viano's paper includes a discussion on the basic principles of biomechanics and emphasizes material properties and failure characteristics of biological tissues subjected to impact loading. The author presents on more than 30 pages what should be known from an engineering point of view about biological tissues. He discusses what types of fibers (collagen), bulky tissue with visco-elastic properties, some of which can consolidate (hyaline cartilage), and crystals in bone tissue (calcium), are responsible for the cohesion of the skeleton (ligaments), the attachment of the muscles to the skeleton (tendons), the transmission of compression forces across the articular surfaces (articular cartilage) and for maintaining the overall shape of the skeleton (bones). For each discussed material the paper describes its biological microstructure and composition, it discusses laboratory setups for material testing, it gives typical stress-strain samples and it outlines possible mathematical models to describe the measured properties up to rupture and fracture.

The textbook by FUNG [1993a] describes "The Mechanical Properties of Living Tissues" (book title). The approach to the description of biomaterials chosen by Fung is the study of the morphology of the organism, the anatomy of the organ, the histology of the tissue, and the determination of the mechanical properties of the materials or tissues in the form of their constitutive equations. The book further deals with setting up the governing differential or integral equations of biomechanical processes, their boundary conditions, their calibration, solution and validation on experiments and predicted results. The constitutive behavior of biomaterials is identified and their equations are defined for the flow properties of blood, blood cells and their interaction with vessel walls, for bio-visco-elastic fluids, for bio-visco-elastic solids, for blood vessels, for skeletal muscle (with a description of Hill's active and passive muscle model), for the heart muscle, for smooth muscles and for bone, cartilage, tendons and ligaments, including the mechanical aspects of the remodeling or growth of certain tissues. The detailed derivation and the mathematical description of the constitutive equations of living tissues is the most distinguishing feature of Fung's textbook. For each treated subject, the book contains extensive lists of references that may be consulted for further reading.

Further collections of biomaterial properties can be found in more recent references, such as the handbook of biomaterial properties by BLACK and HASTINGS (eds.) [1998], which describes in its Part I, from the view point of surgical implants, the properties of cortical bone, cancellous bone, dentin and enamel, cartilage, fibro-cartilage, ligaments, tendons and fascia, skin and muscle, brain tissues, arteries, veins and lymphatic vessels, the intra-ocular lens, blood and related fluids and the vitreous humor. (Part II deals with the properties of surgical implant materials and Part III with the biocompatibility of such materials, not relevant in impact biomechanics.) The cortical bone material is described in its composition (organic, mineral), in its physical properties (density, electromechanical, other) and in its mechanical properties (dry, wet, scatter within the skeleton, stiffness, strength, strain rate and visco-elastic effects). At the end of each chapter additional readings and many references are indicated.

Papers that give detailed stress-strain behavior of biomaterials, including rate effects, are still scarce. Differences between dead and life tissue behavior are seldom described, and data are often inaccessible. The natural scatter between tissues from different individuals is sometimes discussed. The numerical analyst is still constrained to use approximate, incoherent or incomplete data. Many efforts are now undertaken to alleviate this lack of data, a need that was generated only recently by the desire to simulate the biomechanical response the human body using modern computer simulation tools.

Simplest descriptions for biomaterials. Fig. 1.1 shows a selection of some typical biomaterial response curves, as extracted from YANG [1998] (a report of the HUMOS-1 Project, funded by the European Commission under the Industrial and Materials Technologies program (Brite–EuRam III)). The well-known basic elastic, visco-elastic, and elasto-plastic material laws that exist in most dynamic codes and have been applied for biomaterial description. The elastic laws can be linear or nonlinear elastic, isotropic, orthotropic or hyper-elastic. The linear elastic materials are characterized by the elastic moduli, Poisson's ratios, the shear moduli and the mass density. The hyper-elastic materials are characterized by their respective strain energy functions (Mooney, Hart-Smith, etc.). The elastic-plastic material laws are typically defined with the additional hardening modulus, the yield strength, the ultimate strength and strain at failure. The visco-elastic materials need additional coefficients describing the damping, creep and relaxation behavior. The material laws provided with commercial codes are often sufficient to describe hard tissues, such as long bones.

The application to soft tissues is less evident and more research and tests are needed to characterize these materials. In particular, the difference between life and dead tissue behavior is more pronounced in soft than in hard tissues.

In many cases so-called "curve description options" for the standard material laws, as available in the commercial dynamic codes, can be used in order to encode the results directly as obtained from biomaterial tests. These options provide a maximum freedom for the analyst, beyond the usual mathematical descriptions of the materials.

Appendix B gives a summary on the mechanical properties of biomaterials as extracted mainly from YANG [1998].

Bone materials. Fig. 1.2 exemplifies the most frequently studied bone material. Inset (a) shows a cross section through the femur head, with the cortical outer shell of compact bone and the trabecular inner fill of spongeous bone clearly visible. Inset (b) shows the same basic structure in a cross section through the skull bone. Insets (c) and (d) (after RIETBERGEN [1996], RIETBERGEN, MÜLLER, ULRICH, RÜEGSEGGER and HUISKES [1998] and ULRICH [1998]) show so-called "voxel models" of the bony structure, where the trabecular structure of the bone is modeled directly in the optical voxel resolution of micro-scans of the bone. Inset (c), for example, uses several million simplified voxel finite elements to trace (red color) the linear elastic force path through the trabeculae from an axial force loading. Inset (d) shows two different voxel densities, and inset (e) demonstrates that there is practically no visible difference between a real

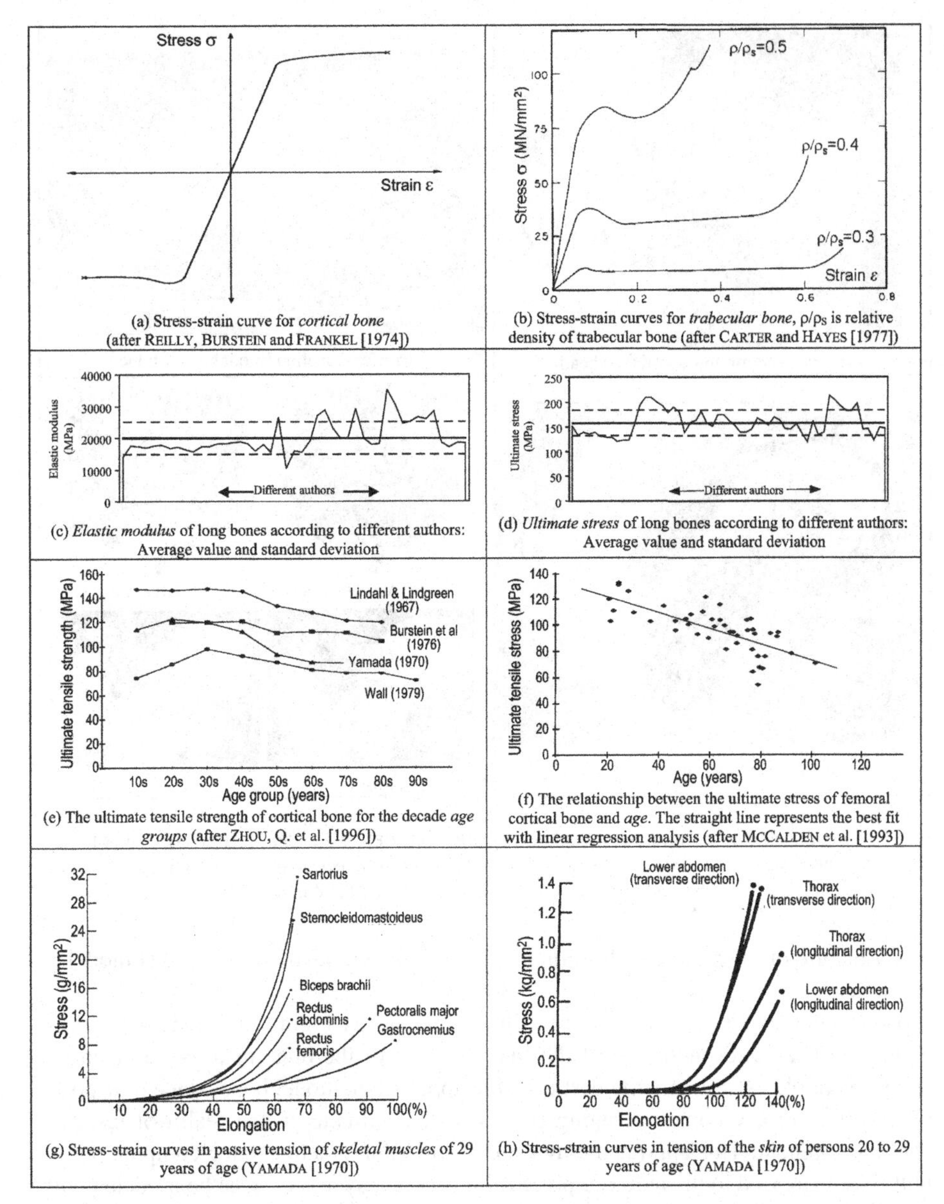

(a) Stress-strain curve for *cortical bone* (after REILLY, BURSTEIN and FRANKEL [1974])

(b) Stress-strain curves for *trabecular bone*, ρ/ρ_S is relative density of trabecular bone (after CARTER and HAYES [1977])

(c) *Elastic modulus* of long bones according to different authors: Average value and standard deviation

(d) *Ultimate stress* of long bones according to different authors: Average value and standard deviation

(e) The ultimate tensile strength of cortical bone for the decade *age groups* (after ZHOU, Q. et al. [1996])

(f) The relationship between the ultimate stress of femoral cortical bone and *age*. The straight line represents the best fit with linear regression analysis (after MCCALDEN et al. [1993])

(g) Stress-strain curves in passive tension of *skeletal muscles* of 29 years of age (YAMADA [1970])

(h) Stress-strain curves in tension of the *skin* of persons 20 to 29 years of age (YAMADA [1970])

FIG. 1.1. Some typical biomaterial responses as a function of material type, inter-individual scatter, age, deformation rate (compiled by YANG [1998]). (Inset (a): Reproduced by permission of Elsevier Health Sciences Rights; Insets (b) and (f): Reproduced by permission of The Journal of Bone and Joint Surgery, Inc.; Insets (c) and (d): Reproduced by permission of Chalmers University of Technology; Inset (e): Reproduced by permission of The Stapp Association; Insets (g) and (h): Reproduced by permission of Lippincott, Williams and Wilkins.)

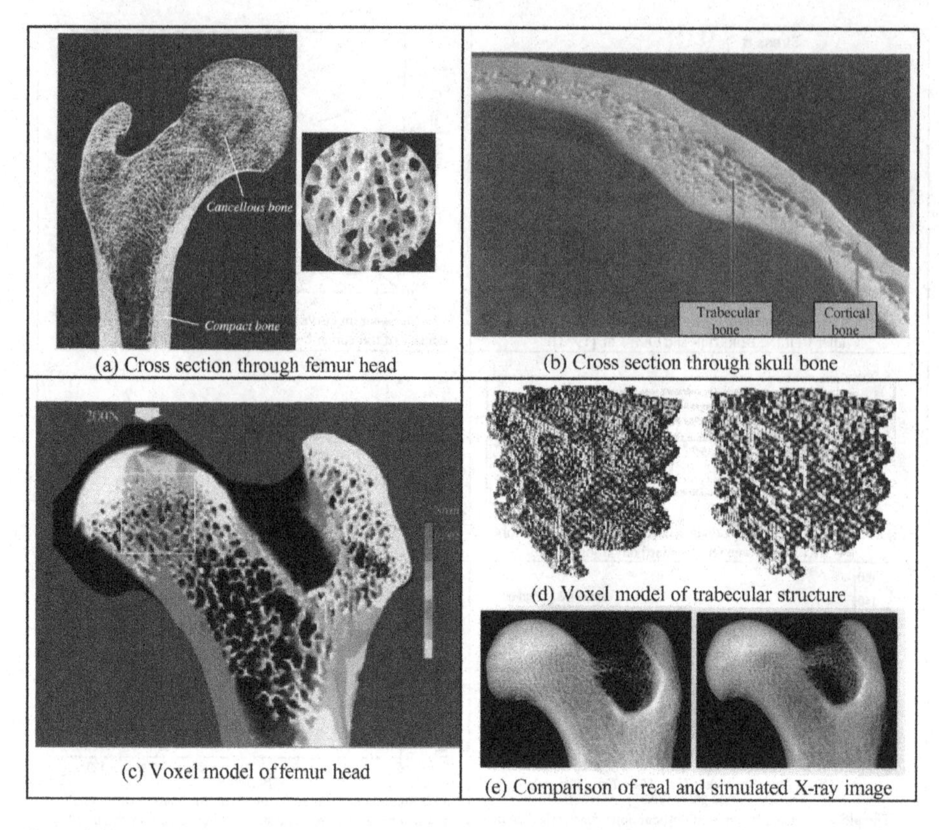

FIG. 1.2. Bone material structure and modeling (after RIETBERGEN, MÜLLER, ULRICH, RÜEGSEGGER and HUISKES [1998], ULRICH [1998]). (Insets (a)–(c), (e): Reproduced by permission of the Journal of Biomechanics; Inset (d): Courtesy Dr. Ulrich of ETH Zurich.)

X-ray picture of the bone and the simulated X-ray picture made from a voxel model of the bone.

This figure stands for the basic need for research concerning the modeling of biomaterials. The direct modeling of the fine structure of the bone material, used in the shown example, is certainly the best possible approach to model bone, since it disposes of uncertain macroscopic averaging processes, but remains in the realm of research and development. This approach may become common practice, once compute power will have increased to the required practical levels. Today, bones must be modeled with "macro" elements, the type of which must be chosen according to the type of bone (e.g., shells for cortical and solids for trabecular bone), and the material densities of which must be evaluated approximately from the average local density of the bone.

Similar remarks can be made for other types of biomaterials, which must be investigated indirectly using "smeared" properties and macro modeling techniques. The brain material, for example, is modeled as solids with homogenized gray or white matter, without taking into account the system of very fine blood vessels that it contains. If this system could be modeled in detail, injury to the vascular system of the brain could be accessed directly, see first attempts made in Fig. 3.3(e)–(h). Today brain injury is accessed indirectly by correlating mechanical field variables to injury through calibration.

References on biomaterial tests, laws, models and simulation. The literature on biomaterials and related subjects is relatively abundant. Appendix B contains references on the biomaterials While many of the indicated references deal with the experimental evaluation of biomaterial properties, others deal with the aspects of their modeling, the use of these materials in biomechanical models and the characterization of trauma and injury.

1.6. Human model validation

Segment and whole body validation. Provided human models can be built and the biomaterials can be calibrated, one of the greatest challenges remains their proper validation. A considerable number of tests on cadaver body segments, whole cadaver bodies and life volunteers were performed in the past, e.g., as listed in the report of the HUMOS-1 project: "Validation Data Base", ROBIN [1999]. As discussed in a later section, the European HUMOS-1 project (Human Models for Safety) produced a first near 50th percentile male European human model in a project funded by the European Commission (HUMOS-1, 1999–2001; HUMOS-2 is under way). The tests listed in this reference comprise the following topics, Fig. 1.3:

Head/neck complex: Frontal tests at 15 g, inset (a); lateral tests at 7 g, inset (b); oblique tests at 10 g, after EWING, THOMAS, LUSTICK, MUZZY III, WILLEMS and MAJEWSKI [1976] (not shown).

Thorax frontal impact: frontal impactor tests by INRETS at low and high velocity (not shown); frontal impactor tests by KROELL, SCHNEIDER and NAHUM [1971], KROELL, SCHNEIDER and NAHUM [1974] at 4.9 m/s, at 6.7 m/s and at 9.9 m/s, inset (c); at 7.0 m/s with seat back (not shown); frontal thorax impactor tests by STALNAKER, MCELHANEY, ROBERTS and TROLLOPE [1973] (not shown).

Thorax belt compression tests: by CESARI and BOUQUET [1990], CESARI and BOUQUET [1994], with a 22.4 kg mass at 2.9 m/s and 7.8 m/s impact velocities, with a 76.1 kg mass at 2.9 m/s impact velocity, inset (d).

Thorax lateral impact: lateral impactor tests by INRETS on the thorax at 3.3 m/s, at 5.9 m/s, inset (e).

Thorax oblique impact: oblique thorax impactor tests by VIANO [1989] at 4.42 m/s, at 6.52 m/s, at 9.32 m/s, inset (f).

Abdomen impact tests: frontal impactor tests on the abdomen by CAVANAUGH, NYQUIST, GOLDBERG and KING [1986], inset (g); oblique impactor tests by Cavanaugh at 31.4 kg and 6.9 m/s (not shown); oblique impactor tests by Viano on the abdomen at 4.8 m/s, at 6.8 m/s, at 9.4 m/s (not shown).

Pelvis impact tests: lateral tests on the pelvis by INRETS at 3.35 m/s, at 6.6 m/s, inset (h); lateral impact tests by Viano at 5.2 m/s, at 9.8 m/s (not shown) (BOUQUET, RAMET, BERMOND and CESARI [1994]).

It is clear that many more tests must be done in order to capture the biomechanical characteristics of humans, not only of the "average" subject (50th percentile male), but also of the inter-individual dispersions that distinguish humans. In order to arrive at

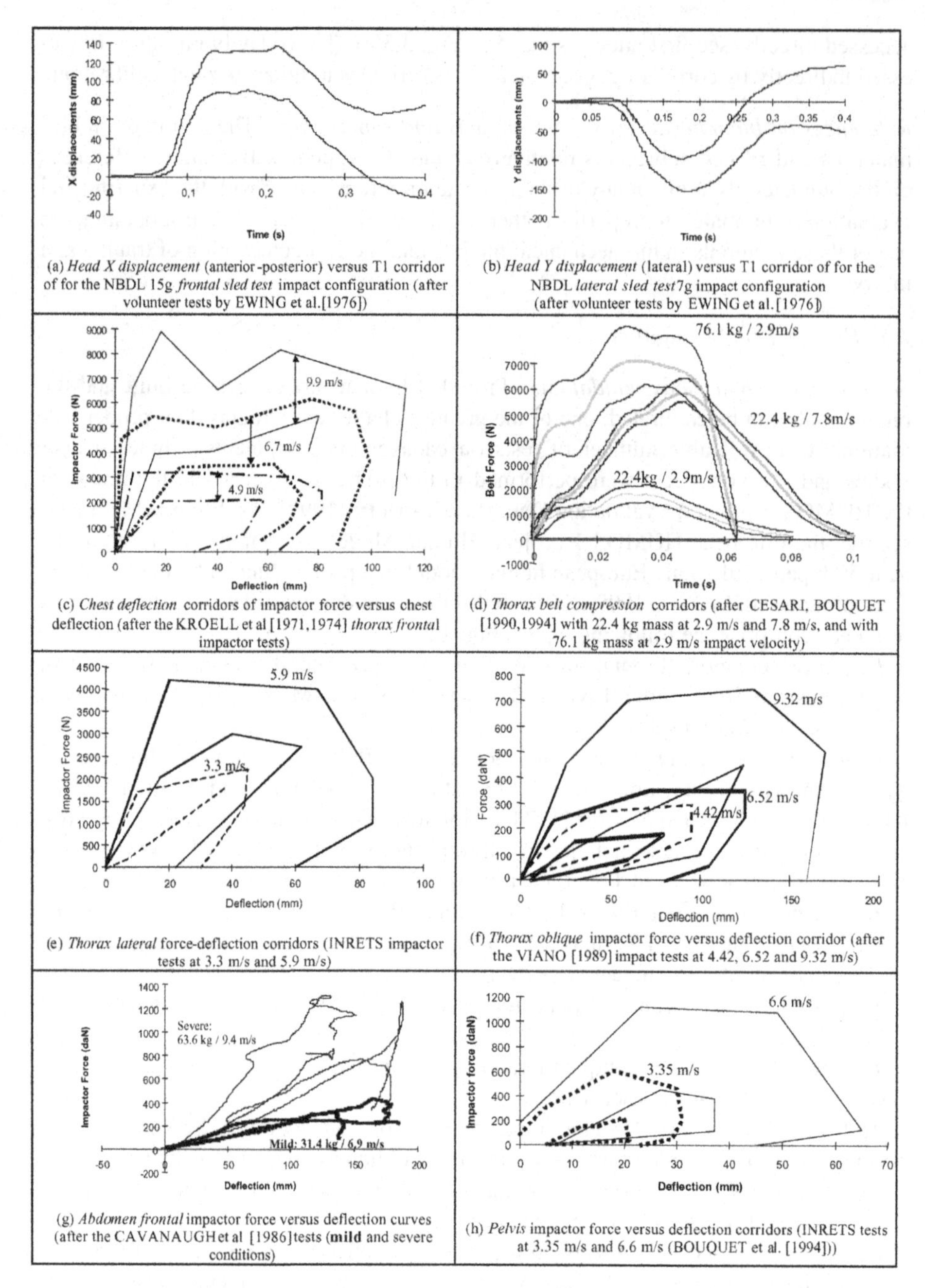

(a) *Head X displacement* (anterior-posterior) versus T1 corridor of for the NBDL 15g *frontal sled test* impact configuration (after volunteer tests by EWING et al.[1976])

(b) *Head Y displacement* (lateral) versus T1 corridor of for the NBDL *lateral sled test* 7g impact configuration (after volunteer tests by EWING et al.[1976])

(c) *Chest deflection* corridors of impactor force versus chest deflection (after the KROELL et al [1971,1974] *thorax frontal* impactor tests)

(d) *Thorax belt compression* corridors (after CESARI, BOUQUET [1990,1994] with 22.4 kg mass at 2.9 m/s and 7.8 m/s, and with 76.1 kg mass at 2.9 m/s impact velocity)

(e) *Thorax lateral* force-deflection corridors (INRETS impactor tests at 3.3 m/s and 5.9 m/s)

(f) *Thorax oblique* impactor force versus deflection corridor (after the VIANO [1989] impact tests at 4.42, 6.52 and 9.32 m/s)

(g) *Abdomen frontal* impactor force versus deflection curves (after the CAVANAUGH et al [1986] tests (**mild** and severe conditions)

(h) *Pelvis* impactor force versus deflection corridors (INRETS tests at 3.35 m/s and 6.6 m/s (BOUQUET et al. [1994]))

FIG. 1.3. Some typical test results for the validation of human models (compiled by ROBIN [1999]). (Insets (a)–(d), (f) and (g): Reproduced by permission of The Stapp Association; Inset (e): Reproduced by permission of INRETS; Inset (h): Material in the public domain by U.S. Department of Transportation.)

"average" responses, and at their likely dispersions, multiple tests should be performed and the results collected in "corridors", which represent best the variable "response" of a given class of human individuals, such as, for example, the 50th percentile male. Some pertinent test results are summarized in Fig. 1.3.

In particular, whole body cadaver sled tests permit to assess the overall response of humans in car accident scenarios. Due to the fact that human models can now be built, calibrated, validated and used in crash simulation or in virtual crash testing, there is a pressing need for reliable data. The necessary tests on human volunteers and cadavers are subject to severe ethical control, which considerably restricts the frequency and number by which such tests can be performed.

2. Overview on mechanical dummies and models

This section may be skipped by readers not interested in mechanical dummies. The material is provided for to give an overview on the mechanical occupant surrogates or legal crash "dummies" as presently used by the auto industry for certification of new car models (KISIELEWICZ and ANDOH [1994]).

Mechanical dummies (occupant surrogates) and their numerical models are used heavily in crash tests and numerical simulations for safe car design. Due to the large number of car crashes each year, crash tests are administered by the National Highway Traffic Safety Administration (NHTSA), an agency within the United States Department of Transportation (DoT). About 35 new model cars have been tested every year since 1979 under the New Car Assessment Program (NCAP). The tests are to see how well different vehicles protect front-seat passengers in a car-to-car head-on collision at equal speeds. The head-on collision is used instead of a rear or a side collision because this is the collision that causes the most deaths and injuries.

The US federal law requires all cars to pass a 30 mph frontal rigid barrier test, so NCAP crash tests on fixed barriers (rigid walls) are performed at 35 mph (56.3 km/h), which corresponds to an impact of two identical cars colliding head-on at a relative velocity of 70 mph (112.6 km/h). These tests show the difference in protection in different car models. The results of the crashes are given on a one-to five star rating, with five being the highest level of protection.

These crash tests are all administered with dummies, the dummies are always wearing seat belts because they are standard equipment on cars today, air bags are used whenever they are available, and test results are only useful in comparing cars of similar weight (within 500 pounds (227 kg) of each other). Dummies heads and knees are painted before a test to see where these areas of the body make contact with the car.

The Hybrid III dummy family is used today in frontal impact tests. For side impact tests, the US DoT SID and the EuroSID special side impact dummies are used. Some of these mechanical occupant surrogates and typical protective measures are shown in Fig. 2.1.

The Hybrid III dummy family. The 50th percentile Hybrid III represents a man of average size. The European standard 50th percentile man is assumed 1.75 meters tall and having a body weight of 75.5 kilograms. Different "standard" sizes may exist in

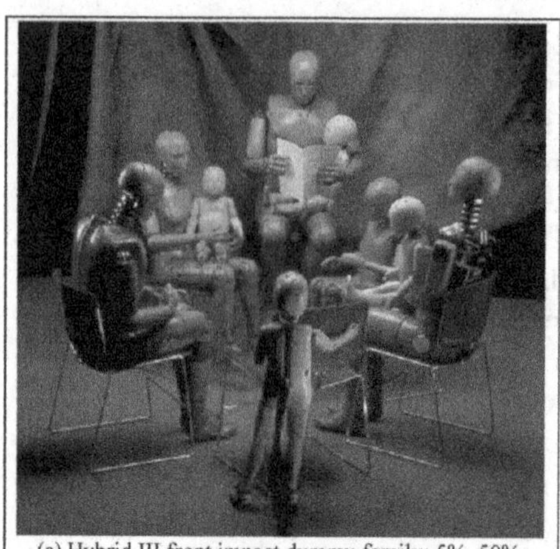

(a) Hybrid III front impact dummy family: 5%, 50%, 95%; 3 and 6 year; CRABI – 6, 12, 18 months; pregnant woman 5% ; EuroSID, BioSID (FTSS)

(b) Airbags and seat belts (schematic models) upper: frontal impact driver and passenger airbags lower: side impact airbags

FIG. 2.1. Front impact dummies and passive safety systems. (Inset (a): Reproduced by permission of First Technology Safety Systems, Inc.; Inset (b) lower right-hand side: Courtesy Autoliv, BMW (FOGRASCHER [1998]); Inset (b) lower left-hand side: Courtesy AUDI.)

TABLE 2.1
Average body heights and weights (subject to variations)

Percentile age	Height		Weight	
	[cm]	[foot′in″]	[kg]	[lbs]
50th Hybrid III dummy	178	~5′10″	77.11	170
50th European adult male	175	~5′9″	75.5	166.45
5th adult female	152	~5′	49.89	110
95th adult male	188	~6′2″	101.15	223
6 year old child	113	$\sim 3'8\frac{1}{2}''$	21.32	47
3 year old child	99	~3′3″	14.97	33

different countries. Born in the USA in the labs of General Motors, the 50th percentile Hybrid III is the standard dummy used in frontal crash tests all over the world. It is called a hybrid, because it was created by combining parts of two different types of dummies. Beside the 50th percentile male, there are the 5th percentile female, the 95th percentile male and the 6 year-old and 3 year-old child dummies, Table 2.1.

Hybrid III dummy models. Fig. 2.2 shows numerical models of members of the Hybrid III dummy family (after FTSS/ESI Software). The models shown under Fig. 2.2(a) have 25 878 (50th percentile), 24 316 (5th percentile), 27 872 (95th percentile), 34 535 (6 year old child) and 13 345 (3 year old child) deformable finite elements, respectively. In (b) a Hybrid III dummy model is shown in a driver position. The 50th percentile male Hybrid III model is shown in Fig. 2.2(c) through (e). Insets (c) and (d) are finite element models, while inset (e) is a section through the simpler multi-body version. Most dummy models are made either as simpler multi-body models, or as more detailed finite element models.

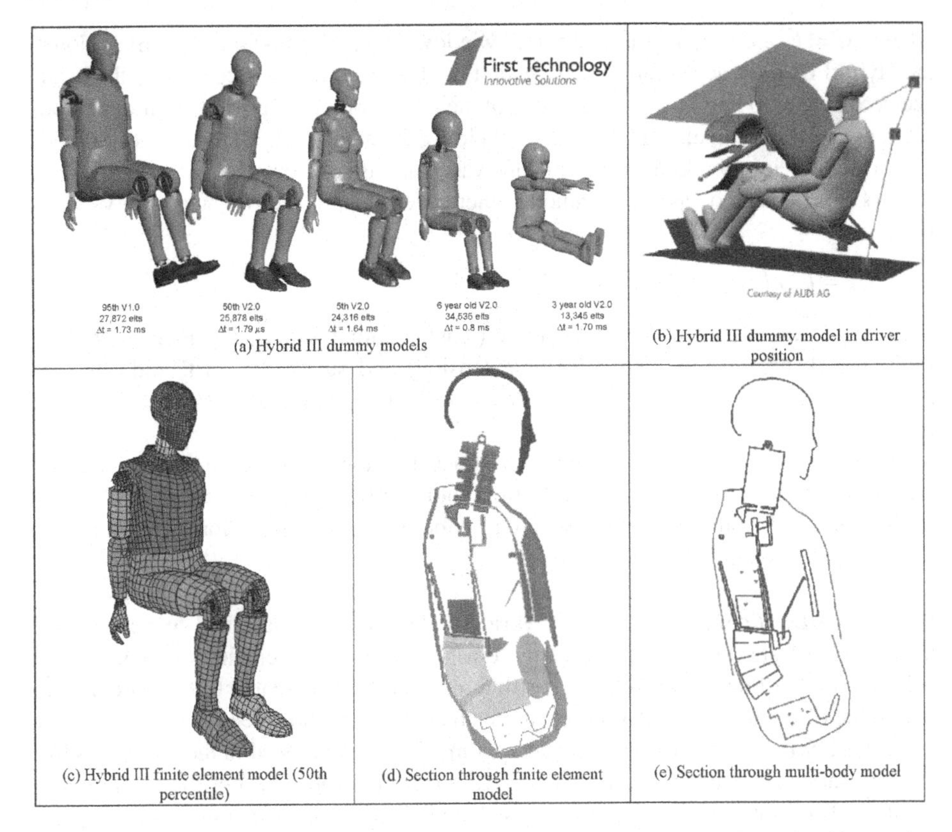

(a) Hybrid III dummy models

(b) Hybrid III dummy model in driver position

(c) Hybrid III finite element model (50th percentile)

(d) Section through finite element model

(e) Section through multi-body model

FIG. 2.2. Numerical models of the Hybrid III dummy family (FTSS/ESI Software). (Reproduced by permission of First Technology Safety Systems, Inc.)

Multi-body modeling techniques comprise linked rigid body tree structures, which contain relatively few deformable parts and which are linked together at the intersections of their anatomical segments. They execute faster but cannot yield detailed injury data. Finite element models are made of the usual standard library of finite elements of the used crash codes (solids, shells, membranes, beams, bars, springs, etc). They take more central processor unit (CPU) computer time, but can yield response data, which are more readily linked to human injuries. Typical solver codes used to analyze car crash scenarios execute the explicit time integration scheme for the set of non-linear equations of motion in the nodal degrees of freedom.

Standard injury criteria. The standard way of assessing injuries of vehicle occupants are heuristic injury coefficients that are calculated from injury criteria defined from the instrumented front or side impact dummy responses in crash tests.

Head injuries of occupants are assessed from the "Head Injury Coefficient" (HIC),

$$\mathrm{HIC} = \max_{t_1 < t_2}\left[(t_2 - t_1)\left\{\frac{1}{t_2 - t_1}\int_{t_1}^{t_2} a(t)\,\mathrm{d}t\right\}^{2.5}\right],$$

where $\Delta t = t_2 \geqslant t \geqslant t_1$ is a normed time window (e.g., 22.5 ms) that is shifted along the Hybrid III head acceleration magnitude time history, $a(t)$, as recorded by the head accelerometers, to find the maximal value of the HIC coefficient over the duration of the crash event. If the calculated HIC-value is below critical (e.g., 1000), then it is assumed that no serious injury (skull fracture; neuro-vascular damage) occurs.

Neck injuries can be assessed (among other criteria) from the N_{ij} neck injury criterion,

$$N_{ij} = F_Z / F_{Z,\mathrm{crit}} + M_Y / M_{Y,\mathrm{crit}},$$

where F_Z is the recorded neck axial force (tension/compression), M_Y is the recorded sagittal neck bending moment (flexion/extension) and subscripts "crit" indicate the respective critical values, set such that $N_{ij} = 1.0$ corresponds to a 30% probability of injury.

Similar criteria exist for the *thorax* (chest acceleration, chest compression, viscous criterion, side impact dummy rib deflection, thoracic trauma index), for the *abdomen* (abdominal peak force, pelvis acceleration, pubic symphysis peak force) and for the *lower extremity* (femur load, tibia index).

SID, EuroSID, BioSID, SID II(s). Hybrid III dummies are designed to be used in frontal crash tests. For tests representing crashes in which a vehicle is struck on the side, a number of dedicated side-impact dummies have been created to measure injury risk to the ribs, spine, and internal organs, such as the liver and spleen, Fig. 2.3.

US DoT SID was the first side-impact dummy. It was developed in the late 1970s by the US National Highway Traffic Safety Administration (NHTSA) of the US Department of Transportation (DoT) and is used in US government-required side-impact testing of new cars.

EuroSID was developed by the European Experimental Vehicles Committee (EEVC) and is used to assess compliance with the European side-impact requirements.

BioSID is based on a General Motors design. It is more advanced than SID and EuroSID, but it is not specified as a test dummy to be used in legal tests.

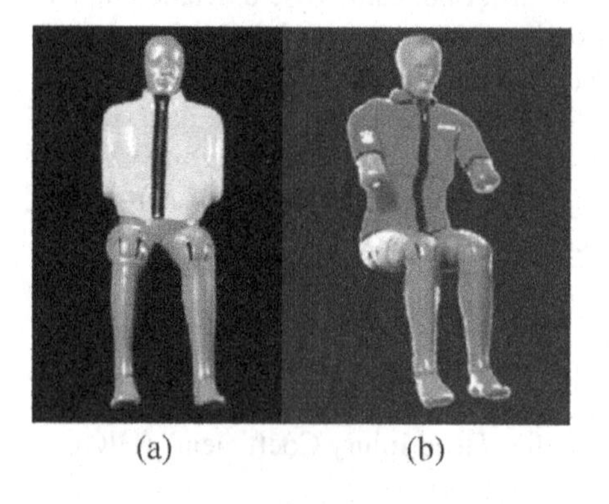

(a) (b) (c) (d)

FIG. 2.3. Side impact dummies (hardware). (a) US DoT SID; (b) EuroSID; (c) BioSID; (d) SID II(s) 5th percentile female. (Reproduced by permission of First Technology Safety Systems, Inc.)

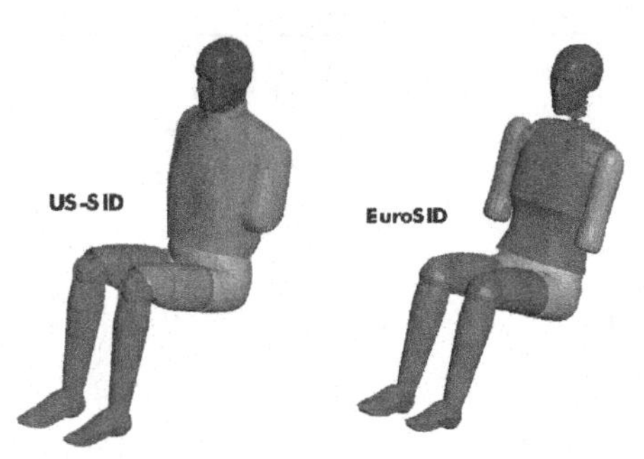

FIG. 2.4. US DoT SID and EuroSID side impact dummy models (FAT/ESI Software).

SID, EuroSID, and BioSID are designed to represent 50th percentile or average-size men 5 feet 10 inches (1.78 m) tall and 170 pounds (77.11 kg).

SID II(s) represents a 5th percentile small female who is 5 feet (1.52 m) tall and weighs 110 pounds (49.89 kg). SID II(s) was created by a research partnership of US automakers. It is the first in a family of technologically advanced side-impact dummies.

SID measures the acceleration of the spine and ribs. Acceleration is the rate of velocity change, and measuring it indicates the forces inflicted on the body during the crash. EuroSID, BioSID, and SID II(s) measure acceleration plus compression of the rib cage. Compression refers to the extent body regions are squeezed during the impact and is used as an indicator of injury to internal organs.

Side impact dummy models. In Fig. 2.4 finite element models of the US DoT SID and of the EuroSID dummy are shown. These models were elaborated on the basis of material, component and whole body tests, performed by the German car manufacturer consortium FAT. The models have about 38 000 deformable finite elements and a time step of 1.5 microseconds in explicit solver codes.

BioRID. A rear-impact dummy has been developed to measure the risk of minor neck injuries, sometimes called whiplash, in low-speed rear-end crashes, which is a big problem worldwide.

BioRID was developed in the late 1990s by a consortium of Chalmers University of Technology in Sweden, restraint manufacturer Autoliv, and automakers Saab and Volvo. It is designed to represent a 50th percentile or average-size man, 5 feet 10 inches tall and 170 pounds in weight, Fig. 2.5, DAVIDSSON, FLOGARD, LÖVSUND and SVENSSON [1999].

BioRID has been designed especially to study the relative motion of the head and torso. For tests representing crashes in which a vehicle is struck in the rear, BioRID can help researchers learn more about how seatbacks, head restraints, and other vehicle characteristics influence the likelihood of whiplash injury.

Unlike Hybrid III dummies, the BioRID spine is composed of 24 vertebra-like segments, so that in a rear-end crash BioRID interacts with vehicle seats and head restraints in a more humanlike way than the Hybrid III. The BioRID segmented neck can take on

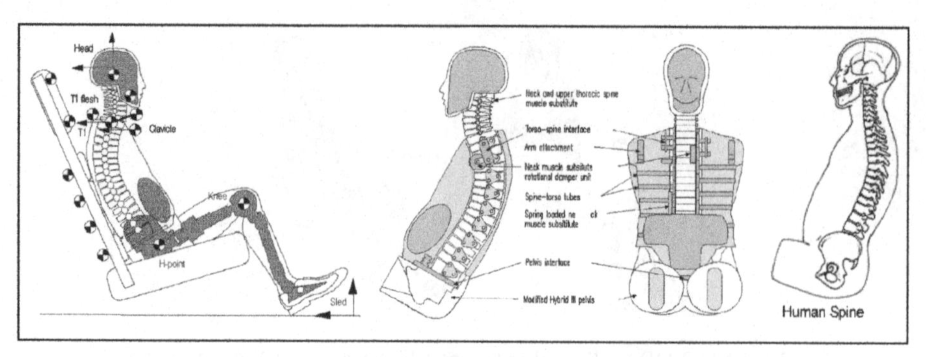

FIG. 2.5. BioRID Rear Impact Dummy and comparative human section. (Reproduced by permission of The Stapp Association.)

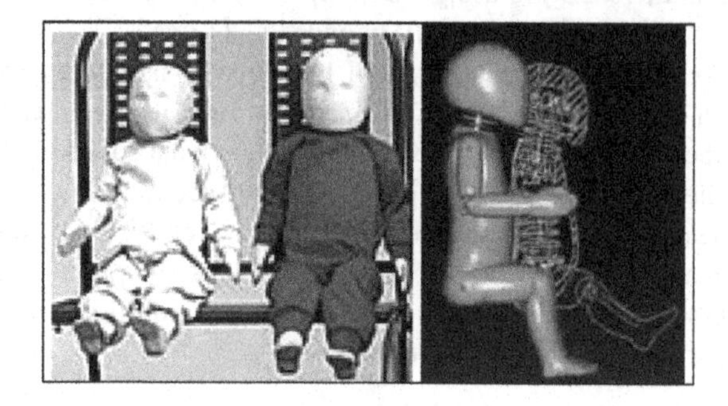

FIG. 2.6. Child Restraint Air Bag Interaction dummy (CRABI). (Reproduced by permission of First Technology Safety Systems, Inc.)

the same shapes observed in human necks during rear-end collisions, an important characteristic for measuring some risk factors associated with whiplash injury. Comparative cross sections through the human spine and BioRID are shown in Fig. 2.5.

CRABI. The Child Restraint Air Bag Interaction dummy was developed by First Technology Safety Systems (FTSS) to represent children, Fig. 2.6. It is used to evaluate child restraint systems, including airbags. There are three sizes: 18 month-old, 12 month-old, and 6 month-old. These dummies have sensors in the head, neck, chest, back, and pelvis that measure forces and accelerations.

THOR. This advanced 50th percentile male dummy is being developed in the United States by NHTSA for use in frontal crash tests, Fig. 2.7 (http://www-nrd.nhtsa.dot.gov/departments/nrd-51/THORAdv/THORAdv.htm). THOR has more human-likefeatures than Hybrid III, including a spine and pelvis that allow the dummy to assume various seating positions, such as slouching, for example, or sitting upright. THOR also has sensors in its face that measure forces so that the risk of facial injury can be assessed, which is not possible with current dummies. In fact, all of THORs standard sensors will provide more injury measurements than those available on Hybrid III.

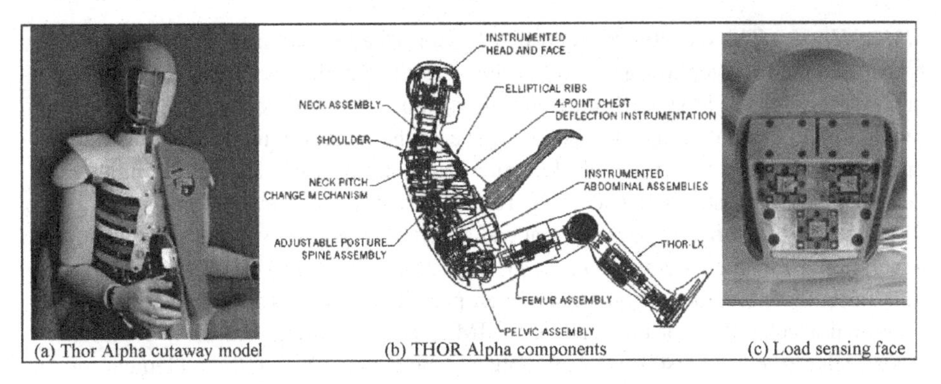

(a) Thor Alpha cutaway model (b) THOR Alpha components (c) Load sensing face

FIG. 2.7. Advanced THOR dummy. (Reproduced by permission of NHTSA, U.S. Government.)

3. Overview of existing human models for occupant safety

The following paragraphs briefly outline the present state of the art in human biomechanical modeling for occupant safety. Most of the information has been drawn from the indicated INTERNET web sites. The interested reader is invited to consult these sites, and more, to get up-to-date information of this rapidly expanding field. The selected examples demonstrate the extensive level of development of human models for impact biomechanics by research institutions and by private industry. The fact that car manufacturers invest actively in human models reflects the need for the use of human models in safe car design. At the same time a unification of these efforts is needed, as reflected, for example, by the pre-competitive European joint development project HUMOS, in which five car manufacturers, several equipment suppliers, several research institutes and three engineering software houses are active partners. While the official approval by the respective National Road Safety Administration authorities for the use of human models for safe car design in "legal virtual crash tests" is pending, new car models are being certified using mechanical dummies in real physical legal crash tests. An interesting concept on the way towards legal virtual crash tests with human models has been introduced recently by the National Highway Traffic Safety Administration (NHTSA), US Department of Transportation (DoT), which can be considered an encouraging step towards the increased use of human models in safe car design. This new concept is discussed first.

3.1. SIMon (Simulated Injury Monitor)

NHTSA experts have developed human models (http://www.nhtsa.dot.gov), which directly simulate bodily injury, unlike dummies or models of dummies, which access injury through equivalent measures. The developed models are meant to help new, advanced, mechanical dummy design (e.g., THOR) on the one hand, and, on the other hand, they serve as vehicles in the numerical interpretation of the enriched output data, harvested from the new generation dummies in crash tests.

SIMon-Head. The recently released first SIMon-Head model is discussed below and it provides the first step towards a new standardized analytical occupant safety analysis methodology, called SIMon (Simulated Injury Monitor). The objective of this particular research is to evaluate injury to the soft tissue of the human brain using finite element models of the brain together with dynamic load data from mechanical dummies, harvested in actual crash tests. The so far released SIMon-Head model consists in a CD Rom with an NT software package that can accept the measured output of nascent new generation mechanical dummies (e.g., THOR). The new dummy head is equipped with nine instead of three accelerometers, which permits to record translational, as well as rotational accelerations of the head. The SIMon-Head package uses these comprehensive acceleration time histories as an input to a built-in calibrated finite element model of the head and brain, to which it applies the recorded accelerations by running an explicit solver code, itself locked into the SIMon-Head package.

The package then analyzes the built-in head/brain model output data and it generates three new brain injury coefficients, namely the "Cumulated Strain Damage Measure (CSDM)", the "Dilatation Damage Measure (DDM)" and the "Relative Motion Damage Measure (RMDM)". The CSDM tells what cumulative volume fraction of the brain matter experienced at some time principal strains larger than a fixed threshold value (15%), known to cause Diffuse Axonal Injury (DAI). The DDM tells what instantaneous volume fraction of the brain matter experienced negative dynamic pressures that can cause vaporization of the cerebral fluids, and contusion. The RMDM tells the percentage of the bridging veins that have stretched beyond a limit curve in a calibrated strain vs. strain-rate diagram, each possibly causing Acute Subdural Hematoma (ASDH) through rupture. The bridging veins connect the soft brain tissues to the skull and may rupture through excessive shearing motions of the brain with respect to the skull.

This new concept will raise the level of precision for injury prediction by directly addressing different types of local injuries, rather than by comparing abstract global coefficients, such as the well-known Head Injury Coefficient (HIC), with calibrated threshold values (cf. NHTSA Federal Motor Vehicle Safety Standard (FMVSS)). The HIC is an integral of translational dummy head accelerations over a moving fixed time window, which tells the maximum acceleration the head experienced over the fixed time interval during the crash event. This measure can be linked to real brain injuries only in a purely heuristic fashion. The SIMon concept will be extended in the near future to further body segments (neck, thorax, femur, etc.), and it will undoubtedly accelerate the widespread use of human models in safe car design.

Fig. 3.1(a)–(h) show a typical (side) impact experiment in a crash laboratory (a), a model of the advanced (frontal) crash test dummy THOR (b), as well as several views of the SIMon Head model, labeled: a mid-coronal section view (c), a coronal-sagittal view, highlighting the boundary between the falx cerebri and the skull (d), a 3D view of the opened SIMon skull model (e), a 3D top view of the brain (f), a top view indicating the location of the parasagittal bridging veins (g), and a 3D view of a brain model (h).

SIMon is designed to provide head (Fig. 3.1), and later neck and thorax (Fig. 3.2), and lower extremity models of occupants, including women and children. These models of human parts can be driven by instrumentation data from advanced crash dummies.

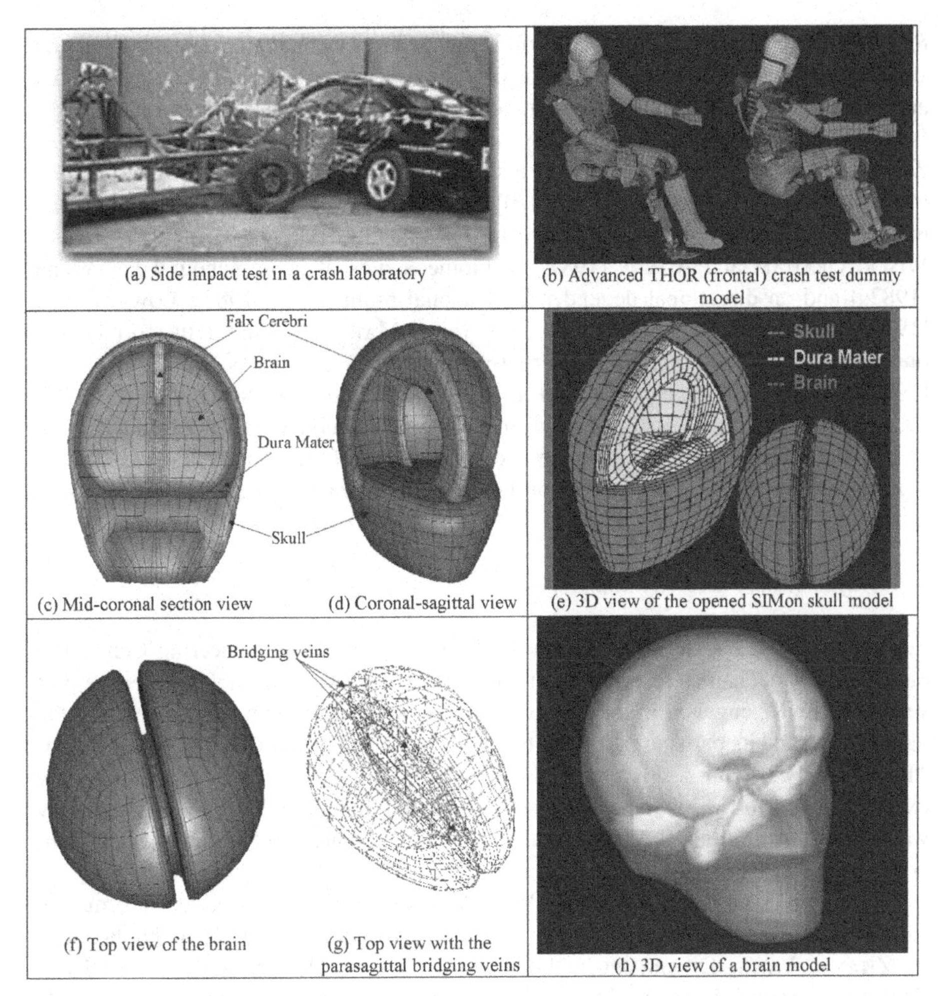

(a) Side impact test in a crash laboratory
(b) Advanced THOR (frontal) crash test dummy model
(c) Mid-coronal section view
(d) Coronal-sagittal view
(e) 3D view of the opened SIMon skull model
(f) Top view of the brain
(g) Top view with the parasagittal bridging veins
(h) 3D view of a brain model

FIG. 3.1. SIMon Head and Brain Models (http://www.nhtsa.dot.gov). (Reproduced by permission of NHTSA, U.S. Government.)

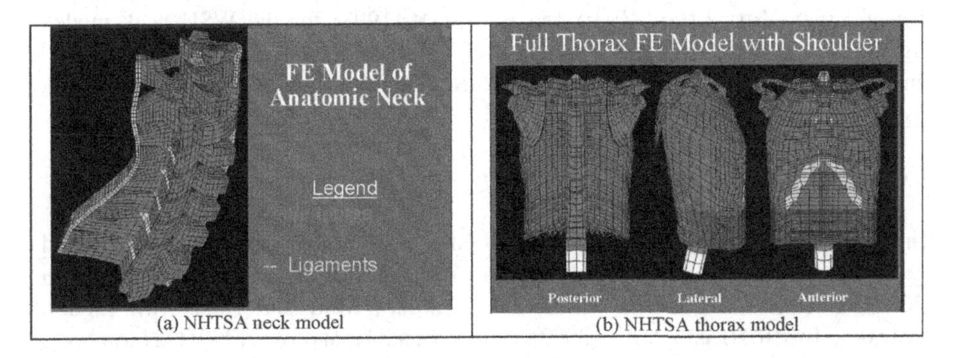

(a) NHTSA neck model
(b) NHTSA thorax model

FIG. 3.2. NHTSA Neck and Chest Models (http://www.nhtsa.dot.gov). (Reproduced by permission of NHTSA, U.S. Government.)

References for SIMon-Head. The following references may be consulted on the subject of the SIMon-Head model: AL-BSHARAT, HARDY, YANG, KHALIL, TASHMAN and KING [1999] on brain/skull relative motions; BANDAK and EPPINGER [1994], BANDAK, TANNOUS, ZHANG, TORIDIS and EPPINGER [1996], BANDAK, TANNOUS, ZHANG, DIMASI, MASIELLO and EPPINGER [2001] on brain FE models and SIMon Head; FALLENSTEIN, HULCE and MELVIN [1970] on dynamic mechanical properties of human brain tissue; GENNARELLI and THIBAULT [1982], GENNARELLI, THIBAULT, TOMEI, WISER, GRAHAM and ADAMS [1987] on biomechanics of acute subdural hematoma (1982a) and on directional dependence of axonal brain injury (1987); LOWENHIELM [1974] on dynamic properties of bridging veins; MARGULIES and THIBAULT [1992] on diffuse axonal injury tolerance criteria; MEANY, SMITH, ROSS and GENNARELLI [1993] on diffuse axonal threshold injury animal tests; NUSHOLTZ, WILEY and GLASCOE [1995] on cavitation effects in head impact model; OMMAYA and HIRSCH [1971] on cerebral concussion tolerance in primates; ONO, KIKUCHI, NAKAMURA, KOBAYASHI and NAKAMURA [1980] on head injury tolerance for sagittal impact from tests.

3.2. *Wayne State University human models*

For over sixty years, the Wayne State University (WSU) Bioengineering Center has pioneered biomechanics research and issued injury tolerance thresholds. Over the past years, the center engaged in a continued activity of the development of models of the human body and its parts, in particular the WSU brain and head model, as shown below. These models are among the most advanced, and their validation is substantiated by experiments performed at the center itself. The WSU human models have served many workers and institutions as a basis for their own development and research (Ford, General Motors, Nissan, Toyota, ESI, Mecalog, etc.).

Fig. 3.3 gives an overview on the WSU human models. The reported information can be found on their web site http://ttb.eng.wayne.edu, as well as in publications by ZHANG, YANG, DWARAMPUDI, OMORI, LI, CHANG, HARDY, KHALIL and KING [2001] and HARDY, FOSTER, MASON, YANG, KING and TASHMAN [2001] for the WSU head injury model, Fig. 3.3(a)–(d); in ZHANG, BAE, HARDY, MONSON, MANLEY, GOLDSMITH, YANG and KING [2002] for the WSU vascular brain model, Fig. 3.3(e)–(h); in YANG, ZHU, LUAN, ZHAO and BEGEMAN [1998] for the WSU neck model, Fig. 3.3(i), (j); after SHAH, YANG, HARDY, WANG and KING [2001] for the WSU chest model, Fig. 3.3(k)–(m); in LEE and YANG [2001] for the WSU abdomen model, Fig. 3.3(n)–(q).

3.3. *THUMS (Total Human Model for Safety)*

The Total Human Model for Safety (THUMS) was recently assembled and tested by Toyota Research Company, see Fig. 3.4(a)–(f), cf. FURUKAWA, FURUSU and MIKI [2002], IWAMOTO, KISANUKI, WATANABE, FURUSU, MIKI and HASEGAWA [2002], KIMPARA, IWAMOTO and MIKI [2002], MAENO and HASEGAWA [2001], NAGASAKA, IWAMOTO, MIZUNO, MIKI and HASEGAWA [2002], OSHITA, OMORI, NAKAHIRA and MIKI

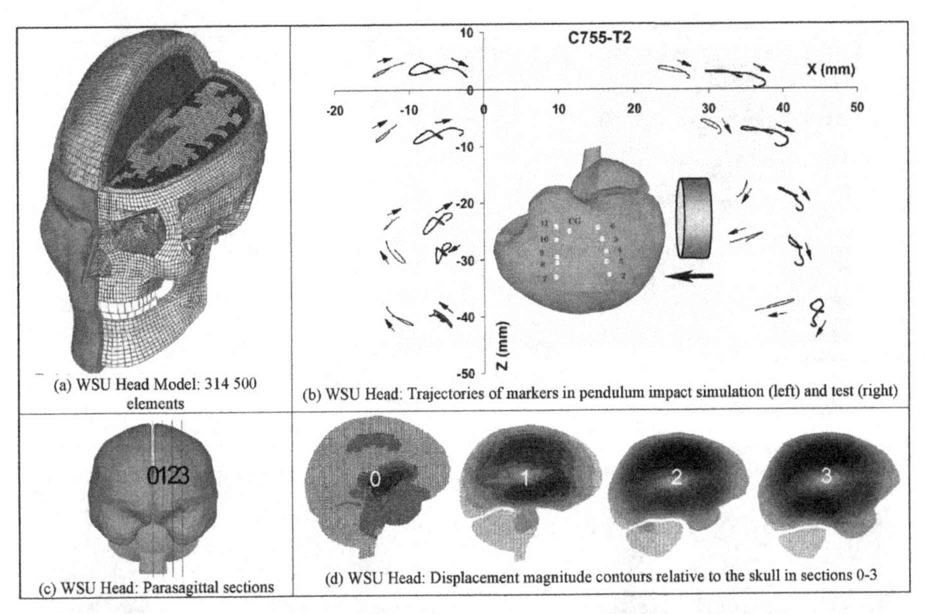

(a) WSU Head Model: 314 500 elements

(b) WSU Head: Trajectories of markers in pendulum impact simulation (left) and test (right)

(c) WSU Head: Parasagittal sections

(d) WSU Head: Displacement magnitude contours relative to the skull in sections 0-3

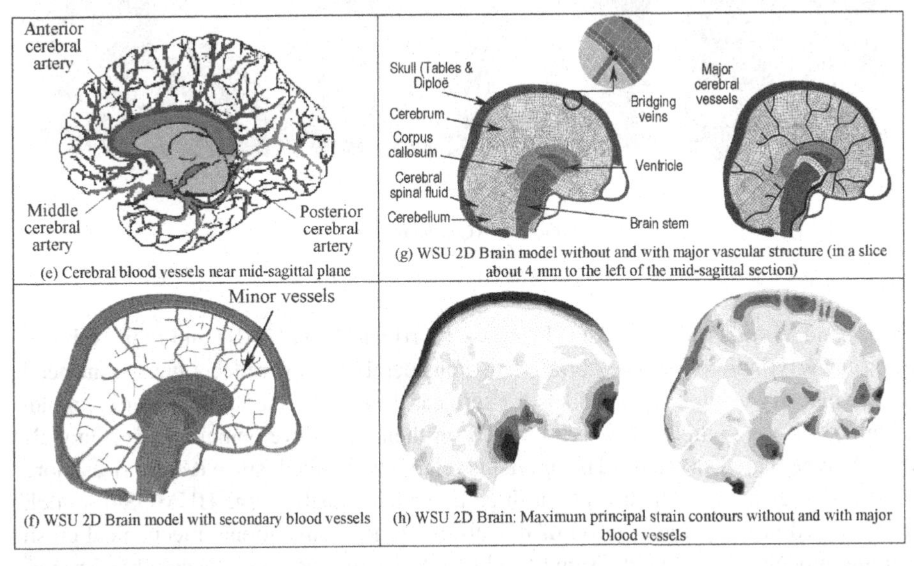

(e) Cerebral blood vessels near mid-sagittal plane

(g) WSU 2D Brain model without and with major vascular structure (in a slice about 4 mm to the left of the mid-sagittal section)

(f) WSU 2D Brain model with secondary blood vessels

(h) WSU 2D Brain: Maximum principal strain contours without and with major blood vessels

FIG. 3.3. The WSU human models. (Insets (a)–(h), (j) and (n)–(q): Reproduced by permission of The Stapp Association; Insets (i) and (k)–(m): Reproduced by permission of King H. Yang, Wayne State University Bioengineering Center; Insets (a)–(d): WSUHIM Head Injury Model (ZHANG, YANG, DWARAMPUDI, OMORI, LI, CHANG, HARDY, KHALIL and KING [2001]) and response under occipital impact test C755-T2 (HARDY, FOSTER, MASON, YANG, KING and TASHMAN [2001]); Insets (e)–(h): WSU 2D Vasculated Brain Injury Model (ZHANG, BAE, HARDY, MONSON, MANLEY, GOLDSMITH, YANG and KING [2002]) and response under impact test nb.37 (NAHUM, SMITH and WARD [1977]); Insets (i) and (j): WSU Neck Model structure and response under whiplash conditions (YANG, ZHU, LUAN, ZHAO and BEGEMAN [1998]); Insets (k) to (q): WSU chest model (k), (l), (m) (SHAH, YANG, HARDY, WANG and KING [2001]), abdomen model (n), (o) (LEE and YANG [2001]) and response (p), (q) under lateral pendulum impact (VIANO [1989]).)

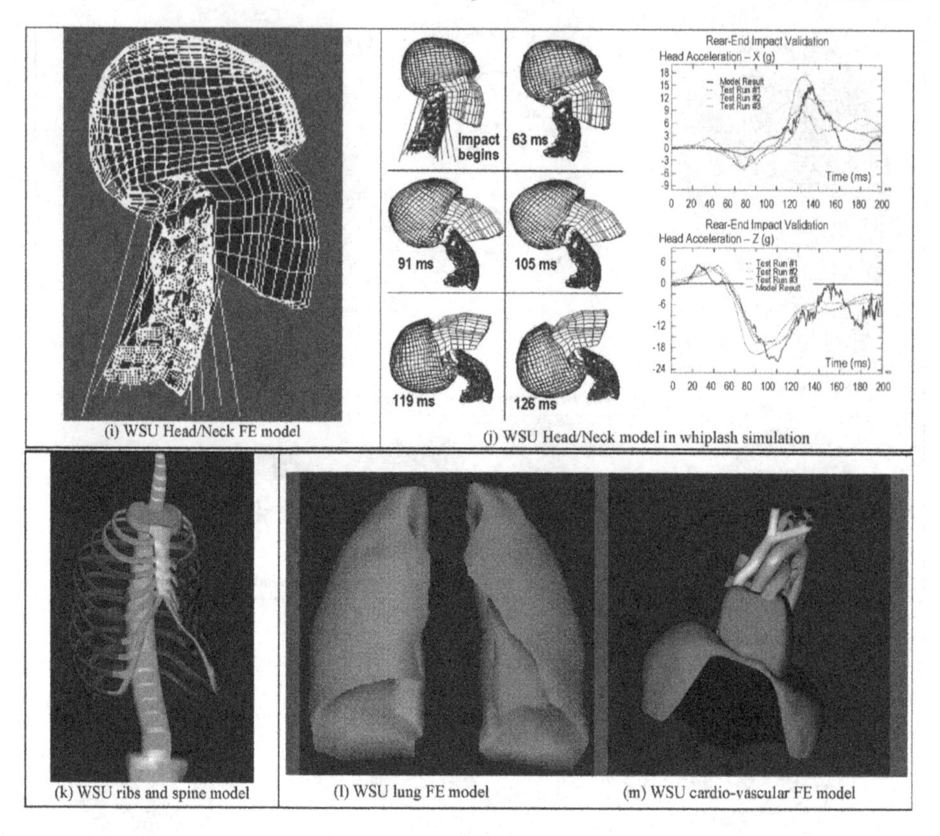

(i) WSU Head/Neck FE model

(j) WSU Head/Neck model in whiplash simulation

(k) WSU ribs and spine model

(l) WSU lung FE model

(m) WSU cardio-vascular FE model

FIG. 3.3. *(Continued.)*

[2002] and WATANABE, ISHIHARA, FURUSU, KATO and MIKI [2001] and their web site http://www.tytlabs.co.jp/eindex.html. This considerable effort reflects the urgent need for car industry using human models for safe car design. A first 50th percentile human male model was completed in 2000, based on their own development and on models from Wayne State University. This model is relatively detailed, since it comprises more than 80 000 elements, which is about three times the density of the HUMOS-1 model, as discussed in a separate section of this article. Since conventional mechanical crash dummy models often have the same level of refinement, the model is suitable for running crash simulations.

THUMS is a family of models, which comprises the AM50 50th percentile male, Fig. 3.4(a), the AF05 5th percentile female and the 6 year old child, (b), and a pedestrian model, (c). The internal organs of the AM50 model are shown in inset (d). Insets (e) and (f) show the deformed shapes of the AM50 and the AF05 models, respectively, under the Kroell chest pendulum impact tests, with a pendulum mass of 23.4 kg, a diameter of 150 mm and an impact velocity of 7.29 m/s. The simulation results were compared with the Kroell tests (KROELL [1971] and KROELL, SCHNEIDER and NAHUM [1971]).

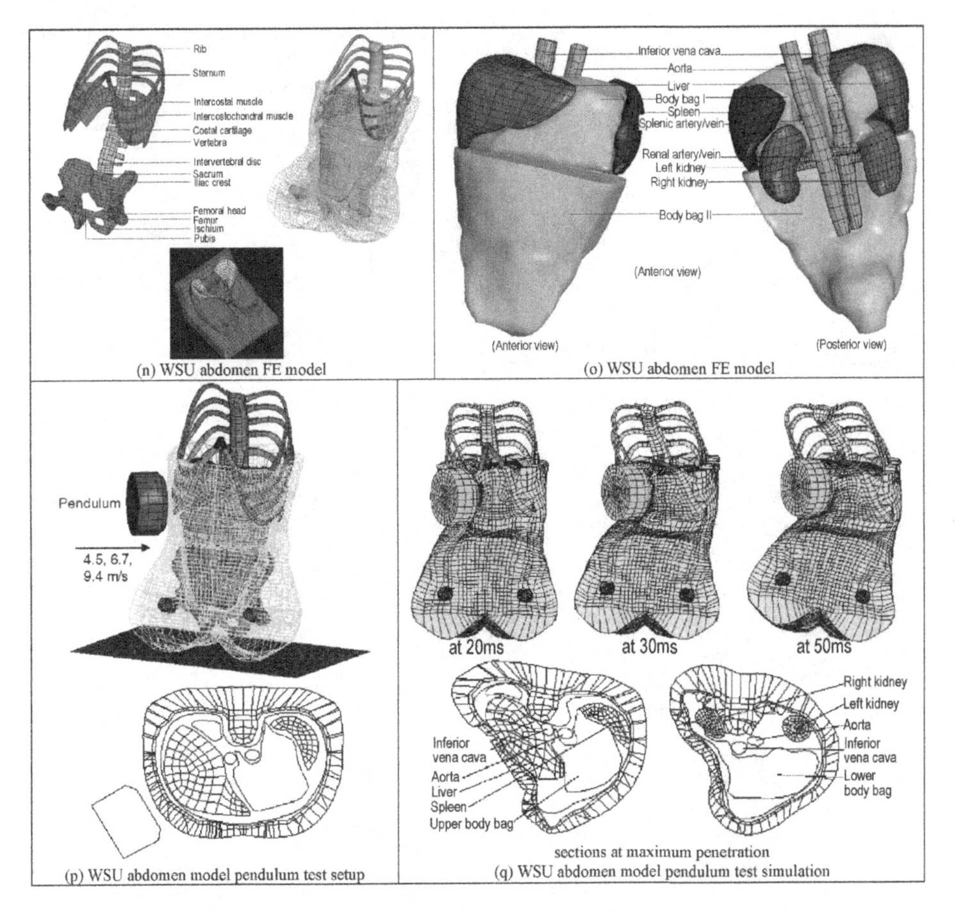

(n) WSU abdomen FE model

(o) WSU abdomen FE model

(p) WSU abdomen model pendulum test setup

(q) WSU abdomen model pendulum test simulation

FIG. 3.3. (*Continued.*)

The pedestrian simulations with 40 km/h bending and shearing tests, inset (c), were compared to tests by KAJZER, SCHROEDER, ISHIKAWA, MATSUI and BOSCH [1997].

3.4. *LAB human model*

The LAB (Laboratoire d'Accidentologie et de Biomécanique of PSA Peugeot, Citroën, RENAULT), in collaboration with CEESAR, ENSAM and INRETS, have developed a complete 50th percentile male human finite element model with 10 000 elements, Fig. 3.5(a)–(i).

The material properties were taken from the literature and a large data base of 30 test configurations and 120 test corridors was compiled and used to validate the model. Comparative studies were performed concerning the differences in the behavior of human models as compared to dummy models in frontal and lateral impact conditions. Fig. 3.5 gives an overview on this pioneering model and its comparisons with models of front impact (HYBRID III) and side impact (EuroSID) dummy models.

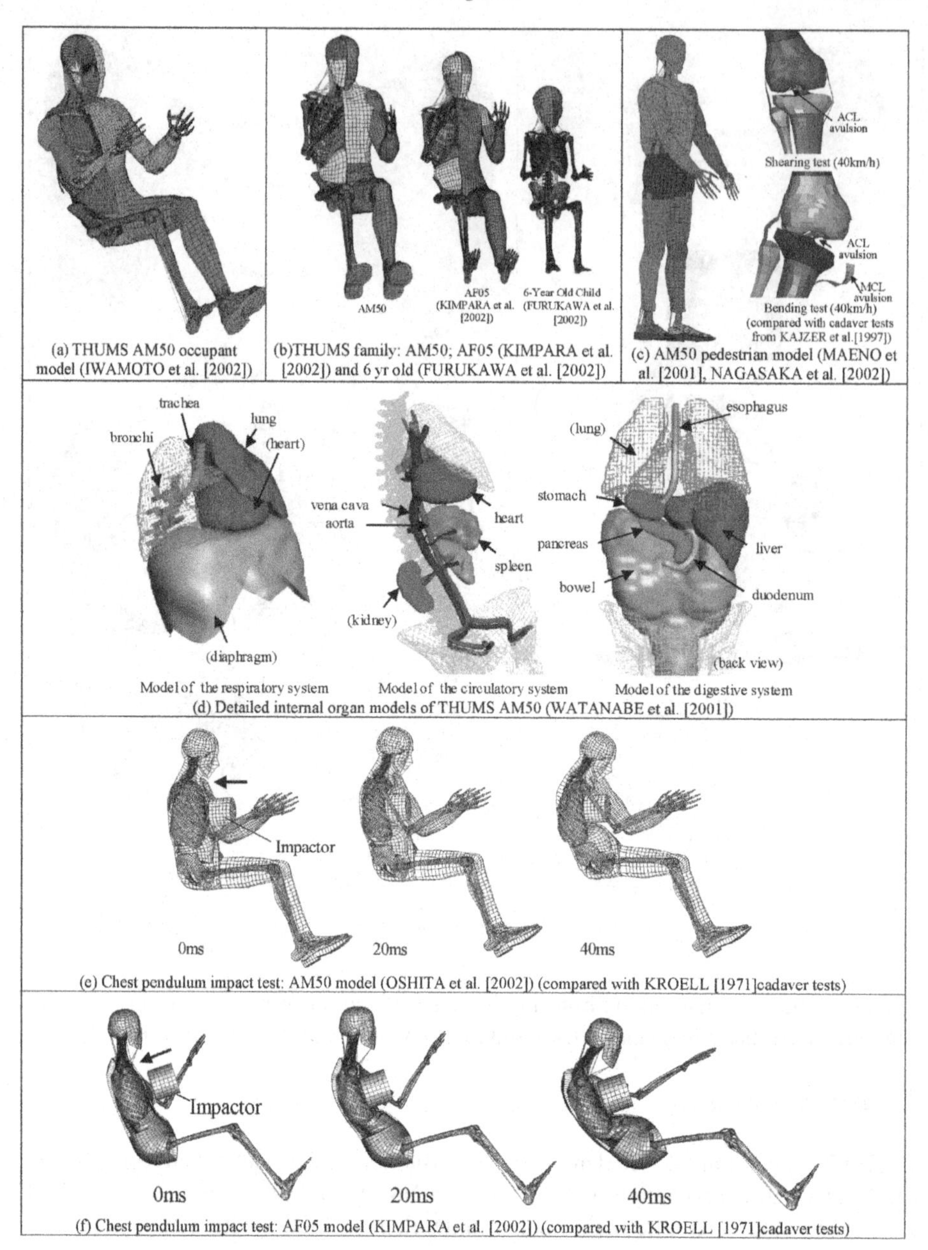

FIG. 3.4. TOYOTA's THUMS Total Model for Human Safety family. (Reproduced by permission of Kazuo Miki, Toyota CRDL.)

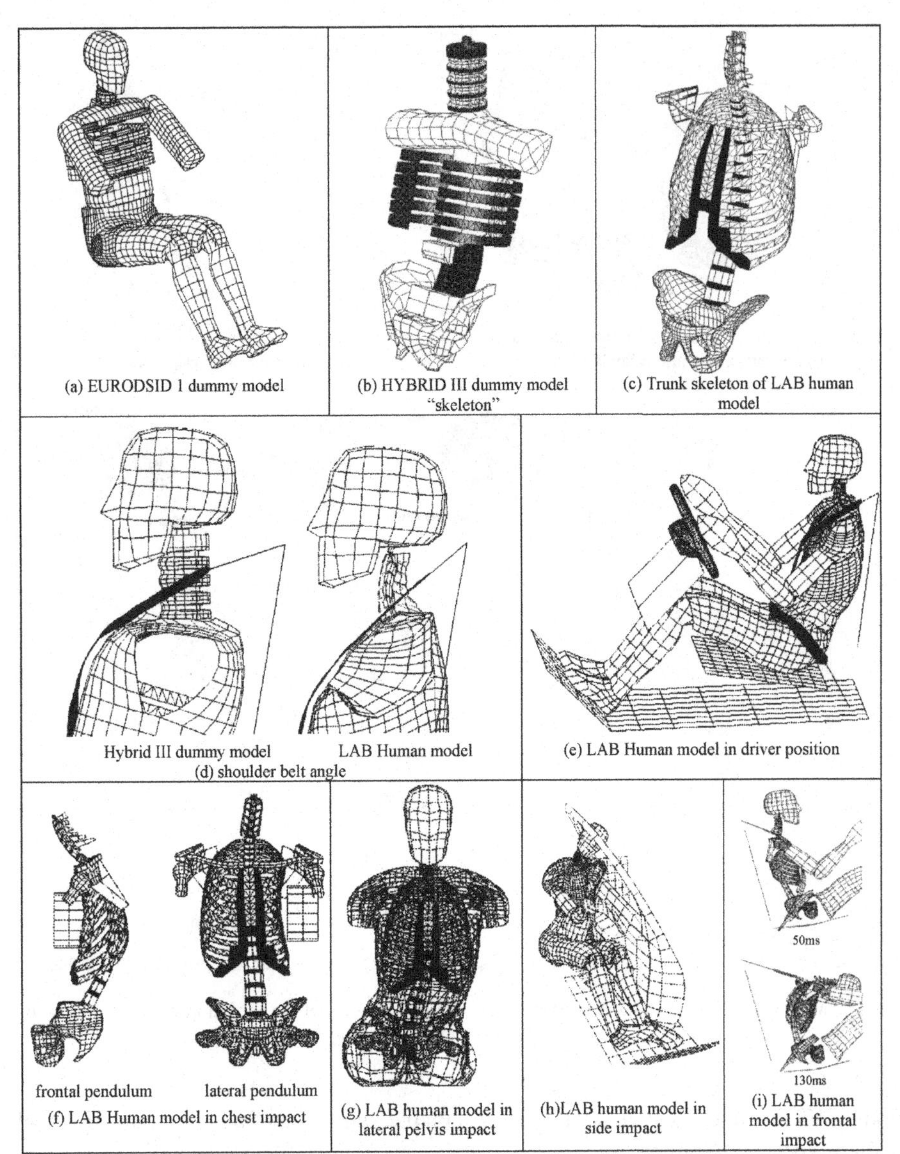

FIG. 3.5. The LAB human and dummy models (BAUDRIT, HAMON, SONG, ROBIN and LE COZ [1999], LIZEE, ROBIN, SONG, BERTHOLON, LECOZ, BESNAULT and LAVASTE [1998]). (Reproduced by permission of The Stapp Association.)

References for the LAB model. The LAB model has been published in the following papers: BAUDRIT, HAMON, SONG, ROBIN and LE COZ [1999] on comparing dummy and human models in frontal and lateral impacts; LIZEE, ROBIN, SONG, BERTHOLON, LECOZ, BESNAULT and LAVASTE [1998] on the development of a 3D FE human body

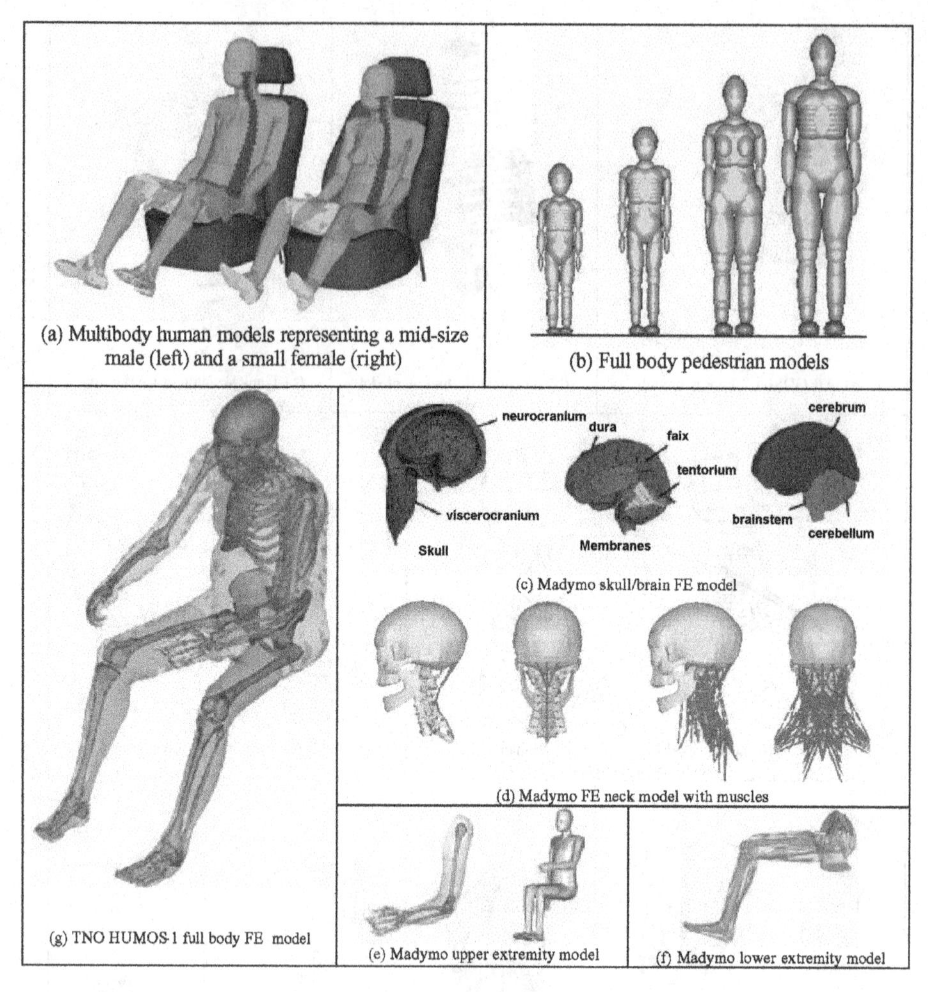

(a) Multibody human models representing a mid-size male (left) and a small female (right)

(b) Full body pedestrian models

(c) Madymo skull/brain FE model

(d) Madymo FE neck model with muscles

(e) Madymo upper extremity model

(f) Madymo lower extremity model

(g) TNO HUMOS-1 full body FE model

FIG. 3.6. Madymo multi-body and deformable finite element models of the human body. (Reproduced by permission of TNO Automotive.)

model; WILLINGER, KANG and DIAW [1999] on the validation of a 3D FE human model against experimental impacts.

3.5. *MADYMO human models*

MADYMO is a TNO Automotive engineering software tool that is used for the design of occupant safety systems. The following extract on their human models is drawn from their web site, http://www.madymo.com, Fig. 3.6(a)–(f).

Human body models have been developed for TNOs software program MADYMO, using a modular approach. Several combinations of detailed multi-body and finite element (FE) segment models are available. The models have been validated for impact

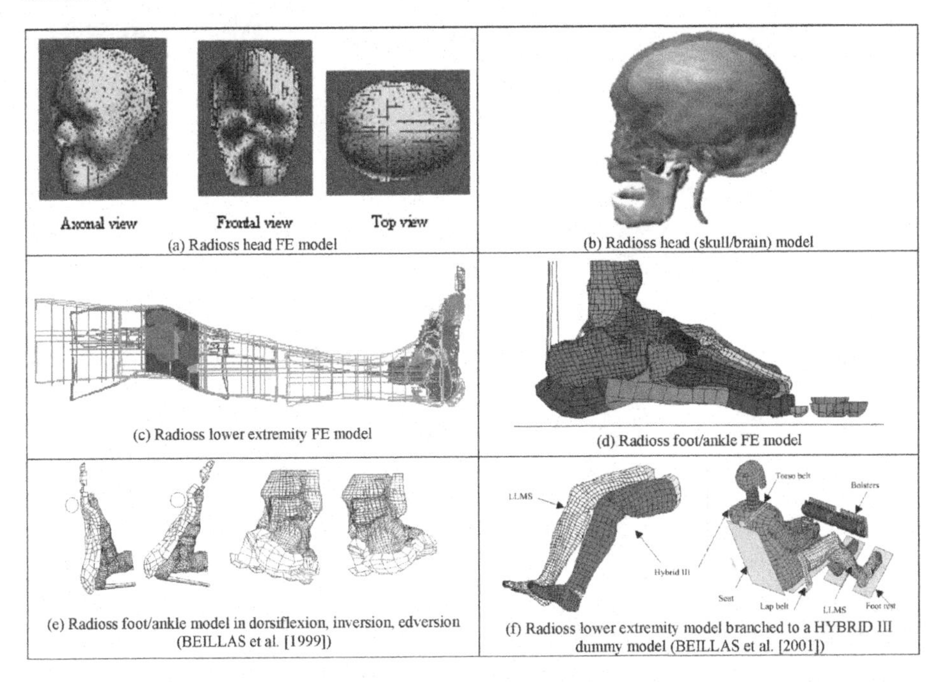

Axonal view
Frontal view
Top view
(a) Radioss head FE model
(b) Radioss head (skull/brain) model
(c) Radioss lower extremity FE model
(d) Radioss foot/ankle FE model
LLMS
Hybrid III
Torso belt
Bolsters
Seat
Lap belt
LLMS
Foot rest
(e) Radioss foot/ankle model in dorsiflexion, inversion, edversion (BEILLAS et al. [1999])
(f) Radioss lower extremity model branched to a HYBRID III dummy model (BEILLAS et al. [2001])

FIG. 3.7. Radioss human models (http://www.radioss.com). (Insets (a)–(d): Reproduced by permission of MECALOG Sarl; Insets (e) and (f): Reproduced by permission of The Stapp Association.)

loading. For their multi-body models, a combination of modeling techniques was applied using rigid bodies for most segments, but describing the thorax as a deformable structure.

A finite element mesh of the entire human body has been developed by TNO, based on the data produced in the EU project HUMOS, Fig. 3.6(g). The mesh was based on a European 50th percentile human in a seated driving position with a detailed 3-D numerical description of the subject's geometry. Apart of this full body model, TNO offers a series of deformable models of various body parts.

3.6. *RADIOSS human models*

As TNO and ESI Software, Mecalog is a partner in the European HUMOS projects. They have developed their own encrypted model from the common HUMOS-1 project data base. Further models of human parts were developed by Mecalog, as shown in Fig. 3.7(a)–(f). Their head and skull model was developed in collaboration with the University of Strassbourg (WILLINGER, KANG and DIAW [1999]). Their lower extremity model was developed in collaboration with WSU (BEILLAS, LAVASTE, NICOLOPOULOS, KAYVANTASH, YANG and ROBIN [1999], BEILLAS, BEGEMAN, YANG, KING, ARNOUX, KANG, KAYVANTASH, BRUNET, CAVALLERO and PRASAD [2001]). More information can be found on their web site http://www.radioss.com.

CHAPTER II

Human Articulated Multi-Body Models

The numerical models and materials presented in this chapter are based on work carried out at ESI Software and the University of West Bohemia (Robby family), and at IPS International and Hong-Ik University (H-Models).

4. Human Articulated Rigid Body (HARB) models

Open tree chain structure. The kinematics of the human body can be described to first order accuracy by the kinematics of a chain of articulated rigid bodies. Such models are computationally efficient, but they provide only limited information. Each member of the human body can be represented as a rigid body, while each skeletal joint is modeled by a corresponding numerical joint element or using non-linear springs. Modern crash codes have multi-body linkage algorithms with a comprehensive set of joint models, by which open-tree linked rigid body structures can be treated effectively. Details about these modeling techniques can be found in the handbooks of these codes. Human Articulated Rigid Body (HARB) models have therefore been introduced, which can describe the overall kinematics of the model under a crash load scenario. Fig. 4.1 shows a fiftieth percentile male HARB model, named "Robby", with its exterior skin, its tree link structure and its skeletal structure (ROBBY1 [1997], ROBBY2 [1998], HYNCIK [1997], HYNCIK [1999a], HYNCIK [2001a], HYNCIK [2002a], HAUG, BEAUGONIN, TRAMECON and HYNCIK [1999], BEAUGONIN, HAUG and HYNCIK [1998]).

Inset (a) of Fig. 4.1 defines the "sagittal", "coronal" (or "frontal") and "transverse" (or "horizontal") planes, used in anatomy to situate the parts of the body. To situate body parts relative to the body, directions are convened as follows: "anterior" pointing towards the front of the body, "posterior" towards the back, "medial" towards the midline, "lateral" away from the midline, "proximal" closer to the trunk and "distal" away from the trunk.

In inset (b), the upright HYBRID III 50th percentile mechanical dummy model is superimposed to the human model Inset (c) shows the upright skeleton with joints. Insets (d) to (f) show the model placed into the posture of a driver. Inset (g) details the articulated spine of the model.

Model geometry. The geometry of the shown fiftieth percentile male HARB model is based on the anatomical data sets available from DIGIMATION (DIGIMATION/VIEWPOINT CATALOG [2002]) and the VISIBLE HUMAN PROJECT [1994]. The exterior skin

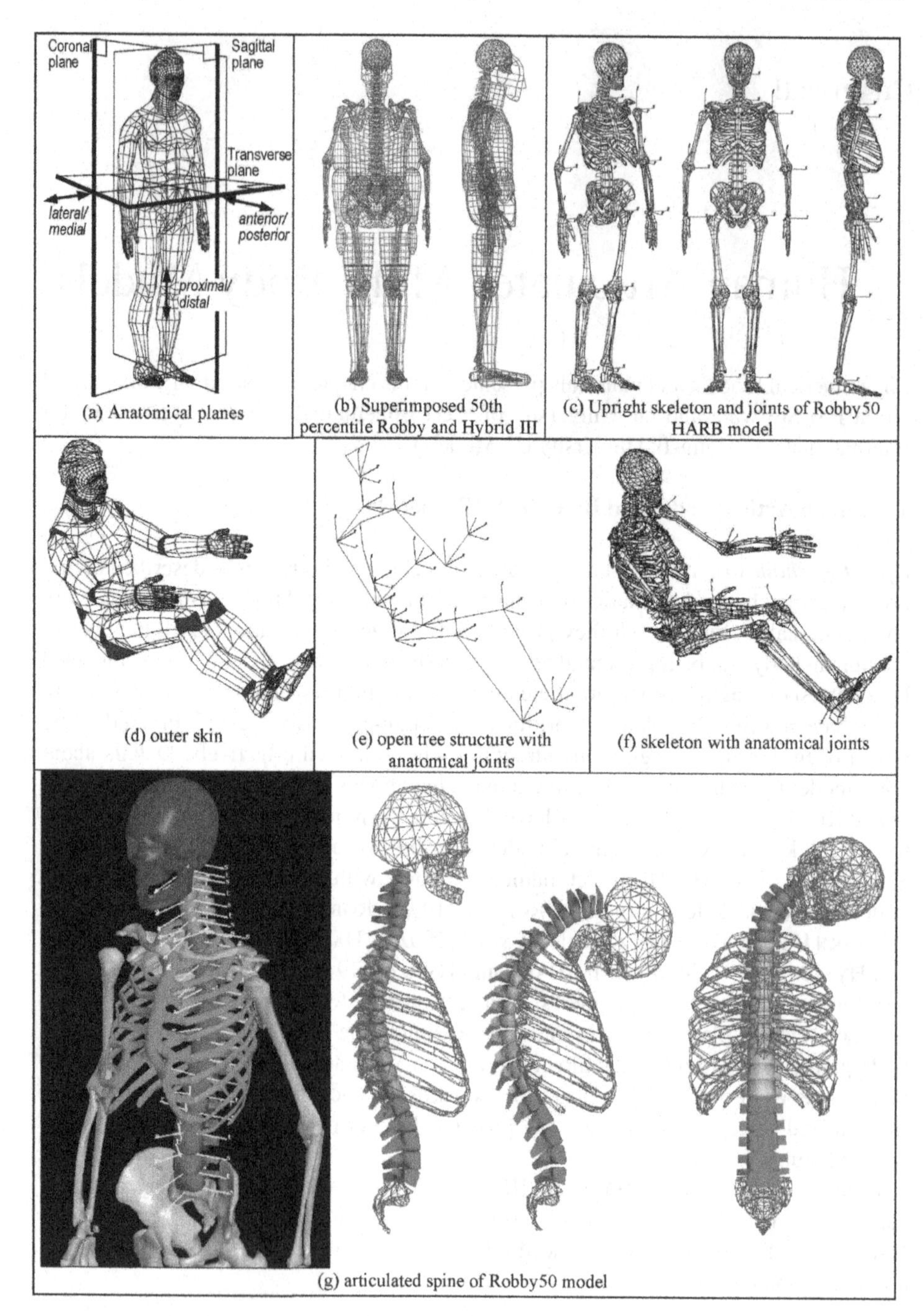

FIG. 4.1. Human Articulated Rigid Body (HARB) model: 50 percentile male (ROBBY1 [1997], ROBBY2 [1998], HYNCIK [1997], HYNCIK [1999a], HYNCIK [2001a], HYNCIK [2002a], HAUG, BEAUGONIN, TRAMECON and HYNCIK [1999], BEAUGONIN, HAUG and HYNCIK [1998]).

and the skeleton of this particular HARB version are modeled with mostly triangular and quadrilateral rigid facets, while the anatomical joints are modeled with nonlinear joint elements that translate the major degrees of freedom and stiffness properties of the skeletal articulations. The body segments head, neck, thorax, upper arms, lower arms, hands, pelvis-abdomen, thighs, lower legs and feet represent each a rigid body made of its skin and skeletal parts (hand and foot articulations beyond the wrist and ankle joints are usually not modeled for crash simulations).

Mass and inertia properties. Each segment has the center of gravity, mass, principal inertias and directions of principal inertias of a fiftieth percentile male, as given in the pivotal UMTRI reports (ROBBINS [1983]), where the anthropometric, the mass and inertia and the basic joint biomechanical properties of the average fiftieth percentile American citizen and of the fifth percentile female and the ninety-fifth percentile male have been collected from an extensive study on human cadavers. The results of this report have led to the construction of the well known Hybrid III mechanical dummy family. One family of the discussed human HARB models is named after the reports by Robbins (Robby, Robina, Bobby).

Contact definition. The modeled body segments can interact with themselves and with the environmental structures and obstacles via their built in and user defined contact interface definitions. Despite the fact that the external skin is rigid, soft nonlinear contact penalty spring definitions permit a realistic treatment of the body contacts. Contact algorithms are described in the handbooks of modern solver codes (cf. Appendix A.3).

Joint modeling. The mechanical joint stiffnesses and maximum excursions of their rotational degrees of freedom are governed by user-specified nonlinear moment-rotation curves. This modeling technique can also allow for joint failure upon over-extension of the joints beyond the anatomical motion ranges. The stiffness and resistance properties are found in the literature (ex: ROBBINS [1983], ROBBINS, SCHNEIDER, SNYDER, PFLUG and HAFFNER [1983]) (see also Appendix B). Fig. 4.2 shows typical curves, based on the Robbins report, for example for the left knee joint.

This model is part of another HARB model, termed H-ARB (Section 9.2). The knee flexion–extension rotation about the joint r-axis has been modeled approximately by a moment-rotation curve with negligible stiffness and resistance in the range between $+1.0$ and -1.2 radians, and with steep slopes at the ends of the flexion–extension excursion range. The internal–external rotation of the knee joint about the t-axis is modeled in similar fashion with a reduced angular range. The varus–valgus rotation about the s-axis is penalized heavily by a stiff slope. The given curves are only approximations to the real motions of the human knee. In reality, the internal–external rotation range depends on the flexion–extension angle, there is a certain forward–backward mobility of the tibia with respect to the femur and the center of rotation of the flexion–extension rotation is mobile. These additional joint mobilities can be captured with more elaborate joint models, or when using detailed finite element models, where the articular surfaces and the stabilizing ligaments and soft tissues are modeled in sufficient detail, which is beyond the range of simple HARB models.

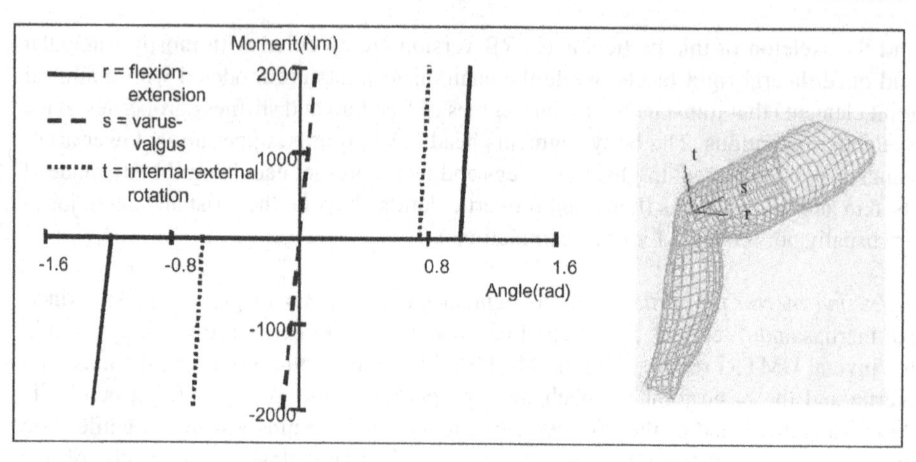

FIG. 4.2. Typical human joint moment-rotation curves (H-Model).

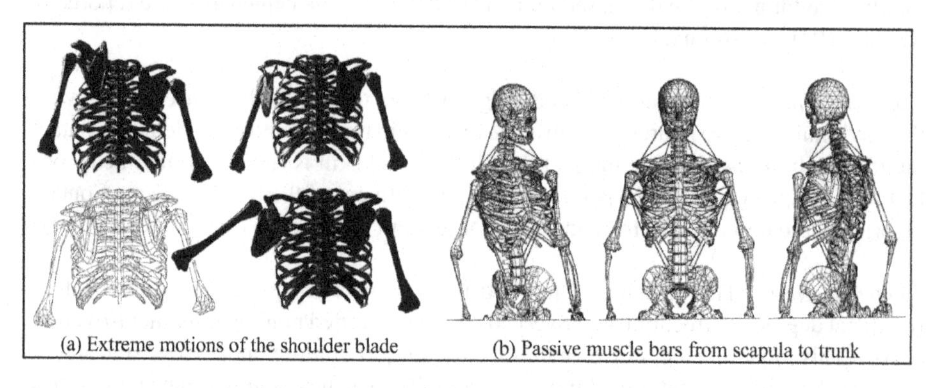

(a) Extreme motions of the shoulder blade (b) Passive muscle bars from scapula to trunk

FIG. 4.3. The attachment of the arm–shoulder complex (ROBBY2 [1998], HYNCIK [1999a]).

Special attention must be given to the articulations of the arm–shoulder complex. The only points of the arm–shoulder complex that are fixed with respect to the trunk are the sterno-clavicular joints, which connect each clavicle to the sternum. The upper arm is connected to the scapula (shoulder blade) by the gleno-humeral joint, while the scapula connects to the clavicula by the acromio-clavicular joint. These joints can be modeled by kinematic joint elements, similar to the one described above. In addition, each scapula can slide about the outer surface of the rib cage inside pockets, which can be modeled as a sliding contact interface.

Fig. 4.3 shows the anatomical mobility of the shoulder blade with the extreme up-down and backward–forward translational motions and inside–outside rotation motions, inset (a), as well as a sub-system of passive Hill muscle bars that help confine these extreme motions by their passive stretch force reactions, inset (b) (HYNCIK [1999a]). If these muscles are absent, the motions of the shoulder blade and the clavicula will be stabilized only by their connecting ligaments, represented in the HARB models by rotational joint elements. A preliminary study by Ludek Hyncik (ROBBY2 [1998]) showed that the passive action of the stretched muscles at the ends of the scapular motion ranges is considerable (more than 30% in average). These muscles therefore help stabilizing the complex mobility of the arm–shoulder complex.

5. The Hill muscle model

The material described in this section is based on work carried out at ESI Software and the University of West Bohemia.

Hill-type muscle bars. The mechanical behavior of the skeletal muscles in the directions of their fibers can be modeled to first order accuracy by Hill-type muscle bars (HILL [1970]). Each Hill-type muscle bar element is characterized by the physiological cross section area of the muscle, cut perpendicular to the fibers, and by the muscle fiber stretch and stretch velocity dependent active and passive mechanical properties of the Hill muscle model, described in Appendix C. The bars cannot, in general, transmit compressive forces.

These bar type finite elements can be arranged in the directions of the active muscle fibers. In that approach each skeletal muscle of interest is subdivided into a sufficient number of segments that can be approximated by bar like elements. For example, each segment of the biceps muscle of the arm can be approximated for most of its gross actions by a single bar element, fixed between its anatomical points of origin and insertion on the skeleton, provided the lever arm topology of the muscle is respected to sufficient precision.

For interactions with the skeleton and other tissues, and with the environment, further numerical modeling devices must be introduced, such as contact sliding interfaces and layers of deformable finite elements. For example, the trapezius muscle, being a flat surface muscle, can be subdivided into several parallel anatomical segments, each of which is represented with its tributary anatomical cross section area as one bar element. In order to facilitate contact of this muscle with the skeleton, each bar element can be further subdivided along its axis into a number of serial bar elements.

In successive further stages of refinement, skeletal muscles can be modeled as parallel assemblies of bars, Fig. 5.1(a), (b) (ROBBY-models, H-UE model: cf. Section 9.6); as bars attached to membrane finite elements which describe the resistance in the direction perpendicular to the muscle fibers, Fig. 5.1(c) (H-UE model); as heterogeneous two-dimensional fiber reinforced composite finite shell or membrane finite elements, where the composite fibers represent the mechanical properties attributed to the active muscle fibers, and where the composite matrix phase provides the passive in-surface stiffness and resistance of the muscle; as three-dimensional composite fiber reinforced solid finite elements, where the composite fiber phase represents the action of the muscle fibers and where the matrix phase provides the volume bulk response of the body of the muscle. It can be noted that the discussed muscle models use deformable finite elements, but the underlying model of the human body can still be a rigid multi-body model, as well as a deformable finite element model.

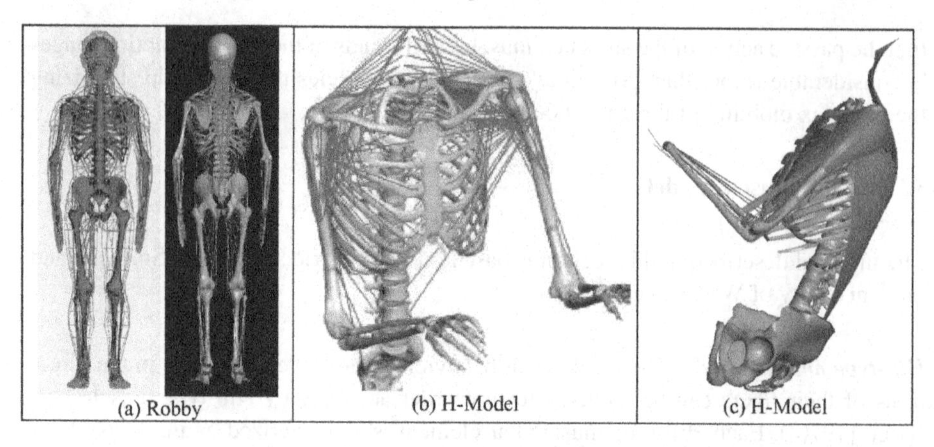

FIG. 5.1. Modeling of skeletal muscles with bars and membranes (ROBBY-models, H-UE model).

This underlines the fact that biomechanical models can be held mixed, incorporating deformable and rigid body models, as required by accuracy and computational efficiency.

The Hill muscle model. The Hill muscle model is one of the simplest phenomenological engineering models of the active and passive biomechanical behavior of skeletal muscles. Its mathematical description is given in Appendix C.

6. Application of Hill muscle bars

Whiplash simulation. The neck models can be equipped with Hill muscle bars for the purpose of simulating the effect of the muscle forces on the head/neck displacements when, for example, a rear car impact victim is bracing the neck muscles in a voluntary or unvoluntary fashion. Fig. 6.1 shows a head/neck model with a number of Hill muscle bars added. The corresponding neck "whiplash" event simulation is discussed in the later section on the H-Neck model (cf. Section 9.4).

Static muscle force distribution. Recently Hill bar models have been used to calculate the distribution of the skeletal muscle forces of human subjects holding a given set of static loads at a given fixed posture (HAUG, TRAMECON and ALLAIN [2001]). To this end, the skeleton of a human model is equipped with Hill muscle bars and the static loads are applied. Since there are many more muscle segments that can be recruited to stabilize the kinematic degrees of freedom inherent in the articulated skeleton, the problem is hyperstatic, i.e., there are more equations than unknowns from which to calculate the equilibrium muscle forces. In fact there is an infinity of equilibrium force patterns for which static equilibrium can be achieved.

Optimization of muscle energy. One rational solution to this problem is given by the assumption that the human body will activate the muscles under a minimum expense

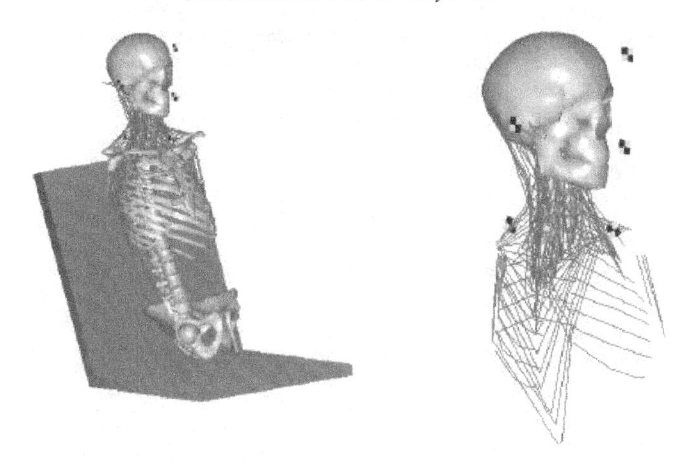

FIG. 6.1. Hill muscle bars added to the H-Neck model for whiplash studies (H-Neck).

of physiological energy. Physiological energy is spent when a muscle must activate to a constant force level over a given interval of time, even in the absence of any motion and external mechanical work ("isometric" conditions). This muscle energy is proportional to the product of force and time and the total energy is the sum of all muscle energies. One solution to the isometric muscle force problem can therefore be found by minimizing the total active muscle energy. This can be done by solving the associated *optimization* problem, where the simplest *objective function*, f, is given by the square root of the sum of the squares of the muscle forces, i.e.,

$$f = \left(\sum \gamma_i (\alpha_i - c)^2\right)^{1/2},$$

where the sum ranges over all participating muscle segments, i, activation level c is the given (average) voluntary level of muscle contraction before the load is applied (0–100%), α_i is the total activation level of the muscle segments that contribute to the task of carrying the load (0–100%) and γ_i is a switch that can have value "1" for the participating groups of muscles and "0", otherwise. This function can be thought to express the least possible overall level of muscle activation, or "energy", to be expended over a time with constant muscle forces. More complex objective functions have been proposed in the literature, for example, SEIREG and ARVIKAR [1989].

The *constraints* for the static optimization process are given by the fact that the accelerations of the links of the kinematic chain, constituted by the involved parts of the skeleton, must all be equal to zero in a position of static equilibrium. These accelerations can be calculated simply by performing an explicit analysis with a dynamic solver code, using the relevant muscled skeleton model, under the applied loads. In fact, one explicit solution time step at time = 0 is enough to determine whether or not the "structure" is in static equilibrium. At equilibrium, the internal muscle forces must balance the applied loads, and the accelerations, calculated by the solver at the centers of gravity of each rigid skeleton link, must be close to or equal to zero.

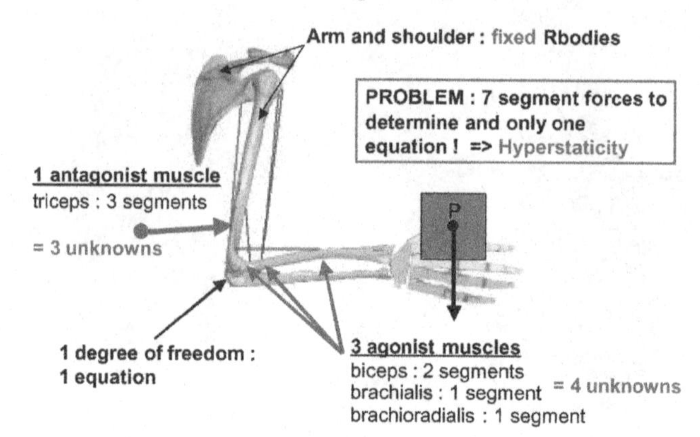

FIG. 6.2. Hill muscle bars activated at minimal energy under static loads (ROBBY models).

The *design parameters* of the optimization problem are given by the activation levels, α_i, of the participating muscle segments, which are the result of the optimization.

The *bounds* on the design parameters are given by $0 < \alpha_i < 100\%$, i.e., the activation level of a muscle cannot be less than zero and not greater than 100%. The outlined optimization procedure can be solved by standard optimization algorithms and it is applied to a simple one degree of freedom system.

Example. Fig. 6.2 shows a first example of an arm with fixed shoulder and upper arm, where a static load is held in place against one elbow rotation degree of freedom. In this example a minimum of seven muscle segments contributes to the task of supporting the load, and the only kinematic degree of freedom is the elbow flexion–extension rotation.

Fig. 6.3(a) shows the model of an upper torso with shoulders and arms, equipped with 132 Hill muscle bars. The model holds a constant static load in the left hand. The muscle forces are calculated for minimal energy. The colors of the muscle bars range from blue for zero activation to red for 100% activation.

In Fig. 6.3(b) a symmetric static load is applied to the same model (cubes), which increases linearly at equal increments. Inset (c) shows the calculated activation levels of some selected muscle bars at loads ranging from zero to 12 kg in each hand. In the diagram the agonist muscles (ex. biceps) are seen to quickly raise their levels of activation, initially in proportion to the load increase, while the less efficient muscles (ex. latissimus dorsi) initially contribute very little to the task. As soon as the first agonist muscles become saturated, i.e., reach 100% activation after about 3 kg loads, the less efficient muscles tend to increase their activation levels faster than linear. After the loads have reached about 8 kg, the inefficient muscles (ex. latissimus dorsi) must compensate for the saturation of the agonists and they raise their levels of activation dramatically. When all muscles reach saturation, the loads can no longer be increased and the limit load has been reached (2×12 kg). Limit loads for different postures are calculated in the examples shown in Appendix C, Fig. C.3.

Fig. 6.3(d) and 6.3(e) show the model pulling the hand brakes in a passenger car and holding the steering wheel while driving. In both postures the needed muscle force

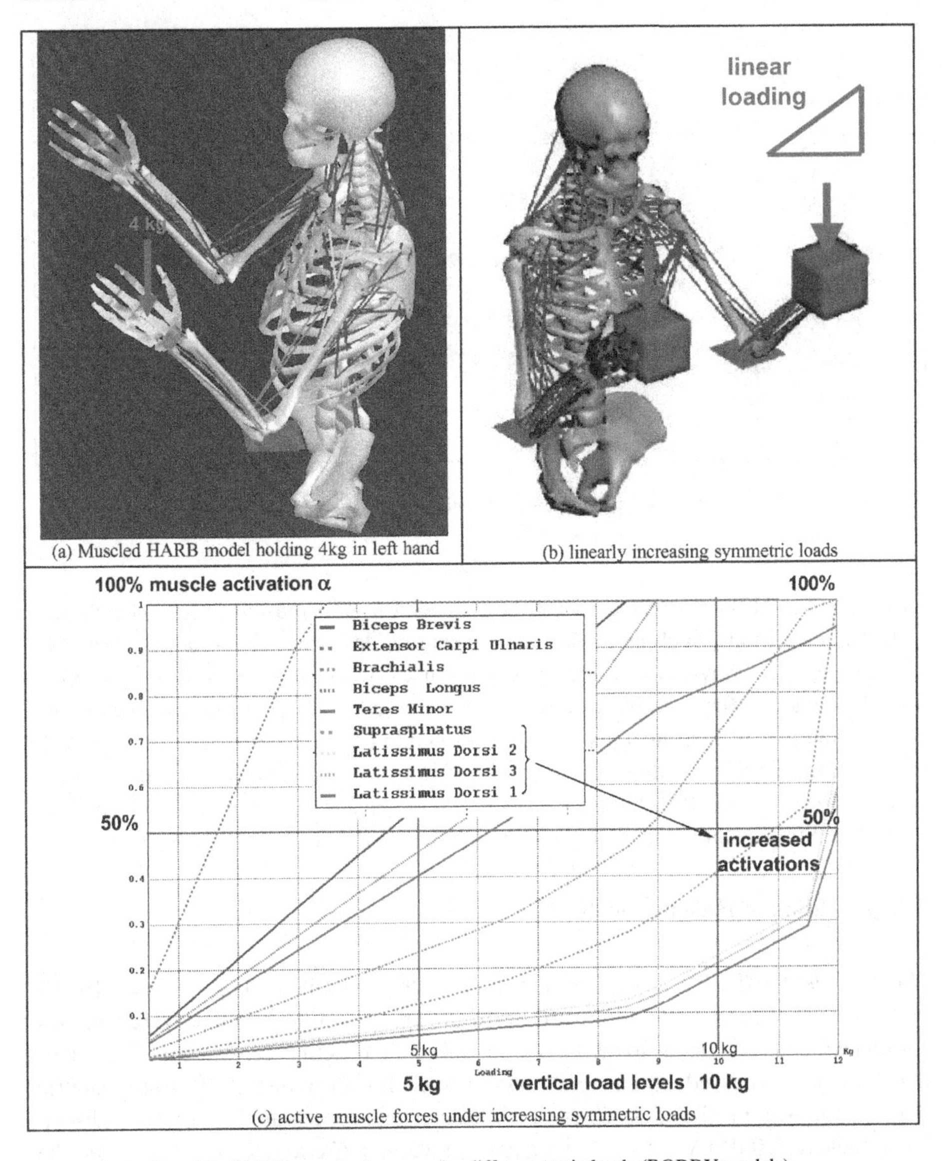

FIG. 6.3. HARB muscle forces under different static loads (ROBBY-models).

pattern was calculated for minimal energy. These examples show that articulated rigid body models can be used together with Hill muscle bars to calculate realistic muscle force distributions. These forces show the limit loads a subject can hold for a short time. For long term static muscle loads, such as in driving a car, the forces are less important, but will lead eventually to muscle fatigue. The levels of fatigue can best be assessed by knowing the forces that cause the fatigue.

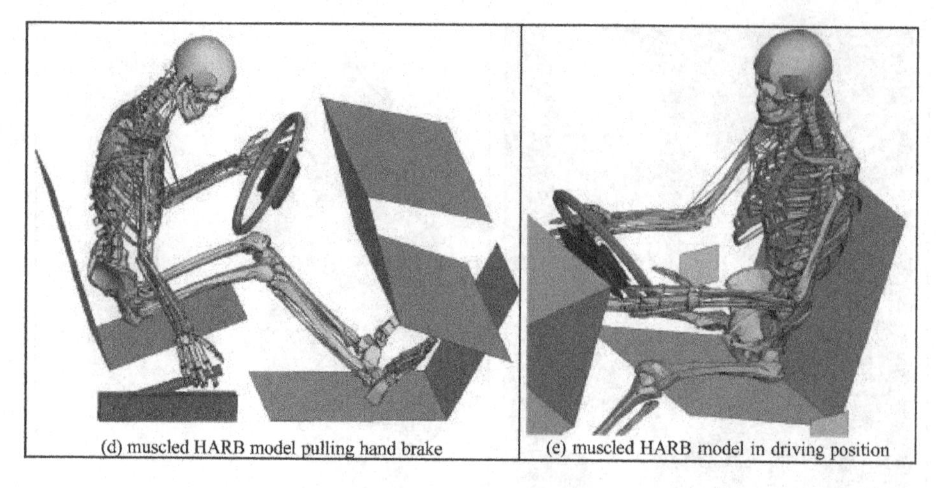

(d) muscled HARB model pulling hand brake

(e) muscled HARB model in driving position

FIG. 6.3. (*Continued.*)

Fatigue. The Hill muscle model can be extended to add muscle fatigue. Under the isometric conditions, such as in the static scenarios described above, a given muscle will fatigue when forced to sustain an active muscle force over a prolonged period of time. If the active force output of a muscle must be constant over time, such as in the simple examples of holding a given load as shown in Fig. 6.2, then muscle fatigue will force the muscle to raise its level of activation in order to compensate for the fatigue-induced drop of force. If the activation level is kept constant, the muscle force will drop over time. Simple static muscle fatigue models are currently under investigation.

7. Application of HARB models

HARB model families. In Fig. 7.1 models of a fifth percentile female (ROBINA [1998]) and a six year old child (BOBBY – ESI Software) are shown. These models are built in a fashion which is analogous to the HARB model of the fiftieth percentile male, and their details are not discussed separately. It must be noted, however, that basic anthropometric data exist mostly for the adult humans, while such data for children are close to absent. Nevertheless, DIGIMATION (DIGIMATION/VIEWPOINT CATALOG [2002]) provide the external geometry, the skeleton geometry and the organ geometry of male and female adults, as well as some muscle surface geometry. The external geometry of pregnant woman and children is also supplied, Fig. 7.1(a). The skeletal geometry and the organ geometry of children is not available, however.

Fig. 7.1(b) shows the external and skeletal geometry of the 5th percentile female (Robina). Fig. 7.1(c) shows the external and skeletal geometry of a six year old child (Bobby). The external geometry (skin) was hand-scaled from an available eight year old, while the skeleton was hand-scaled from the adult geometry to the anthropometric dimension of the six year old child, where different body parts were scaled individually.

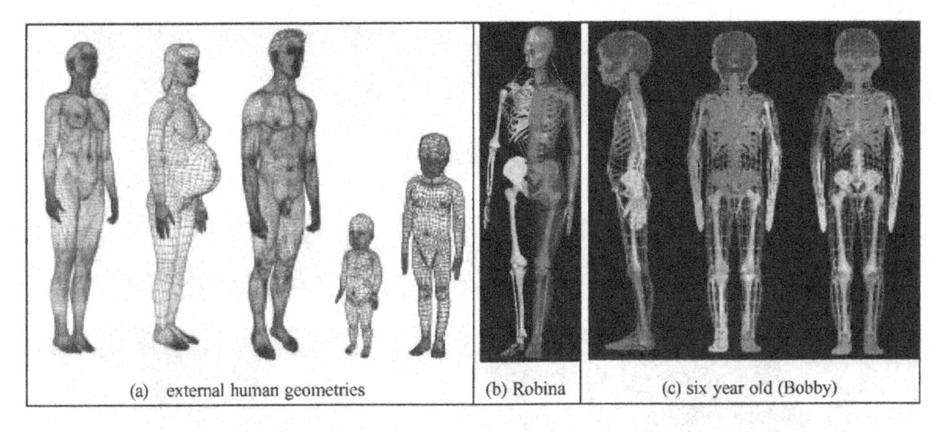

FIG. 7.1. Different HARB models (male, female, children). (Inset (a): Reproduced by permission of Digimation/Viewpoint, http://www.digimation.com.)

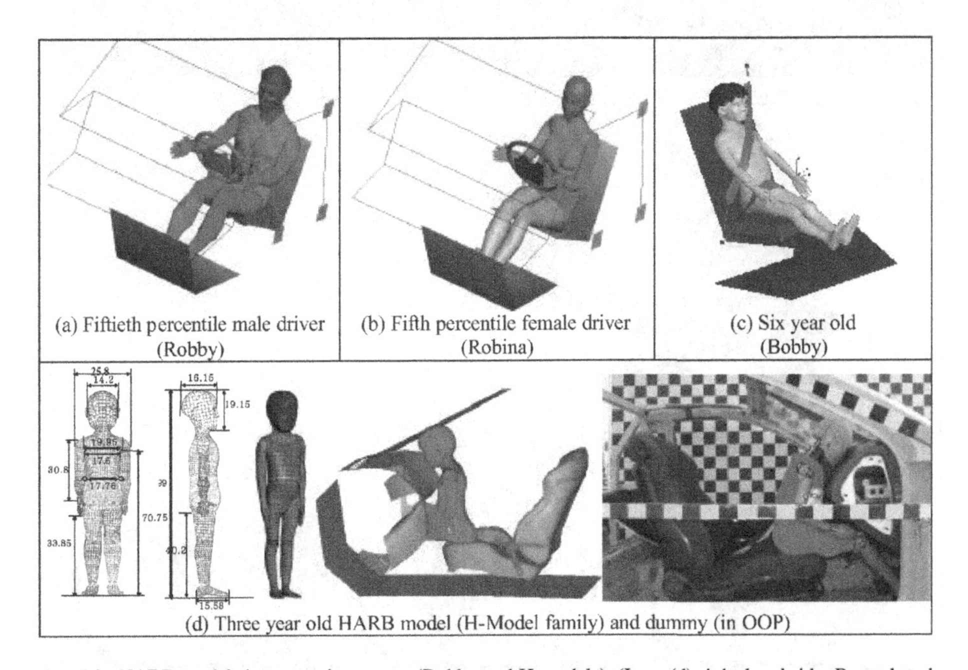

(a) Fiftieth percentile male driver (Robby)
(b) Fifth percentile female driver (Robina)
(c) Six year old (Bobby)
(d) Three year old HARB model (H-Model family) and dummy (in OOP)

FIG. 7.2. HARB models in car environments (Robby and H-models). (Inset (d) right-hand side: Reproduced by permission of Prof. Kim, Kwangwon University.)

Crash test simulations with HARB models. Fig. 7.2 presents the different HARB models (Robby family) in a passenger car environment, with and without airbags in driver and passenger positions, including a three year old in passenger airbag out-of-position (OOP) posture (H-Model family).

Fig. 7.3(a) shows displaced shape snapshots of a frontal impact sled test simulation for the seat belted 50th percentile male (Robby) with a deploying driver airbag.

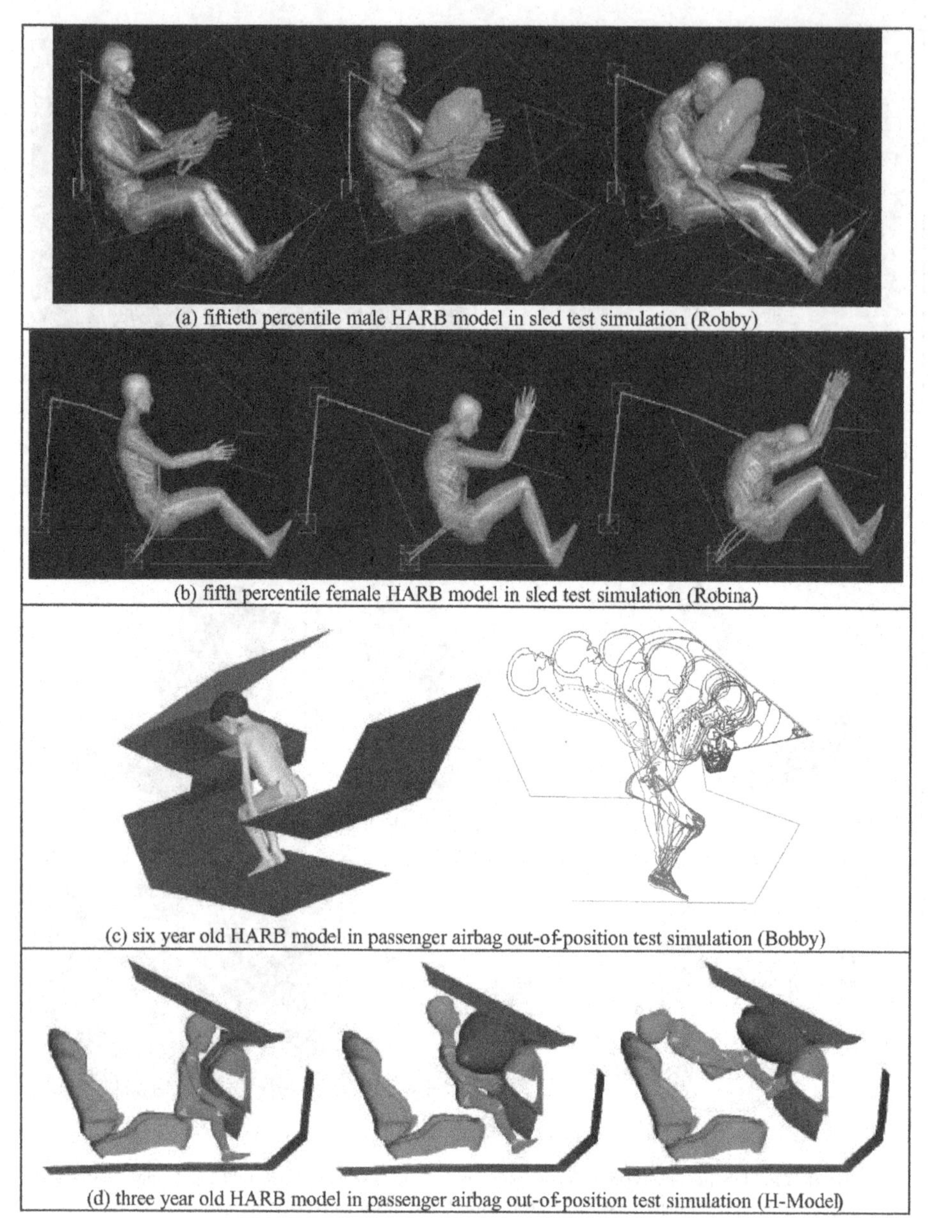

(a) fiftieth percentile male HARB model in sled test simulation (Robby)

(b) fifth percentile female HARB model in sled test simulation (Robina)

(c) six year old HARB model in passenger airbag out-of-position test simulation (Bobby)

(d) three year old HARB model in passenger airbag out-of-position test simulation (H-Model)

FIG. 7.3. HARB models in frontal crash test situations (Robby and H-Models).

Fig. 7.3(b) repeats this scenario for the belted 5th percentile female (ROBINA [1998]) who is not protected by an airbag. Fig. 7.3(c) shows the six year old HARB model (Bobby) in an OOP posture on the passenger seat. If the passenger side airbag fires at that moment, the subject is hit by the bag during its deployment phase, which can cause "bag slap" injuries and overextension and torsion of the neck.

Although the six year old should normally not be in the front seat, unbelted and leaning forward, car industry is interested in studying such events in an effort to render the airbag deployment phase as little aggressive to out-of-position subjects as possible, while keeping the airbag effective under standard conditions. Since the standard mechanical dummies are not biofidelic for such events, simulation with human models is recognized as a welcome alternative. Fig. 7.3(d), finally, shows a three year old HARB model (H-Model family) undergoing the same passenger airbag OOP scenario.

Validation of HARB models. Fig. 7.4 summarizes validations performed on the example of the model of the 50th percentile Robby model of the HARB family (ROBBY1 [1997], ROBBY2 [1998]). Validation of the other models were performed in a similar way.

Validation of biomechanical simulation techniques (algorithms, codes, models) in general can best be performed against physical test results. Directly exploitable physical results of crash events involving humans are rarely available. Access to physical results can be gained through accidentological studies, accident reconstruction analyses, cadaver test simulations and, rarely, volunteer test results for low energy impacts (rear impact).

The physical validation test results of Fig. 7.4 were obtained from an existing data base of cadaver tests performed at the University of Heidelberg (e.g., SCHMIDT, KALLIERIS, BARZ, MATTERN, SCHULZ and SCHÜLER [1978]). The subject selected from this data base was close to a 50th percentile male. The figure compares global head, thorax and pelvis accelerations with the corresponding cadaver test results.

In view of the large scatter expected from different human subjects, the comparison is considered satisfactory. In fact, validations should be made against experimental test corridors obtained from a sufficiently large subset of near 50th percentile human subjects. If the numerical simulation result curves remain within the given test corridors, the validation is successful. Moreover, the numerical results should not be performed on one "average" model of a given percentile slot, but simulations with stochastic variants of the average model should match the physical test corridors. This requirement is absent when working with mechanical dummies, because mechanical dummies of a given percentile category are all alike.

References on multi-body and muscle models. The following references may be consulted on the subjects discussed in this chapter: BAUDRIT, HAMON, SONG, ROBIN and LE COZ [1999] on dummy vs. human comparison; BEAUGONIN, HAUG and HYNCIK [1998] on the Robby family of articulated human models; CHOI and LEE [1999b] on a deformable human model; COHEN [1987] on the analysis of frontal impacts; DENG [1985], DENG and GOLDSMITH [1987] on the head/neck/upper torso response to dynamic loading; DIGIMATION/VIEWPOINT CATALOG [2002] contains CAD data of the human body; GRAY'S ANATOMY [1989]; HAUG, LASRY, GROENENBOOM, MUNCK, ROGER, SCHLOSSER and RÜCKERT [1993], HAUG [1995] and HAUG, BEAUGONIN, TRAMECON and HYNCIK [1999] on biomechanical models for vehicle accident simulation; HILL [1970] on muscle tests; HUANG, KING and CAVANAUGH [1994a], HUANG, KING and CAVANAUGH [1994b] on human models for side impact; HYNCIK

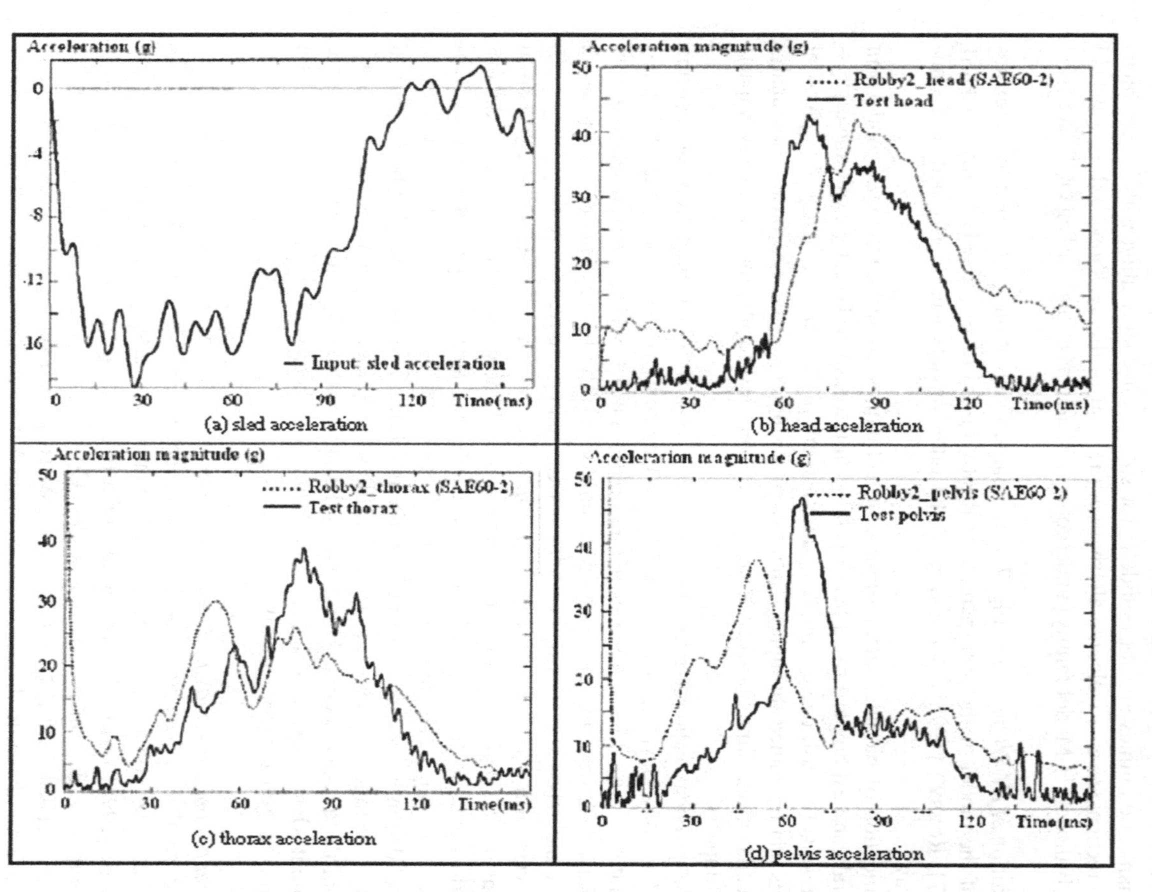

FIG. 7.4. 50th percentile HARB model validation with Heidelberg sled test results (Robby model).

[1997], HYNCIK [1999a], HYNCIK [1999b], HYNCIK [2000], HYNCIK [2001a], HYNCIK [2001b], HYNCIK [2002a], HYNCIK [2002b], HYNCIK [2002c] on human articulated rigid body models and deformable abdominal organ models; IRWIN and MERTZ [1997] on child dummies; KISIELEWICZ and ANDOH [1994] on critical issues in biomechanical tests and simulation; KROONENBERG, A. VAN DEN, THUNNISSEN and WISMANS [1997] on rear impact human models; LIZEE and SONG [1998], LIZEE, ROBIN, SONG, BERTHOLON, LECOZ, BESNAULT and LAVASTE [1998] on deformable human thorax and full body models; MA, OBERGEFELL and RIZER [1995] on human joint models; MAUREL [1998] on a model of the human upper limb; ROBBINS [1983], ROBBINS, SCHNEIDER, SNYDER, PFLUG and HAFFNER [1983] contains basic human percentile data; ROBBY1 [1997], and ROBBY2 [1998] on the 50th percentile Robby multi-body models; ROBIN [2001] on the HUMOS model; ROBINA [1998] on the 5th percentile female multi-body model of the Robby family; SCHMIDT, KALLIERIS, BARZ, MATTERN, SCHULZ and SCHÜLER [1978] on loadability limits of human vehicle occupant; SEIREG and ARVIKAR [1989] on muscle models; STÜRTZ [1980] on biomechanical data of children; VISIBLE HUMAN PROJECT [1994] is a CD ROM with anatomy data of the human body; WILLINGER, KANG and DIAW [1999] on 3D human model validation; WINTERS and STARK [1988] (muscle properties); WITTEK and KAJZER [1995], WITTEK and KAJZER [1997], WITTEK, HAUG and KAJZER [1999], WITTEK, KAJZER and HAUG [1999], WITTEK, ONO and KAJZER [1999], WITTEK, ONO, KAJZER, ÖRTENGREN and INAMI [2001] on active and passive muscle models.

CHAPTER III

Deformable Human Models

8. The HUMOS human models for safety

8.1. Introduction

International projects, like HUMOS-1, pave the way for standardizing the modeling methodology of the human body for impact biomechanics and occupant safety. The major foreground information produced by the first HUMOS project is therefore not specific computer models, but a database for the modeling of the human body with commercially available dynamic crash software. The encryption of the contents of the data base into specific versions of the individual commercial crash codes is a secondary result of the project. If models made with commercial software make use of the data published in the HUMOS database, if their modeling techniques are conform to the rules fixed in the base, and if they pass the standard tests laid down in this base, then the models could obtain the "HUMOS label". This label would certify the model to be conform to the standards laid down in the data base.

The HUMOS project embodies the fact that multi-disciplinary and multi-national efforts are needed to eventually achieve the goal of the industrial use of internationally standardized human models for safe car design. Putting humans into crash tests is only feasible in the computer. The methodology of "virtual testing" using compute models and codes, therefore, must be "certified" when it should complement or replace certain legal crash dummy tests. There is at present no officially binding certification procedure for simulation, neither for mechanical dummy nor for human models. The HUMOS projects can help to establish such procedures.

The following paragraphs are extracted from, and prepared after, reports of the HUMOS-1 project and the paper written by the HUMOS-1 project coordinator:

Robin, S. (2001), HUMOS: Human Model For Safety – A Joint Effort Towards the Development of Refined Human-Like Car Occupant Models, *17th ESV Conference*, Paper Number 297.

The selected paragraphs and material intend to briefly summarize the contents and major results of the first HUMOS project. The extracted material, figures and tables are reproduced by permission of the ESV Conference Organizers.

In the first European HUMOS program (1999–2001) the development of commonly accepted human car occupant models and computer methods was attacked. It was clear from the outset, that the results of the first HUMOS program must be extended into a second program, where other percentile human models and models with different human morphologies should be created by appropriate modeling, scaling and morphing techniques, and where the material description base should be enlarged.

Fourteen partners were involved, including six car manufacturers and several suppliers, software houses, public research organizations and universities. The program dealt with the synthesis and further development of the current knowledge of the human body in terms of geometry, cinematic behavior, injury threshold and risk, the implementation of this knowledge in new human body models, the development of the utilities for the design office use, and delivering of the models to be available for their integration in the car design process.

A wide bibliographical review supported those major goals. The geometry acquisition of a mid-sized 50th percentile male in a car driver seated position was achieved. The main human body structures were then reconstructed using a CAD method. The meshing of the different structures was based on the CAD definition and has led to models accounting for skin, bones, muscles as well as the main organs (lungs, heart, liver, kidneys, intestine etc.). The validation process was undertaken on a segment basis, each main part of the human body being confronted with the available literature results. The assembly of the whole model was the conclusion of this program.

The first problem that had to be solved was the acquisition of the inner and outer geometry of a human, seated in a car occupant posture (the driver posture was chosen). The publication made by Robbins, Schneider, Snyder, Pflug and Haffner [1983] on the seated posture of vehicle occupants served as a basis for many developments of human substitutes. This work mainly qualified the position of external anatomical landmarks on many different car occupants. But there were some limitations in this work, particularly concerning the relative position of the different bony structures and of the different organs, that had to be explored by the project.

8.2. *Geometry acquisition*

Selection of the subject. The geometry acquisition of a seated near 50th percentile adult male is one of the original achievements of the HUMOS-1 program. Commonly available human geometrical databases (e.g., GEBOD) furnish the main external dimensions of the human being. Little information is available concerning the geometry of the different organs and their relative positions of the different structures in a seated position. Table 8.1 summarizes the main characteristics of the selected HUMOS subject and those of the 50th percentile European male.

Subject frozen in driving position and physical slicing. The method consisted of placing the chosen cadaver in a driving position, Fig. 8.1.

Fig. 8.2 gives an overview on the applied methodology. The selected subject was installed in a standard full-size car cockpit with the hands placed on the steering wheel, and it was frozen in this position, Fig. 8.2, inset(1). A reference frame related to the cockpit was defined and the subject was embedded in a polymer block, Fig. 8.2(2). The method of physical slicing of a frozen cadaver was chosen, Fig. 8.2(3). Each slice was 5 mm thick, and the saw was 2.5 mm thick (268 slices were made). The different slices

TABLE 8.1
Main characteristics of the HUMOS subject and 50th percentile adult male (ROBIN [2001])

	Sitting height [mm]	Standing height [mm]	Weight [kg]
HUMOS subject	920	1730	80
50th percentile European male	915	1750	75.5

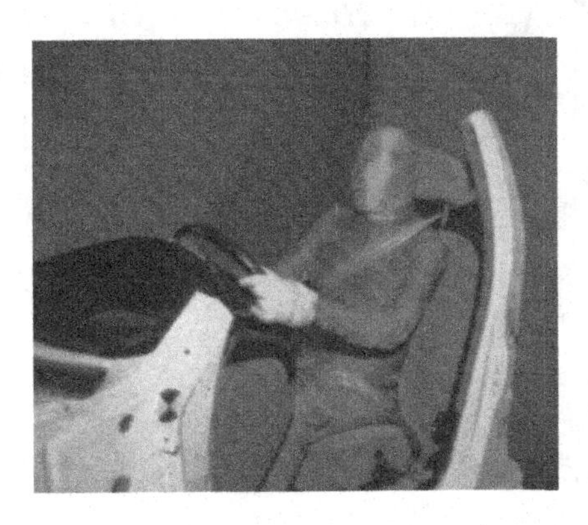

FIG. 8.1. HUMOS subject in its driving position. (Reproduced by permission of ESV Conference Organizers.)

were photographed on each side (491 images) and then each image was hand-contoured by anatomists, Fig. 8.2(4). This acquisition phase generated a set of about 13.000 files representing 300 different organs. Each file was composed of a set of points characterized by their 2D coordinates related to the slice, for example, Fig. 8.2(5). The position of each slice with regard to the reference frame was also available in the different files. A 3D visualization, Fig. 8.2(6), based on the nodes, was carried out in order to validate the acquisition process and, if need be, to modify some of the points describing the organs.

3D geometrical CAD reconstruction process. Results of the 3D CAD reconstruction process are shown on Fig. 8.3. During this process the point-by-point description files were transformed into CAD geometrical files. Standard commercially available CAD software was used. The main organs were reconstructed using available mathematical surface definitions. A back and forth process was set up between anatomists and CAD engineers in order to double-check at each step the shape of the reconstructed surfaces. When some assumptions had to be made during this phase, they were thoroughly discussed with the anatomists.

At this point it should be kept in mind that human occupant models for safety should be "generic", rather than "specific" in the sense that "average" people should be placed in cars during crash simulations. In reality, only specific subjects are available for data acquisition, with particular defaults and deviations from average. It is therefore legitimate to correct the acquired data for obvious deviations from the presumed anatomical

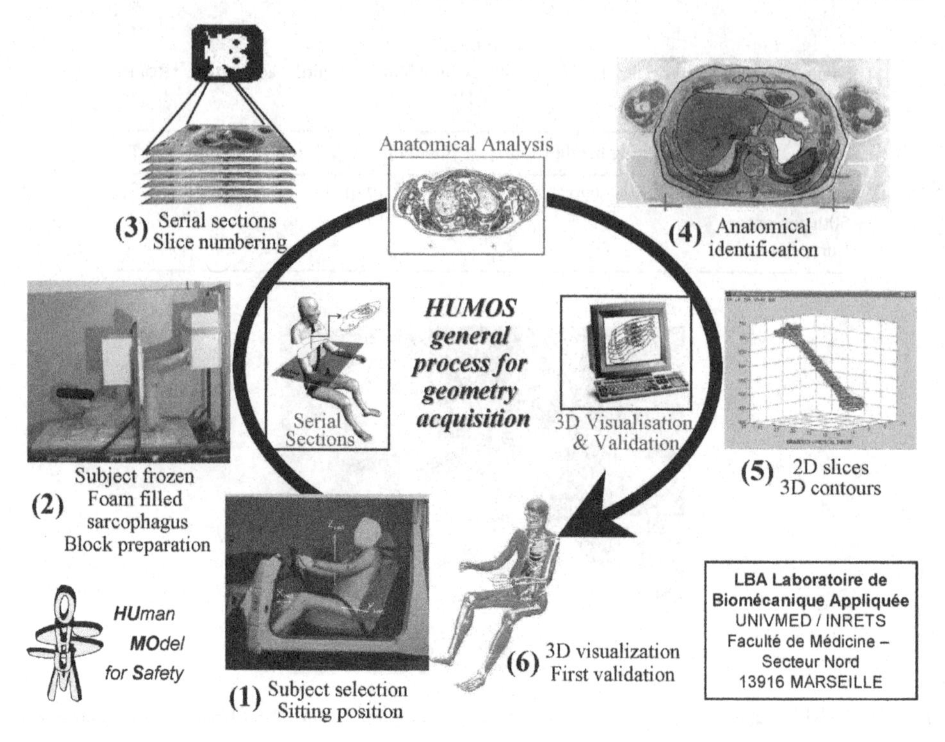

FIG. 8.2. Overview on the HUMOS-1 geometry acquisition process (HUMOS-1 project).

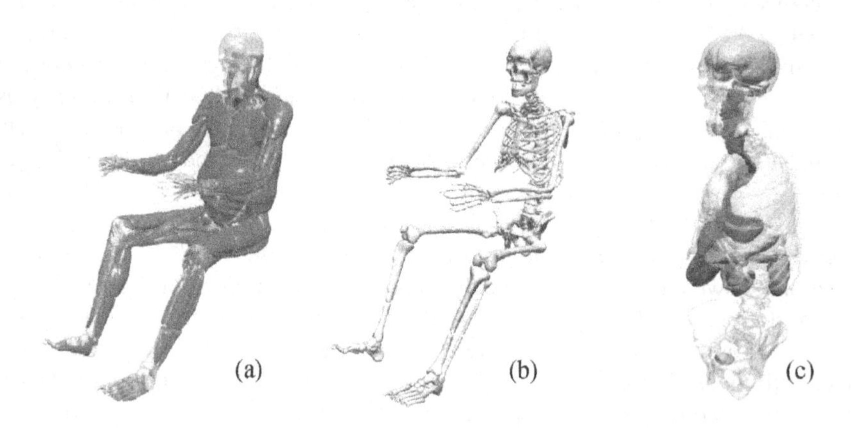

FIG. 8.3. Geometry acquisition result of (a) the whole body (without skin), (b) the skeleton and (c) the major organs. (Reproduced by permission of ESV Conference Organizers.)

standard. In fact, human models for safety should be as standardized as possible. Definitions of standard human anatomies have yet to be made. These goals must be achieved in follow-up projects.

8.3. Meshing process

During meshing, a close collaboration was maintained with the anatomists in order to double-check the shape of the reconstructed organs as well as the validity of the organ connectivity assumptions. Only half of the skeleton was meshed and sagittal symmetry was assumed. Further assumptions were made for some parts of the human body, which were not described accurately enough. For example, the posterior arcs of the different vertebrae were not always visible, and not all the ligaments and muscles were digitized during the acquisition process. Thus, some assumptions based on anatomical descriptions were made in order to add some muscle parts and some ligaments.

For reasons of computational efficiency it was decided to try to keep the number of deformable finite elements below 50 000, and requirements were issued on the final time step (cf. Appendix A) of the complete model, which put a lower limit on the size of the elements. Fig. 8.4 illustrates different segment meshes developed during this process.

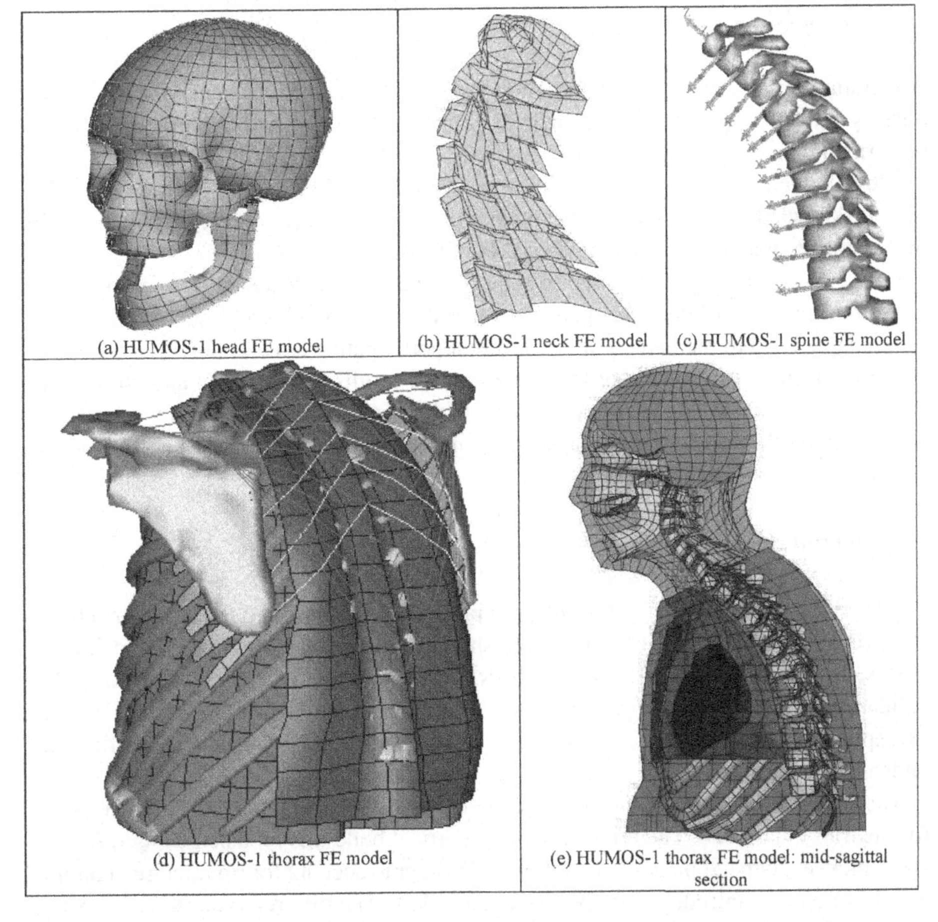

(a) HUMOS-1 head FE model

(b) HUMOS-1 neck FE model

(c) HUMOS-1 spine FE model

(d) HUMOS-1 thorax FE model

(e) HUMOS-1 thorax FE model: mid-sagittal section

FIG. 8.4. Mesh of skeleton and soft tissue parts. (Reproduced by permission of ESV Conference Organizers.)

It should be noted that the resulting mesh is relatively rough, and that finer meshes of parts might be substituted in the future for refined studies.

The bony part of the skull was modeled with standard thin shells. The cervical spine was modeled with shells and solids, some muscles and ligaments with bars. The remaining spinal column has rigid vertebrae and nonlinear joint elements for the inter-vertebral discs.

The final mesh of the whole thorax is a good example for the chosen mesh quality. The clavicle, sternum, ribs and connective cartilaginous structures were modeled using standard shell and solid finite elements. The scapula and the different vertebrae were modeled using shell elements for the definition of the shape of those structures. In the mechanical model, these bony parts are described as rigid elements. The muscular structures were meshed using both bar and solid elements. The flesh and connective tissues were modeled using solid elements.

The other body parts were meshed with comparable quality.

8.4. Material laws

A literature review enabled to gather the available knowledge, cf. Appendix B. Due to missing information, a limited number of experimental static and dynamic tests were performed within the program.

Ribs, clavicles, sternum and pelvic bone were tested extensively in order to gain specific knowledge about their mechanical properties. The results were associated with the definition of a parameterized material law (a power law was used).

Furthermore, static and dynamic muscle compression experiments were carried out. The strain-rate sensitivity was quantified for the soft tissues. Existing material laws were adapted and implemented in the different participating crash codes.

For the different organs, very few experimental results are available and there is an obvious lack of knowledge in this field. Available linear visco-elastic laws were chosen according to other modeling publications. Soft tissue characterization is still regarded as a largely uncovered area.

8.5. Segment validation process

Single rib model validation. The material properties were mainly derived from a literature review. The response corridors published by LIZEE and SONG [1998] and LIZEE, ROBIN, SONG, BERTHOLON, LECOZ, BESNAULT and LAVASTE [1998] were used for validation. Some material tests were carried out during the HUMOS-1 project. As an example, the rib models were first checked against static and dynamic tests carried out on isolated human ribs.

The ribs were modeled using shell elements for the cortical bone and solid elements for the trabecular bone. The thickness of the cortical bone was modeled using different shell thickness depending upon the location. In order to account for rib fracture, a failure plastic strain was introduced in the rib material law. The different values used for the material behavior definition of the ribs are reported in Table 8.2.

TABLE 8.2
Material properties of the rib cortical bones (ROBIN [2001])

Rib cortical bone material model: elastic-plastic behavior with failure	
Elasticity modulus	14 GPa
Yield Stress	70 Mpa
Ultimate stress	70 Mpa
Maximum deformation	4%
Poisson's ratio	0.3
Density	6000 kg/m^3

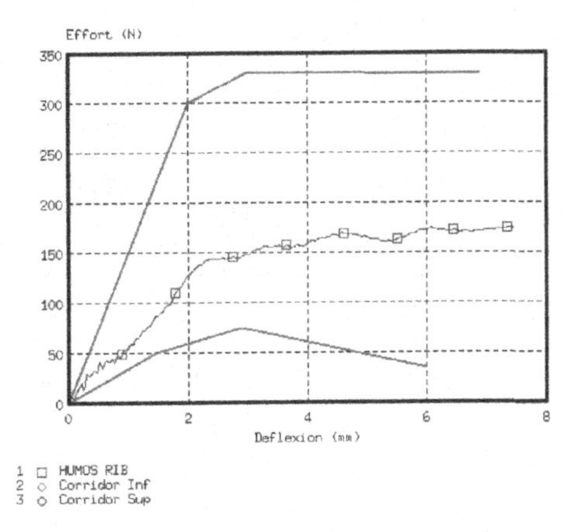

FIG. 8.5. Computer result compared with experimental corridor of an isolated rib subjected to a quasi-static loading. (Reproduced by permission of ESV Conference Organizers.)

These values were used for both the static and dynamic validation of the ribs. Fig. 8.5 illustrates the static behavior of the so-called "HUMOS rib" compared with the results obtained on an isolated human rib by the University of Heidelberg.

Thorax model validation. For the bones and cartilaginous parts of the thorax for which experimental data were available, the same process was used for the first description of the model. The assembled segment models were validated against the published experimental results. Figs. 8.6 and 8.7 represent the results obtained for the HUMOS-1 thorax model (ESI Software version) as compared to the experimental pendulum impact test results published by KROELL, SCHNEIDER and NAHUM [1971], KROELL, SCHNEIDER and NAHUM [1974] for frontal impact and by VIANO [1989] for lateral impact. In these experiments, the thorax was impacted with a 23.4 kg cylindrical impactor at 4.9 m/s and 4.6 m/s, in frontal and lateral directions, respectively.

The organs of the thorax were modeled with the airbag modeling technique (lungs) or as "bio-bags" (heart), cf. Appendix D. A bio-bag model is an adaptation of a standard airbag model, for which the mechanical properties of the enclosed gas are modified to be close to incompressible, like a fluid. This feature permits to approximately simulate

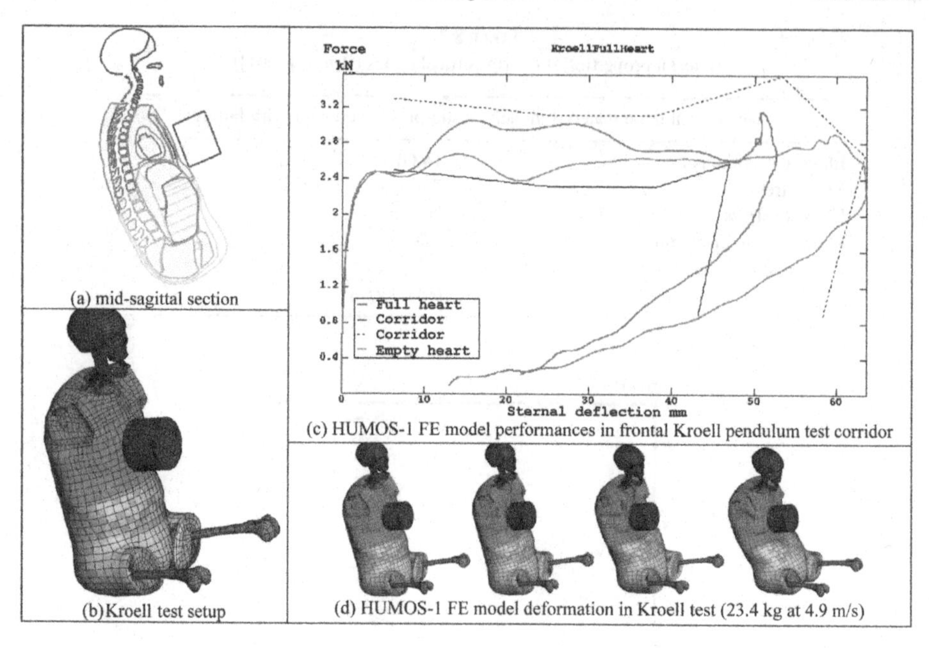

(a) mid-sagittal section

(b) Kroell test setup

(c) HUMOS-1 FE model performances in frontal Kroell pendulum test corridor

(d) HUMOS-1 FE model deformation in Kroell test (23.4 kg at 4.9 m/s)

FIG. 8.6. HUMOS-1 FE model: frontal impact test (Kroell test) (HUMOS-1 project/ESI).

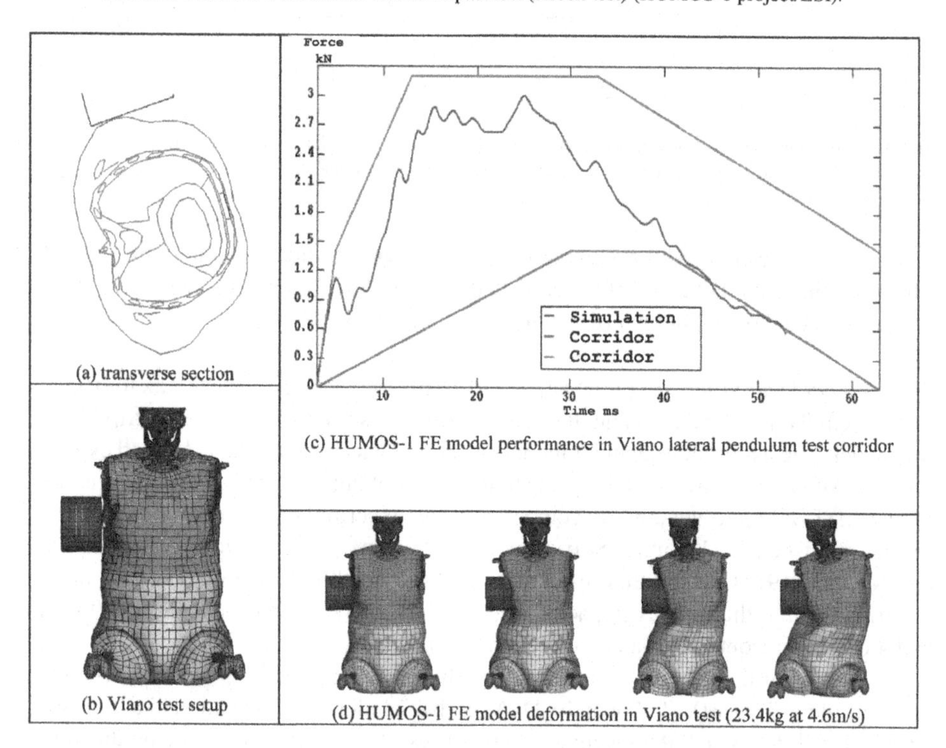

(a) transverse section

(b) Viano test setup

(c) HUMOS-1 FE model performance in Viano lateral pendulum test corridor

(d) HUMOS-1 FE model deformation in Viano test (23.4kg at 4.6m/s)

FIG. 8.7. Lateral impact test (Viano test) (HUMOS-1 project/ESI).

the impact response of hollow internal organs, filled partly or fully with air and liquid contents. It also allows organ contents to escape (vent out) under compression of the organ, when the standard venting options of the used airbag models are invoked. Standard airbag models are supported by all dynamic crash codes. They model a compressible gas contained in a flexible and extensible hull, made of membrane, shell or solid finite elements. More elaborate modeling of the gas or fluid contents of the hollow organs will require the techniques of fluid or particle dynamics, not discussed here.

The importance of at least approximate modeling of the blood contents of the heart is demonstrated in Fig. 8.6, where the impact force versus sternum deflection curves are shown both, for an empty and for a full heart.

Abdomen model validation. The test reported by CAVANAUGH, NYQUIST, GOLDBERG and KING [1986] was simulated with the HUMOS model (ESI Software version), Fig. 8.8. In this test, a bar (⌀25 mm, length 381 mm) with total mass of 32 kg hits the abdomen with a velocity of 6.9 m/s. The simulation outcome depends largely on the way the abdominal cavity and the intestines are modeled, Fig. 8.8(a).

The abdominal cavity is treated like an equivalent bio-bag, closed by the abdominal walls, and filled with a fluid and with sub-models of the intestines and internal organs. The intestines can either be modeled as solids, Fig. 8.8(a)(1), as an inner simple (2) or "meandered" (3) bio-bag, each placed within the outer bio-bag of the abdominal cavity.

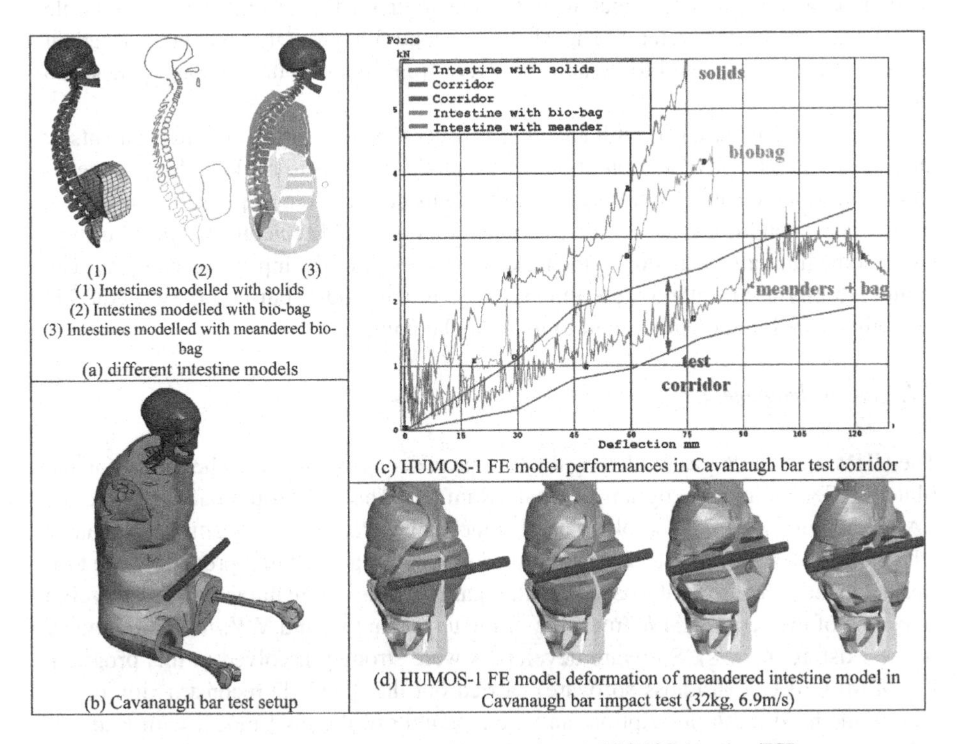

FIG. 8.8. Abdomen bar impact test (Cavanaugh test) (HUMOS-1 project/ESI).

The solid model of the intestines gave too much shear and compressive stiffness at large deformations, Fig. 8.8(c). The model with the bio-bag responded well initially, but developed too high internal fluid pressures at large deformations. The intestine model with a meandered bio-bag gave the best results and the response curve falls in between the experimental test corridor. The results show that even for a faithful reproduction of gross results, such as external load-deformation curves, the models should follow as closely as possible the anatomical realities, which allow for major shifts and relative displacements of the inner organs.

8.6. *Discussion and conclusions*

The HUMOS-1 program led to a first definition of a finite element model of the human body in a *seated driving posture*. In an effort to create a European standard human model, useful for car in industry, the model was implemented with three commercially available dynamic crash codes (Madymo, Pam–Crash and Radioss). The mesh of the model is shared by the different software packages, but the validation of the different models was carried out separately for the different codes.

From the beginning of this research work, it was foreseen that some major limitations would be encountered. First, the geometrical definition of the model comes from a unique human subject, i.e., is not generic. It is expected in a follow-up program to develop scaling techniques, which would enable to first define a generic 50th percentile model from the current reference mesh, and second 5th and 95th percentile occupant models. It is also expected to be able to derive from this first model some pedestrian models.

A validation database was built and used in order to validate the different segments of the model. The global validation of the whole model remains to be done. Some investigations need to be carried out on the muscle tonus contribution, especially for the low speed impact conditions that can be encountered in real field accident analysis. Furthermore, some limitations are due to the lack of knowledge of the injury mechanisms. The main currently used criteria were implemented in the model, but its injury prediction capabilities are limited with regard to its limited complexity.

8.7. *Acknowledgements*

The HUMOS project was funded by the European Commission under the Industrial and Materials Technologies program (Brite–EuRam III). The consortium partners were the LAB (Laboratory of Accidentology and Biomechanics PSA Peugeot, Citroën, Renault), who coordinated the work and was involved mainly in the meshing process of the thorax and in the validation database. The other participating car manufacturers were Volvo (meshing of the neck), BMW (meshing of the upper limbs), and VW (literature review of the existing models). Software developers were strongly involved in this program. ISAM/MECALOG (Radioss software) carried out the 3D CAD reconstruction of the model, the head mesh description, and the assembly of the final model with Radioss. ESI Software (Pam–Crash software) modeled the lower limbs and was in charge of the homogeneity of the different segment models. ESI Software also performed the assembly of the final model with Pam–Crash. TNO (Madymo Software) carried out modeling work and coordinated the soft tissue behavior activities. The supplier FAURECIA carried out the pelvis and abdomen modeling and coordinated the geometry acquisition process. INRETS LBMC (Laboratoire de Biomécanique et Mécanique des Chocs) carried out full-scale sled tests with human substitutes and contributed to the extension of the validation database. Marseille University was in charge of the geometrical acquisition of the seated human body, carried out by INRETS LBA (Laboratory of Applied Biomechanics). Athens University defined the physical material laws for the different human soft tissues and carried out experiments on some muscle properties. Heidelberg University was in charge of experimental investigations on different human tissues (bones and some cartilage structures) and produced a set of new experiments for the extension of the validation database. Chalmers University performed a bibliographical study on the current knowledge about human tissue behavior and identified the fields of missing knowledge. Chalmers University also contributed in the validation of the neck model.

8.8. References for the HUMOS model

The following references were found essential for the work reported in this section: CAVANAUGH, NYQUIST, GOLDBERG and KING [1986] on the impact response and tolerance of the human lower abdomen; KROELL, SCHNEIDER and NAHUM [1971], KROELL, SCHNEIDER and NAHUM [1974] on the impact response and tolerance of the human thorax; LIZEE and SONG [1998], LIZEE, ROBIN, SONG, BERTHOLON, LECOZ, BESNAULT and LAVASTE [1998] on 3D FE models of the thorax and whole body; MERTZ [1984] on a procedure for normalizing impact response data; ROBBINS, SCHNEIDER, SNYDER, PFLUG and HAFFNER [1983] on the seated posture of vehicle occupants; ROBIN [2001] on the HUMOS model for safety; VIANO [1989] on biomechanical responses and injuries in blunt lateral impact.

9. The fiftieth percentile male H-Model

9.1. Introduction

Preamble. This section gives an overview on the structure of the fiftieth percentile male H-Model of the human body. This family of models is presently distributed and developed by the private companies ESI Software (Paris) and IPS International (Seoul), and by Hong-Ik University (Seoul). The presently available basic model represents a 50th percentile male human body and it was conceived primarily to study injury mechanisms and to assess injuries of the human skeleton and organs, which result from car accidents, including pedestrian injuries, e.g., CHOI and LEE [1999b]. A fifth percentile female and a 3 and 6 year old child model are under development. The basic model permits a fast and ongoing absorption of the rapidly growing biomechanical research results. The model can assist the conception of improved crash protection measures and devices for the human driver and passenger. It allows the omni-directional analysis for different impact directions (front, side, rollover, etc.) with a single model with global and local responses to bags, belts, head restraints, knee bolsters, etc. The analysis of different body postures, and the effect of muscle activity an be studied using the model.

From the H-Models other models for the analysis of passenger sitting comfort and riding comfort can be derived. The models for sitting comfort permit the assessment of seat and backrest pressure and they are whole body models equipped with elaborate representations of the flesh. The models for riding comfort permit to assess the effect of seat vibrations and they have well adapted spines and organ masses to detect resonance.

An attempt is made to systematically present the material for each body part in paragraphs which discuss anatomy, injury, model structure and calibration, and validation. The discussed items are the whole body H-ARB models and models of the head, neck, torso, upper extremity, lower extremity and the foot/ankle complex.

Features. Major features of the H-Model are given by the possibility of modular assembly of the deformable external and internal components into an underlying multi-body H-ARB (Human Articulated Rigid Body) model. The H-ARB model assures a correct overall kinematic behavior in a simulated crash scenario. Modularity permits to study each body part of interest in detail. While assuring the good overall kinematics, detailed studies can so be made for the head in head impacts, the neck in whiplash events, the thorax and abdomen for belt injuries, the lower extremity in knee bolster impacts, the foot/ankle complex in toe panel intrusions and the upper extremity and shoulder complex in side impacts and airbag aggressions. Fig. 9.1 shows an overall view of the model with skin and skeleton.

Model structure. The H-Model is a recent combination of two previous biomechanical models of the human body, namely the "ROBBY" (ESI Software) and the "H-Dummy" (IPSI/Hong-Ik) model families. The overall anthropometric properties of both models were mainly extracted from the report edited by ROBBINS [1983], after which the ROBBY family had been named. The new model family is now co-developed by

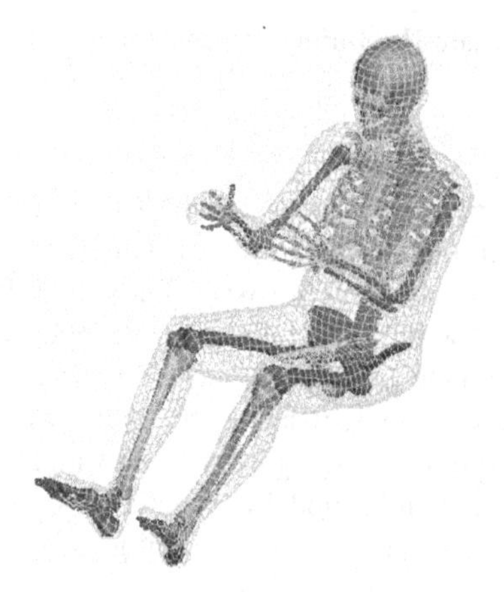

FIG. 9.1. H-Model with skin and skeleton (fiftieth percentile male).

the Seoul and Paris biomechanics teams. The development of the H-Model family was started prior to and is independent of the HUMOS-1 model described in Section 8 of this article.

The geometry of all listed models and sub-models has been derived mostly from the anatomy data found in the CAD geometry data base issued by DIGIMATION (DIGIMATION/VIEWPOINT CATALOG [2002]) and in the CD ROM issued by the VISIBLE HUMAN PROJECT [1994]. Further inspiration, concerning mainly the anatomical details of the ligaments, tendons, muscles, blood vessels and other soft tissues and internal organs was found in anatomical atlases, such as GRAY'S ANATOMY [1989]. Where necessary, the data were scaled to the presumed dimensions of the 50th percentile male. The Viewpoint CAD data were transformed into surface finite element meshes, and three-dimensional bulk FE meshes were generated in a compatible fashion, using commercially available FE mesh generator packages. The Visible Human CT data were digitized manually. The digitized anatomical cross sections were used to generate surface and bulk FE meshes.

The HUMOS-1 project (cf. Section 8) used a specific slicing technique, applied on a PMHS (post mortem human subject) in a sitting driver position. This procedure enabled the project to produce unique data of a seated subject. Since only one subject could be treated, scaling to the fiftieth percentile was necessary.

Calibration. All listed models and sub-models have been given material properties found in the open literature. A description of bio-materials is given in Section 1.5 and in Appendix B of this article.

Validation. All listed models and sub-models have been validated with available literature results. The basic validation tests and results are described in the available H-Model Reference Manuals. All validation test cases are relatively simple, representative, repeatable and well documented and controlled experiments, which facilitates modeling under well identified conditions.

9.2. *The H-ARB Human Articulated Rigid Body model*

Model structure and calibration. The H-ARB versions of the H-Models consist of articulated rigid body segments with flexible joints. The fiftieth percentile male model, H-ARB50, was built after ROBBINS [1983] and is documented in the H-Model Reference Manuals. It represents the basic platform for the modular assembly of detailed deformable and frangible skeleton components with soft tissues and internal organs. It serves for the evaluation of the overall kinematic and kinetic behavior of the occupant for omni-directional impacts. Fig. 9.2 shows the H-ARB50 version of the H-Model (76 kg) with the joint tripods (a) indicating the location and the orientations of the anatomical joints.

The outer geometry of the skin is taken from the clay model published in ROBBINS [1983], Fig. 9.2(b). The major dimensions of the outer surface of the model are shown in inset (c). The neck of the H-ARB model has several joints (skull-C1, C1-C2, etc.) as shown in inset (c), but no muscles. The various joints have rotational moment/angle

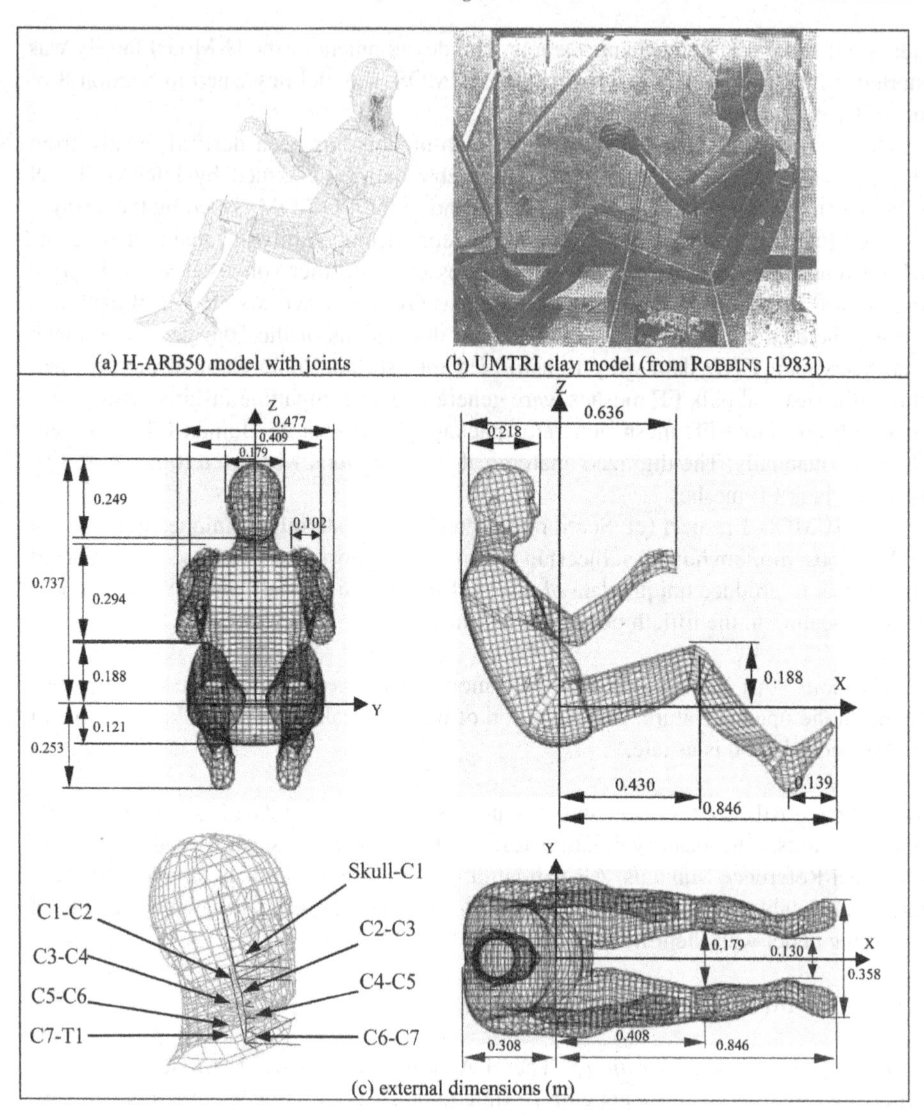

FIG. 9.2. H-ARB50 version of the H-Model (fiftieth percentile male). (Inset (b): from ROBBINS [1983] (UMTRI-83-53-2, US DoT NHTSA public domain report).)

relationships indicated in ROBBINS [1983] (see also Appendix B). Two typical curves are shown in Fig. 9.3. Some joints are given damping coefficients in order to stabilize their dynamic response. The joints are modeled with standard joint finite elements with linear and nonlinear moment-angle curves.

Accelerometers. Three accelerometers, which can be used to evaluate global motions of the H-ARB models, are defined using three additional nodal points. The locations of the accelerometers is adopted in analogy to the Hybrid III 50% male dummy model.

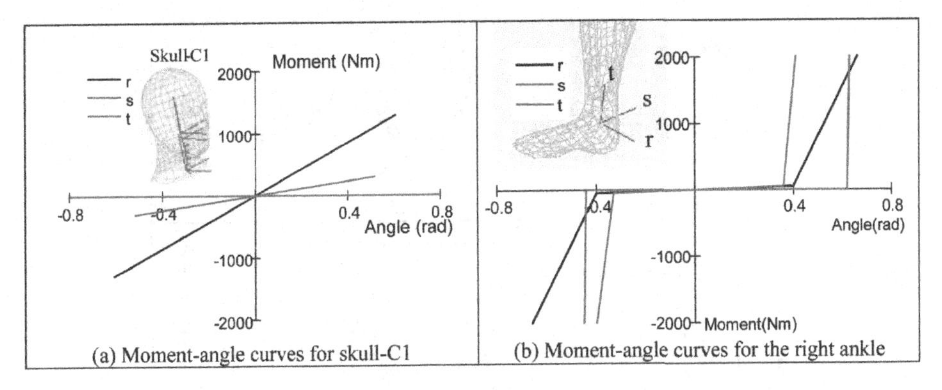

FIG. 9.3. H-Model H-ARB50 joint moment-angle curves.

The solver code will specifically output the acceleration time histories at these precise locations.

Contact. Contact surfaces in are defined between different segments of the H-Model, such as arms to thorax, between upper legs, etc. Note that in any application problem, additional contact interfaces involving the H-Models should be defined, such as H-Model to airbag, to belt, to car seat, etc. Standard contact algorithms are used, as implemented in the commercial crash codes (cf. Appendix A, Section A.3).

Validation and performance. First, the *extension motion* of the neck due to rear impact between two cars is presented. The geometry of the animated neck motion is shown in Fig. 9.4 at different times. The acceleration pulses of the striking and the hit car, the input acceleration pulse applied at T1 (first thorax vertebra), and the linear acceleration responses of the human volunteer and of the H-ARB50 head model are also shown in the figure. Any active muscle action is not considered in this simple H-ARB model. The passive muscle action, and the action of the discs and neck ligaments, is considered in a global fashion via the modeled standard joint finite elements with calibrated linear and nonlinear moment-angle curves.

Second, for the *lateral bending motion* of the neck due to the side impact, the animated neck motion for different times is shown in Fig. 9.5. The input acceleration pulse given at T1 (first thorax vertebra) is shown in the figure. The responses of the H-ARB50 neck model is compared with corridors of human volunteers.

Third, for a *frontal sled test simulation*, Fig. 9.6 represents the overall response of the H-ARB50 model to the medium intensity acceleration diagram indicated in inset (a). The H-ARB response is compared at different times to the response of a Hybrid III 50th percentile dummy model, inset (b). The head and upper torso linear acceleration versus time diagrams of both articulated rigid body models are compared in inset (c), where the differences in the responses reflect the differences between humans and crash dummies.

Finally, for an *arbitrary passenger out-of-position* (OOP) scenario, Fig. 9.7 shows the overall response of the H-ARB version of the H-Model.

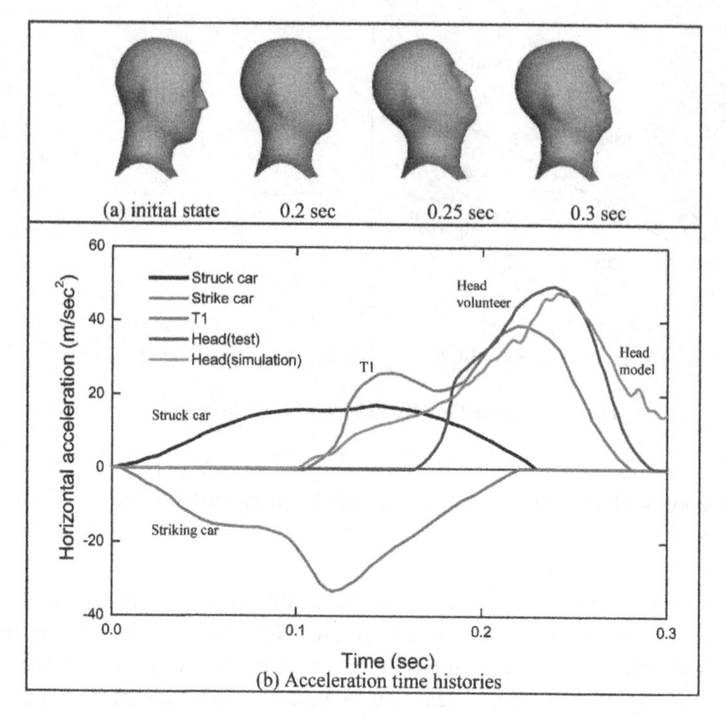

FIG. 9.4. H-ARB50 Validation: Horizontal acceleration histories due to rear impact.

Zooms with FE sub-models. For detailed studies ("zooms") of the impact response of the individual body parts, it is possible to substitute their detailed finite element models into the H-ARB versions of H-Model. The following sections outline such detailed sub-models. This modular approach saves CPU time while the overall model response correctly represents the kinematics of the body to generate the dynamic boundary conditions for the required zoom.

References on multi-body human models. To obtain further information on the shown HARB models, the following references may be consulted (in addition to the references listed at the end of Chapter II in the paragraph "References on multi-body and muscle models"): JAGER, SAUREN, THUNNISSEN and WISMANS [1994], JAGER [1996] on head/neck models; PRASAD [1990] on comparison between Hybrid dummy performances; SAE ENGINEERING AID 23 [1986] is the Hybrid III User's Manual; THUNNISSEN, WISMANS, EWING and THOMAS [1995] on human volunteers in whiplash loading; ZINK [1997] on 6 year old out-of-position passenger airbag simulation.

9.3. *H-Head: Skull and brain*

An overview on the H-Head model structure and application is given in CHOI [2001].

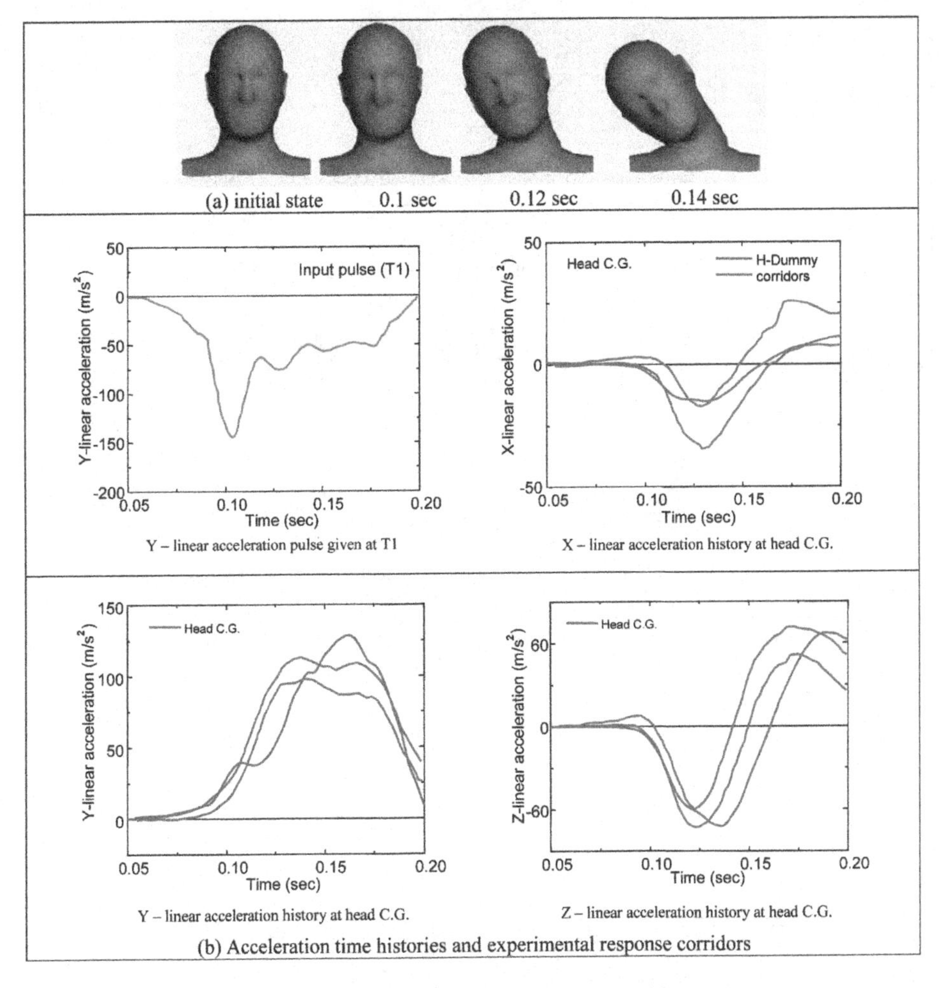

FIG. 9.5. H-ARB50 Validation: Neck motion due to side impact.

Anatomy. The complex anatomy of the skull and brain is summarized in Fig. 9.8. More details can be found in anatomical atlases and textbooks, e.g., GRAY'S ANATOMY [1989] and in the VISIBLE HUMAN PROJECT [1994]. The finite element model of the human head (H-Head) has been constructed by evaluating cross sections through the head. Three such sections (top-down) and the corresponding traces made of the brain matter are shown in Fig. 9.8(a) (SPITZER and WHITLOCK [1998]).

Cranial structure. The skull consists of three layers referred to as the outer table, diploe, and inner or vitreous table, Fig. 9.8(b) (PIKE [1990]) and (e) (PUTZ and PABST [2000]). The diploe consists of trabecular bone, and is located between the other two that are made up of compact (cortical) bone. The inside surface of the cranial cavity is lined by a layer of dense fibrous irregular connective tissue and enclosing venous sinuses. This layer adheres for the most part tightly to the cranial bones.

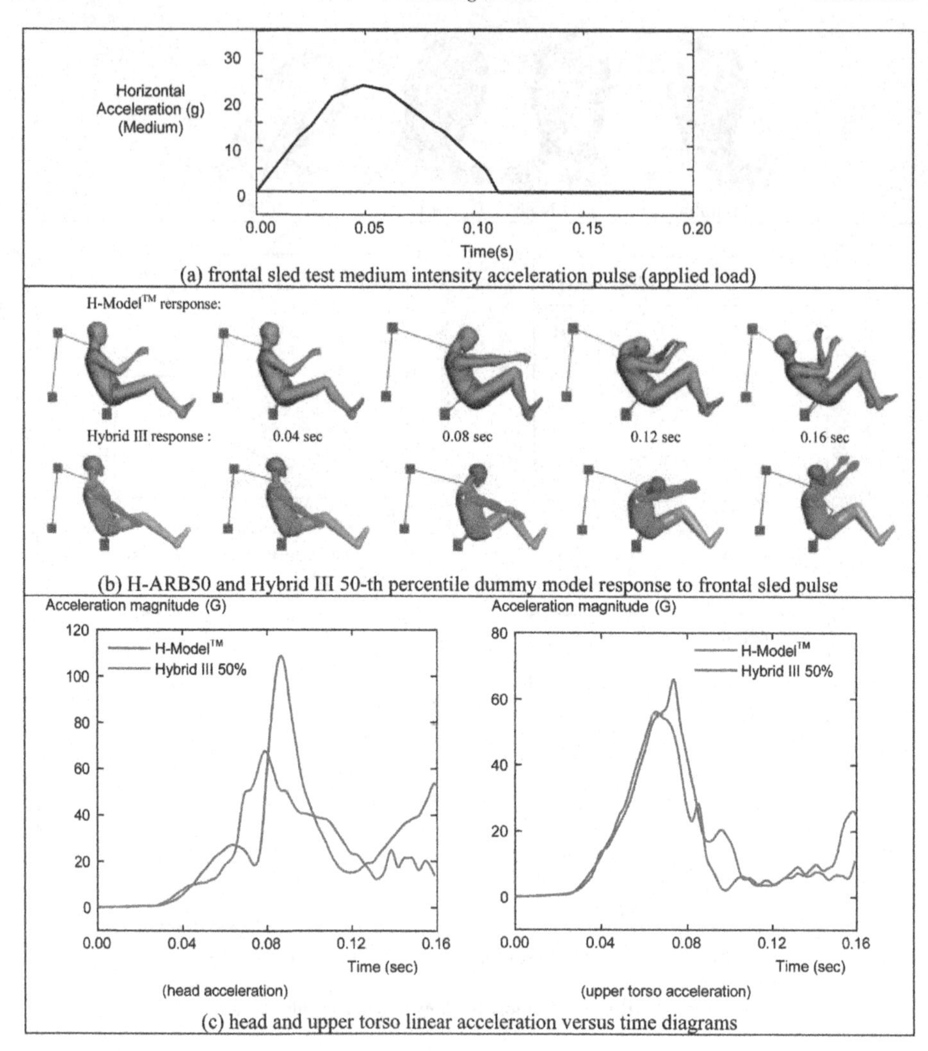

FIG. 9.6. H-ARB50 Validation: Sled test for frontal impact.

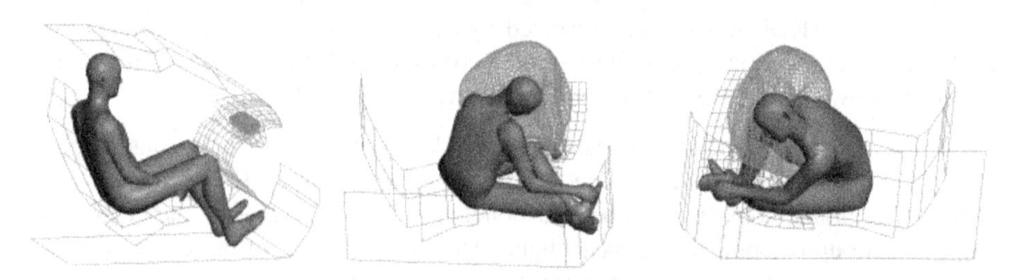

FIG. 9.7. OOP passenger airbag with H-ARB50.

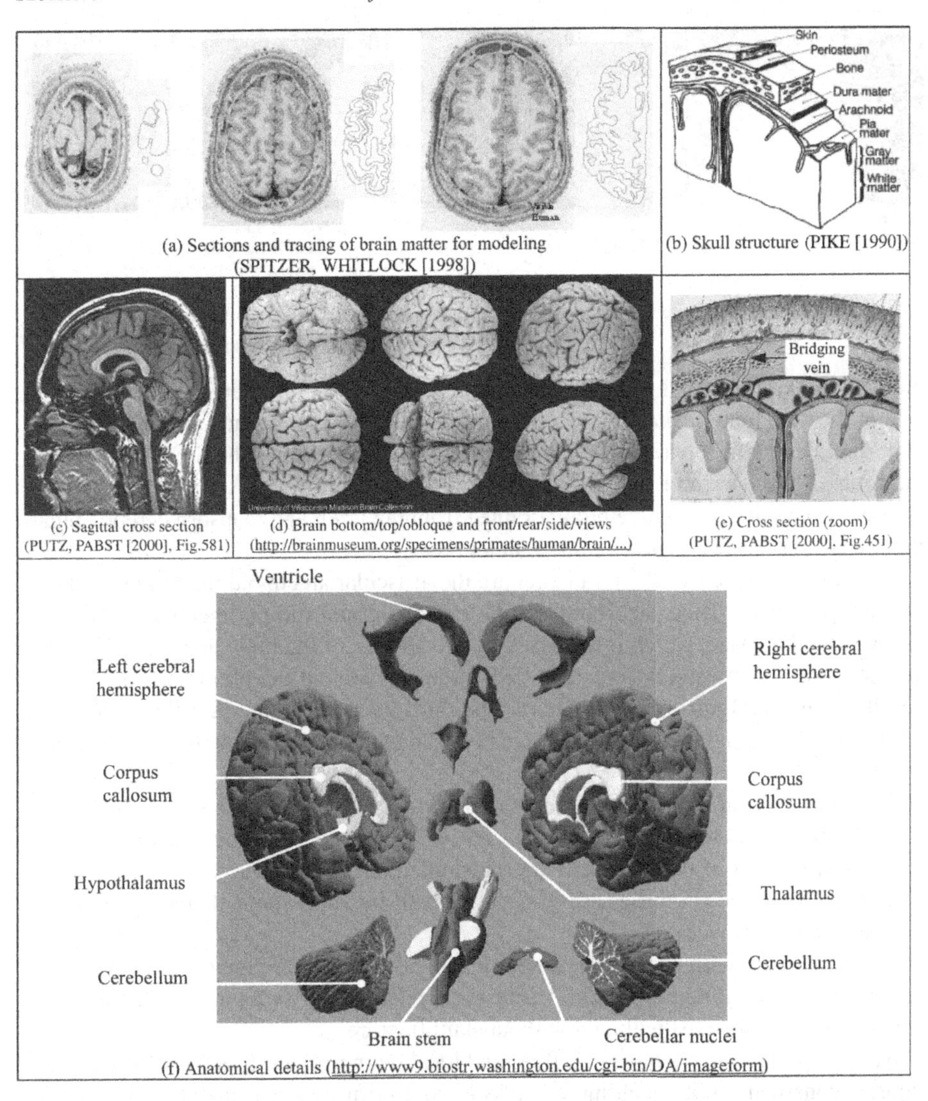

FIG. 9.8. Anatomy of the human head (overview). (Inset (a): Reproduced by permission of Jones & Bartlett Publishers Inc.; Inset (b): Reprinted with permission from "Automotove Safety" ©1990 SAE International; Insets (c) and (e): Reproduced by permission of Urban & Fischer Verlag; Inset (d): Reproduced by permission of Wally Welker at The Department of Neurophysiology, The University of Wisconsin; Inset (f): Graphic by John Sundsten, courtesy of the Structural Informatics Group at the University of Washington.)

Intra-cranial structure. The intra-cranial structure contains mainly the brain, the cerebrospinal fluid (CSF), the vascular structure, and the membranous coverings (dura mater, pia mater) and partition structures, Fig. 9.8(c)–(f) (PUTZ and PABST [2000] and web sites http://brainmuseum.org/specimens/primates/human/brain/human8sect6.jpg and http://www9.biostr.washington.edu/cgi-bin/DA/imageform). The brain occupies

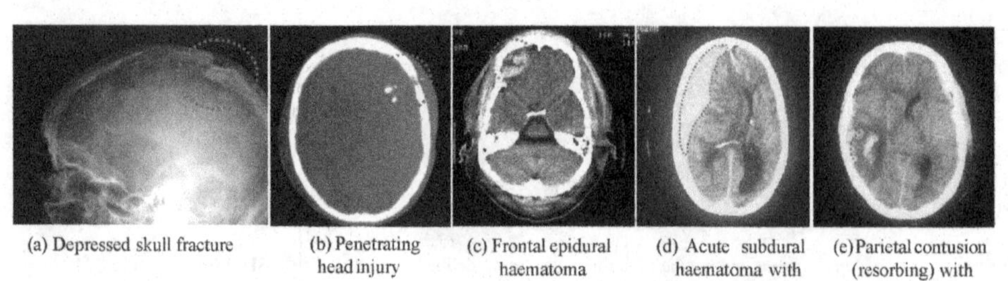

(a) Depressed skull fracture (b) Penetrating head injury (c) Frontal epidural haematoma (d) Acute subdural haematoma with midline shift (e) Parietal contusion (resorbing) with midline shift

FIG. 9.9. Brain injuries (http://www.trauma.org/imagebank/imagebank.html). (Reproduced by permission of Trauma.Org.)

most of the cranium and is composed of right and left hemispheres, separated by a fold of the dura mater called the falx cerebri. Another fold of the dura mater, called the tentorium cerebelli, separates the cerebrum and the cerebellum. The junction between the folds of the dura and the inner surface of the cranium forms some of the venous sinuses. These sinuses receive blood drained from the brain and reabsorb cerebrospinal fluid in regions knows as arachnoid granulations.

The next layers below the dura mater are the avascular arachnoid membrane and the delicate pia mater. The space between the arachnoid and the pia mater is called the subarachnoidal space, which is traversed by the arteries of the brain and cranial nerves and contains cerebrospinal fluid. The CSF circulates around the brain and spinal cord and through four ventricles (two lateral, the third ventricle at the mid-line below the lateral ventricles, and the fourth ventricle between the brain stem and the cerebellum).

Head injury. Head injuries can be caused by external or internal loads. External loads result from impacts with objects or obstacles, while internal dynamic loads result from motions of the rest of the body and are transmitted to the head through the neck (e.g., whiplash). Major head injuries are considered skull fractures (bone damage) and brain injuries (neural damage).

Skull fractures with increasing severity can be linear, comminuted (fragmented), depressed and basal fractures. Skull fractures correlate weakly with neural damage, i.e., neural damage may exist with and without skull fracture.

Neural damage can result from diffuse and focal brain injuries, Fig. 9.9. Diffuse brain injuries consist in brain swelling, concussion and diffuse axonal injury (DAI), while focal brain injuries lead to subdural hematomas (SDH), epidural hematomas (EDH), intra-cerebral hematomas (ICH), and contusions (from coup and countercoup pressure). For example, upon high angular head acceleration associated with an impact, the bridging veins at the top of the brain can rupture, causing a subdural hematoma or blood clot to develop. The clot exerts large pressures on the brain and if not quickly relieved can deprive the brain of its blood supply, causing brain death. The stretch in some of the bridging veins can reach 200%! (according to Viano: "humans are made to fall from trees").

Model structure and calibration. The detailed FE head model (H-Head) of the 50th percentile H-Model consists of deformable bone and brain components, Fig. 9.10. The modeling of the head comprises the scalp (can attenuate blunt impacts), the skull and

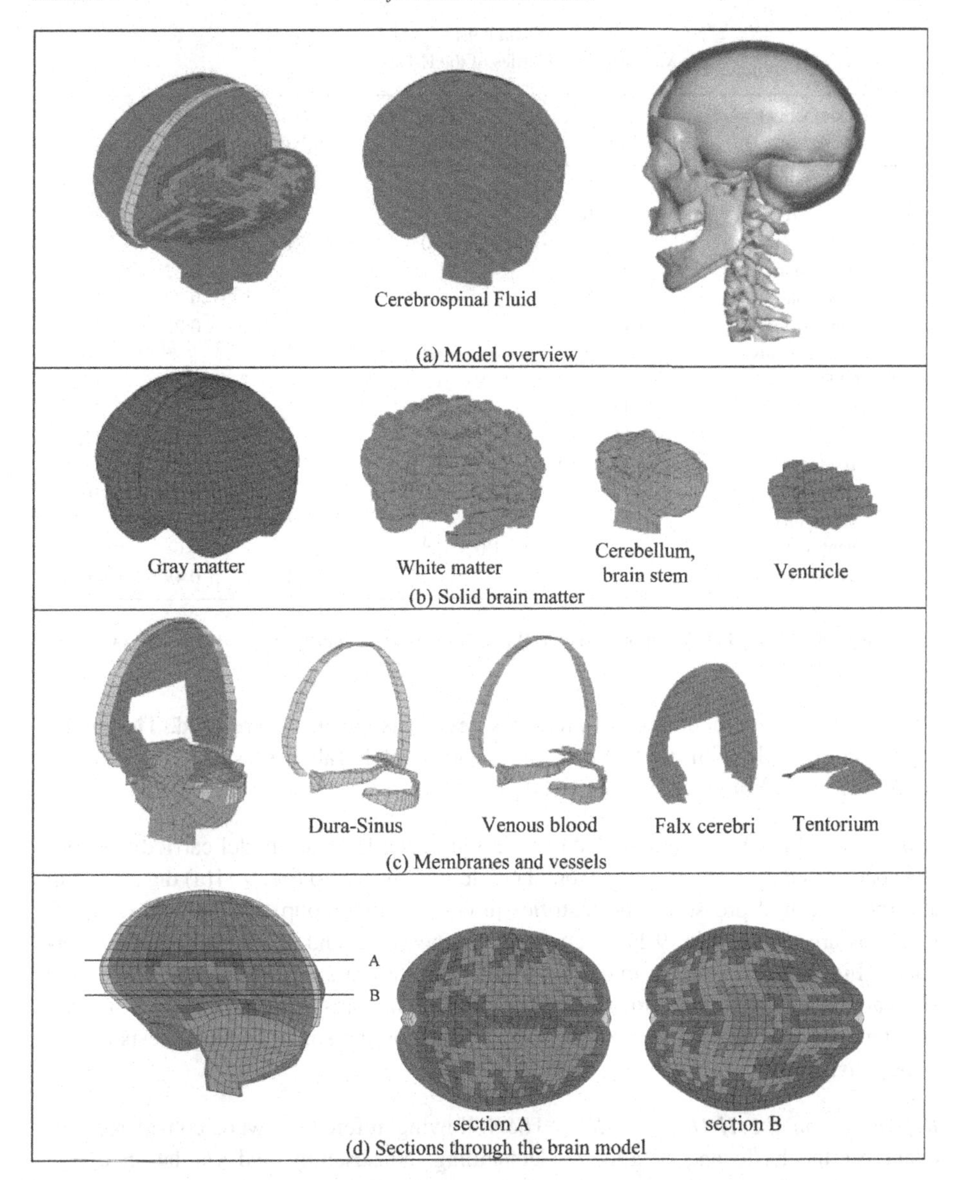

FIG. 9.10. H-Head model overview.

facial bone (the latter can be held more approximate), the cerebro-spinal fluid (CSF) layers and the ventricles, the brain membranes (dura mater), the falx, the tentorium, the gray and the white brain matter and the cerebellum. The CSF is presently modeled with the Murnaghan equation of state, $p = p_0 + B((\rho/\rho_0)^\gamma - 1)$, where p is the hydrostatic pressure, ρ is the mass density and B and γ are material constants. For the modeling of the vacuum cavity that might develop in the CSF at the interface between the skull and

TABLE 9.1
Mechanical properties of the H-Head model

	Component	E [kPa]	K [kPa]	G [kPa]	ν	ρ [kg/m^3]
Skull	Outer table	7.3×10^6			0.22	3000
	Inner table	7.3×10^6			0.22	3000
	Diploe		2.02×10^5	1.39×10^5	0.22	1410
	Facial bone	7.3×10^6			0.22	2700
	Mandible	7.3×10^6			0.22	2700
	Dura mater	3.15×10^4			0.22	1133
	Venous sinus		1.0×10^5			1000
	CSF		1.0×10^5			1000
	Falx	3.15×10^4			0.45	1133
	Pia	3.15×10^4			0.45	1133
	Tentorium	3.15×10^4			0.45	1133
Brain	Gray matter		7.96×10^3	7.96×10^1	~ 0.499	1040
	White matter		1.27×10^4	1.27×10^2	~ 0.499	1040
	Ventricle		1.0×10^5			1000
	Cerebellum and stem		1.27×10^4	1.27×10^2	~ 0.499	1040

E = Young's modulus [kPa], K = Bulk modulus [kPa], G = Shear modulus [kPa], ν = Poisson's ratio, ρ = Mass density [kg/m^3].

brain from negative countercoup pressures, ideal gas equations are used. The model is fully compatible with the H-Neck spinal cord model. Table 9.1 shows the material coefficients as calibrated for the H-Head model.

Validation. Fig. 9.11 summarizes a validation of the H-Head model carried out after tests performed by Nahum, using frontal pendulum impact. In Fig. 9.11(a) the test setup and the measured pressure time histories in coup, countercoup, parietal and occipital locations are shown. Fig. 9.11(b) contains the measured and calculated head accelerations. Fig. 9.11(c) gives an overview on the surface pressures as calculated with the H-Head model at different times. In Fig. 9.11(d)–(f) the pressure time histories in coup, countercoup and occipital locations are shown and comparison to Nahum's tests is seen to be satisfactory.

References on the H-Head model. The following references were considered relevant for the basic and detailed understanding, construction and validation of the head model: ABEL, GENNARELLI and SEGAWA [1978] on incidence and severity of cerebral concussion in rhesus monkeys from sagittal acceleration; BANDAK and EPPINGER [1994], BANDAK [1996] on brain FE models and impact traumatic brain injury; CHAPON, VERRIEST, DEDOYAN, TRAUCHESSEC and ARTRU [1983] on brain vulnerability from real accidents; CHOI and LEE [1999b] on deformable FE models of the human body; CLAESSENS [1997], CLAESSENS, SAUREN and WISMANS [1997] on FE modeling of the human head under impact conditions; COOPER [1982a], COOPER [1982b] on injury of the skull, brain and cerebro-spinal fluid related; DIMASI, MARCUS and EPPINGER [1991] on a 3D anatomical brain model for automobile crash loading;

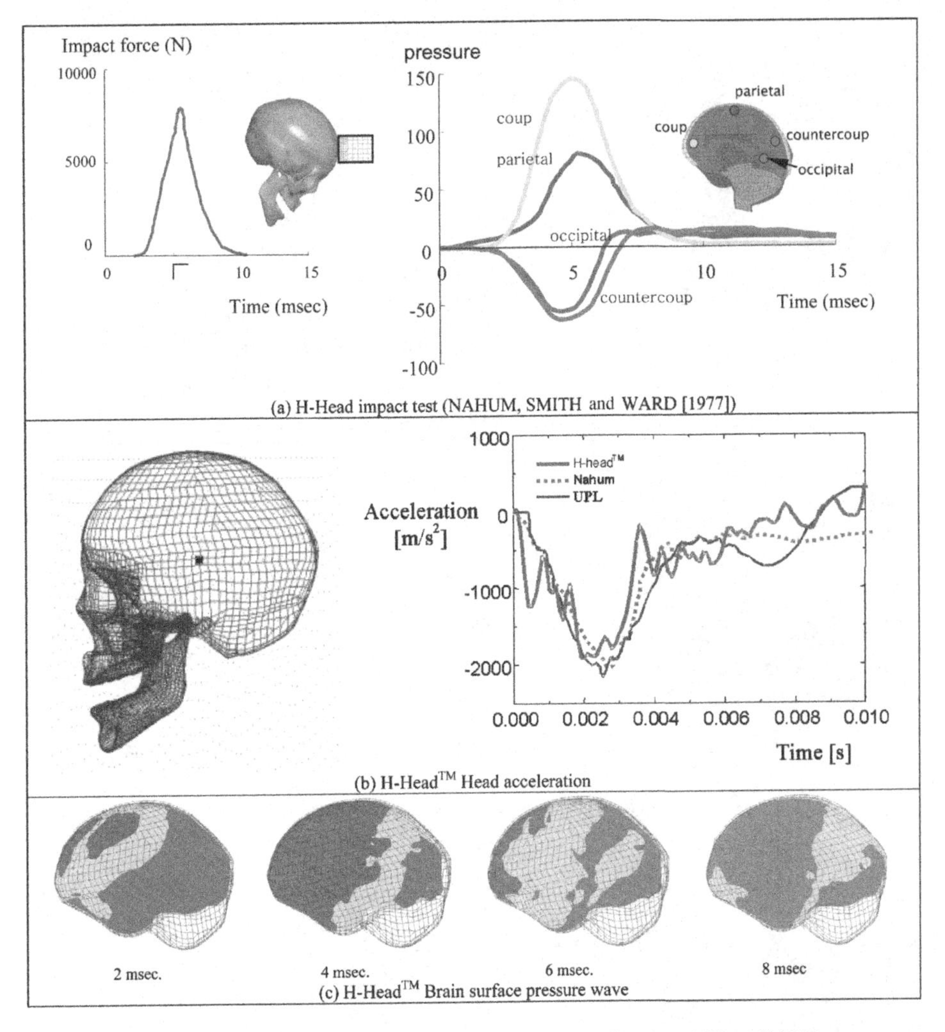

FIG. 9.11. H-Head model validations (tests: after NAHUM, SMITH and WARD [1977]).

DONNELLY and MEDIGE [1997] on shear properties of human brain tissues; EWING, THOMAS, LUSTICK, MUZZY III, WILLEMS and MAJEWSKI [1978] on the effect of the initial position on the head/neck response in sled tests; GENNARELLI [1980], GENNARELLI, THIBAULT, ADAMS, GRAHAM, THOMPSON and MARCINCIN [1982] on the analysis of head injury severity by AIS-80 (1980) and on diffuse axonal injury and traumatic coma in primates (1982); GRAY'S ANATOMY [1989]: Atlas of Anatomy; GURDJIAN and LISSNER [1944], GURDJIAN, WEBSTER and LISSNER [1955], GURDJIAN, ROBERTS and THOMAS [1966] on mechanisms of head injury and brain concussion and tolerance of acceleration and intra-cranial pressure; HOLBOURN [1943] on the mechanics

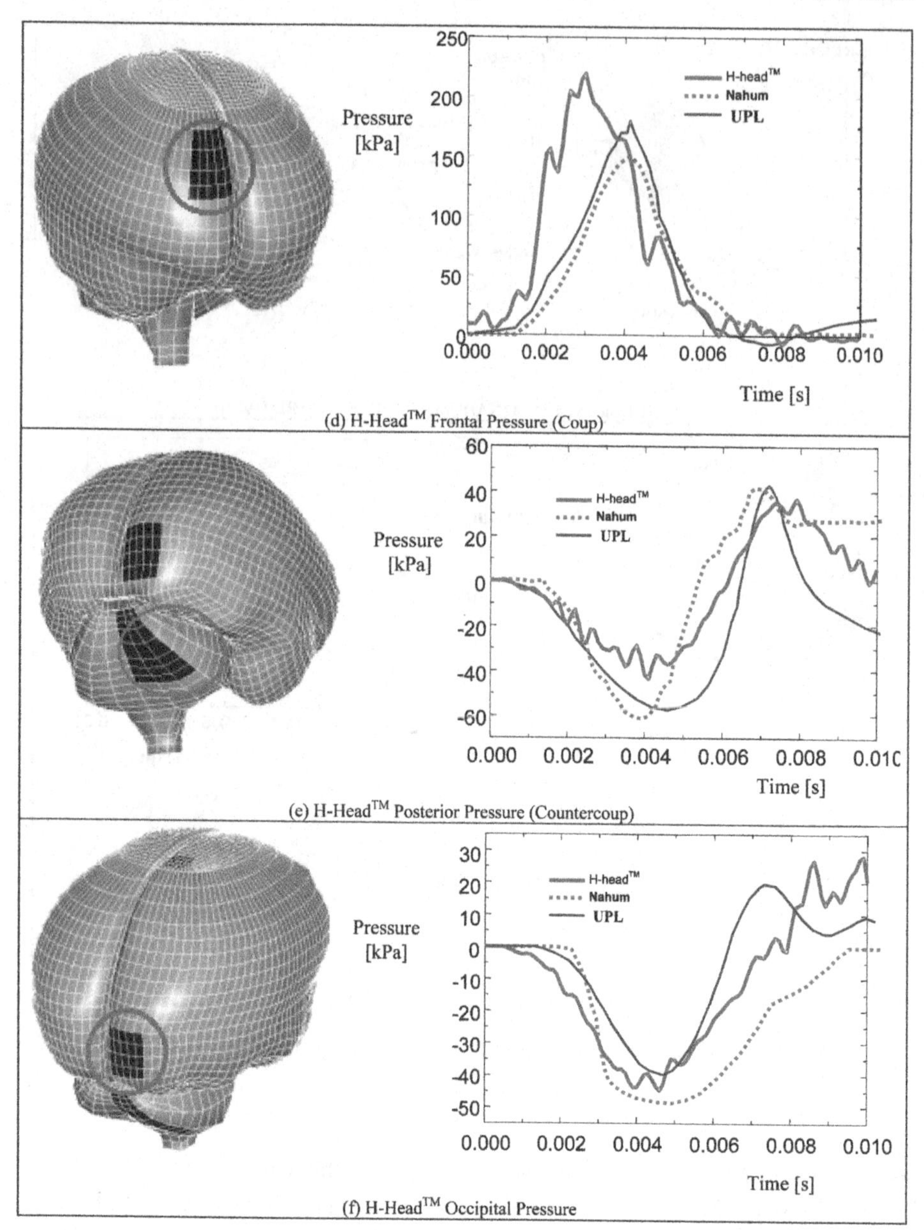

FIG. 9.11. (*Continued.*)

of head injury; KANG, WILLINGER, DIAW and CHINN [1997] on the validation of a 3D FE model of the human head in motorcycle accidents; LEE, MELVIN and UENO [1987], LEE and HAUT [1989] on FE analysis of subdural hematoma and bridging vein failure characteristics; LISSNER, LEBOW and EVANS [1960] on experiments on intra-cranial pressure; MARGULIES, THIBAULT and GENNARELLI [1990] on modeling brain injury in primates; MILLER and CHINZEI [1997] on constitutive modeling of brain tissue; NAHUM, SMITH and WARD [1977] on intra-cranial pressure dynamics during head impact; NEWMAN [1993] on head protection; OMMAYA, HIRSCH, FLAMM and MAHONE [1966], OMMAYA and HIRSCH [1971], OMMAYA and GENNARELLI [1974] on cerebral concussions in the monkey (1966), on their tolerances (1971) and on their clinical/experimental correlation (1974); ONO, KIKUCHI, NAKAMURA, KOBAYASHI and NAKAMURA [1980], ONO [1999] on head injury tolerance for sagittal impact from tests (1980) and on spine deformation and on vertebral motion from whiplash test volunteers (1999); PENN and CLASEN [1982] on traumatic brain swelling and edema; PUTZ and PABST [2000]: Sobotta Atlas of Human Anatomy; RUAN, KHALIL and KING [1991] on human FE head model response in side impacts; RUAN and PRASAD [1994] on head injury assessment in frontal impacts by mathematical modeling; SANCES ET AL. [1982] on head and spine injuries; SCOTT [1981] on the epidemiology of motor cyclist head and neck trauma; SPITZER and WHITLOCK [1998]: Atlas of the Visible Human Male; TARRIERE [1981] on investigation of the brain with CT-scanners; TORG [1982], TORG and PAVLOV [1991] on athletic injury on the head, neck and face; TURQUIER, KANG, TROSSEILLE, WILLINGER, LAVASTE, TARRIERE and DÖMONT [1996] on the validation of a 3D FE head model against experiments; UENO, MELVIN, LUNDQUIST and LEE [1989] on 2D FE analysis of human brain under impact; VOO, KUMARESAN, PINTAR, YOGANANDAN and SANCES [1996] on a finite element model of the human head; WALKE, KOLLROS and CASE [1944] on the physiological basis of concussion; WILLINGER, KANG and DIAW [1999] on the validation of a 3D FE human model against experimental impacts; WISMANS ET AL. [1994] on injury biomechanics; ZEIDLER, STÜRTZ, BURG and RAU [1981] on injury mechanisms in head-on collisions; ZHOU, KHALIL and KING [1996] on the visco-elastic brain FE modeling for sagittal and lateral rotation acceleration. The following sites were used: http://www9.biostr.washington.edu/cgi-bin/DA/imageform; http://brainmuseum.org/specimens/primates/human/brain/human8sect6.jpg; http://www.trauma.org/imagebank/imagebank.html.

9.4. H-Neck: Cervical spine with active muscles

Overviews on the H-Neck model structure and applications are given in CHOI and EOM [1998], LEE and CHOI [2000] and CHOI, LEE and HAUG [2001b]. A study on the whiplash injury due to the low velocity rear-end collision is presented in CHOI, LEE, EOM and LEE [1999].

Anatomy. The anatomy of the cervical spine is summarized in Fig. 9.12 (after http://www.rad.washington.edu/RadAnat/Cspine.html and PUTZ and PABST [2000]). The human cervical spine is composed of 7 vertebrae, C1–C7, with five similar lower vertebrae (C3–C7) and two dissimilar upper vertebrae (C1 "atlas" and C2 "axis"). Approximately 47% of the bending and stretching of the cervical spine occurs between the head and C1, while over 50% of the rotational motions occur between C1 and C2. Between each pair of adjacent vertebrae except for between C1 and C2 exist a structure known as the disc of cervical spine. Each disc is composed of the nucleus pulposus,

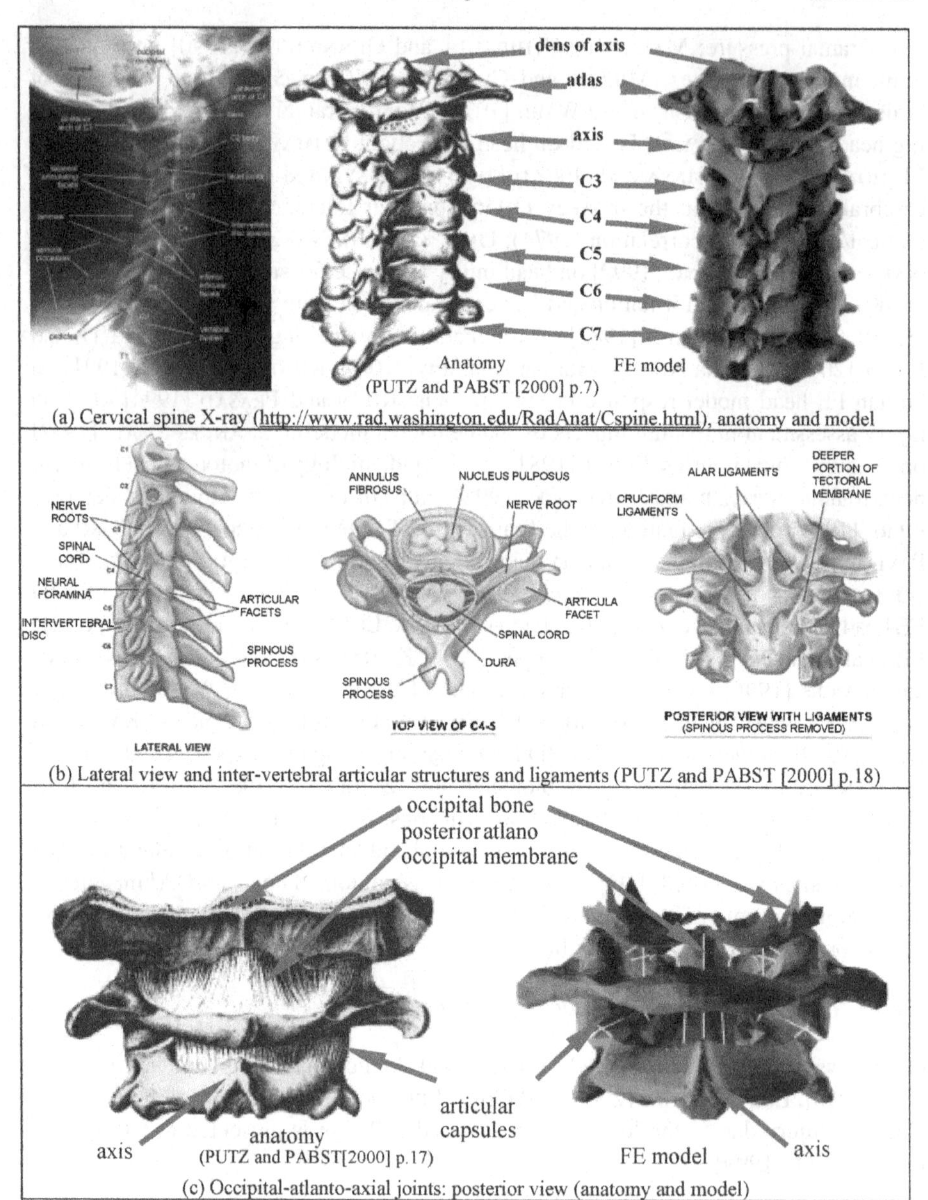

FIG. 9.12. Anatomy of the human cervical spine. (Inset (a) left: Reproduced by permission of Michael L. Richardson, University of Washington, Deparment of Radiology; Insets (a) center, (b) and (c) left: Reproduced by permission of Urban & Fischer Verlag.)

the annulus fibrosus and a cartilageous end-plate. Between 70 to 90% of the nucleus pulposus by weight is water, and takes up as much as 40 to 60% of the disc area.

The annulus fibrosus is a laminated and hence anisotropic structure composed of several layers in which each layer maintains a 30° angle of inclination from the horizontal plane. The inner boundary of the annulus fibrosus is attached to the cartilageous end-plate, and the outer surface is directly connected to the vertebra body. The discs play a dominant role under compressive loads.

The discs show greater stiffness for the front/rearward motion than for the side/side motion. Spinal ligaments (anterior, posterior, ...) connect the vertebrae of the cervical spine. They ensure that the spinal motions occur within physiological bounds, thus preventing spinal code injury due to excessive spinal motion. The ligaments are usually situated between two adjacent vertebrae, and in some cases connects several vertebrae.

Neck injury. Neck injury from car accidents has been studied extensively and is summarized hereafter according to McELHANEY and MYERS [1993] (eds. Nahum and Melvin). Fig. 9.13, inset (a), summarizes the anatomical head motions. Inset (b) shows the modes of loading on the neck. Inset (c) lists 10 distinct injury mechanisms and identifies a total of 25 different neck injuries. Inset (d) shows three tension-extension injury mechanisms, including the important whiplash injury (B). Inset (e) shows three flexion-compression neck injury mechanisms.

For example, the tension–extension injury (A) of inset (d) to Fig. 9.13 is described as "*Fixation of the head with continued forward displacement of the body. This occurs commonly in unbelted occupants hitting the windshield, and as a result of falls and dives*". The tension–extension whiplash injury (B) is described as "*Inertial loading of the neck following an abrupt forward acceleration of the torso as would occur in a rear-end collision*". The tension–extension injury (C) is described as "*Forceful loading below the chin directed postero-superiorly (as in a judicial hanging)*.

The frequent whiplash injury occurs as a result of more or less mild rear-end collisions. It is considered a hyperextension injury and it may cause muscle stiffness and neck pain. Larger accelerations, in addition to producing whiplash symptoms, may produce disruption of the anterior longitudinal ligament and intervertebral disk, and horizontal fractures though the vertebra. The remaining neck injuries are described in the above reference. The survivable neck injuries from car accidents are often injuries of the soft connecting tissues, while bone fractures are less common.

Model structure and calibration. The modeling aspects of the human cervical spine are summarized in Figs. 9.14 and 9.15. The H-Neck model is fully compatible and interfaced with the H-Head model. Based on 3D CAD data of 50% male cervical vertebrae supplied by DIGIMATION (DIGIMATION/VIEWPOINT CATALOG [2002]), the cervical spine in the current H-Neck model is built as a chain of articulated rigid vertebrae. This reduces computational cost, all the while soft tissue injury plays a more important role than bony fracture does in neck injury. Fully deformable vertebra bodies can easily be modeled, for example, as demonstrated earlier by NITSCHE, HAUG and KISIELEWICZ [1996].

The H-Neck model contains important elements such as vertebrae, discs, ligaments, and stabilizing muscles. The inter-vertebral contacts are modeled with sliding interfaces and contact stiffnesses. They can also be modeled by appropriate layers of finite elements. The vertebrae of the H-Neck model are interconnected by the neck ligaments,

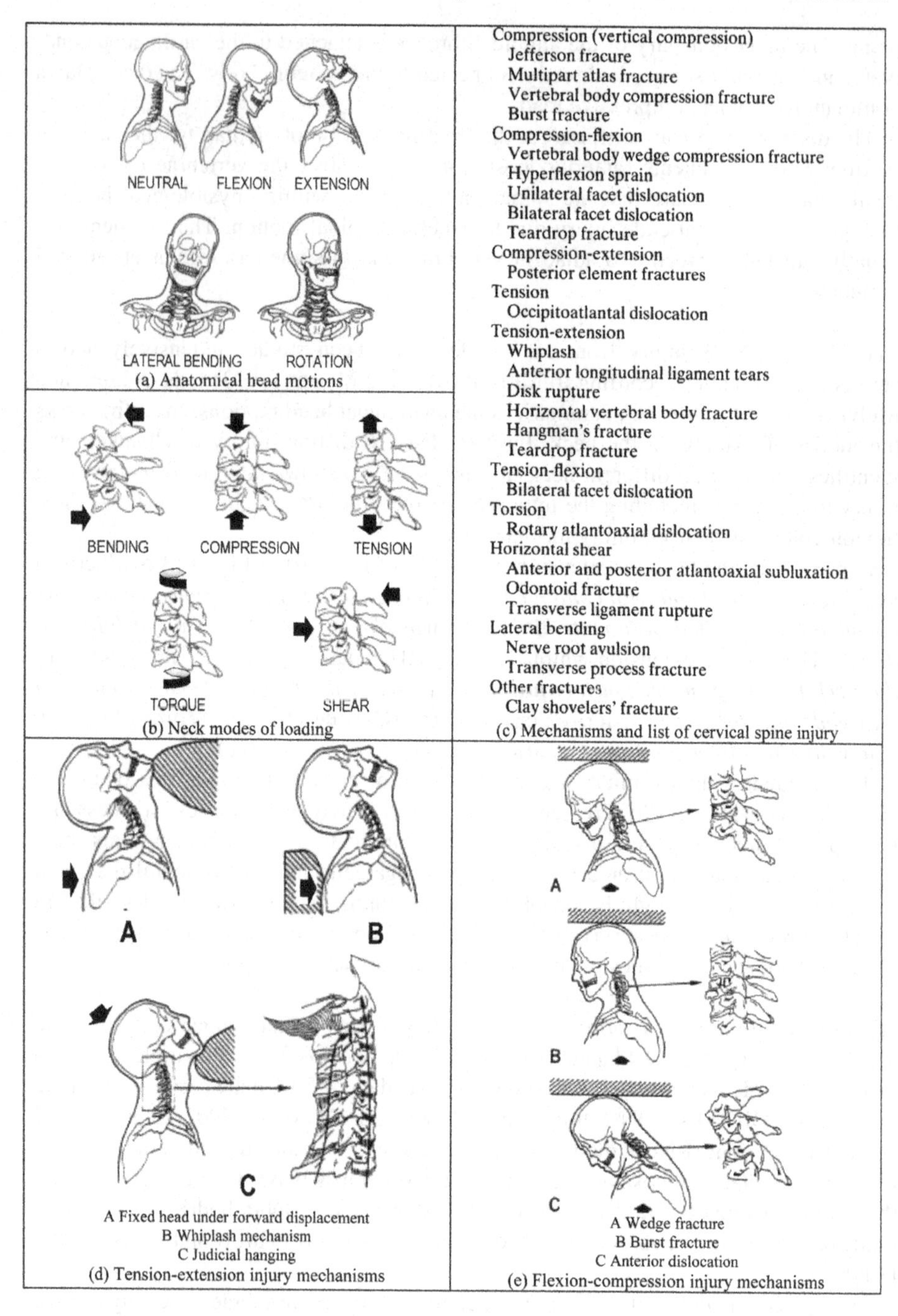

FIG. 9.13. Neck injury overview (after MCELHANEY and MYERS [1993]). (Reproduced by permission of Springer Verlag from: Nahum and Melvin eds., Accidental Injury, Biomechanics and Prevention: Insets (a) and (b): Figure 14.5, p. 318; Inset (c): Table 14.5, p. 319; Inset (d): Figure 14.10, p. 324; Inset (e): Figure 14.8, p. 320.)

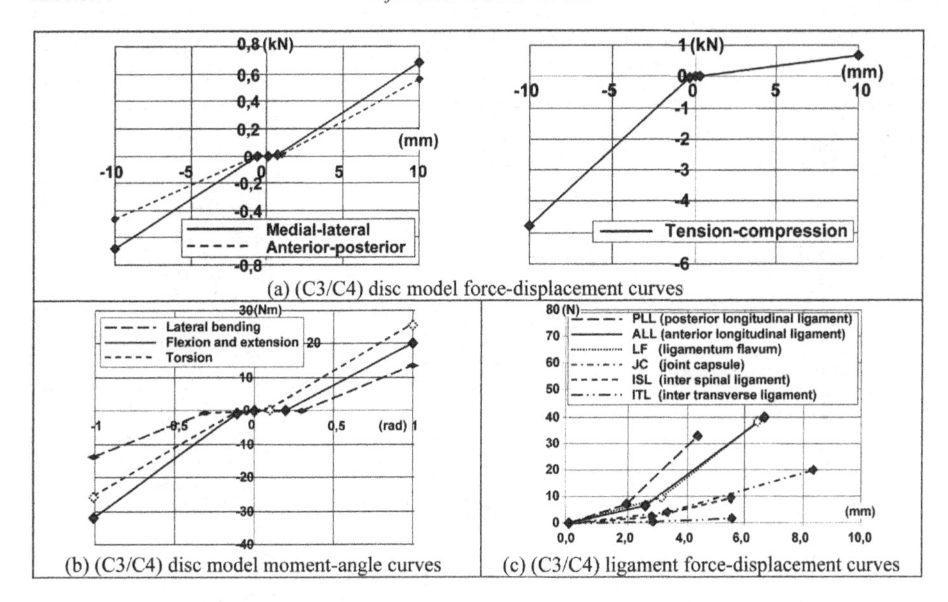

(a) (C3/C4) disc model force-displacement curves

(b) (C3/C4) disc model moment-angle curves

(c) (C3/C4) ligament force-displacement curves

FIG. 9.14. (C3–C4) disc 6 DOF kinematic joint element and ligament calibration (H-Neck).

modeled as bars or membranes. To account for the visco-elastic and anisotropic biomechanical properties of the ligaments, nonlinear one-dimensional tension-only bar elements are used. For cruciform, transverse, and tectorial ligaments in the atlanto-axial joint however, membrane elements are used to allow for surface contact (see also Fig. 9.12(c)). Each pair of adjacent vertebrae (except C1 and C2) is connected by the elastic discs of the cervical spine.

In the H-Neck model, the complex deformation of the discs is modeled by 6-DOF nonlinear spring elements with stiffness calibrated as shown in Fig. 9.14(a), (b). The ligament force-displacement calibrations are shown in Fig. 9.14(c).

These calibrations were made with test results found in the literature, for example: CHAZAL, TANGUY and BOURGES [1985] on biomechanical properties of the spinal ligaments; MCCLURE, SIEGLER and NOBILINI [1998] on in vivo 3D flexibility properties of the human cervical spine; MORONEY, SCHULTZ, MILLER and ANDERSSON [1988], MORONEY, SCHULTZ and MILLER [1988] on lower cervical spine properties and neck loads; MYKLEBUST and PINTAR [1988] on the tensile strength of spinal ligaments; NIGHTINGALE, WINKELSTEIN, KNAUB, RICHARDSON, LUCK and MYERS [2002] on upper/lower cervical spine strength comparison in flexion/extension; PANJABI, CRISCO, VASAVADA, ODA, CHOLEWICKI, NIBU and SHIN [2001] on cervical spine load-displacement curves; YOGANANDAN, SRIRANGAM and PINTAR [2001] on the biomechanics of the cervical spine.

Fig. 9.15(a) shows the anatomical details (PUTZ and PABST [2000]), the H-Neck ligaments and the assembled H-Head/H-Neck model, where the head can be modeled effectively as a rigid body for most whiplash simulations. The muscle forces obey the well known active and passive Hill muscle law, Fig. 9.15(b). The muscles are modeled by nonlinear bar elements, which can be curved in space, assuring a correct introduction

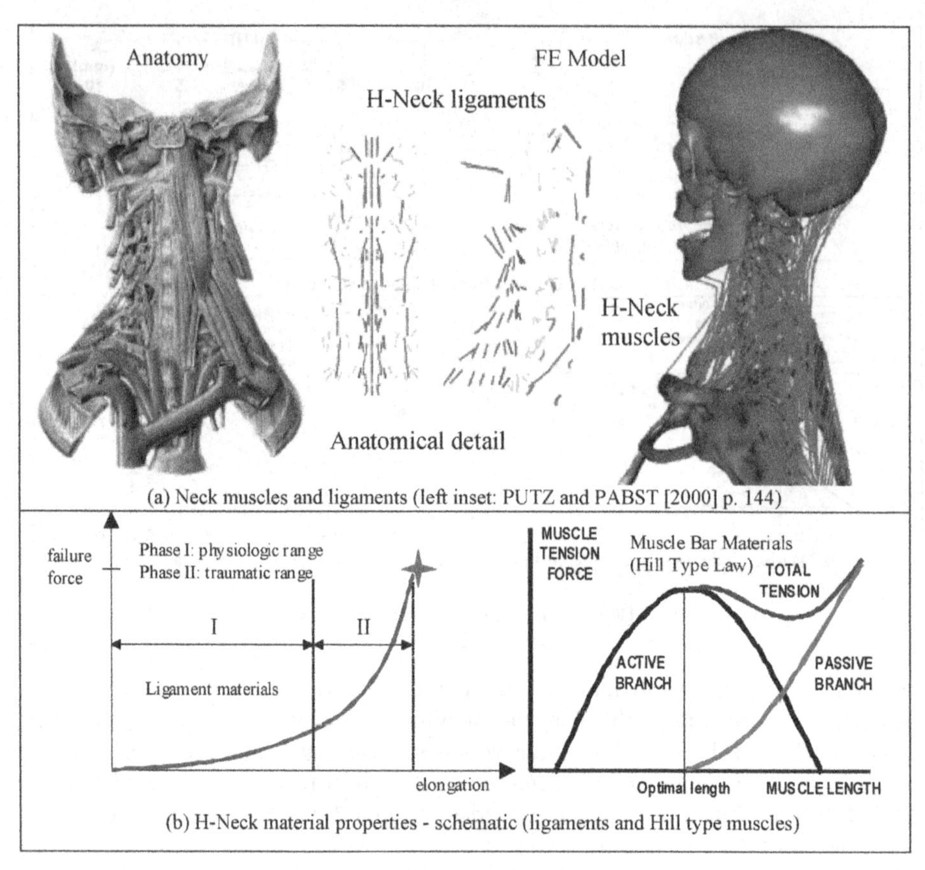

FIG. 9.15. H-Neck model overview. (Inset (a) left: Reproduced by permission of Urban & Fischer Verlag.)

of the muscle forces along the curved neck line. The active muscle force can contribute to conscious bracing and motion. The passive muscle forces contribute to the resistance of the connected body parts near the end of their respective anatomical motion ranges. It increases with muscle length in a highly nonlinear fashion (cf. Appendix C).

Validation. Fig. 9.16 shows in inset (a) the flexion response of the H-Neck model in a frontal passenger car impact simulation. In inset (b) the extension response in whiplash for a rear impact is shown. Inset (c) shows pictures of the deformed neck in lateral bending as from a side impact. The muscle bars are seen to follow the neck as it flexes and extends.

Fig. 9.17 contains several results of the H-Head/H-Neck response to a frontal car crash event, as compared to the response corridors obtained with volunteers in live tests (after THUNNISSEN, WISMANS, EWING and THOMAS [1995] and HORST, THUNNISSEN, HAPPEE, HAASTER and WISMANS [1997]).

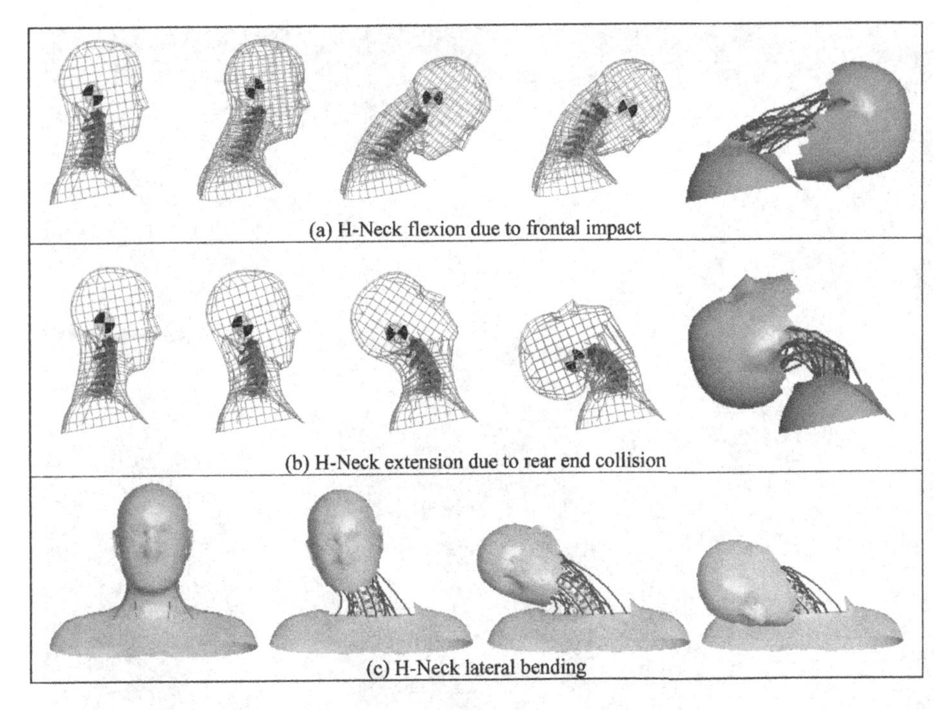

(a) H-Neck flexion due to frontal impact
(b) H-Neck extension due to rear end collision
(c) H-Neck lateral bending

FIG. 9.16. H-Neck model validations: Flexion, extension and torsion motions.

Fig. 9.18 shows the trajectories of several points of the H-Neck model in extension from rear impacts (whiplash). The results compare well with corridors from tests performed with life volunteers (ONO, KANEOKA, WITTEK and KAJZER [1997]).

Fig. 9.19(a)–(c) compare the responses of the H-Neck model mounted on the H-ARB version in a rear impact sled test simulation, both, with and without a headrest. Fig. 9.19(d) compares the H-Head/H-Neck response to a rear impact with a headrest in a low and a high position. Fig. 9.20, finally, shows the head and the cervical spine skeleton response with a deformable headrest to such a rear impact. The results demonstrate the utility of such models to study the effects of preventive measures in the design of passive safety restraints of passenger cars.

JARI volunteer tests. Recently the H-Neck™ model is being validated against volunteer neck whiplash tests, Fig. 9.21, in which human volunteers underwent equivalent mild frontal and rear head accelerations in the sagittal plane, as they are typical in mild frontal and rear impacts of passenger cars, ONO, KANEOKA, SUN, TAKHOUNTS and EPPINGER [2001]. While the evaluations of the tests are still ongoing, some simulations of the scenarios have been made with an attempt to study the influence of voluntary and involuntary contraction of the neck muscles which counteract the accidental neck/head motions in the sagittal plane, Fig. 9.21(a). The neck muscles are divided into groups of neck flexors and neck extensors in the sagittal plane, insets (b) and (c), which are activated separately in a parametric study in different assumed reflex time scenarios to their

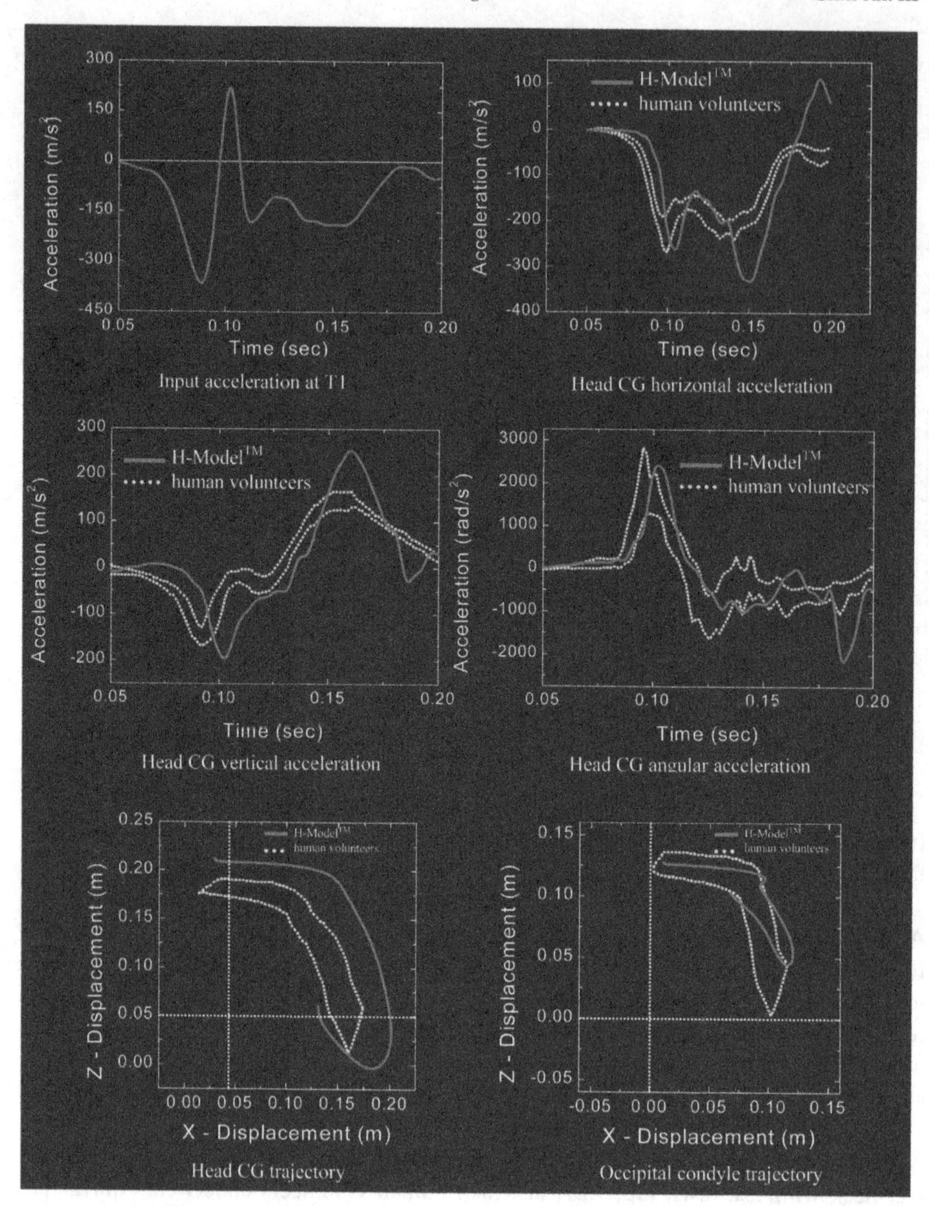

FIG. 9.17. H-Neck model validations: Volunteer tests in frontal crash (flexion). (Test results after Stapp papers THUNNISSEN, WISMANS, EWING and THOMAS [1995] and HORST, THUNNISSEN, HAPPEE, HAASTER and WISMANS [1997].)

highest level of activation (100%). The flexor group pulls the head forward (agonists) and the extensor group backward (antagonists).

For example, the subject's head is being suddenly pulled rearward with the help of a string attached to a mask, which contains accelerometers and optical gauges, as

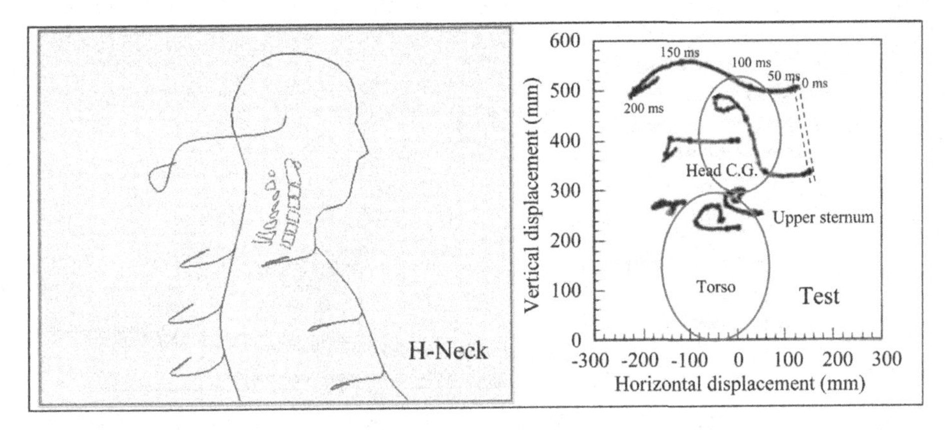

FIG. 9.18. H-Neck model validations: Trajectories in extension for a rear impact (whiplash). (Right inset: Reproduced by permission of The Stapp Association.)

shown in inset (a). The volunteer is either asked to relax, or to fully stiffen its neck muscles prior to the sudden pull. An intermediate case is when the subject involuntarily stiffens its neck muscles during the event (stretch reflex). In inset (d) deformed shape snapshots of a simulation of a relaxed subject with no activation of the active muscle forces are shown. In (e) the neck flexor muscles are assumed to be fully tensed, starting at 100 ms after the pull, followed by the extensor muscles, starting at 150 ms. As expected, the maximal neck flexion occurs for the relaxed subject, (d), with trajectories, (f), and this motion is reduced for the subject with the (arbitrarily) assumed pattern of full muscle activation, (g), (h). Since the exact pattern of involuntary muscle activation is unknown, such activation patterns can be found by back calculation from various assumed simple activation patterns and comparison with volunteer tests. Pattern (i) was not a good assumption, since the late activation of the flexors allowed for large neck flexion motions.

The active muscle forces were built up and applied according to Hill's law, using the physiological cross section area of each muscle. Although it is presently unknown in precisely what pattern a living subject activates the neck muscles in a whiplash event, such parametric studies can help clarify the open questions. Light can also be shed on the different loads and injuries of the cervical spine, which are influenced by the muscle actions.

References on the H-Neck model. The following references were considered useful for the preparation of the H-Neck model: BOKDUK and YOGANANDAN [2001] on minor injuries of the cervical spine; CUSICK and YOGANANDAN [2002] on major injuries of the cervical spine; DOHERTY, ESSES and HEGGENESS [1992] on odontoid fracture; DVORAK, HAYEK and ZEHNUDER [1987], DVORAK, PANJABI and FROEHLICH [1988] on rotary instability of the upper cervical spine; FOREMAN and CROFT [1995] (whiplash injuries); GRAY'S ANATOMY [1989]: Atlas of Anatomy; HIRSCH [1955] (disc compression reaction); HORST, THUNNISSEN, HAPPEE, HAASTER and WISMANS [1997] on the influence of muscle activity on head/neck impact response; HUELKE, NUSHOLTZ and KAIKER [1986] on an overview on cervical fracture and dislocation; JAGER [1996] on the mathematical

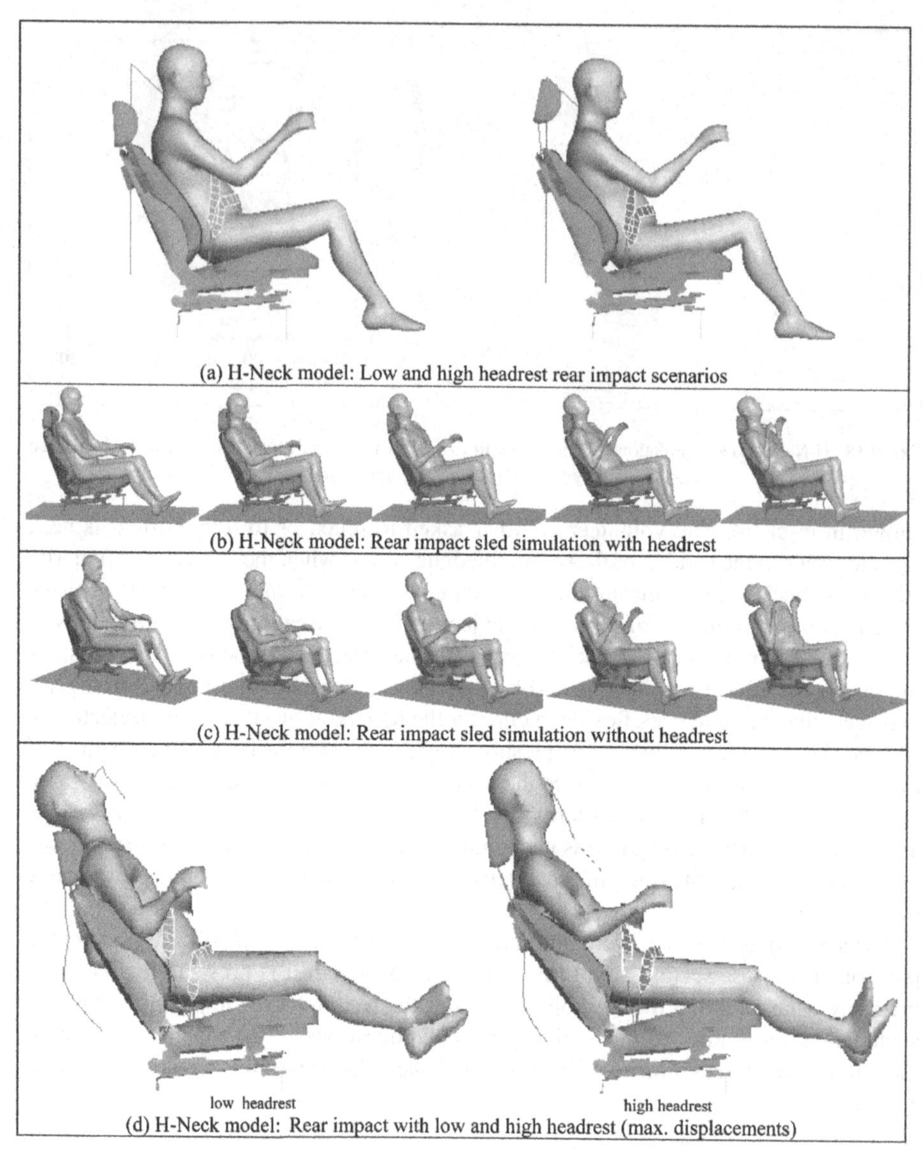

(a) H-Neck model: Low and high headrest rear impact scenarios
(b) H-Neck model: Rear impact sled simulation with headrest
(c) H-Neck model: Rear impact sled simulation without headrest
low headrest
high headrest
(d) H-Neck model: Rear impact with low and high headrest (max. displacements)

FIG. 9.19. H-Neck model validations: Low and high headrest scenarios.

modeling of the cervical spine (overview); JAGER, SAUREN, THUNNISSEN and WISMANS [1994] on head/neck models; KULAK, BELYTSCHKO, SCHULTZ and GALANTE [1976] on non-linear behavior of discs under axial load; KALLIERIS and SCHMIDT [1990] on side impact neck response from cadaver tests; MARKOLF and MORRIS [1974] on components of intervertebral discs; MCELHANEY, PAVER, MCCRACKIN and MAXWELL [1983] on cervical spine compression response; MYERS, MCELHANEY, RICHARDSON, NIGHTINGALE and DOHERTY [1991], MYERS, MCELHANEY, DOHERTY, PAVER and GRAY [1991] on the influence of end conditions (a) and torsion (b) on cervical spine injury; NOVOTNY

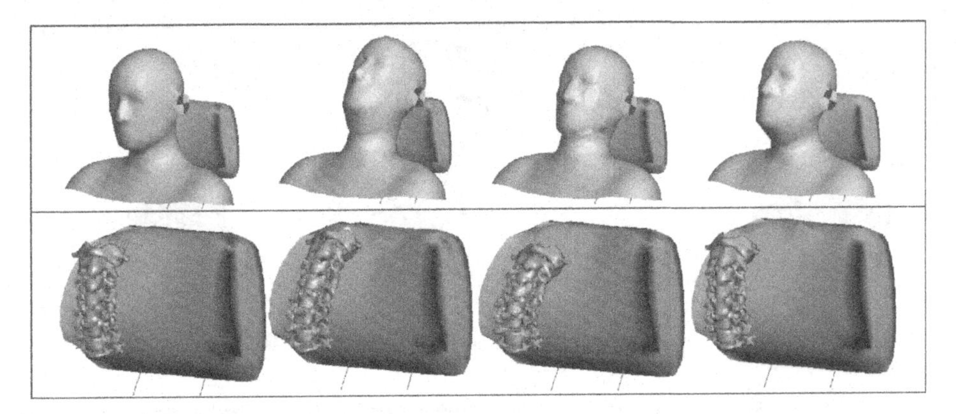

FIG. 9.20. H-Neck model validations: Rear impact whiplash response with deformable headrest.

[1993] on spinal biomechanics; ONO [1999] on spine deformation and vertebral motion from whiplash test volunteers; PANJABI, DVORAK and DURANCEAU [1988] on 3D movements of the upper cervical spine; PENNING [1979], PENNING and WILMARK [1987] on movements and rotations of the cervical spine; PUTZ and PABST [2000]: Sobotta Atlas of Human Anatomy; SONNERUP [1972] on cervical disc analysis in compression; SVENSSON and LÖVSUND [1992] on new dummy neck for rear end collision; TENNYSON and KING [1976a], TENNYSON and KING [1976b] on biodynamic model of the spinal column and measure of muscle action; THUNNISSEN, WISMANS, EWING and THOMAS [1995] on human volunteers in whiplash loading; VIRGIN and LUDHIANA [1951] on physical properties of intervertebral discs; WHITE and PANJABI [1990] on clinical biomechanics of the spine; WINTERS and STARK [1988] on muscle properties; WISMANS, VAN OORSCHOT and WOLTRING [1986] on omnidirectional human head-neck response; WITTEK and KAJZER [1995], WITTEK and KAJZER [1997], WITTEK, HAUG and KAJZER [1999], WITTEK, KAJZER and HAUG [1999], WITTEK, ONO and KAJZER [1999], WITTEK, ONO, KAJZER, ÖRTENGREN and INAMI [2001] on active and passive muscle models; YOGANANDAN, SANCES and PINTAR [1989b] on axial compression of human cadaver and manikin necks.

The following web sites were consulted: http://www.rad.washington.edu/RadAnat/Cspine.html.

9.5. *H-Torso: Rib cage, spine and thoracic organs*

An overview on the H-Torso model is given in CHOI and LEE [1999a].

Anatomy of the spine and torso. The anatomy of the human spine and thorax is summarized in Fig. 9.22(a)–(f), prepared with the web sites http://www9.biostr.washington.edu/cgi-bin/DA/imageform, (a) and (b), and http://www.nlm.nih.gov/research/visible/image/abdomen_mri.jpg, (c), of the National Library of Medicine (NLM).

The human spine consists in the flexible cervical spine (neck; seven vertebrae), the more rigid thoracic spine (twelve vertebrae with attached ribs), the flexible lumbar spine (five vertebrae), the rigid sacrum (five fused vertebrae) and the coccyx (tail bone; four

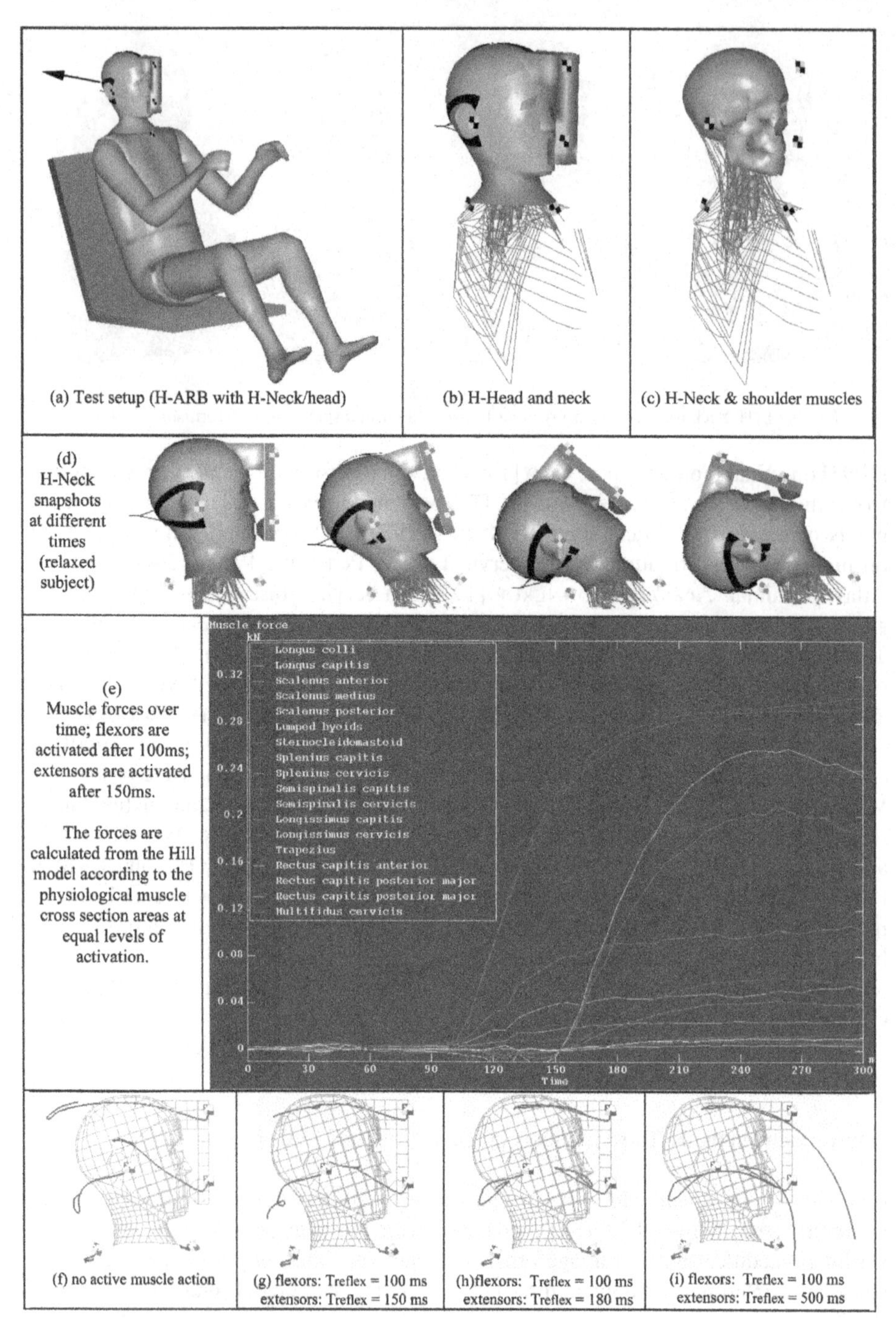

FIG. 9.21. Volunteer test simulation for sudden rearward pulling on subjects head mask (H-Neck).

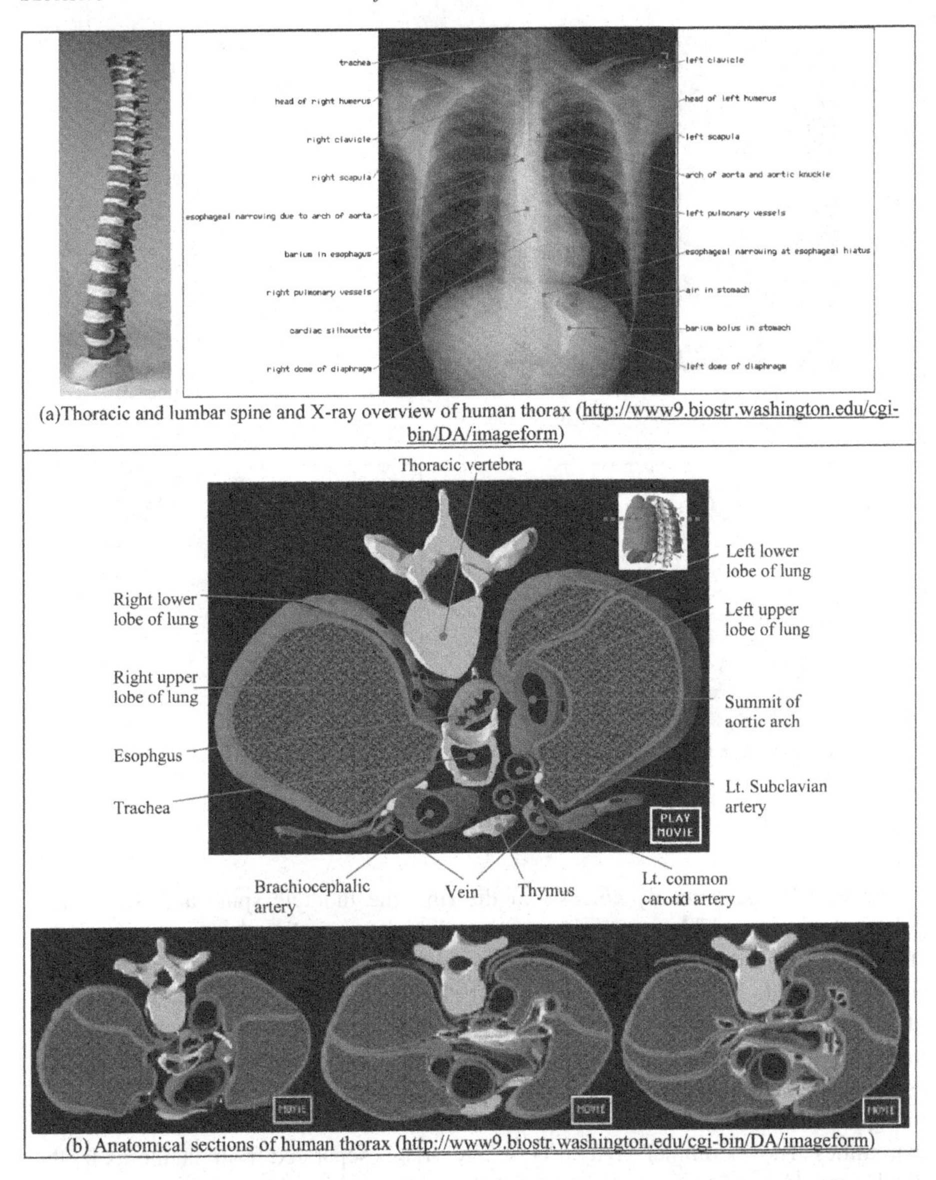

(a)Thoracic and lumbar spine and X-ray overview of human thorax (http://www9.biostr.washington.edu/cgi-bin/DA/imageform)

(b) Anatomical sections of human thorax (http://www9.biostr.washington.edu/cgi-bin/DA/imageform)

FIG. 9.22. Anatomy of the human torso. (Insets (a) right and (b): Reproduced by permission of the University of Washington Digital Anatomist; Inset (c): National Library of Medicine (NLM) public domain information.)

fused segments of bone). The spine supports the upper body, it protects the spinal chord and it provides mobility via its twenty four movable and nine fixed vertebrae. The vertebrae are hinged together by their facet joints and by the inter-vertebral discs. The facet joints, discs and ligaments of the spine limit its overall deformation in bending, torsion and elongation.

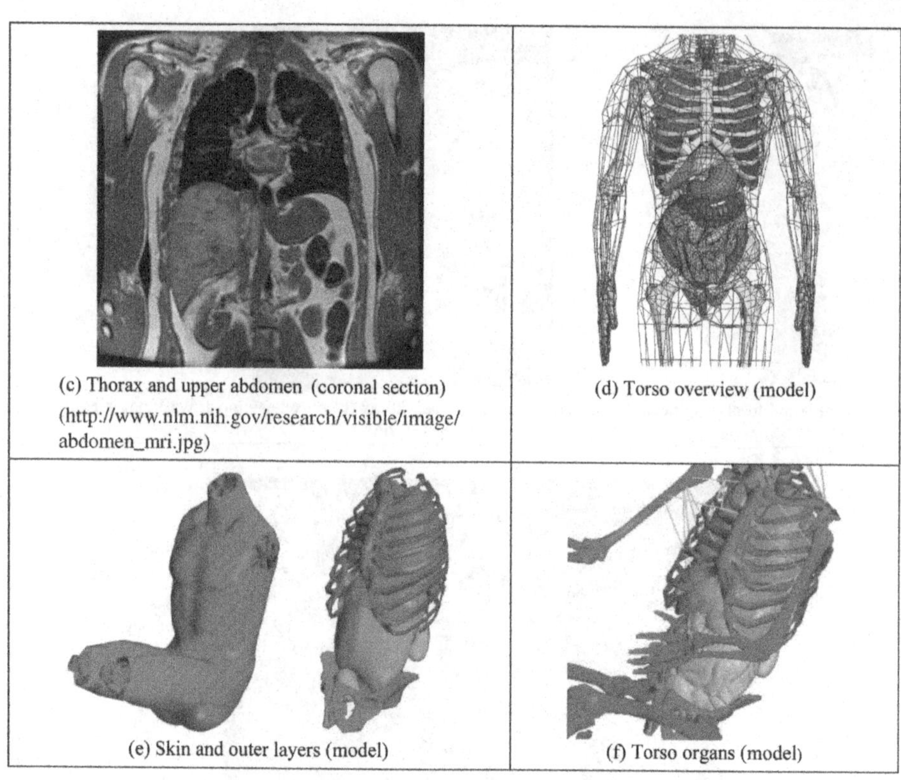

(c) Thorax and upper abdomen (coronal section) (http://www.nlm.nih.gov/research/visible/image/abdomen_mri.jpg)

(d) Torso overview (model)

(e) Skin and outer layers (model)

(f) Torso organs (model)

FIG. 9.22. *(Continued.)*

The human chest mainly consists of the ribs, the thoracic spine and the cardio-pulmonary organs. The upper ten ribs are directly or indirectly connected to the sternum via their costal cartilages, while the lowest two ribs are floating. The twelve ribs of each side articulate posterior with the twelve thoracic vertebrae. Three layers of intercostal muscles connect the ribs. The mediastinum forms the space between the cardio-pulmonary organs, consisting of trachea, lungs, aorta and the heart. The esophagus transverses the thorax. The diaphragm separates the thorax region from the upper abdomen. The thorax serves as a support for the shoulders and the upper extremities. The abdominal cavity, Fig. 9.22(c)–(f), is separated from the thorax by the diaphragm. The abdomen can be subdivided into the upper and lower abdomen. It contains the "solid" organs: liver, spleen, pancreas, kidneys, adrenal glands and ovaries; and the "hollow" organs: stomach, small and large intestines, urinary bladder and the uterus (ROUHANA [1993], HYNCIK [1999b], HYNCIK [2001b], HYNCIK [2002b], HYNCIK [2002c]). These organs are less well protected for impact than the organs of the thorax.

Injury to the thoraco-lumbar spine. According to KING [1993], injury to the bony part of the thoraco-lumbar spine is rare in frontal car accidents. Injury can be classified into anterior wedge fractures and burst fractures of vertebral bodies, dislocations

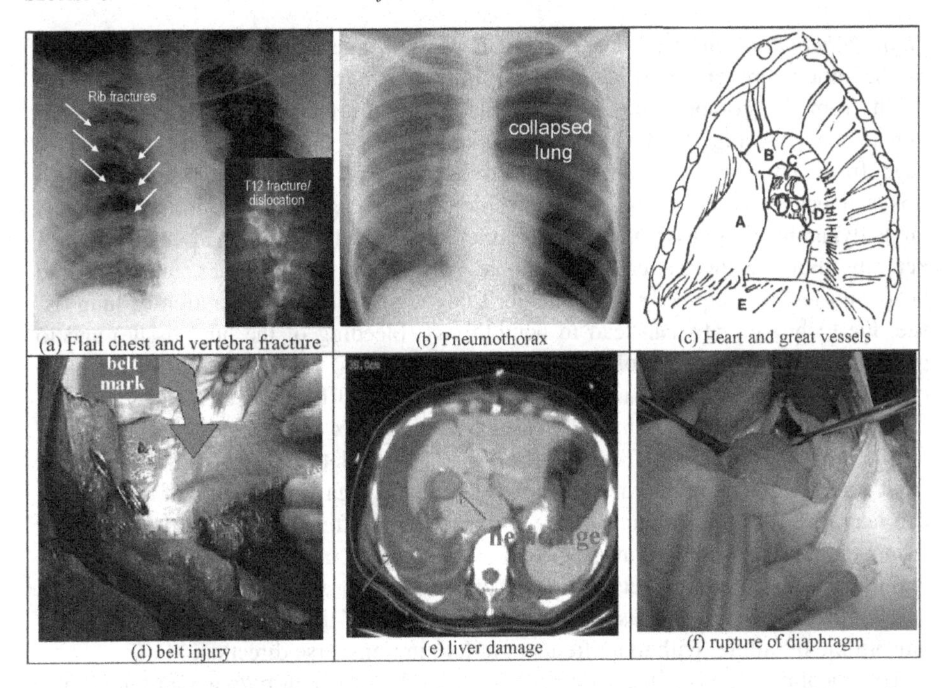

(a) Flail chest and vertebra fracture (b) Pneumothorax (c) Heart and great vessels
(d) belt injury (e) liver damage (f) rupture of diaphragm

FIG. 9.23. Chest and upper abdomen injury. (Insets (a) and (f): Reproduced by permission of Trauma.Org; Inset (b): Reproduced by permission of the Yale University School of Medicine; Inset (c): CAVANAUGH [1993], reproduced by permission of Springer Verlag from: Nahum and Melvin eds., Accidental Injury, Biomechanics and Prevention, Figure 15.4, p. 367; Insets (d) and (e): from CIREN William Lehman Injury Research Center at the Ryder Trauma Center Case # 97-018 (public domain material, U.S. Department of Transportation, NHTSA), see NHTSA/CIREN [1997].)

and fracture dislocations, rotational injuries, Chance fractures, hyper-extension injuries and soft tissue injuries. A typical vertebra fracture is shown in the inset to Fig. 9.23(a). Anterior wedge fractures occur under combined axial compression and bending (pilot ejection) and consist in a lighter injury. Burst fractures threaten the integrity of the cord and occur at higher loads. Dislocations can produce misalignments of the facets between adjacent vertebrae and occur under combined flexion, rotation and postero-anterior shear and can involve bone fracture and ligament rupture, threatening the cord. Rotational injuries come from excessive torsion of the spine under simultaneous axial and shear loads, with lateral wedge fractures and a danger of paraplegia. Chance fractures (CHANCE [1948]) occur by over-flexion of the lumbar spine when wearing only a lap belt in a frontal collision and it splits a lumbar vertebra in its transverse plane. If the upper body is properly fixed, excessive flexion over the restraining lap belt cannot occur. Hyper-extension injuries occur in pilot ejection and rarely in car accidents. They may produce bone fracture and ligament rupture. Soft tissue injuries involve the intervertebral discs, the ligaments, the joint capsule, the facet joints and the muscles and tendons attached to the spinal column.

Thorax/abdominal injury. For occupants of passenger cars, most crash injuries of the thorax are given by single and multiple rib fracture ("flail chest"), which result from the compression of the rib cage and the ensuing bending and breaking of the ribs, Fig. 9.23(a) (http://www.trauma.org). Flail chest prevents the rib cage from acting as a coherent protective cage and impairs respiration. Broken ribs can cause injury to the organs inside the rib cage, either from unrestrained large dynamic compression, or from punctures caused by the sharp ends of broken ribs. For the thorax region, pulmonary artery and lung injuries, and liver and heart injuries are frequently encountered. Lung contusions and laceration of lung tissue due to broken ribs can lead to hemothorax (bleeding of the lung tissue) and to pneumothorax (pneumatic collapse of one or both lungs) from punctures of the pleural sac between the lungs and the rib cage, Fig. 9.23(b), respectively (http://info.med.yale.edu/intmed/cardio/imaging/cases/pneumothorax_tension/graphics/rad1.gif).

The heart, Fig. 9.23(c) (from CAVANAUGH [1993]), letter A, can suffer contusion from compression and high rates of compression, and high compression of the sternum can lead to laceration. Rupture of the aorta, B, occurs frequently at its root, near the ligamentum arteriosum (isthmus), C, and at its insertion into the diaphragm, E. In lateral impacts, aorta rupture between its unsupported upper and its more firmly anchored descending branch, D, can result from lateral motions of the heart. The aorta walls act anisotropic in failure, with more frequent failure in transverse direction.

Trauma of the organs of the abdomen, Fig. 9.23(d) and (e) (http://www-nrd.nhtsa.dot.gov/departments/nrd-50/ciren/ciren1.html) and (f) (http://www.trauma.org/~imagebank/chest/images/chest0023c.jpg), may stem from penetrating objects, or blunt impacts, the latter being predominant in transport vehicle accidents, including injuries from the lap belt. The abdominal cavity is filled by the solid organs (liver, spleen, pancreas, kidneys, adrenal glands, ovaries) and by the hollow organs (stomach, large and small intestines, urinary bladder, uterus). These organs are only lightly tethered.

They can slide easily within their membrane envelopes (peritoneum) and they are held in place by their blood supply vessels (kidneys), or by peritoneal folds and ligaments. The large relative mobility of the abdominal organs makes injury sensitive to the attitude of the subject at the time of the accident. Blunt trauma affects the solid organs more than the hollow organs. Side impacts favor renal injury. The action of lap belts frequently leads to minor injuries, while preventing from serious injuries.

Model structure and calibration. The modeling aspects of the human torso are summarized in Figs. 9.24 and 9.25.

The *thoracic and lumbar spine* model is shown in Fig. 9.24.

The *thorax* of the H-Model mainly consists in deformable and damageable rib and rigid spine bones and in the thoracic organs and great vessels with internal air (lungs) or blood modeling, Fig. 9.25(a) (http://www.bartleby.com/107/illus490.html) and (b). The blood-filled vessels are simulated with an incompressible fluid model, while the lungs can compress like airbags. Both models can leak out fluid or compressible air ("bio-bags", cf. Appendix D). If impacts on the thorax are simulated, the abdominal cavity can be represented approximately as an incompressible bio-bag (see Fig. 9.27(b), 9.27(d)). In Section 8.5 on the HUMOS model validation, Fig. 8.6(c) demonstrates the effect of

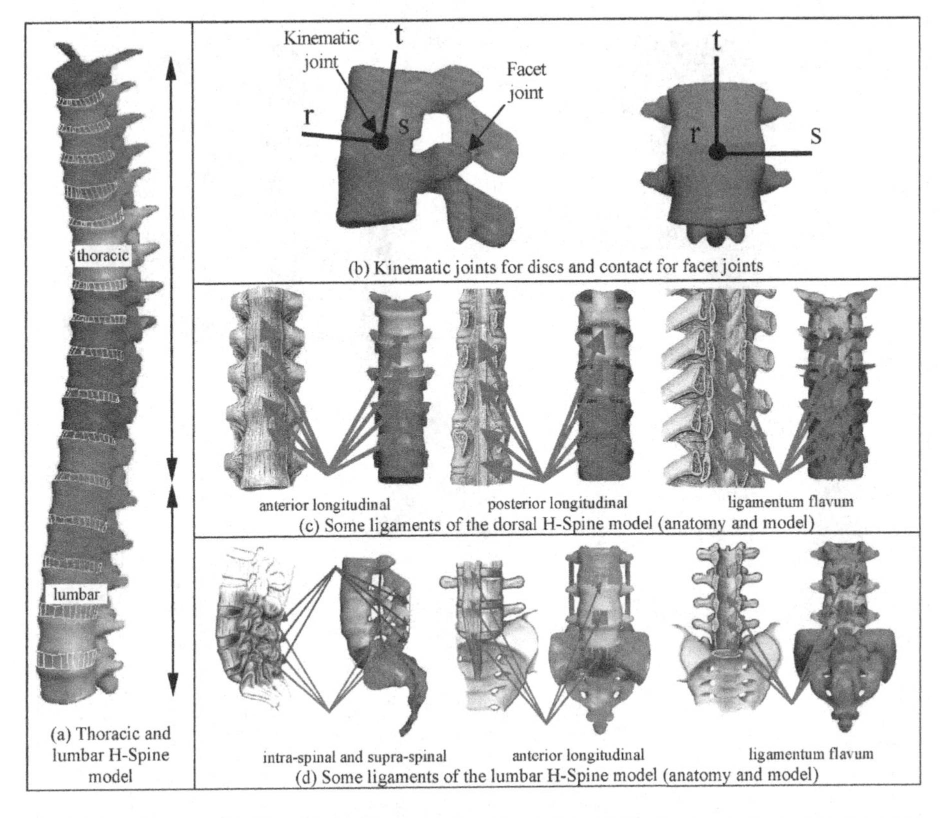

FIG. 9.24. H-Spine model: Thoracic and lumbar spine. (Insets (c) and (d): Anatomy after PUTZ and PABST [2000]: Reproduced by permission of Urban & Fischer Verlag.)

simulating the blood fill of the heart on the Kroell frontal pendulum impact test. If the heart was modeled as an empty bag with elastic skin, the resulting chest deformation was over-predicted.

Fig. 9.25(c) depicts the finite element models of the upper abdominal organs of the H-Model. The hollow organs (stomach, etc.) are modeled with bio-bags, while the solid organs (liver, etc.) are modeled with solid finite elements. An ad hoc modeling of the viscera was introduced in Section 8.5, Fig. 8.8(a), (d). Further investigation will clarify to which detail the visceral organs must be modeled for various purposes. If the upper abdominal organs are studied, the viscera can often be modeled approximately as bio-bags. Similar models were established in collaboration with the University of West Bohemia, Prof. Rosenberg, by HYNCIK [1999b], HYNCIK [2001b], HYNCIK [2002b], HYNCIK [2002c].

Fig. 9.26 summarizes the modeling and calibration aspects of the ribs of the H-Torso model. Fig. 9.26(a) shows the more complex thin shell solid models and the simpler tapered beam model of the ribs. The rib bone material properties, including for rib fracture, were calibrated against test results, Fig. 9.26(b) and (c).

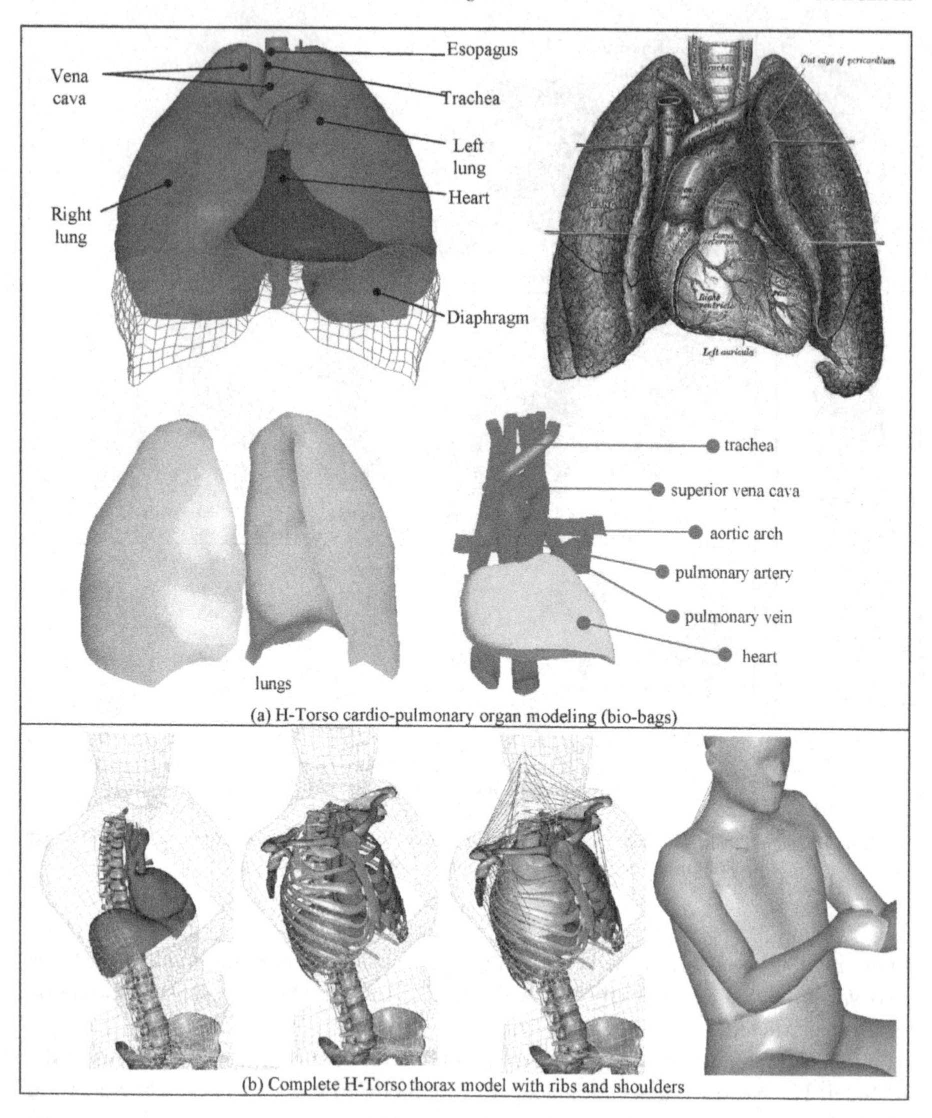

FIG. 9.25. H-Torso model (thorax and upper abdomen). (Inset (a) anatomical drawing: Reproduced by permission of Bartleby.com, Inc.)

Validation. For a *frontal pendulum impact loading,* Fig. 9.27 shows the response of the H-Torso model after the cadaver (and animal) tests done by KROELL [1971], KROELL, SCHNEIDER and NAHUM [1971], KROELL, SCHNEIDER and NAHUM [1974], KROELL, ALLEN, WARNER and PERL [1986]. Fig. 9.27(a) represents the test setup and the resulting pendulum force-chest deflection corridor. Fig. 9.27(b)–(d) display the H-Torso model response to the applied frontal pendulum impact loading. In (c) and (d) the deformations of the heart, lungs and blood vessels are clearly visible. Fig. 9.27(e)

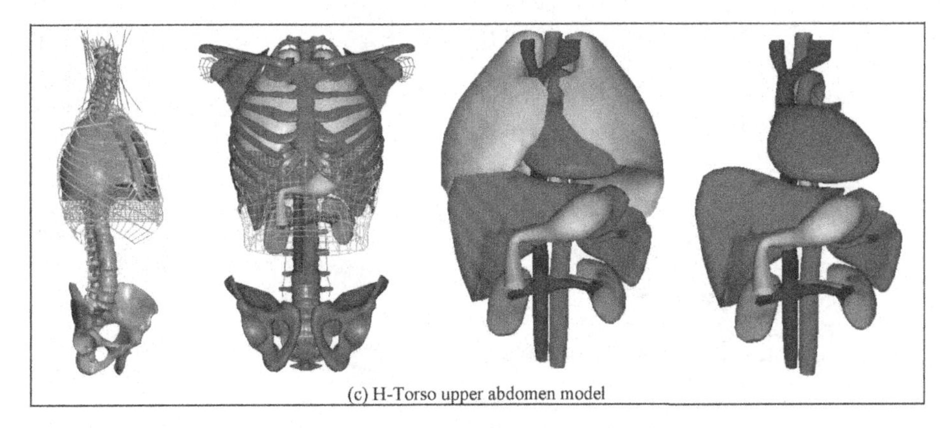

(c) H-Torso upper abdomen model

FIG. 9.25. (*Continued.*)

shows the response of the H-Torso rib cage with von Mises surface stress contours added on the ribs and with the failure of some ribs ("flail chest"). Fig. 9.27(f), finally, shows the H-Torso model pendulum force versus thorax deflection curve to fall well into the response corridor obtained from many cadaver tests by Kroell. Similar good responses (not shown here) were obtained for lateral pendulum impacts (Viano tests), similar as shown in Section 8.5, Fig. 8.7, for the HUMOS model segment validation.

For *chest belt loading*, Fig. 9.28 summarizes the response obtained with the H-Torso model.

Fig. 9.28(a) shows the corresponding cadaver test setup (CESARI and BOUQUET [1990], CESARI and BOUQUET [1994]) with the measured cadaver and Hybrid III dummy chest band and thorax deflection versus impact energy results. The chest band device is an elastic steel band that is wrapped tightly around the chest. Strain gauges measure the flexural deformation from which the deflected shape of the tape, and hence the thorax, can be deduced. Fig. 9.28(b) shows the H-Torso model global deformation response to the applied belt loading. The response of the ribs, including fracture near maximum deformation, are shown in Fig. 9.28(c). Fig. 9.28(d) and (e) compare chest band results from the cadaver tests with the response of the H-Torso model, where the chest band results are approximated by plotting the external model section contours at the locations of the upper and lower chest bands, respectively. The agreement is deemed satisfactory, although the model should have duplicated the real chest bands with their true physical properties and contact with the chest for a closer representation.

For *abdomen bar impact tests*, validations of the H-Model similar as the ones carried out in Section 8.5 for the HUMOS model were performed (not repeated here).

References on the H-Torso model. The following references were consulted for the establishment of the H-Torso model (first authors): ALLAIN [1998] on thorax model calibration; ALLEN, FERGUSON, LEHMANN and O'BRIEN [1982] on fracture/dislocation of the lower cervical spine; BEGEMAN, KING and PRASAD [1973] on spinal loads from front/rear acceleration; BELYTSCHKO, KULAK, SCHULTZ and GALANTE [1972], BELYTSCHKO, KULAK, SCHULTZ and GALANTE [1974] on FE analysis of intervertebral

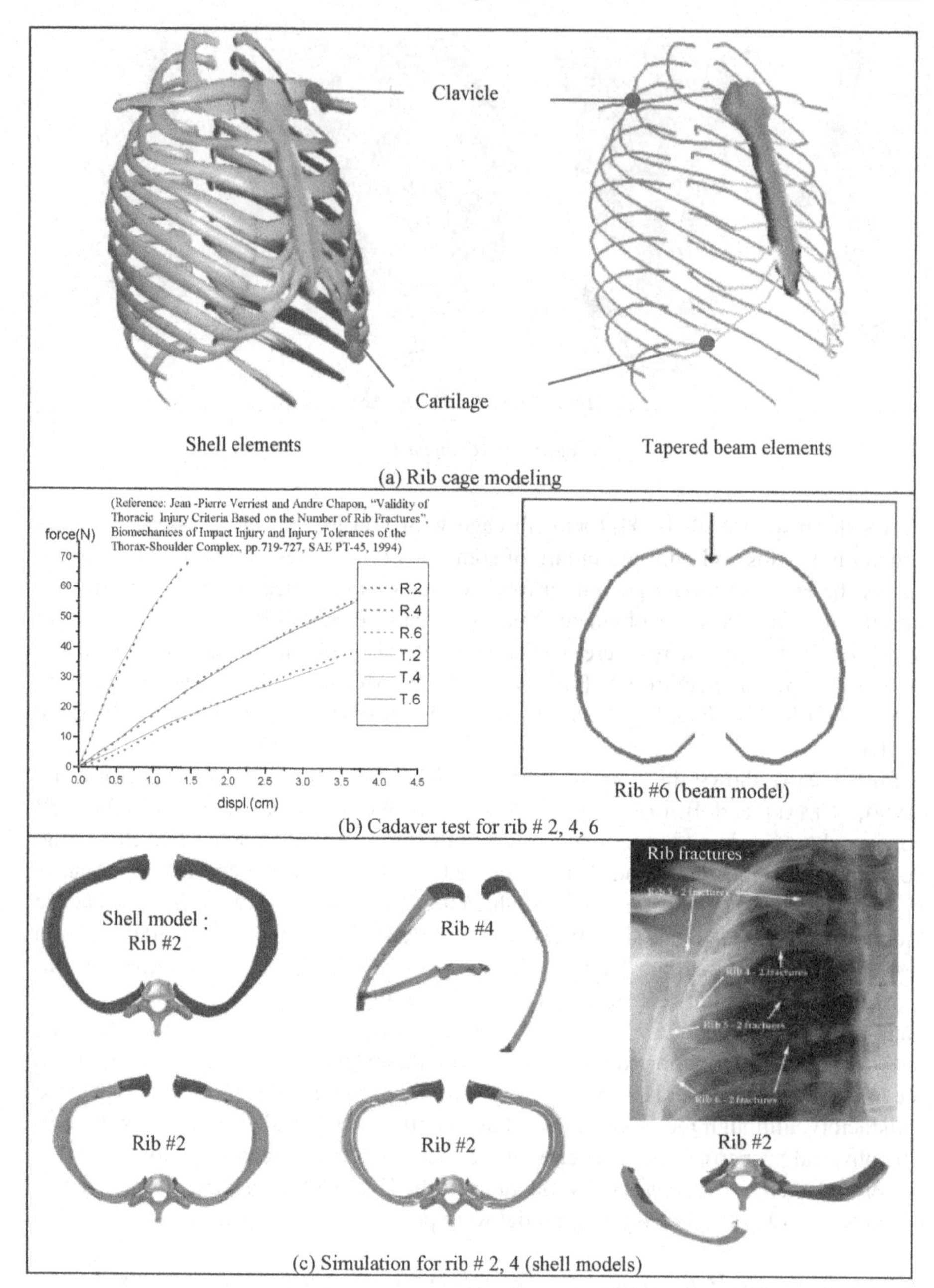

FIG. 9.26. H-Torso rib modeling and calibration. (Inset (b) left diagram: Reprinted with permission from SAE paper number 856027 © 1985 SAE International; Inset (c): X-ray picture: Reproduced by permission of Trauma.Org.)

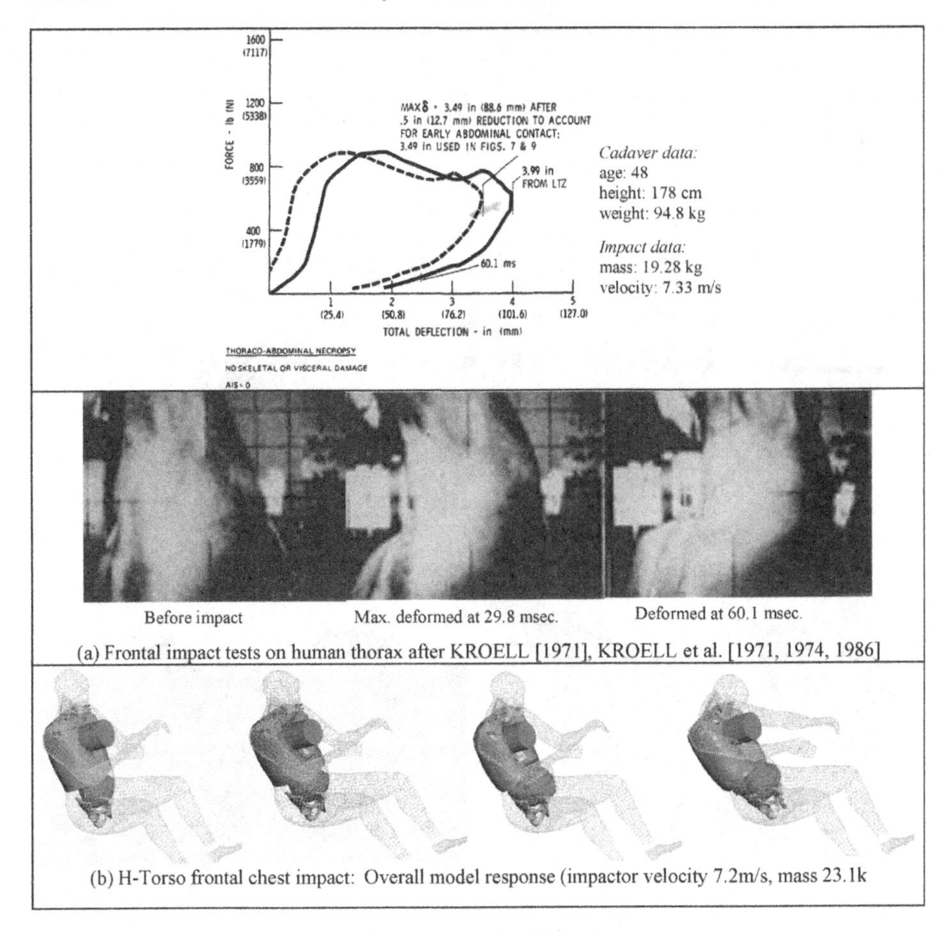

(a) Frontal impact tests on human thorax after KROELL [1971], KROELL et al. [1971, 1974, 1986]

(b) H-Torso frontal chest impact: Overall model response (impactor velocity 7.2m/s, mass 23.1k

FIG. 9.27. H-Torso frontal chest pendulum impact validation. (Inset (a): Reproduced by permission of The Stapp Association.)

discs; BOUQUET, RAMET, BERMOND and CESARI [1994] on thorax/pelvis response to impact; CESARI and BOUQUET [1990] and CESARI and BOUQUET [1994] on thoracic belt loading; CHAZAL, TANGUY and BOURGES [1985] on spinal ligament properties; CLEMENTE [1981] (anatomy atlas); COOPER, PEARCE, STAINER and MAYNARD [1982] on thorax trauma and cardiac injury; EPPINGER [1976], EPPINGER, MARCUS and MORGAN [1984] on thoracic injury and dummy development; FUNG and YEN [1984] on lung injury experiments; GOEL, GOYAL, CLARK, NISHIYAMA and NYE [1985] on lumbar spine kinematics; HUELKE, NUSHOLTZ and KAIKER [1986] on thoraco-abdominal biomechanics research; KAPANDJI [1974c] (joint physiology of the trunk and vertebral column); KAZARIAN, BEERS and HERNANDEZ [1979], KAZARIAN [1982] on the spine injuries; KEITHEL [1972] on thoraco-lumbar intervertebral joint deformation due to loads; KLEINBERGER, SUN, EPPINGER, KUPPA and SAUL [1998] on improved injury criteria; KULAK, BELYTSCHKO, SCHULTZ and GALANTE [1976] on non-linear

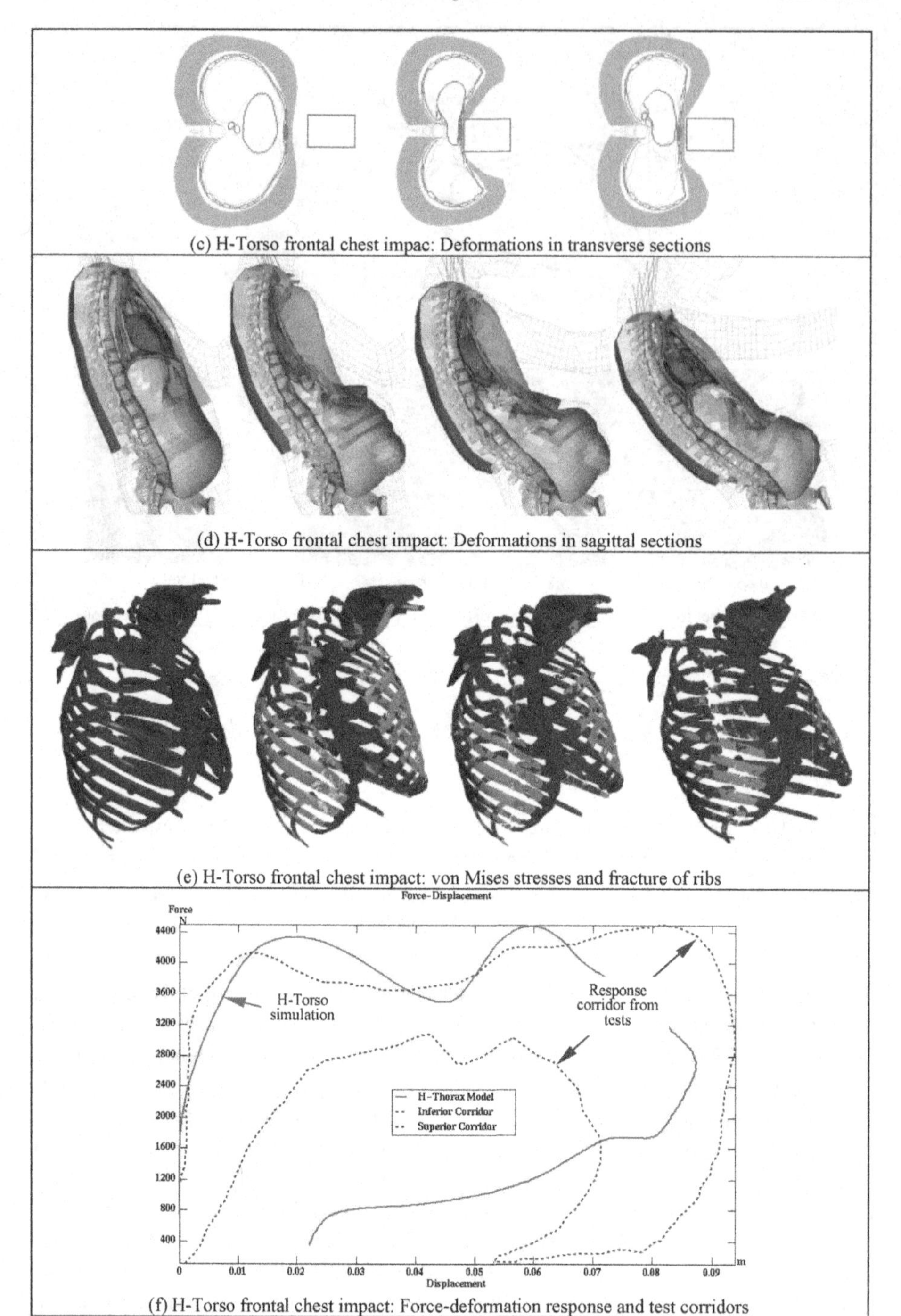

(c) H-Torso frontal chest impac: Deformations in transverse sections

(d) H-Torso frontal chest impact: Deformations in sagittal sections

(e) H-Torso frontal chest impact: von Mises stresses and fracture of ribs

(f) H-Torso frontal chest impact: Force-deformation response and test corridors

FIG. 9.27. (*Continued.*)

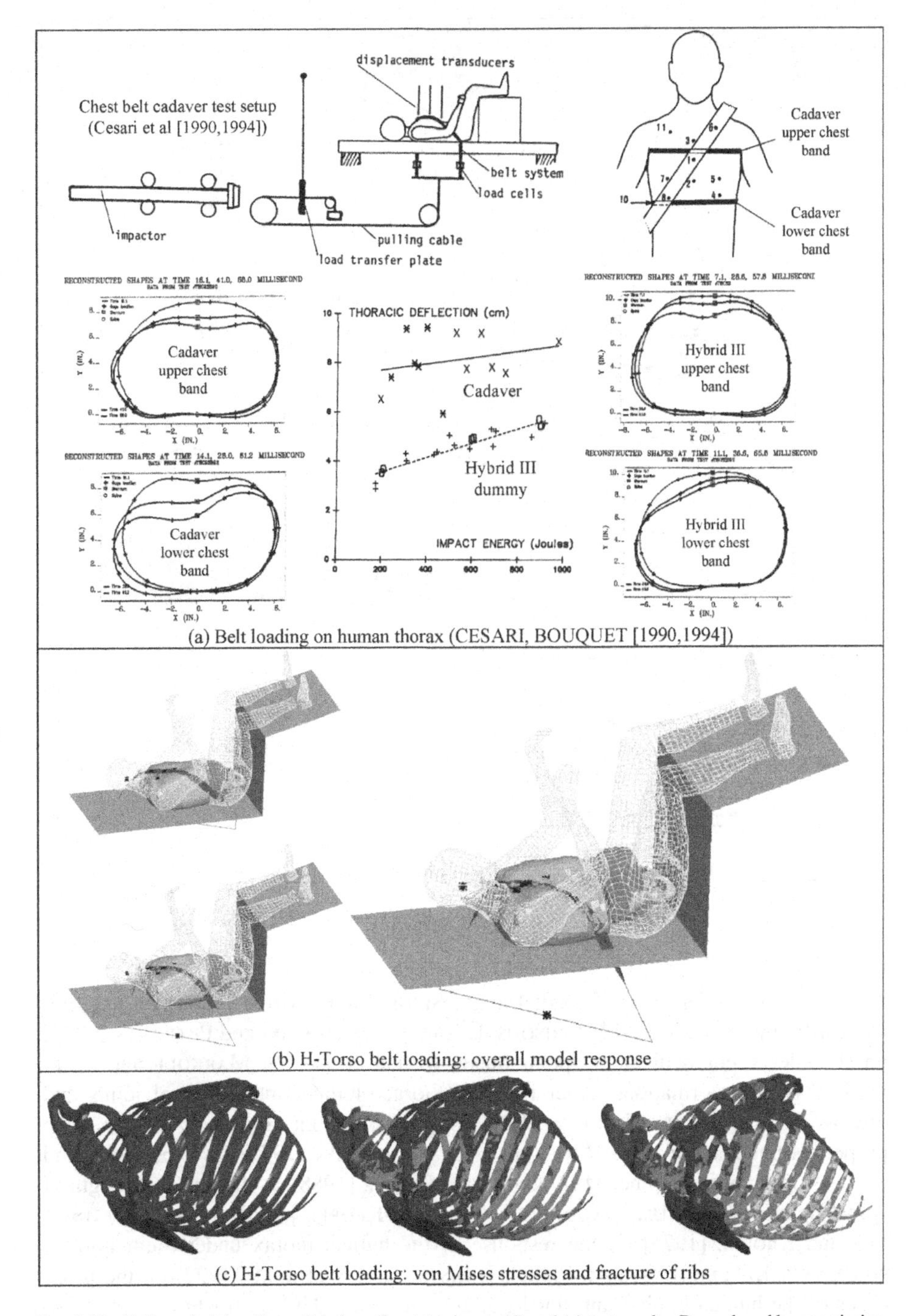

(a) Belt loading on human thorax (CESARI, BOUQUET [1990,1994])

(b) H-Torso belt loading: overall model response

(c) H-Torso belt loading: von Mises stresses and fracture of ribs

FIG. 9.28. H-Torso belt loading validation. (Inset (a), insets (d) and (e) test results: Reproduced by permission of The Stapp Association.)

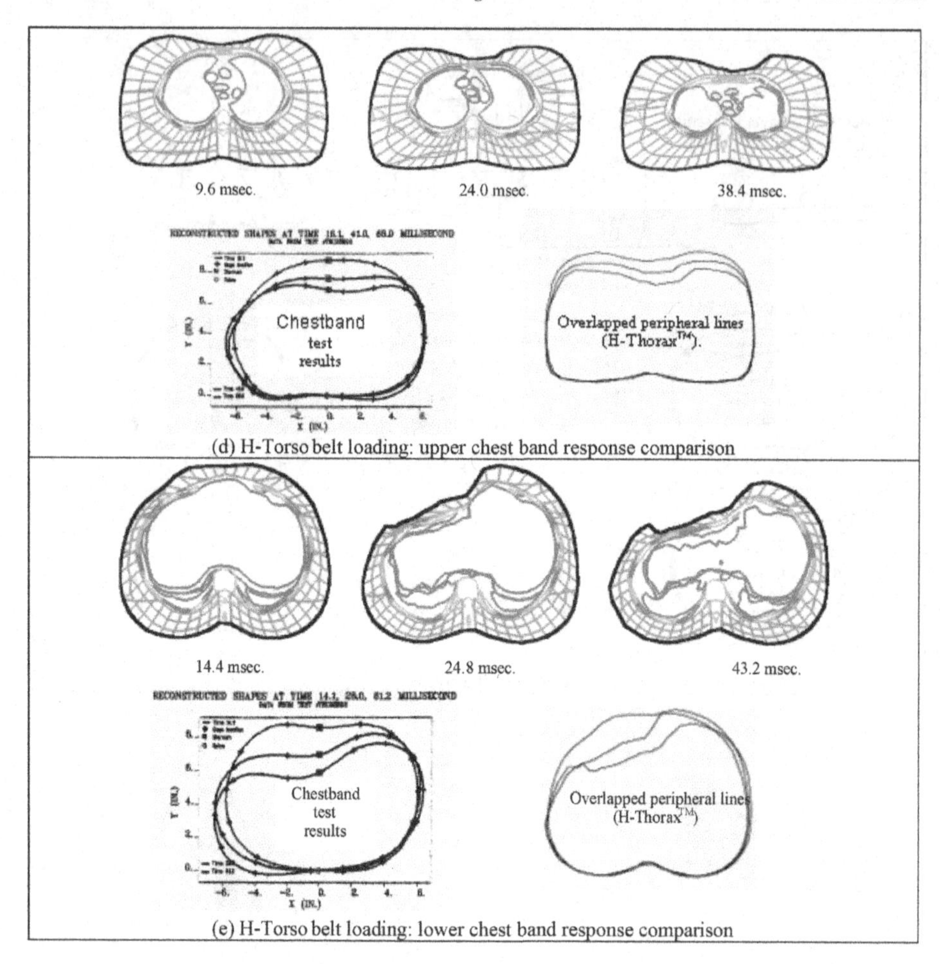

FIG. 9.28. (*Continued.*)

behavior of discs under axial load; LASKY, SIEGEL and NAHUM [1968] on automotive cardio-thoracic injuries; MAKHSOUS, HÖGFORS, SIEMIEN'SKI and PETERSON [1999] on shoulder strength in the scapular plane; MARKOLF [1972], MARKOLF and MORRIS [1974] on deformations under loads of thoraco-lumbar intervertebral joints and disc structure; MILLER, SCHULTZ, WARWICK and SPENCER [1986] on lumbar spine properties under large loads; MOFFATT, ADVANI and LIN [1971] on experiment and analysis of the human spine; MYKLEBUST and PINTAR [1988] on the tensile strength of spinal ligaments; NAHUM, GADD, SCHNEIDER and KROELL [1970], NAHUM, SCHNEIDER and KROELL [1975] on the response of the human thorax under blunt impact; NEUMANN, KELLER, EKSTROM, PERRY, HANSSON and SPENGLER [1992] on the properties of the human lumbar anterior ligament; NICOLL [1949] on fracture of the torso lumbar spine; NOVOTNY [1993] on spinal biomechanics; PANJABI and BRAND [1976] on the mechanical properties of the human thoracic spine; PLANK, KLEINBERGER and EPPINGER [1994] on FE analysis of thorax/restraint system interaction; POPE, KROELL, VIANO, WARNER and ALLEN [1979] on the postural influence on thoracic impact; PRASAD and KING [1994] on an experimentally validated dynamic spine model; ROAF [1960] on the mechanics of spinal injury; SCHNEIDER, KING and BEEBE [1990] on thorax-abdomen trauma assessment; STALNAKER and MOHAN [1974] on human chest impact protection criteria; STOCKIER, EPSTEIN and EPSTEIN [1969] on seat belt trauma to the lumbar spine; VERRIEST and CHAPON [1994] on thoracic injury criteria and rib fractures; VIANO and LAU [1983], VIANO [1989] on the influence of impact velocity and chest compression in thorax injury (1983) and on the response and injuries blunt lateral impact (1989); WHITE and PANJABI [1978], WHITE and PANJABI [1990] on spine biomechanics; YOGANANDAN, HAFFNER, MALMAN, NICHOLS, PINTAR, JENTZEN, WEINSHEL, LARSON and SANCES [1989], YOGANANDAN and PINTAR [1998] on the trauma of the human spine (1989a) and on the biomechanics of thoracic ribs (1998).

The following web sites were consulted: http://info.med.yale.edu./intmed/cardio/imaging/cases/pneumothorax_tension/graphics/rad1.gif; http://www.bartleby.com/107/illus490.html; http://www.bionetmed.com/; http://www.nlm.nih.gov/research/visible/image/abdomen_mri.jpg; http://www.trauma.org; http://www9.biostr.washington.edu/cgi-bin/DA/imageform.

9.6. H-UE: Shoulder and arms

An overview on the human upper extremity model structure and application is given in CHOI, LEE and HAUG [2001a].

Anatomy. The anatomy of the human upper extremity is summarized in Fig. 9.29. The upper extremity consists of the arm and the shoulder complex, Fig. 9.29(a) (after http://www.rad.washington.edu/RadAnat and PUTZ and PABST [2000]). The shoulder (scapula, clavicula) is connected to the thorax via the scapula-thoracic (sliding) joint and via the sterno-clavicular (ball) joint. The upper arm connects to the scapula by the gleno-humeral joint. The scapula connects to the clavicula by the acromio-clavicular joint.

The skeleton of the upper extremity is connected by several systems of muscles, Fig. 9.29(b) (after http://www.fitstep.com/Advanced/Anatomy/Shoulders.htm) and Fig. 9.29(c). These muscles connect the upper arm to the body (latissimus dorsi, pectoralis major), the upper arm to the scapula (supraspinatus, infraspinatus, teres, subscapularis, deltoid, biceps, triceps, coraco-brachialis, brachialis), the scapula to the thorax (trapezius, levator anguli, rhomboids, serratus, pectoralis minor), the upper arm to the clavicula (deltoid, pectoralis major) and the clavicula to the thorax (subclavius, trapezius). The bones of the upper extremity are further connected by ligaments, Fig. 9.29(d) (after http://eduserv.hscer.washington.edu/hubio553/atlas/shjointlig.html). The ligaments that are connecting the clavicula to the ribs are not shown.

Injury. The injury of the upper extremity is discussed in part in LEVINE [1993]. The upper extremity is most frequently hurt through side impacts between two vehicles,

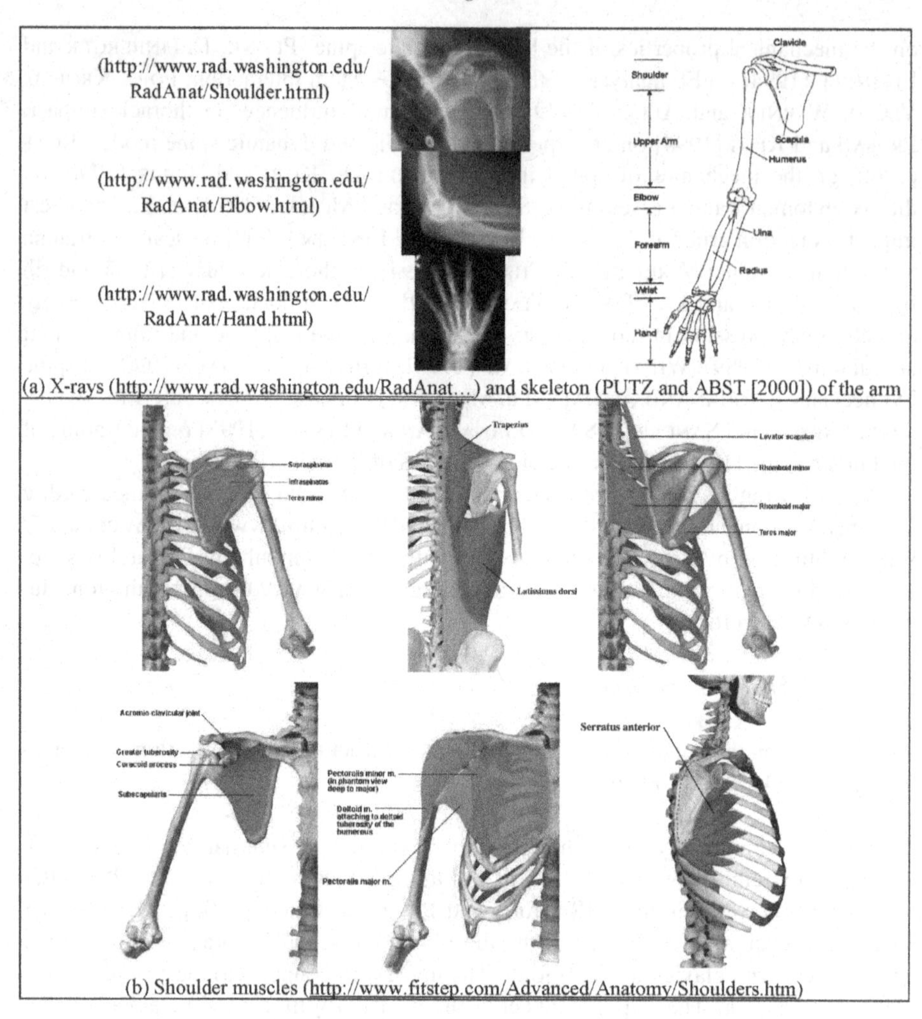

FIG. 9.29. Anatomy of the human upper extremity. (Inset (a) X-ray pictures: Reproduced by permission of Michael L. Richardson, University of Washington Medical Center, Department of Radiology; Inset (a) anatomical drawing: Reproduced by permission of Urban & Fischer Verlag; Insets (b) and (c): Reproduced by permission of BetterU, Inc.; Inset (d): Reproduced by permission of Carol C. Teitz, University of Washington.)

where the occupant may hit the door in a lateral motion. The forearm may be affected when it is positioned over the steering wheel while the driver side airbag deploys. The upper extremity may also be injured from violent projection by side impact airbags. Because they must fire faster, side impact airbags turn out to be more aggressive than front impact airbags. Upon side impact, the clavicle with its acromio-clavicular and sterno-clavicular joints may be injured through fracture and dislocations. When the lower arm is placed over a driver side airbag deploying from the steering wheel, there may be fractures of the arm. Upon inflation of side airbags located in the back rest, the

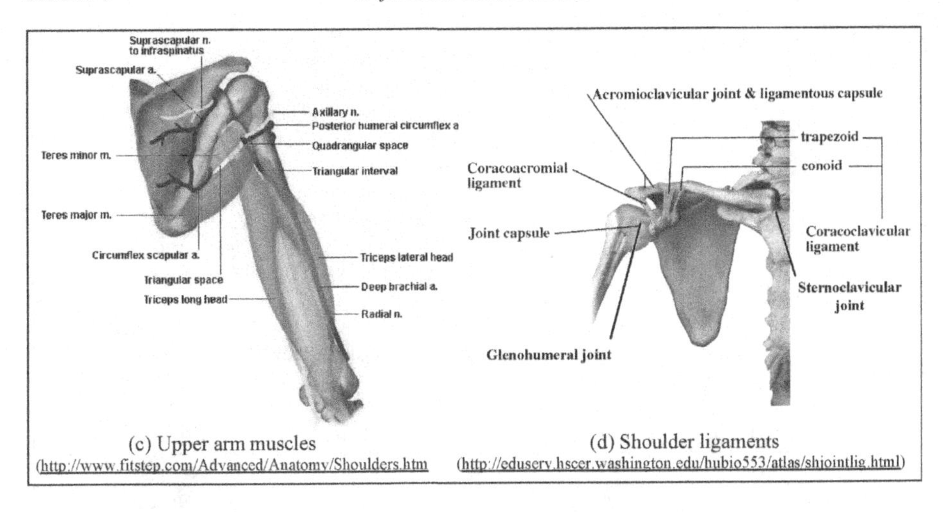

(c) Upper arm muscles
(http://www.fitstep.com/Advanced/Anatomy/Shoulders.htm

(d) Shoulder ligaments
(http://eduserv.hscer.washington.edu/hubio553/atlas/shjointlig.html)

FIG. 9.29. *(Continued.)*

upper arm may experience fracture and the gleno-humeral joint may become dislocated. The injuries can be more or less severe, ranging from dislocations to closed fractures to severe open fractures.

Model structure and calibration. The modeling aspects of the human upper extremity are summarized in Fig. 9.30. The upper extremity (H-UE) of the H-Model mainly consists of deformable and damageable bones and flesh padding. The flesh padding uses solid elements (not shown). Active and passive muscle forces are modeled with bar elements using the Hill muscle model. Nonlinear contact interfaces model the cartilage layers on the shoulder and elbow joints. Fig. 9.30(a) shows the H-UE model in its bodily context, while Fig. 9.30(b) shows the skeleton and the attached long muscle bars. Fig. 9.30(c) contains the modeling of the major joint ligaments, which are modeled with short nonlinear bars. Fig. 9.30(d) shows the modeling of surface muscles with two-dimensional finite elements and superimposed muscle bars. Here membrane elements are used as a "matrix" onto which a number of Hill type muscle bars are attached like "fibers" in series and in parallel, as required by the surface curvature and the surface area occupied by each muscle segment. The membranes represent the passive muscle material properties perpendicular to the fibers, while the bars represent the active and passive longitudinal muscle fiber properties. This way the curvature and the sliding contact of the muscles with the rib cage and other parts of the skeleton is well modeled. Fig. 9.30(e) shows the behavior of this model under imposed motions of the arm and shoulder. The next step might be to adapt "composite" materials, with the transverse and bulk properties of the muscles represented by the composite matrix, and the active and passive muscle fiber properties assigned to the fiber phase of the composite materials.

Fig. 9.31 shows, for example, the arm bone material calibration from *three point bending tests* on cadaver arms, after PINTAR, YOGANADAN and EPPINGER [1998] (public domain). The tests, Fig. 9.31(a), were simulated up to fracture, Fig. 9.31(b), and

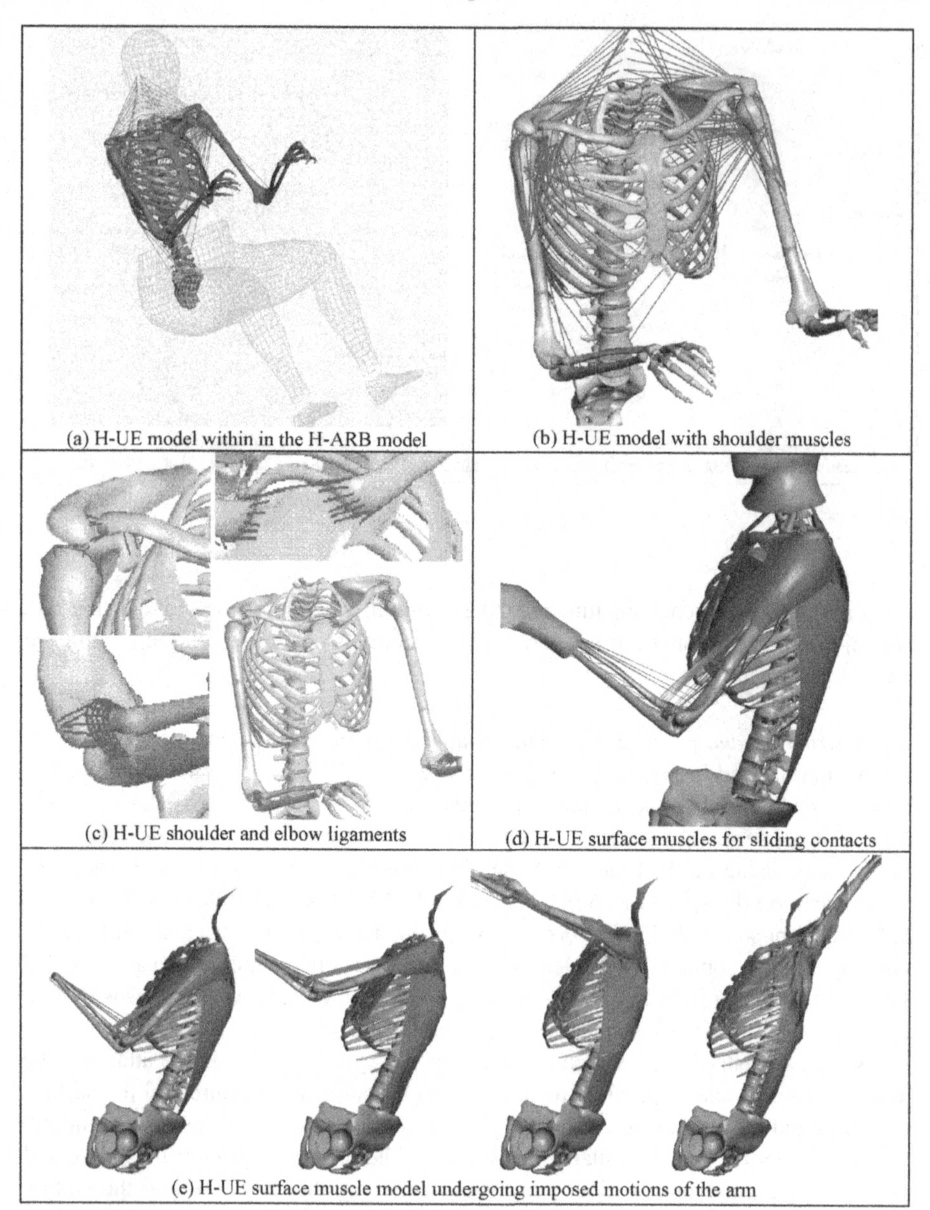

(a) H-UE model within in the H-ARB model
(b) H-UE model with shoulder muscles
(c) H-UE shoulder and elbow ligaments
(d) H-UE surface muscles for sliding contacts
(e) H-UE surface muscle model undergoing imposed motions of the arm

FIG. 9.30. H-UE modeling.

the calculated calibrated and the measured force-displacement curves are compared in Fig. 9.31(c).

Validation. Based on PALANIAPPAN, WIPASURAMONTON, BEGEMAN, TANAVDE and ZHU [1999], Fig. 9.32 shows the calibration of an earlier version of the model on a

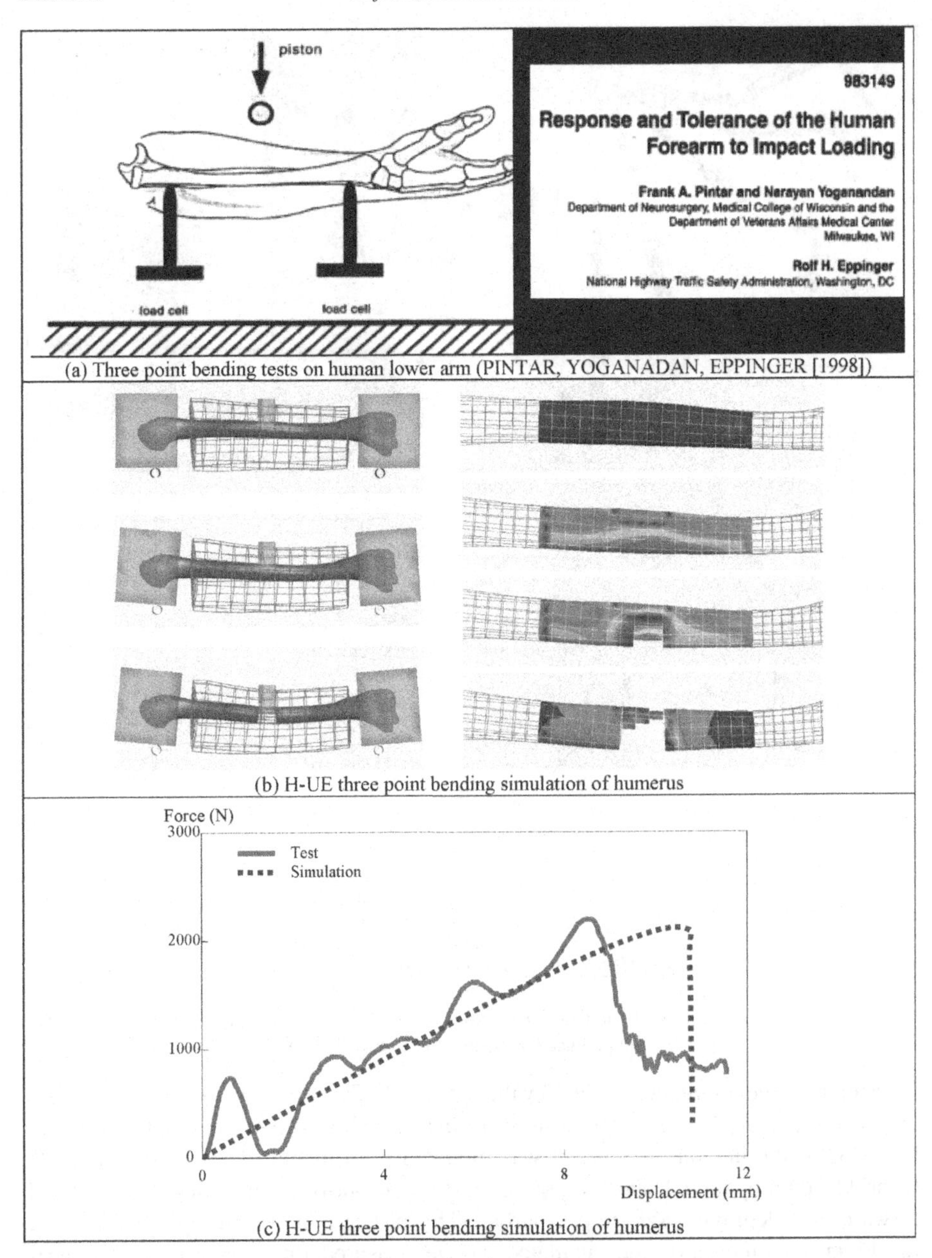

(a) Three point bending tests on human lower arm (PINTAR, YOGANADAN, EPPINGER [1998])

(b) H-UE three point bending simulation of humerus

(c) H-UE three point bending simulation of humerus

FIG. 9.31. H-UE arm bone material calibration.

pendulum impact experiment. The hands were modeled rigid, the deformable bones were modeled with shells and the flesh was modeled with deformable solid elements. The flesh model was calibrated on the pendulum impact tests. The attenuation of the shocks through the deformable flesh was found to be an important element to achieve the required accuracy of the model.

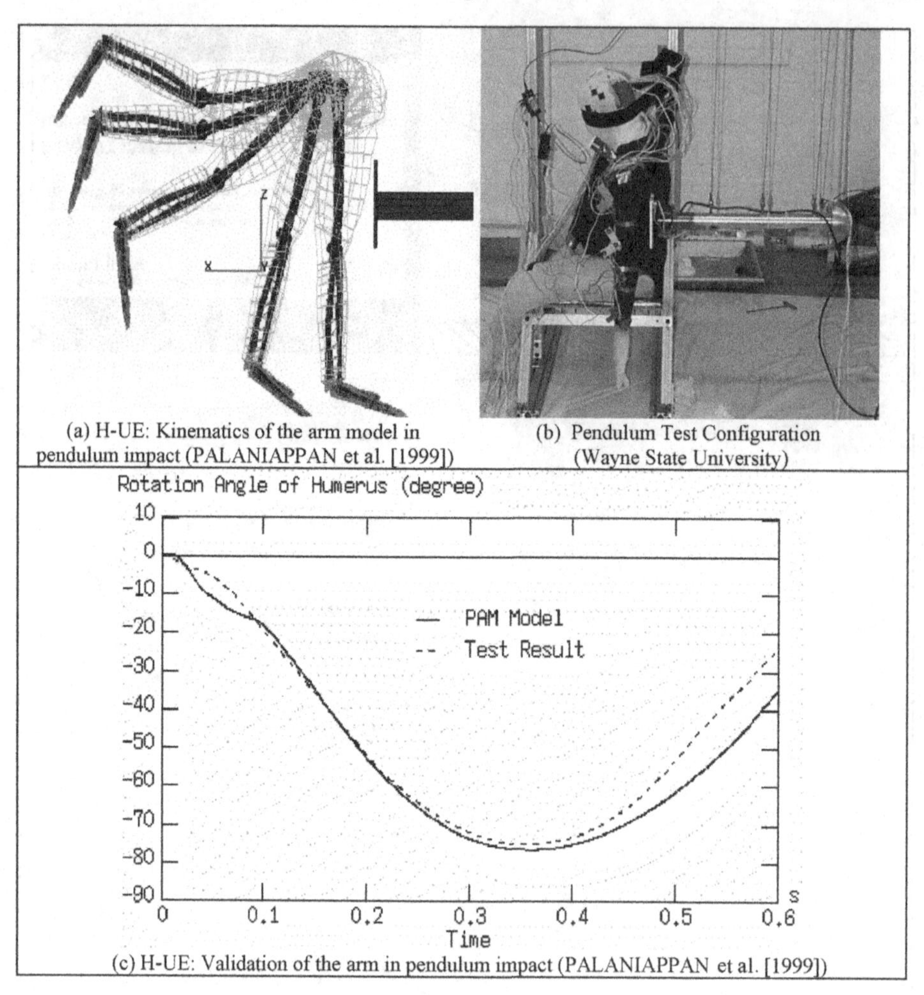

FIG. 9.32. H-UE pendulum impact validation (PALANIAPPAN, WIPASURAMONTON, BEGEMAN, TANAVDE and ZHU [1999]). (Reproduced by permission of The Stapp Association.)

From the same reference, Fig. 9.33 validates the H-UE model on a *side impact airbag out of position deployment*. The tests were performed at Wayne State University. In Fig. 9.33(a) the human arm model was mounted on a model of a fiftieth percentile Hybrid III dummy model. The left arm of the driver is out-of-position and it is projected forward by a deploying side impact airbag. This airbag is mounted in the backrest of the driver seat. In inset (b) the calculated and the measured arm displacement responses are compared.

References on the H-UE model. The following references were consulted when establishing the H-UE model (first authors): ENGIN [1979], ENGIN [1980], ENGIN [1983], ENGIN [1984], ENGIN and PEINDL [1987], ENGIN and CHEN [1989], ENGIN and T'MER [1989] on works on the biomechanics of the human shoulder complex,

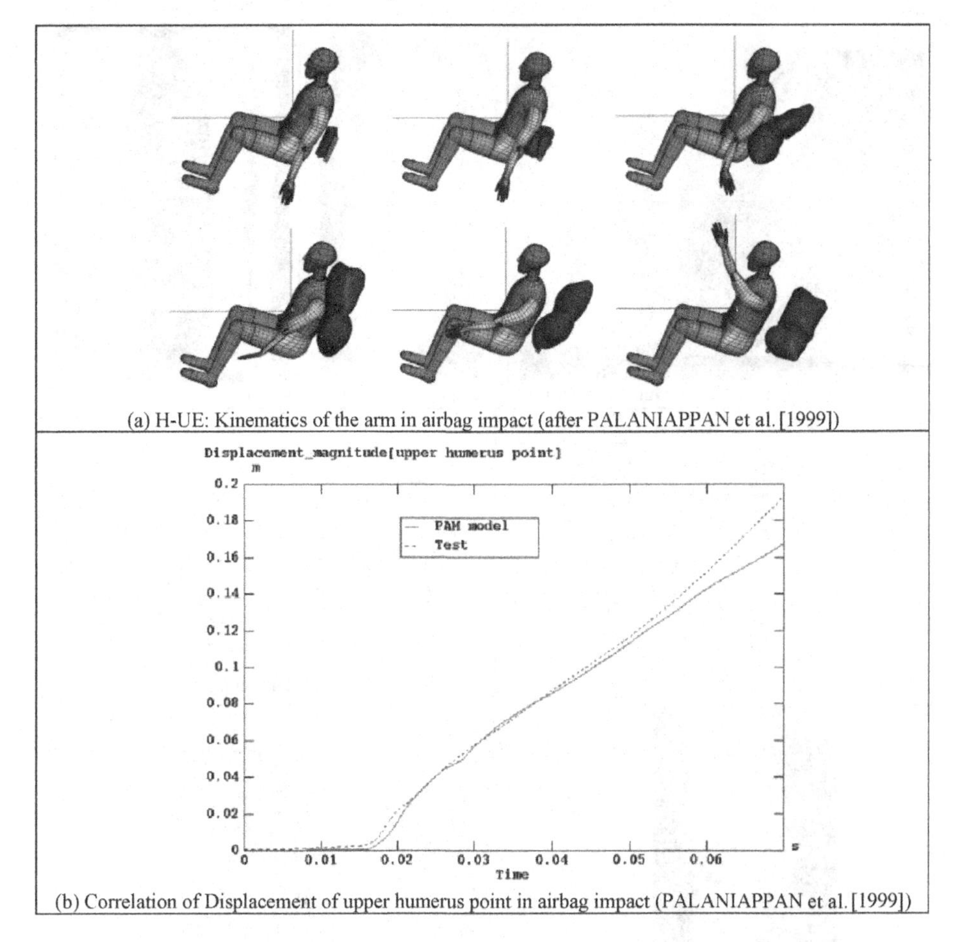

(a) H-UE: Kinematics of the arm in airbag impact (after PALANIAPPAN et al. [1999])

(b) Correlation of Displacement of upper humerus point in airbag impact (PALANIAPPAN et al. [1999])

FIG. 9.33. H-UE airbag impact validation (PALANIAPPAN, WIPASURAMONTON, BEGEMAN, TANAVDE and ZHU [1999]). (Reproduced by permission of The Stapp Association.)

bone and joint resistance; GOLDSTEIN, FRANKENBURG and KUHN [1993] on the biomechanics of bone; IRWIN [1994] on shoulder and thorax response cadaver tests and their analysis; KAPANDJI [1974a] (upper limb joint physiology); PEINDL and ENGIN [1987] on passive resistive properties of the shoulder complex; PRADAS and CALLEJA [1990] on the non-linear visco-elastic properties of human hand flexor tendon; PUTZ and PABST [2000]: Sobotta Atlas of Human Anatomy; T'MER and ENGIN [1989] on a 3D model of the shoulder complex. The following web sites were used: http://www.rad.washington.edu/RadAnat; http://www.fitstep.com/Advanced/Anatomy/Shoulders.htm; http://eduserv.hscer.washington.edu/hubio553/atlas/shjointlig.html.

9.7. H-LE: Knee-thigh-hip complex

An overview on the H-LE model and applications is provided in YOO and CHOI [1999].

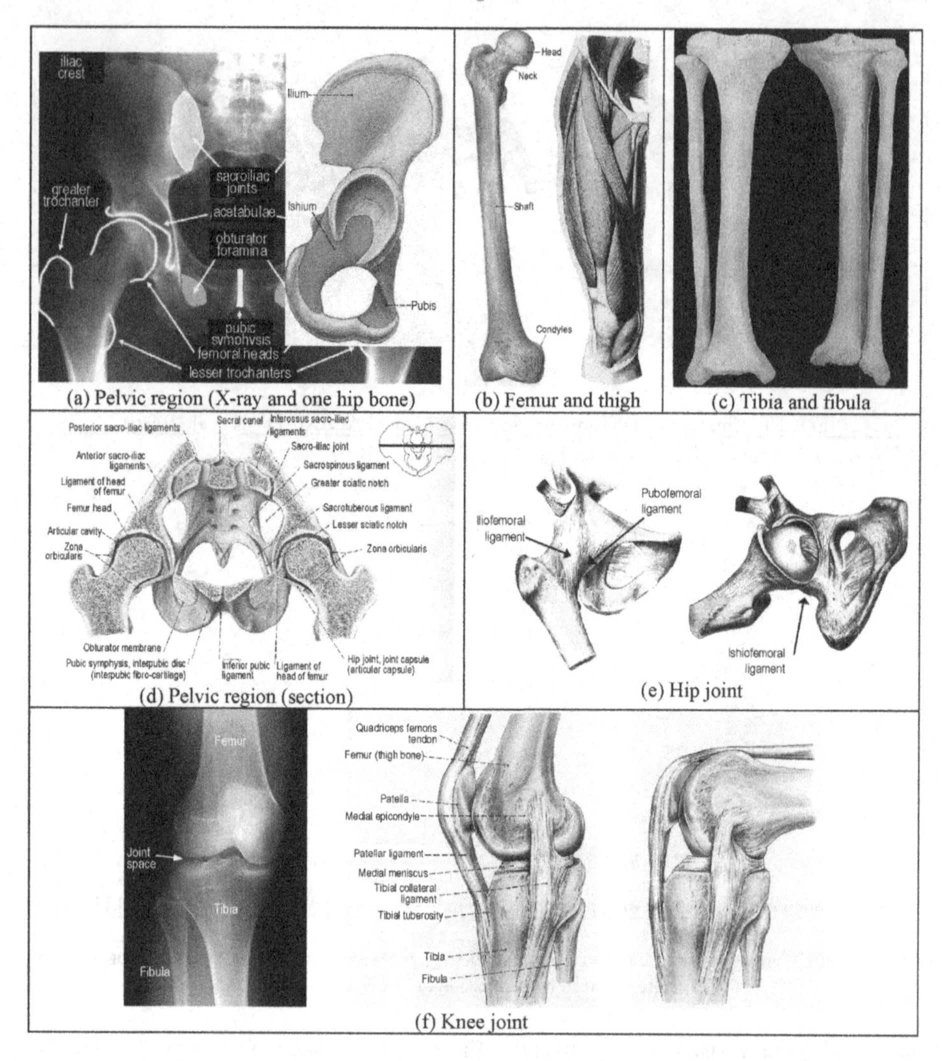

FIG. 9.34. Anatomy of the human lower extremity (without foot/ankle). (Insets (a) and (f) X-ray pictures: Reproduced by permission of Michael L. Richardson, University of Washington Medical Center, Department of Radiology; Insets (a) and (f), anatomical drawings, and insets (b) and (d): Reproduced by permission of Urban & Fischer Verlag; Insets (c) and (e): Figs. 7.16a, p. 205, 7.16d, p. 206 and 8.14c, p. 231 from HUMAN ANATOMY, 4th ed. by Frederic H. Martini, Michael J. Timmons and Robert H. Tallitsch, Copyright © 2003 by Frederic H. Martini, Inc. and Michael J. Timmons.)

Anatomy. The anatomy of the human lower extremity and pelvis is summarized in Fig. 9.34. It consists of hip, upper and lower legs, ankles and foot, articulated by the hip, knee, and ankle joints (the foot/ankle complex is discussed in a separate section). The pelvis supports the spinal column and it contains the sacrum, the coccyx, and the two hip bones, each made of the three fused ilium, ischium and pubic bones, Fig. 9.34(a) (http://www.rad.washington.edu/RadAnat/pelvis.html and PUTZ and PABST

[2000]), Fig. 9.34(d) (PUTZ and PABST [2000]) and Fig. 9.34(e) (MARTINI, TIMMONS and TALLITSCH [2003]). These three pelvic bones converge at the acetabulum, the articulation for the head of the femur, (a). The femur (thigh bone), (b) (after PUTZ and PABST [2000]), is the longest bone in the body. Its lower end joins the tibia (shin) to form the knee joint. Its upper end is rounded into a ball (or "head" of the femur) that fits into a socket in the pelvis (the acetabulum) to form the hip joint, (d), (e).

The neck of the femur gives the hip joint a wide range of movement, but it is a point of weakness and a common site of fracture.

The tibia is the inner and thicker of the two long bones in the lower leg. The tibia runs parallel to the smaller and thinner fibula to which it is attached by ligaments. The upper end of the tibia joins the femur to form the knee joint. The lower ends of tibia and fibula form the ankle joint with the medial and lateral malleolus, Fig. 9.34(c) (MARTINI, TIMMONS and TALLITSCH [2003]).

The knee joint, Fig. 9.34(f) (http://www.rad.washington.edu/RadAnat/knee.html and PUTZ and PABST [2000]), is held together by flexible ligaments. The collateral ligaments run along the sides of the knee and limit sideways motion. The anterior cruciate ligament (ACL) limits rotation and relative forward motion of the tibia. The posterior cruciate ligament (PCL) limits relative backward motion of the tibia. The lateral meniscus and medial meniscus are pads of cartilage that cushion the joint, acting as shock absorbers between the bones. The patella is the roughly triangular-shaped bone at the front of the knee joint. It transmits redirecting forces from the quadriceps muscle to the knee joint, which it protects.

Injury. The injury of the lower extremity, Fig. 9.35, is discussed in part in LEVINE [1993]. The most frequent and severe injuries of the knee-thigh-hip complex are skeletal bone fractures and hip dislocations. Fractures of the skeletal bone can be classified roughly in shaft or diaphysis fracture of the long bones and in crushing or compression of the short bones and of the meta- and epiphysis (articular surface) near the articulations of the long bones. The injuries can range from dislocations to closed fractures to severe open fractures.

In front collision, knee or pelvic dislocation can occur if contact is made between the femur and the dash board, Fig. 9.35(a), (b), (d) (Profs. Choi, Poitout) and (e) (http://www.sicot.org/: Library: Online Report E006, February 2002, Figure 1 by: COSTA-PAZ, RANALLETTA, MAKINO, AYERZA and MUSCOLO [2002]). Because of crash forces transmitted along the axis of the femur to the hip joint, knee injuries, such as femoral condyle split and patella fracture, Fig. 9.35(d) and (e), would more likely occur with relatively harder surfaces of the dash board, while hip dislocation, (b), and femur shaft fracture, (c), occur with softer knee padding.

Hip dislocation, Fig. 9.35(b) and femoral neck and acetabulum fractures, (f), are typically seen in a collision of the pelvis with lateral components, causing the femoral head to punch through the acetabulum through direct force application via the greater trochanter (Fig. 9.34(a)). These injuries can occur as a result of direct contact with side structures. This produces a more complex injury pattern, as shown in Fig. 9.35(f)

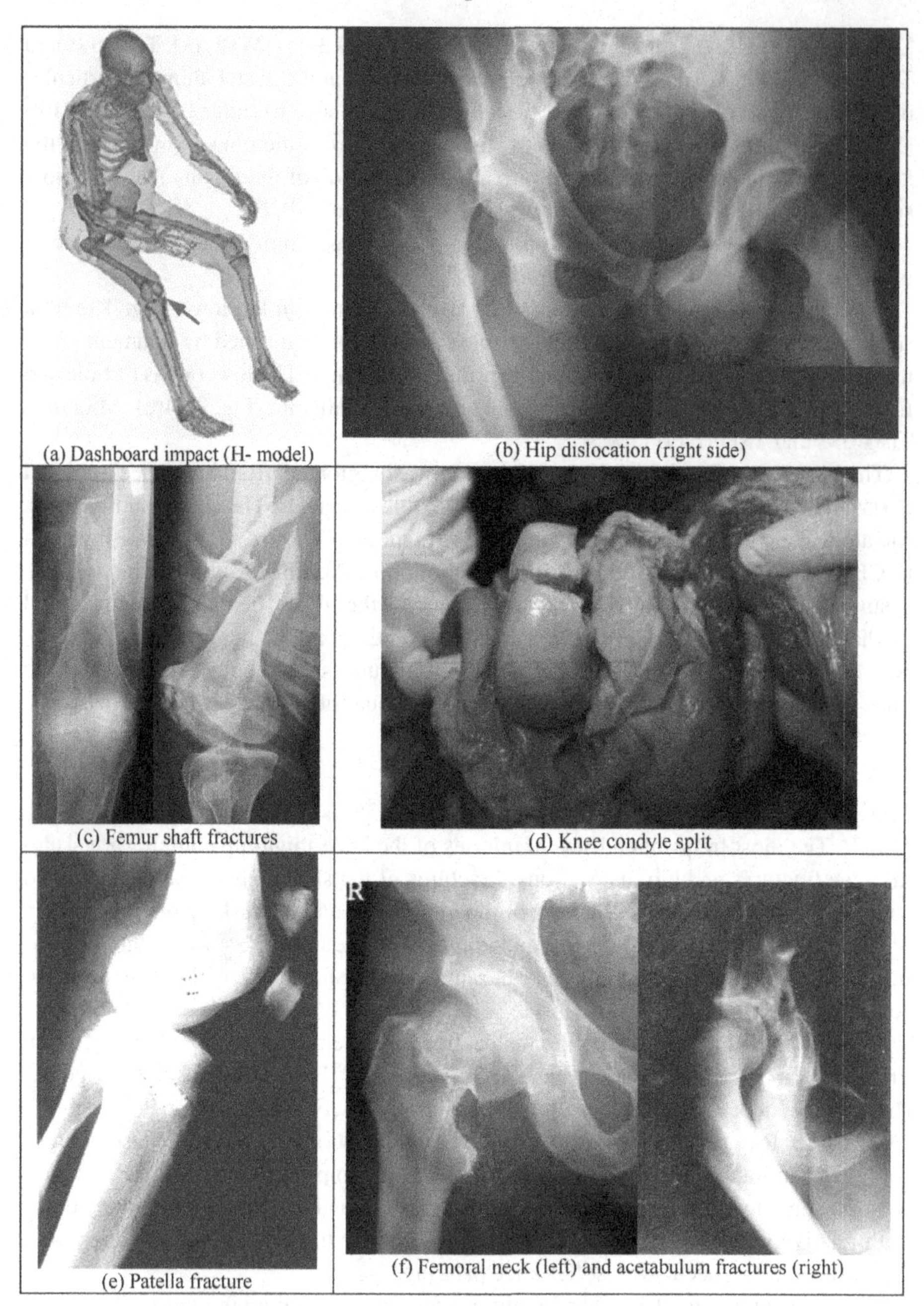

FIG. 9.35. Lower extremity injury. (Insets (b), (c) and (f) left-hand side: Courtesy Prof. D. Poitout, Service de Chirurgie Orthopédique et Traumatologique, Hôpital Nord, Marseille; Inset (d): Private photograph (Prof. H.Y. Choi); Inset (e): Reproduced by permission of SICOT Sociéte Internationale de Chirurgie Orthopedique et de Traumatologie; Inset (f) right-hand side: Reproduced by permission of Glacier Valley Medical Education.)

(Prof. D. Poitout and http://www.glaciermedicaled.com/bone/bonesc11p3.html), that not only involves the articular surfaces of the acetabulum, but also the iliac wing and the pubic rami (arch-like structures) of the pelvis. While bony disruption of the lower limb constitutes a severe injury, additional complications can result from the possibility of soft tissue injury.

Model structure and calibration. The modeling aspects of the human lower extremity are summarized in Fig. 9.36. The H-LE model mainly consists of deformable and damageable bones, including the pelvis bones, and flesh padding. Active and passive muscle forces are modeled with bar elements using the Hill muscle model. Nonlinear contact interfaces model the cartilage layers on hip and knee joints. Fig. 9.36(a) shows the H-LE model in its context, with the attached muscle bars and with the flesh paddings. Fig. 9.36(b) shows the modeling of the cartilage layers of the hip joint with nonlinear contact interfaces. Fig. 9.36(c) contains the modeling of the major hip joint ligaments, which are modeled with nonlinear bars, while Fig. 9.36(d) represents the major knee ligaments. The materials are chosen similar to the materials of the upper extremity.

Validations. From *knee bolster impact* in car frontal crash events, Fig. 9.37 summarizes an investigation made on the injury of the femur (HAYASHI, CHOI, LEVINE, YANG and KING [1996]). Fig. 9.37(a) shows the H-LE model response with distal femur fracture: ("condyle split") due to a hard knee bolster padding material. In that case, the impact force magnitude is high and leads to the observed injury mode. In Fig. 9.37(b) a soft knee bolster padding material was applied, which leads to the observed hip dislocation, because the femur axial force magnitude is too low to cause fracture, but too high over too much time for the hip joint to remain intact. Fig. 9.37(c) represents the same impact using a knee bolster padding material of intermediate stiffness, that leads to the observed femur shaft fracture. The latter injury is deemed to be the least damaging and best healing femur/hip injury. Fig. 9.37(d), finally, compares experimental and calculated impact force time histories.

For a *side impact* scenario, Fig. 9.38 shows the displacements and the deformations of the pelvis (upper pictures) and the pelvic bone with von Mises stress contours and fracture of the pelvic bone (lower pictures).

References on the H-LE model. The H-LE model was built under consultancy of the following references: BACH, HULL and PATTERSON [1997] on the measurement of strain in the anterior cruciate ligament; BEDEWI, MIYAMOTO, DIGGES and BEDEWI [1998] on the human femur FE impact and injury analysis; CAVANAUGH, WALILKO, MALHOTRA, ZHU and KING [1990] on the biomechanical response and injury tolerance of the pelvis in side impact tests; CESARI, BERMOND, BOUQUET and RAMET [1994] on testing and simulation of impacts on the human leg; DALSTRA and HUISKES [1995] on the load transfer across the pelvic bone; DOSTAL [1981] on a 3D biomechanical model of the hip musculature; ENGIN [1979], ENGIN and CHEN [1988a], ENGIN and CHEN [1988b] on the kinematics and passive resistances of the hip joint; FUKUBAYASHI and KUROSAWA [1980] on the contact area and the pressure distribution of the knee joint; HAYASHI, CHOI, LEVINE, YANG and KING [1996] on the

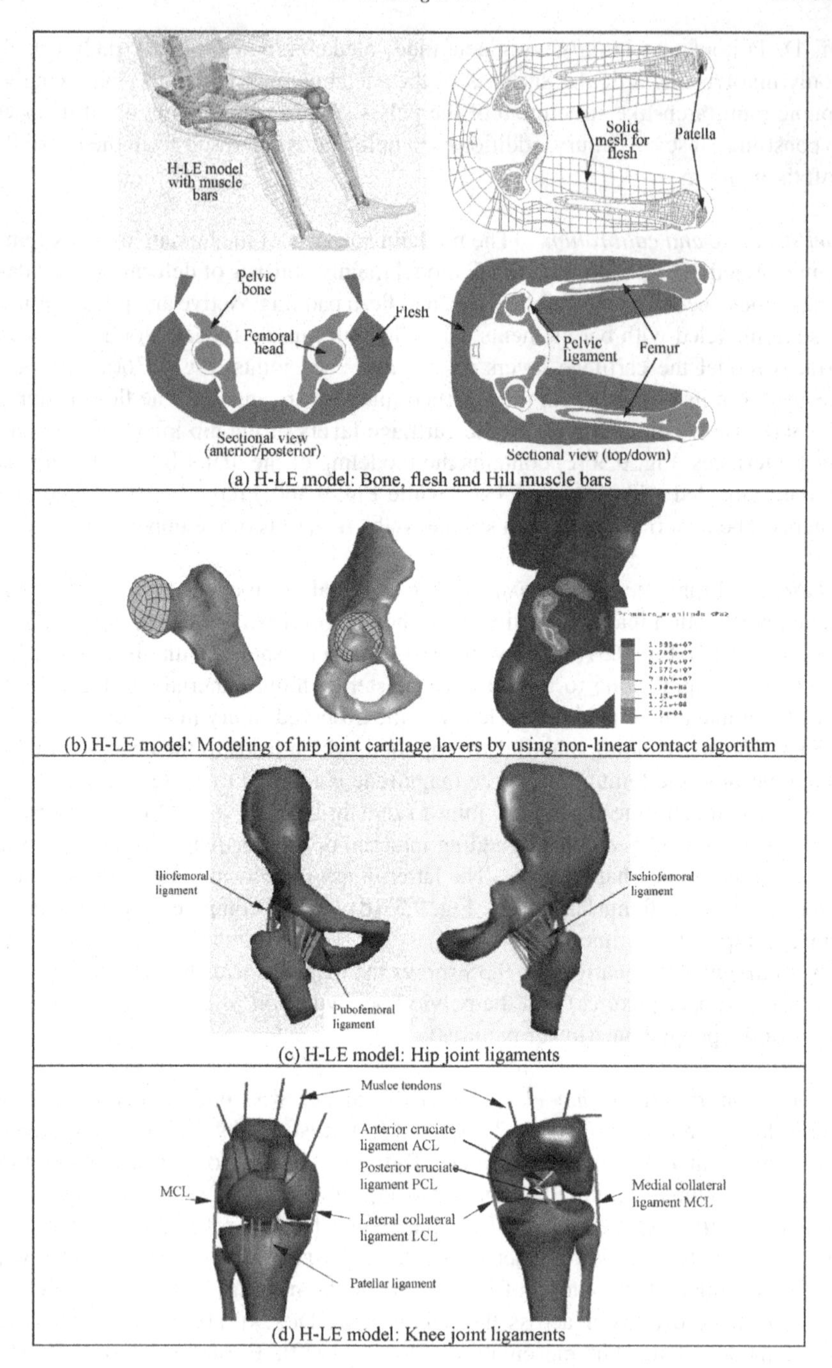

(a) H-LE model: Bone, flesh and Hill muscle bars

(b) H-LE model: Modeling of hip joint cartilage layers by using non-linear contact algorithm

(c) H-LE model: Hip joint ligaments

(d) H-LE model: Knee joint ligaments

FIG. 9.36. H-LE model overview.

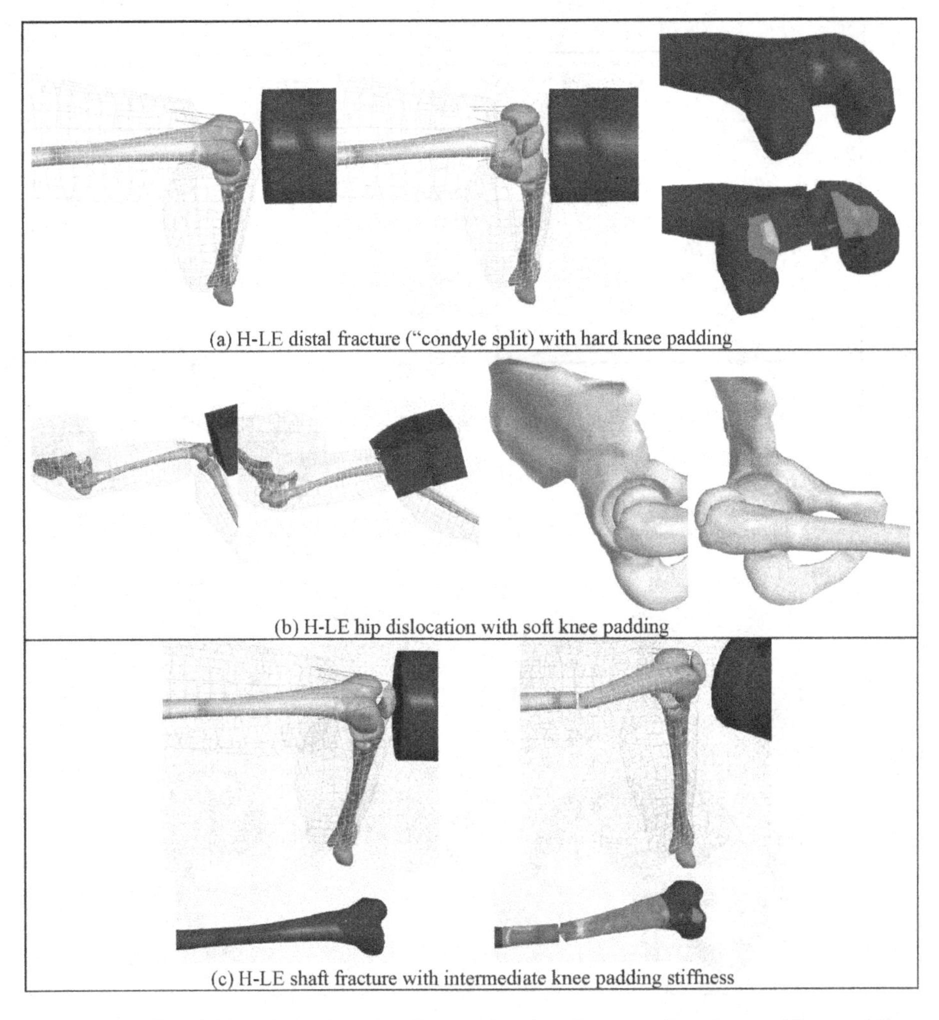

(a) H-LE distal fracture ("condyle split) with hard knee padding

(b) H-LE hip dislocation with soft knee padding

(c) H-LE shaft fracture with intermediate knee padding stiffness

FIG. 9.37. H-LE model knee bolster impact validations (test from HAYASHI, CHOI, LEVINE, YANG and KING [1996]).

experimental and analytical study of frontal knee impact; KAJZER [1991] on the impact biomechanics of knee injury; KAPANDJI [1974b] (lower limb joint physiology); KING [1993] on the injury of the thoraco-lumbar spine and pelvis; KRESS, SNIDER, PORTA, FULLER, WASSERMAN and TUCKER [1993] on the human femur response to impact loading; LEVINE [1993] on injury to the extremities; MARTIN and THOMPSON [1986] on Achilles tendon rupture; MARTINI, TIMMONS and TALLITSCH [2003]: Human Anatomy; MOMERSTEEG, BLASKEVOORT, HUISKES, KOOLOOS and KAUER [1996] on the mechanical behavior of human knee ligaments; NYQUIST, CHENG, EL-BOHY and KING [1985] on the tibia bending strength and response; PATTIMORE, WARD, THOMAS and BRADFORD [1991] on the nature and causes of lower

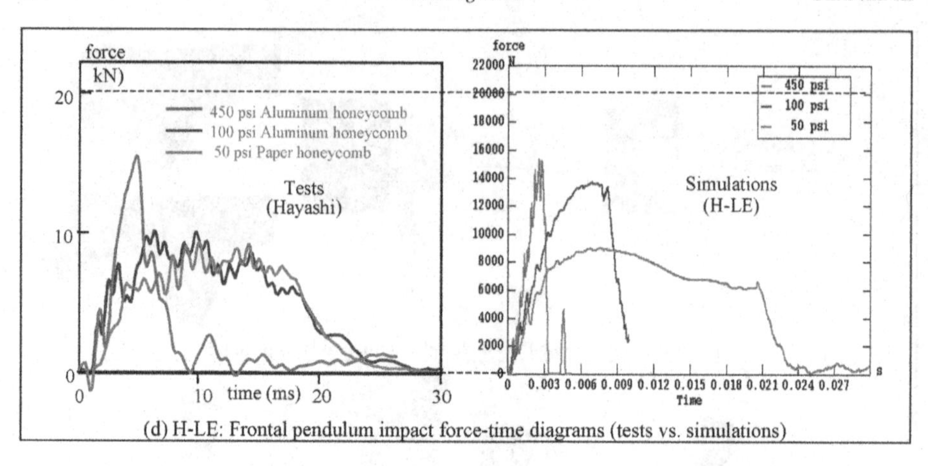

(d) H-LE: Frontal pendulum impact force-time diagrams (tests vs. simulations)

FIG. 9.37. (*Continued.*)

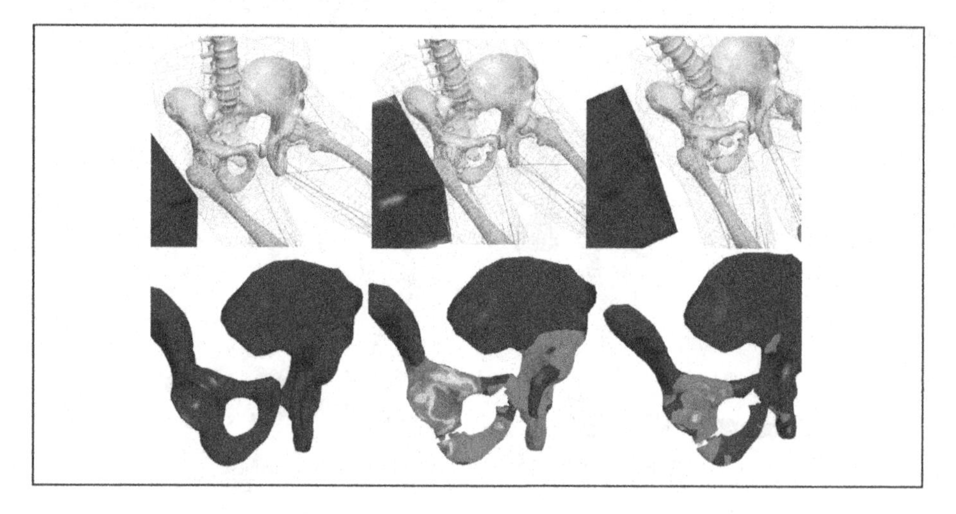

FIG. 9.38. H-LE model side impact validations (pelvis deformation and stress and fracture locations).

limb injuries in car crashes; PORTIER, TROSSEILLE, LE COZ, LAVASTE and COLTAT [1993] on lower leg injuries in real-world frontal car accidents; PUTZ and PABST [2000]: Sobotta Atlas of Human Anatomy; RENAUDIN, GUILLEMOT, PÉCHEUX, LESAGE, LAVASTE and SKALLI [1993] on a 3D FE model of the pelvis in side impacts; ROHEN and YOKOCHI [1983]: Color Atlas of Anatomy; STATES [1986] on adult occupant injuries of the lower limb; WYKOWSKI, SINNHUBER and APPEL [1998] on a finite element model of the human lower extremity in frontal impact; YANG and LÖVSUND [1997] on a human model for pedestrian impact simulation. The following web sites were consulted: http://www.rad.washington.edu/RadAnat/pelvis.html; http://www.rad.washington.edu/RadAnat/knee.html; http://www.sicot.org/; http://www.glaciermedicaled.com/bone/bonesc11p3.html.

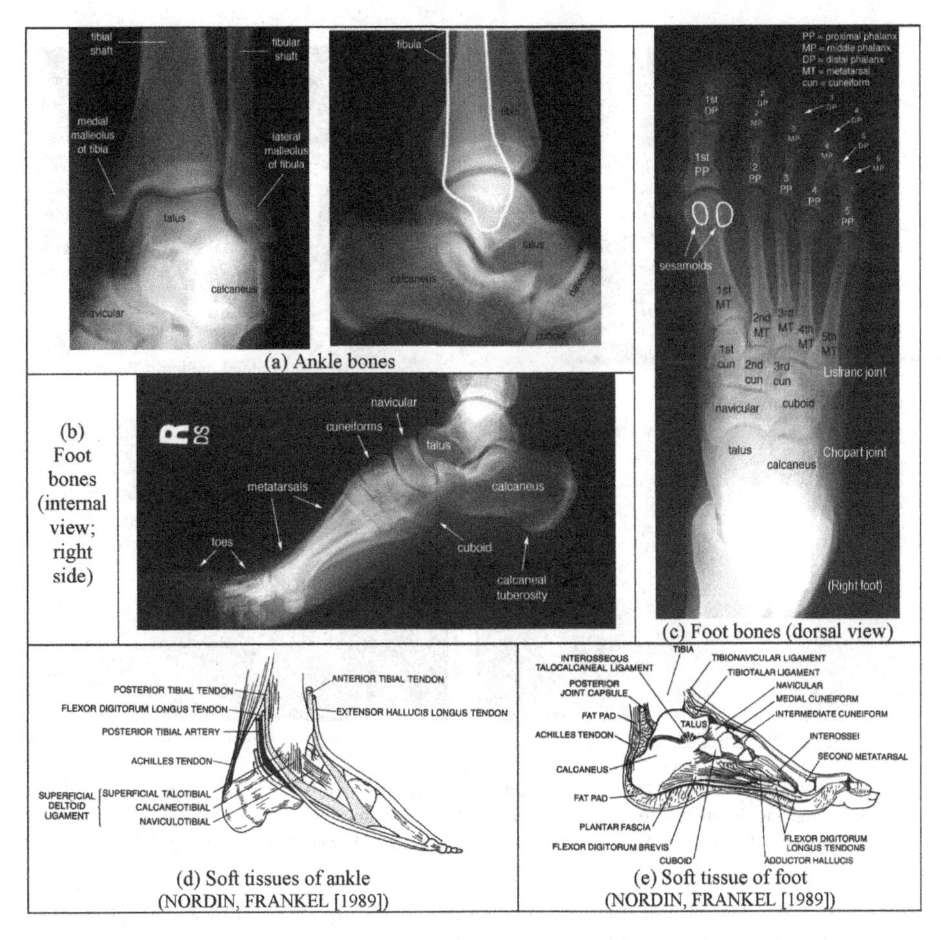

(a) Ankle bones

(b) Foot bones (internal view; right side)

(c) Foot bones (dorsal view)

(d) Soft tissues of ankle (NORDIN, FRANKEL [1989])

(e) Soft tissue of foot (NORDIN, FRANKEL [1989])

FIG. 9.39. Anatomy of the human foot/ankle complex. (Insets (a)–(c): Reproduced by permission of Michael L. Richardson, University of Washington Medical Center, Department of Radiology; Insets (d) and (e): Reproduced by permission of Lippincott, Williams & Wilkins.)

9.8. *H-Ankle&Foot*

Overviews on the H-Ankle&Foot model structure and applications are given in BEAUGONIN, HAUG, MUNCK and CESARI [1995], BEAUGONIN, HAUG and CESARI [1996], BEAUGONIN, HAUG, MUNCK and CESARI [1996], BEAUGONIN, HAUG and CESARI [1997].

Anatomy. The anatomy of the human foot/ankle complex is summarized in Fig. 9.39. The skeleton of the foot consists in the short bones of the tarsus (a) (http://www.rad.washington.edu/RadAnat/AnkleMortiseLabelled.html and http://www.rad.washington.edu/RadAnat/AnkleLaterallLabelled.html), namely the talus (or astragalus), which establishes the articulated connection between the foot and the leg (tibia, fibula), the calcaneous, the navicular (or scaphoid), the cuboid and the cuneiform bones,

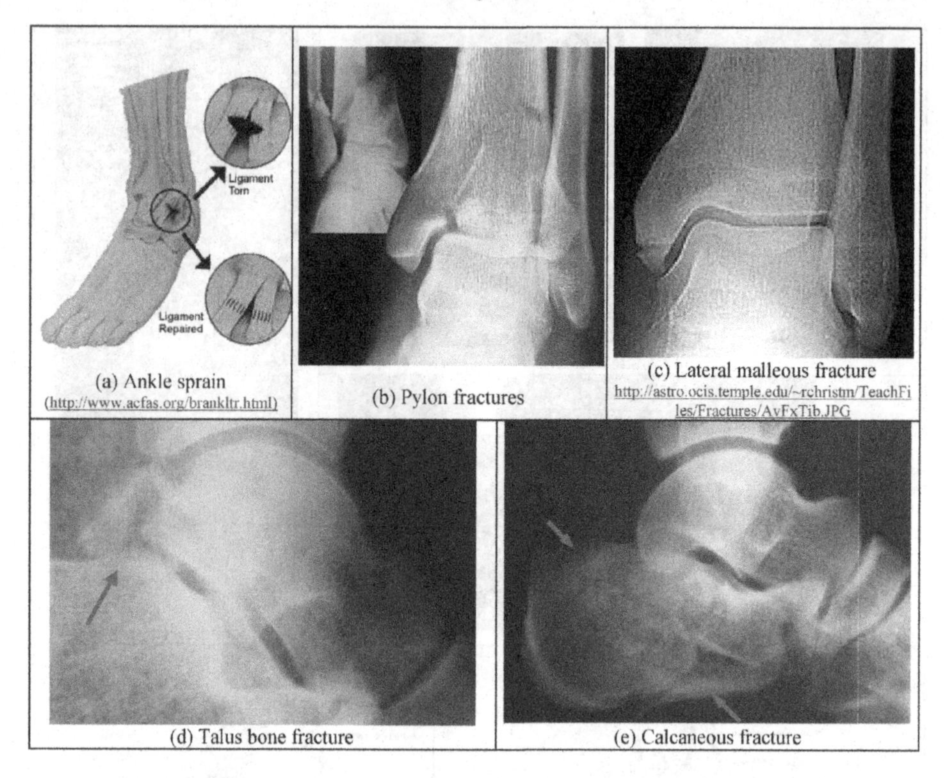

(a) Ankle sprain (http://www.acfas.org/brankltr.html)
(b) Pylon fractures
(c) Lateral malleous fracture http://astro.ocis.temple.edu/~rchristm/TeachFiles/Fractures/AvFxTib.JPG
(d) Talus bone fracture
(e) Calcaneous fracture

FIG. 9.40. Foot and ankle injury. (Inset (a): Reproduced by permission of ACFAS American College of Foot and Ankle Surgeons; Insets (b), (d) and (e): Courtesy Prof. D. Poitout, Service de Chirurgie Orthopédique et Traumatologique, Hôpital Nord, Marseille; Inset (c): Courtesy of Robert A. Christman, D.P.M., Philadelphia, PA.)

and of the metatarsal and the phalanx bones (b) (http://www.rad.washington.edu/RadAnat/FootLateralLabelled.html), (c) (http://www.rad.washington.edu/RadAnat/FootAPLabelled.html). Besides the major articulations of the talus bone, the Lisfranc and the Chopart joints provide minor mobility (c). These bones are connected by numerous ligaments and soft tissues, (d) and (e) (NORDIN and FRANKEL [1989]).

Injury. Injury of the foot/ankle complex, Fig. 9.40 (see indicated web sites), is discussed in part in LEVINE [1993]. As much as one third of all surviving vehicle crash victims sustain lower limb injury, where belt use does not alter significantly the risk. These injuries are not life threatening, but cause extensive health care cost and long periods of recovery. Ankle and foot injuries are mainly attributed to foot well intrusion in frontal crashes. They can be classified as *skeletal injury*, such as ankle fractures (malleolar or bimalleolar fractures; tibial pylon fractures; talar fractures) and foot fractures (e.g., the metatarsal bones; the calcaneus; the cuboid) and *internal injury,* such as joint injury (Lisfranc and Chopart joints), ligamentous injury (sprains, tears), tendon injury (Achilles tendon damage or rupture).

MORGAN, EPPINGER and HENNESSEY [1991] described six mechanisms which they consider the most frequent in vehicle crash (leg trapped between floor and instrument

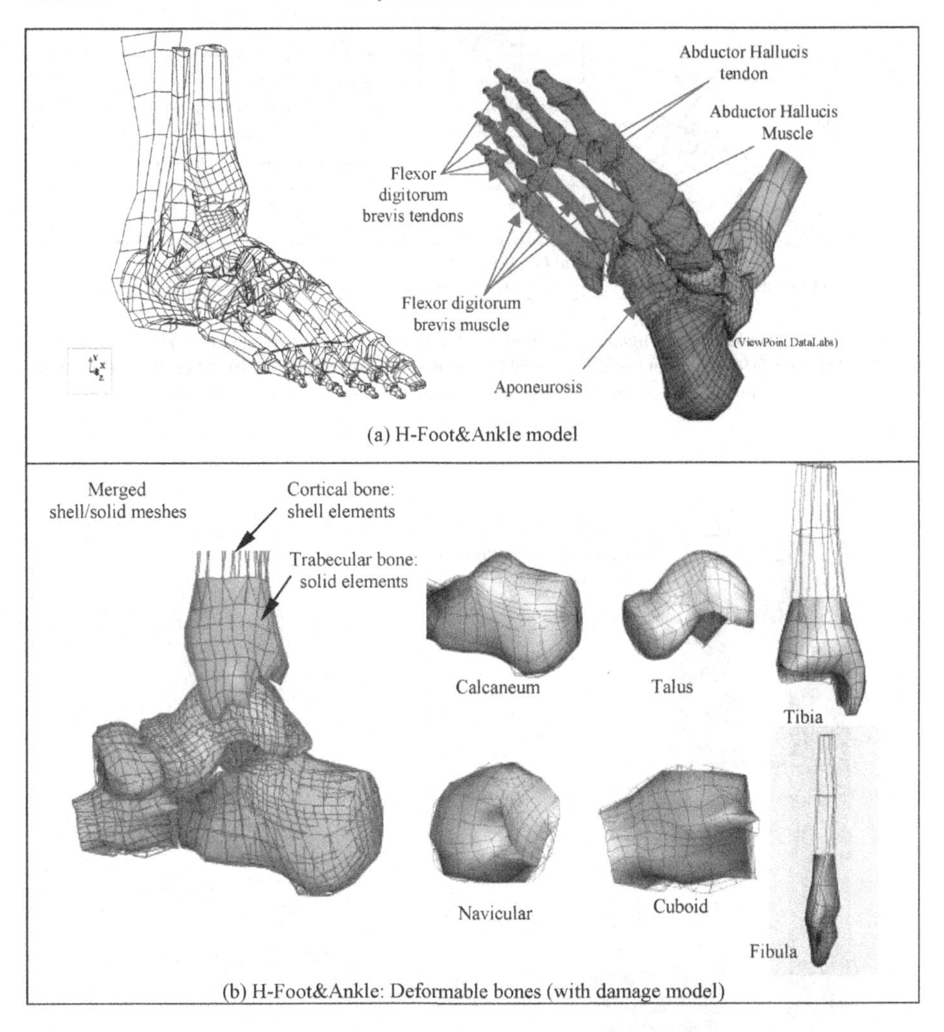

FIG. 9.41. H-Foot&Ankle model overview (BEAUGONIN, HAUG, MUNCK and CESARI [1995], BEAUGONIN, HAUG and CESARI [1996]). (Reproduced by permission of The Stapp Association.)

panel with pocketing of instrument panel; contact with foot controls; wheel well intrusion; contact with floor; collapse of leg compartment; foot trapped under pedals). These mechanisms correspond to a combination of ankle/foot simple movements like dorsiflexion, plantarflexion, pronation and supination. These movements can be associated with direct or indirect loading conditions.

Model structure and calibration. The foot/ankle complex (H-Ankle&Foot) of the H-Model mainly consists of deformable and damageable bone models with ligaments modeled with membranes and bars, Fig. 9.41 (BEAUGONIN, HAUG, MUNCK and CESARI [1995], BEAUGONIN, HAUG and CESARI [1996]).

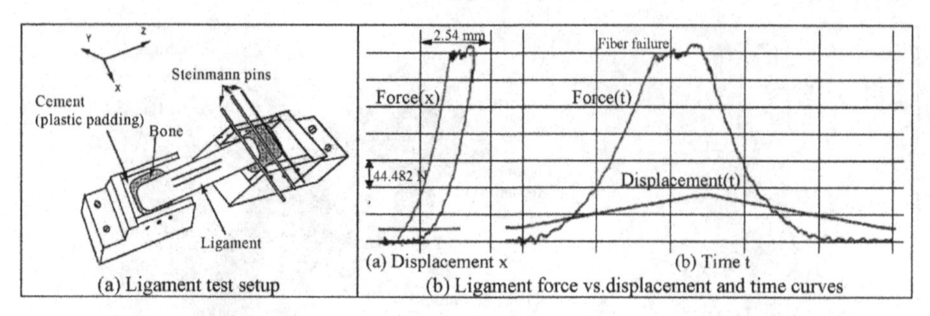

FIG. 9.42. H-Foot&Ankle model ligament calibration (after PARENTEAU, VIANO and PETIT [1996] and BEGEMAN and AEKBOTE [1996]). (Inset (a): Reproduced by permission of ASME; Inset (b): After material received from Prof. Begeman at Wayne State University, Bioengineering Center.)

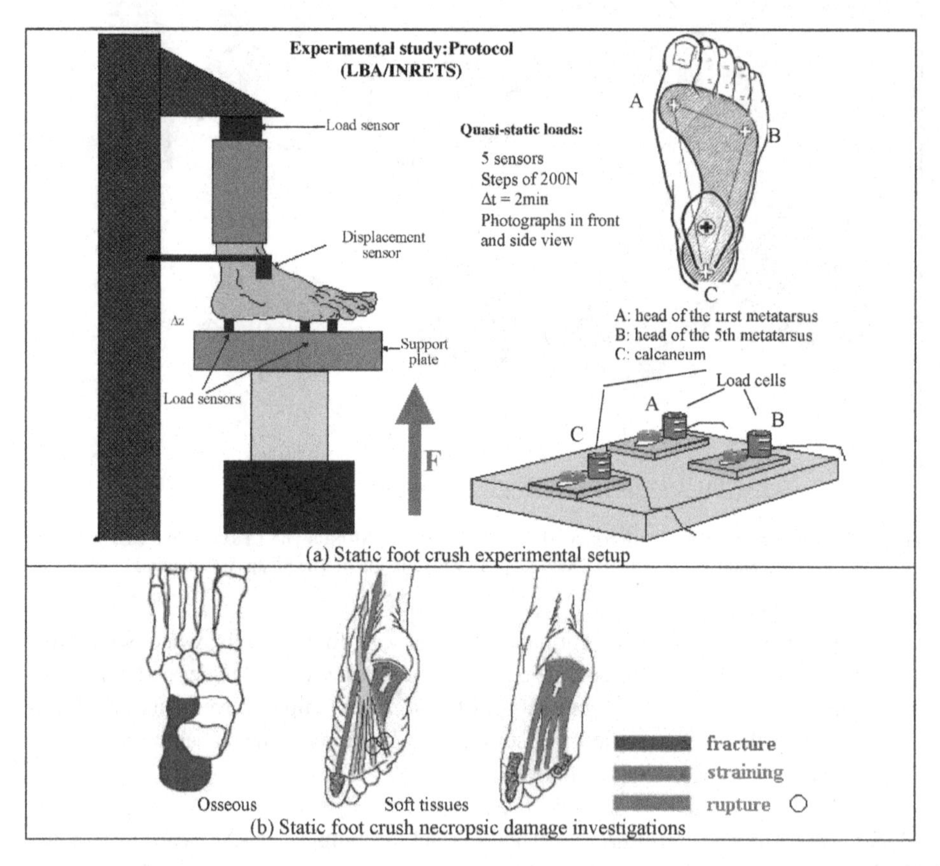

FIG. 9.43. H-Foot&Ankle validation: Static crush behavior: Experimental tests and numerical simulations (MASSON, CESARI, BASILE, BEAUGONIN, TRAMECON, ALLAIN and HAUG [1999]). (Reproduced by permission of IRCOBI.)

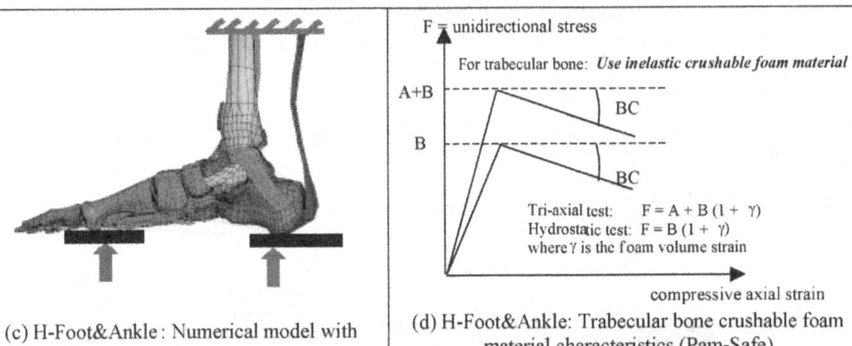

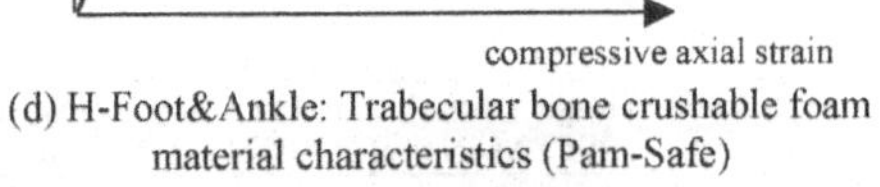

(c) H-Foot&Ankle: Numerical model with damageable bones

(d) H-Foot&Ankle: Trabecular bone crushable foam material characteristics (Pam-Safe)

Two competing damage modes:

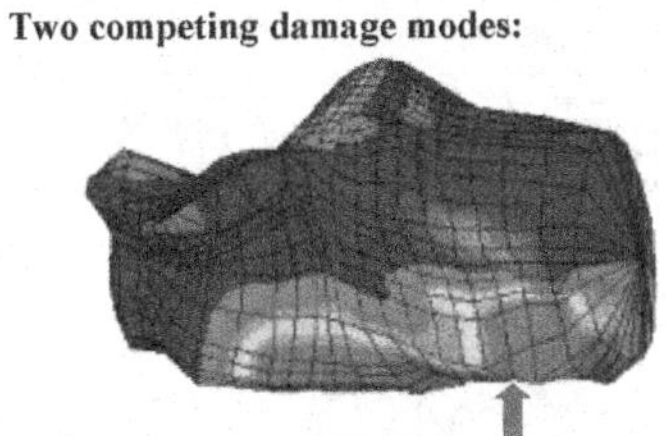

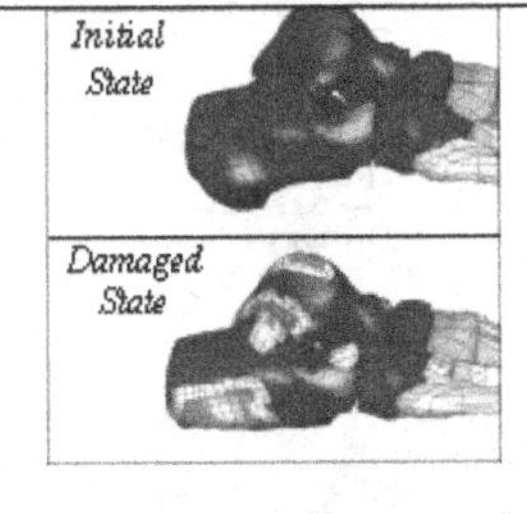

Cortical bone low yield stress (50MPa)
(calcaneum crushing)

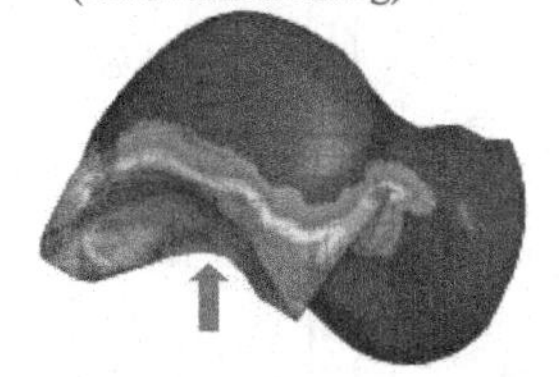

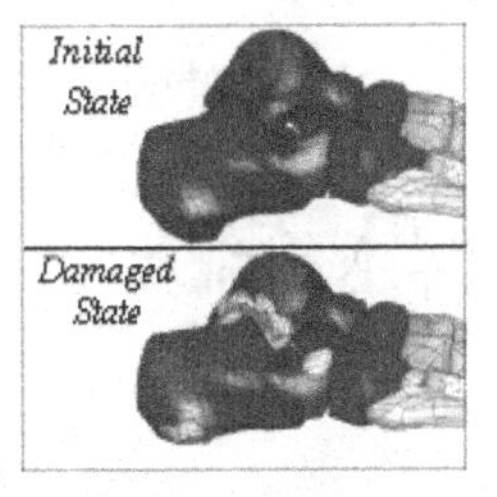

Cortical bone high yield stress(110MPa)
(talus crushing)

(e) H-Foot&Ankle: Numerical simulation results: Influence of cortical bone characteristics

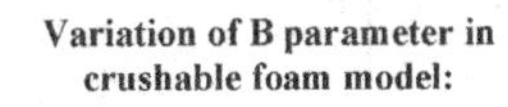

Variation of B parameter in crushable foam model:

If B is low (3MPa and 5MPa) ➡ *crush mode*

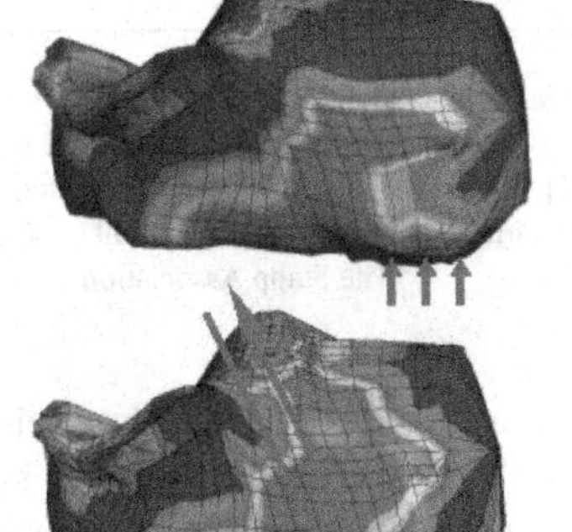

If B is high (7MPa and 10MPa) ➡ *shear mode*

(f) H-Foot&Ankle: Numerical simulation results: Influence of trabecular bone characteristics

FIG. 9.43. (*Continued.*)

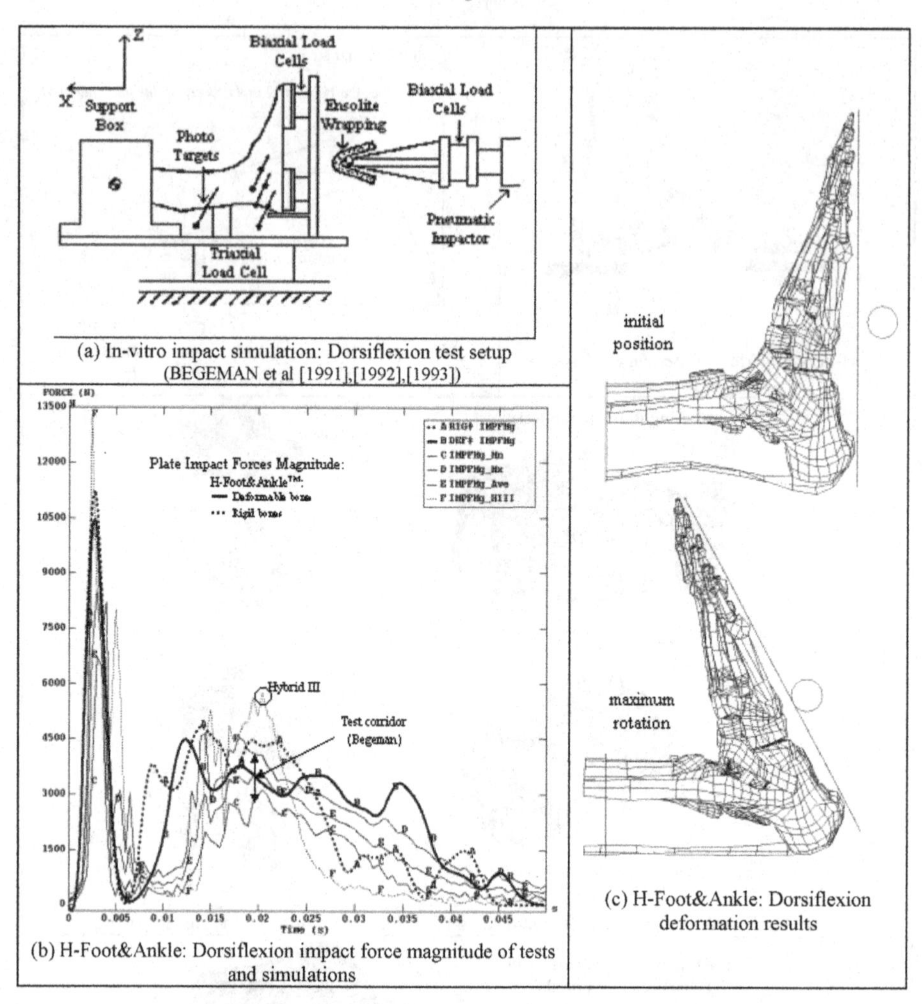

FIG. 9.44. H-Foot&Ankle validation: In vitro dynamic impacts: Experimental tests and numerical simulations (BEAUGONIN, HAUG and CESARI [1996], BEAUGONIN, HAUG, MUNCK and CESARI [1996], BEAUGONIN, HAUG and CESARI [1997], BEGEMAN and KOPACZ [1991], BEGEMAN, BALAKRISHNAN, LEVINE and KING [1992], BEGEMAN, BALAKRISHNAN, LEVINE and KING [1993]). (Insets (a)–(g) and (i): Reproduced by permission of The Stapp Association; Inset (h): ESI Software.)

Nonlinear contact interfaces model the cartilage layers of the major foot/ankle joints. Nonlinear joint elements connect the lesser bones. Fig. 9.41(a) and 9.41(b) show the H-Ankle&Foot model globally and with its major bones in detail. The latter are modeled with damageable thin shell elements for the cortical bone and with crushable foam solids for the trabecular bone.

Fig. 9.42 shows a typical ligament *calibration* test, where a ligament is isolated with its bony insertions, potted in grips and tested in tension (PARENTEAU, VIANO and PETIT [1996]). Typical force-displacement and force-time response curves are shown

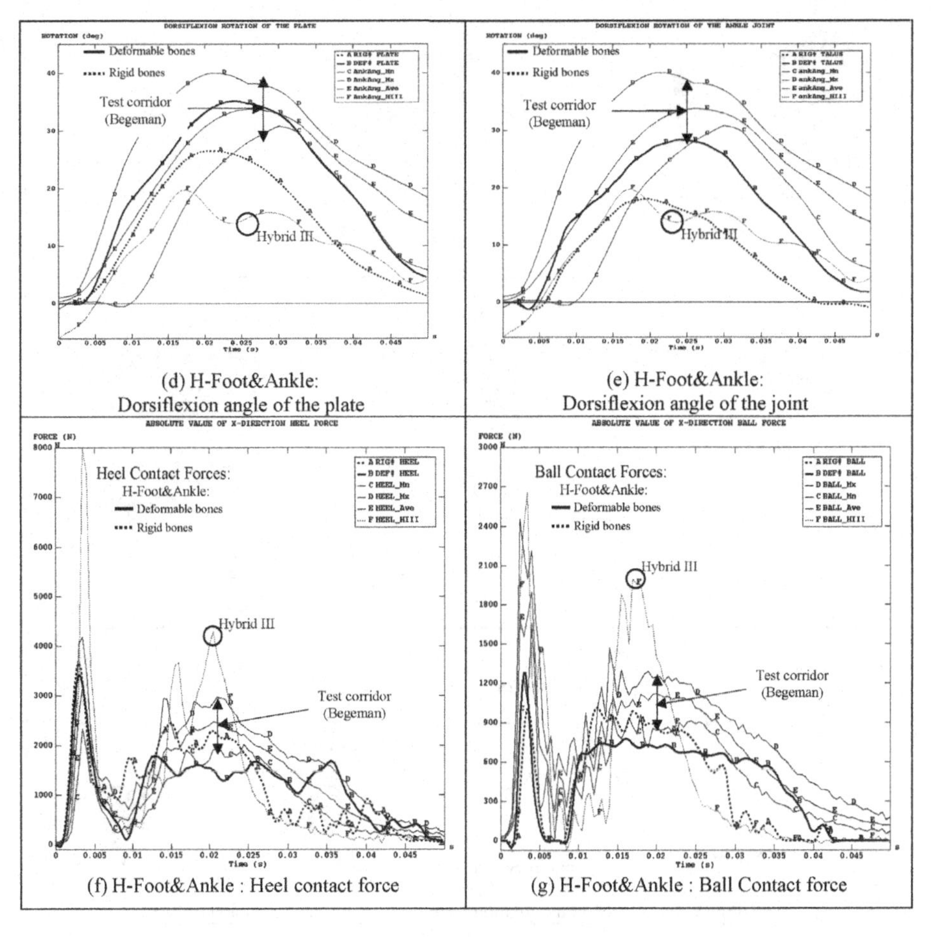

FIG. 9.44. *(Continued.)*

(BEGEMAN and KOPACZ [1991], BEGEMAN, BALAKRISHNAN, LEVINE and KING [1992], BEGEMAN, BALAKRISHNAN, LEVINE and KING [1993]), which exhibit unloading hysteresis and fiber failure. This behavior is easily simulated using bars with nonlinear elastic-plastic material behavior.

Validations. For *static crush behavior* of the H-Foot&Ankle model, Fig. 9.43 shows experimental tests and numerical simulations, where different calcaneum and talus bone failure modes are obtained when the material properties of the cortical and trabecular bones are subject to parametric variations (MASSON, CESARI, BASILE, BEAUGONIN, TRAMECON, ALLAIN and HAUG [1999]).

For *dynamic plantar impacts*, Fig. 9.44 depicts the in vitro experimental tests (BEGEMAN and KOPACZ [1991], BEGEMAN, BALAKRISHNAN, LEVINE and KING [1992], BEGEMAN, BALAKRISHNAN, LEVINE and KING [1993]) and numerical simulations for the H-Foot&Ankle model validation (BEAUGONIN, HAUG, MUNCK and CESARI [1995],

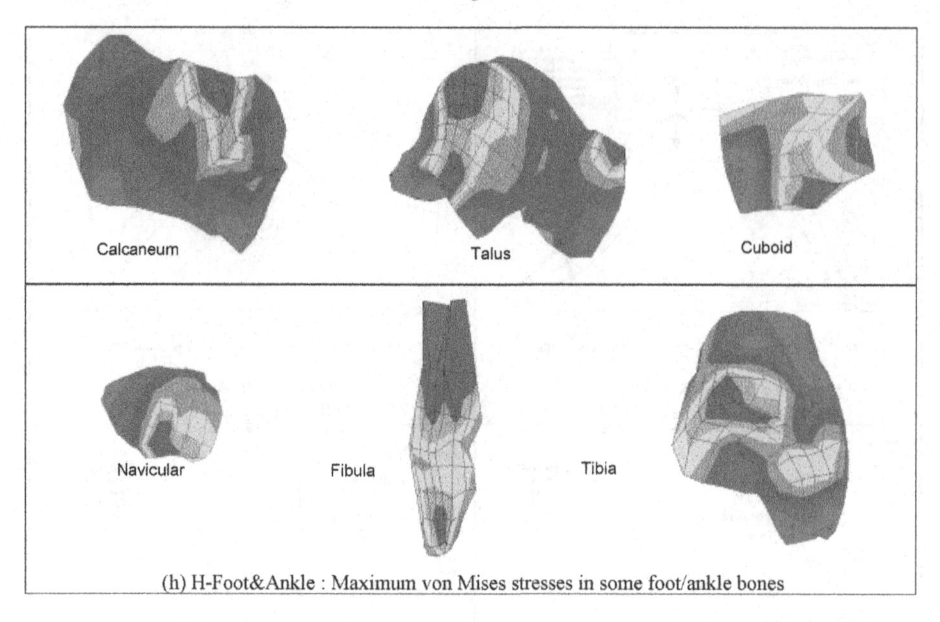

(h) H-Foot&Ankle : Maximum von Mises stresses in some foot/ankle bones

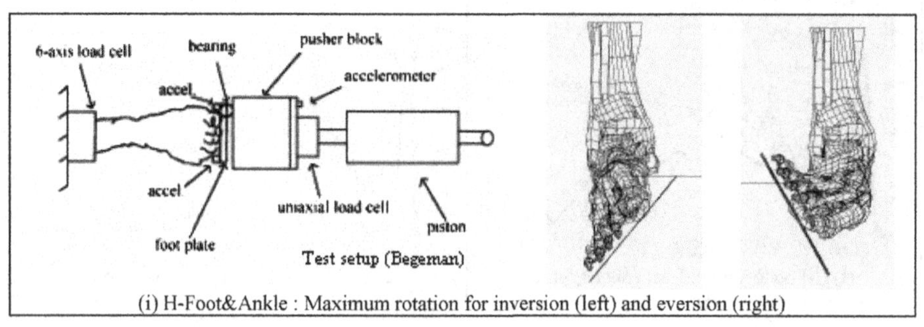

(i) H-Foot&Ankle : Maximum rotation for inversion (left) and eversion (right)

FIG. 9.44. *(Continued.)*

BEAUGONIN, HAUG and CESARI [1996], BEAUGONIN, HAUG, MUNCK and CESARI [1996], BEAUGONIN, HAUG and CESARI [1997]).

While Fig. 9.44(a)–(h) treat the dorsiflexion case, Fig. 9.44(i) treats the cases of inversion and eversion due to plantar impacts (BEGEMAN, BALAKRISHNAN, LEVINE and KING [1993], BEAUGONIN, HAUG and CESARI [1996]). The validations are described in more detail in the cited literature. The comparisons with the response curves of a Hybrid III mechanical crash dummy model, indicated in the diagrams, shows a considerable deviation from the response curves obtained with the human model, which demonstrates the limited biofidelity of crash dummies.

References on the H-Ankle&Foot model. The following references were consulted for the establishment of the H-Foot&Ankle model: ATTARIAN, MCCRACKIN, DEVITO, MCELHANEY and GARRETT [1985] on the biomechanical characteristics of the human ankle ligaments; BEGEMAN and KOPACZ [1991], BEGEMAN, BALAKRISHNAN, LEVINE and KING [1992], BEGEMAN, BALAKRISHNAN, LEVINE and KING [1993] on the biomechanics of the human ankle impact in dorsiflexion (1991), on the human ankle dynamic response in dorsiflexion (1992) and for inversion/eversion (1993); CESARI, BERMOND, BOUQUET and RAMET [1994] on testing and simulation of impacts on the human leg; KAPANDJI [1974b] (lower limb joint physiology); LESTINA, KUHLMANN, KEATS and MAXWELL ALLEY [1992] on fracture mechanisms of the foot/ankle complex in car accidents; LUNDBERG, GOLDIE, KALIN and SELVIK [1989], LUNDBERG, SVENSSON, BYLUND, GOLDIE and SELVIK [1989] on the kinematics of the ankle/foot complex in dorsiflexion (a) and in pronation and supination (b); NAHUM, SIEGEL, HIGHT and BROOKS [1968] on lower extremity injuries in front seat occupants; NORDIN and FRANKEL [1989] on the Basic Biomechanics of the Musculoskeletal System; OTTE, VON RHEINBABEN and ZWIPP [1992] on the biomechanics of ankle/foot joint injury; PARENTEAU and VIANO [1996] on the kinematics and PARENTEAU, VIANO and PETIT [1996] on the biomechanical properties of the ankle subtalar joints in quasi-static loading up to failure; PATTIMORE, WARD, THOMAS and BRADFORD [1991] on the nature and causes of lower limb injuries in car crashes; PORTIER, TROSSEILLE, LE COZ, LAVASTE and COLTAT [1993] on lower leg injuries in real-world frontal car accidents; STATES [1986] on adult occupant injuries of the lower limb; WILSON-MACDONALD and WILLIAMSON [1988] on severy ankle joint injuries; WYKOWSKI, SINNHUBER and APPEL [1998] on a finite element model of the human lower extremity in frontal impact. The following web sites were consulted: http://www.rad.washington.edu/RadAnat; http://www.acfas.org/brankltr.html; http://astro.ocis.temple.edu/~rchristm/TeachFiles/Fractures/AvFxTib.JPG.

10. The fifth percentile female H-Model

Model structure and calibration. The fifth percentile female H-Model (weight 50 kg; height 1.52 m) is being built according to the same basic principles than the fiftieth percentile male H-Model (weight 75.5 kg; height 1.75 m), where the anatomy of the body was adjusted to the dimensions of the fifth percentile woman, Fig. 10.1. Inset (a) compares the shape of the model to the shape of the 5th percentile female Hybrid III mechanical dummy model. Inset (b) gives an overview of the model of the thorax. Inset (c) gives details of this model, concerning the deformable neck, breasts, rib cage, thoracic organs and the heart. All sub-models were built and calibrated in a way similar to the corresponding 50th percentile male sub-models, described earlier. The breasts were modeled as “bio-bags”, Appendix D, which consist in a non-linear elastic outer skin with a quasi-incompressible filling in the volume created by the inner thorax lining and the outer skin. The modeling of the breasts is not considered final at that stage, and models using solid finite elements for the bulk (fatty tissue) and membranes for the outer envelope (skin) are built.

Calibration for thorax pendulum impact. Fig. 10.2(a) shows the Kroell pendulum impact test setup for female cadavers (KROELL, SCHNEIDER and NAHUM [1971], KROELL, SCHNEIDER and NAHUM [1974]). In inset (b) the measured, idealized, high velocity 6.71 m/s (solid lines) and low velocity 4.27 m/s (dashed lines) impact force versus chest compression corridors for pendulum impacts of 23.15 kg mass are shown.

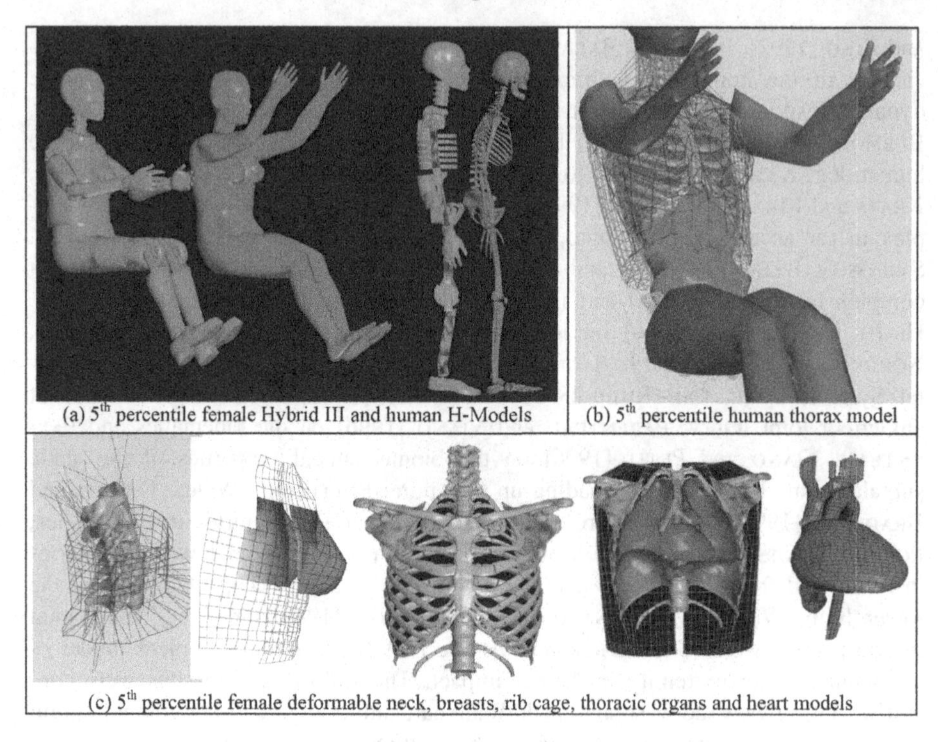

Fig. 10.1. Fifth percentile deformable female H-Model.

Inset (c) demonstrates the final calibrated response of the female model at high impact velocity (6.71 m/s) to fall well within the test corridor. Inset (d) demonstrates at low impact velocity (4.27 m/s) the influence of various stiffness parameters, and notably of the presence and absence of the breasts during calibration.

Inset (e) contains sections at 10 millisecond intervals of the compression of the chest and organs, while inset (f) shows overall pictures and details on predicted rib fracture during this test.

During the calibration process, a first model used raw material data from the male model, leading to Fig. 10.2(d), curves (1a) and (1b). This model was generally too stiff. A preliminary evaluation of the modeling of the breasts with that model showed a considerable influence of their presence or absence. If breasts were modeled, curve (1a), the chest deformations were under-predicted, as compared to the case when the breasts were removed, curve (1b). After calibration of all material data, the response of the model with breasts was close to the measured test corridors, curve (2).

Validation for out-of-position airbag inflation. In Fig. 10.3(a) the fifth percentile female H-Model is exposed to a typical "out-of-position" (OOP) driver side airbag inflation scenario, cf. Rudolf, Fellhauer, Schaub, Marca and Beaugonin [2002]. Contrary to a standard driver position (Fig. 10.1 insets (a), (b)), Fig. 10.3 shows the 5th percentile female H-Model positioned close to the steering wheel, being rejected

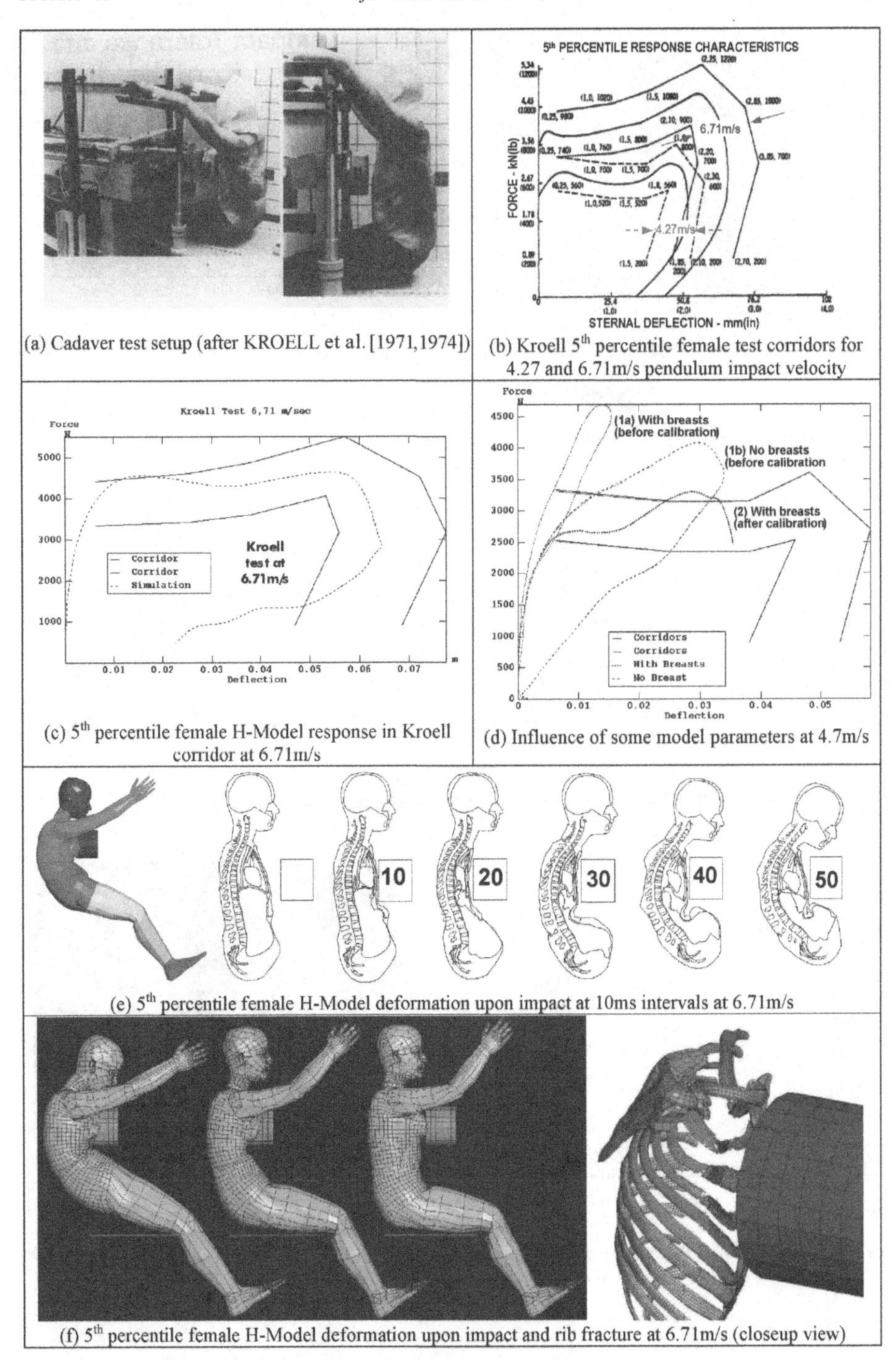

(a) Cadaver test setup (after KROELL et al. [1971,1974])

(b) Kroell 5th percentile female test corridors for 4.27 and 6.71m/s pendulum impact velocity

(c) 5th percentile female H-Model response in Kroell corridor at 6.71m/s

(d) Influence of some model parameters at 4.7m/s

(e) 5th percentile female H-Model deformation upon impact at 10ms intervals at 6.71m/s

(f) 5th percentile female H-Model deformation upon impact and rib fracture at 6.71m/s (closeup view)

FIG. 10.2. Calibration of the 5th percentile deformable female H-Thorax (tests after KROELL, SCHNEIDER and NAHUM [1971], KROELL, SCHNEIDER and NAHUM [1974]). (Insets (a), (b): Reproduced by permission of The Stapp Association.)

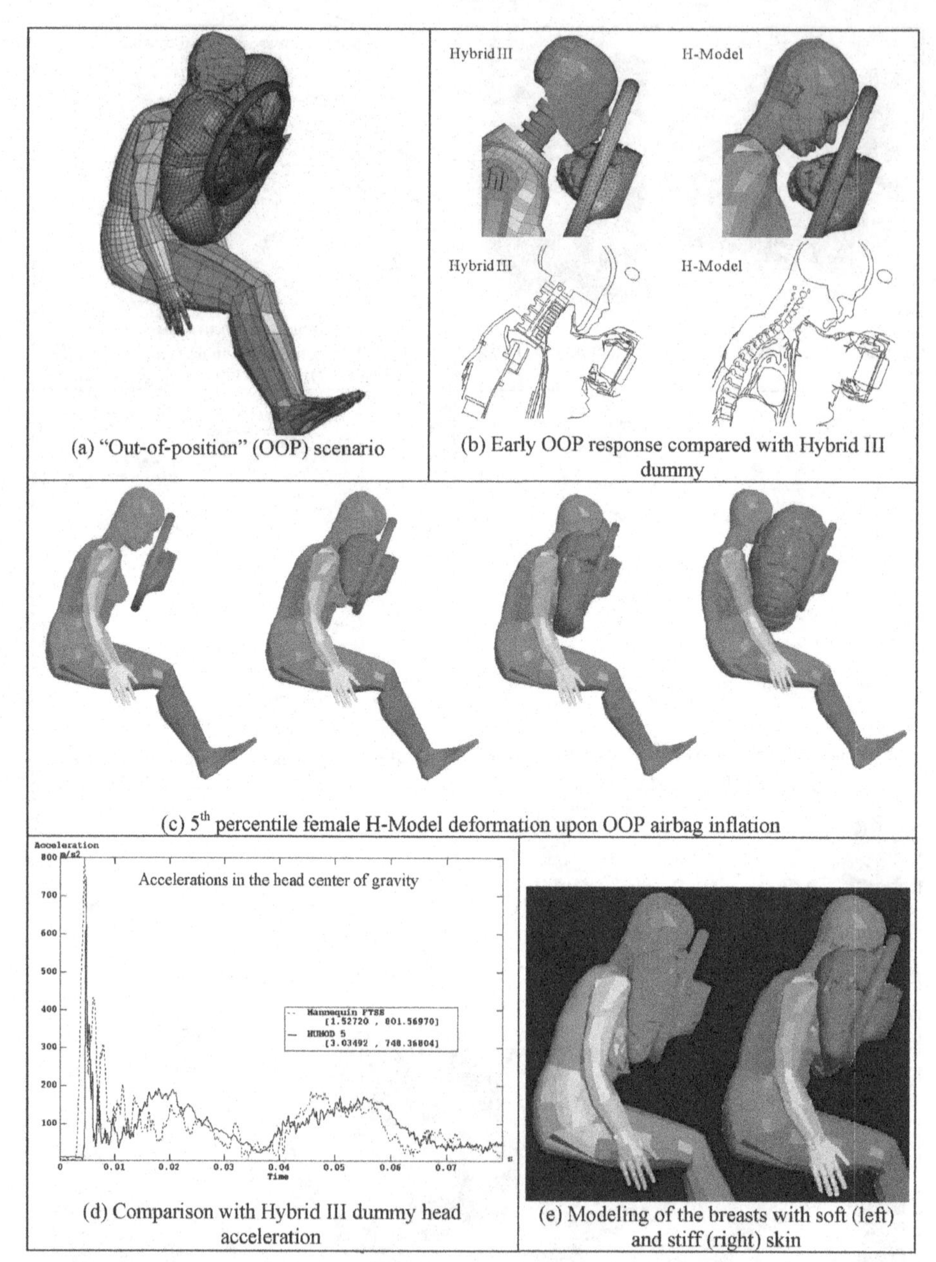

(a) "Out-of-position" (OOP) scenario

(b) Early OOP response compared with Hybrid III dummy

(c) 5th percentile female H-Model deformation upon OOP airbag inflation

(d) Comparison with Hybrid III dummy head acceleration

(e) Modeling of the breasts with soft (left) and stiff (right) skin

FIG. 10.3. Validation of the 5th percentile deformable female H-Model (OOP case) (RUDOLF, FELLHAUER, SCHAUB, MARCA and BEAUGONIN [2002]).

to the rear by the inflating driver side airbag, inset (c). Whereas in a normal position the occupant comes into contact with a fully inflated airbag, the OOP occupant is hit by the airbag while it inflates, which can case injury through the phenomenon called "bag-slap". Further, the inflating bag may be trapped under the chin, inset (b), and thus project the head and neck towards the rear with an over extension of the neck. The mid-sagittal section plots of inset (b) suggest that injury of the head/neck/thorax complex can be assessed much more readily from the human model than from the dummy model. Using human models and models of the standard 5th percentile female Hybrid III mechanical dummy, one can appreciate the differences between the response of the dummy and the human, inset (d). Inset (e), finally, gives an impression on the influence of the modeling if the outer skin of the breasts, which, when not correctly stiffening at large skin strain, may be reacting in an overly soft manner.

In the out-of-position airbag validation test case, the outer skin of the breasts was initially modeled with the isotropic material properties as calibrated from the pendulum impact tests. This led to large deformations in the OOP case, as observed in Fig. 10.3(e) (left). Such deformations were not seen previously in the pendulum impact simulations, because the pendulum impacted in a more concentrated fashion, between the breasts. Then the skin was re-modeled with an orthotropic fiber reinforced material, with stiffening fibers at larger strains, as it is characteristic of skin (YAMADA [1970] and Fig. 1.1(h)). If this stiffening effect was neglected, the skin was too soft and the final deformations were too large. It was found that the volume and the mechanical resistance of the breasts can play an important role in the energy absorption characteristics of the thorax from frontal impacts. It is therefore important to further evaluate their influence in future studies.

Further investigations on the small female dummy and cadaver tests under static airbag out-of-position deployment tests can be found, e.g., in CRANDALL, DUMA, BASS, PILKEY, KUPPA, KHAEWPONG and EPPINGER [1999].

APPENDIX A

Basic Theory of Crash Codes

The following paragraphs are based mainly on Pam–Crash documentation of ESI Software. See also HAUG, CLINCKEMAILLIE and ABERLENC [1989a], HAUG, CLINCKEMAILLIE and ABERLENC [1989b].

A.1. Overviews on solution methods and finite elements

Solution method overview. Modern crash and impact simulation codes are three-dimensional, Lagrangian, finite element, explicit and implicit, vectorized and multi-tasked *solid codes* for the non-linear dynamic and large deformation analysis of solid structures in the realm of computational structural mechanics (CSM). They analyze crash phenomena at discrete points in space and time.

Space discretization. Space is discretized with the most often used *finite element methodology*, which is based on the *displacement method* of structural analysis, where the discrete nodal displacements and rotations constitute the unknowns of the problem. A comprehensive account on the theory and application of the finite element method in engineering is given in the textbook by BATHE [1996]. Some solid codes are coupled to *flow codes*, which treat problems in the realm of computational fluid dynamics (CFD), in order to treat fluid-structure interaction (FSI) dominated problems. Most solid codes are provided with alternative spatial discretization schemes called *particle methods*, where a solid, fluid or gaseous medium is represented by *mesh-less* discrete (mass or integration) points, such as in the most common mass-point smooth particle hydrodynamics (SPH) scheme. This permits to solve certain classes of FSI problems with the solid codes without the necessity to couple with a separate flow code.

The solid codes allow to model 3D structures of arbitrary geometry using solid elements, membrane elements, plate and shell elements, beam and bar elements, and discrete spring and joint elements. In typical crashworthiness and impact simulations, plates and shells are used to model thin-walled metal or plastic components. Beams and bars are used for stiffening frames, wheel suspensions, shafts, special connections or secondary components. Solid elements may be used for modeling the bulk of crushable foams. In typical finite element simulations of the human body, plates and shells are used for the simulation of cortical bone, beams and bars are used for the modeling of long bones, tendons and muscles, and solid elements are used for the modeling of the bulk of soft tissues and for the spongeous bone.

Time discretization. The application of the displacement finite element method leads to the discretized, coupled and nonlinear equations of motion in each displacement and rotation degree of freedom. These equations can be integrated in the time domain by using either implicit or explicit methods. Both methods use time discretization operators which permit to solve for the unknown displacements, rotations, linear and angular velocities and accelerations of each degree of freedom, at a given discrete point in time, from the known states of the structure at previous points in time. An overview on the explicit and implicit solution methods can be found in HUGHES, PISTER and TAYLOR [1979].

Standard *implicit methods* require linearization of the set of nonlinear equations of motion and lead to sets of coupled algebraic equations which, for non-linear problems, must be solved at a considered point in time in an iterative fashion, in order to achieve dynamic equilibrium at that time. The degree of coupling of the equation set is measured by the "bandwidth" of the linearized system matrix, which envelopes the extent of non-zero matrix elements of each of its rows. In a "diagonal" equation matrix the bandwidth is minimal (one) while in a "full" matrix the bandwidth is maximal (number of equations). Implicit methods can be used together with explicit solution, when the explicit solution is costly, such as in quasi-static problems, which arise, for example, when seating an occupant model into a deformable car seat before simulating a crash scenario, or by calculating the elastic spring-back of stamped parts.

Standard *explicit methods* do not require repeated solution of linearized coupled equation systems, and lead to a set of uncoupled algebraic equations when a lumped (diagonal) mass matrix is assumed. Solution of diagonal equation systems is trivial and computer time per discrete solution time step is negligible as compared to the computer time needed to repeatedly solve the potentially huge coupled systems of algebraic equations of the implicit methods. To date, crash models with up to one million thin shell finite elements and more have been built and solved successfully on powerful computers with the explicit methods.

The time increment of explicit methods, however, is restricted for solution stability, while, in principle, the time increment in unconditionally stable implicit methods is not restricted in size. In typical crashworthiness and impact studies over relatively short durations and involving large distortions of the structural parts, this advantage of the implicit methods has no bearing, however, because the structural states must be known at many discrete points in time in order to allow for an accurate tracing of the complex physical phenomena, including "contact", and to account for material and geometrical non-linearity. Time increments of the order of one microsecond and less are typical in crash simulations with explicit time integration.

In crashworthiness and in higher velocity dynamic impact studies, therefore, the explicit time integration methods have proven computationally advantageous. Implicit solution algorithms, on the other hand, may be applied with advantage to quasi-static or slow vibration dynamic problems with limited non-linearity.

Finite elements overview. Standard finite elements are defined by a set of nodes and a connectivity array, which relates their geometric topology to these nodes. In nonlinear crash and impact analysis, experience shows that "higher order" elements are not beneficial. This is mainly due to the fact that sharp stationary or traveling plastic hinge folding lines of crushed thin shell structures, made of elastic-plastic material, are not well accommodated even by complex interpolation shape functions of higher order elements, and that plastic hinge lines therefore require a dense nodal point spacing in the finite element models of thin walled structures, even when computationally expensive higher order elements are used.

Similarly, high gradients of three-dimensional stress and of plastic deformation are better accommodated by dense finite element models of simple 8-node 3D solid elements, than by using higher order solids that, for computational competitiveness, must have coarser meshes. Similar remarks apply for the simulation of fracture, where the small scale of the physics of fracture requires even higher mesh densities, sometimes achieved by automatic adaptive local mesh refinement schemes. Recently, so-called "mesh-less" methods are applied to simulate fracture, such as the EFG (element-free Galerkin), DE (discrete elements), SPH (smoothed particle hydrodynamics) (MONAGHAN and GINGOLD [1983] and MONAGHAN [1988]), FPM (finite point methods), etc. These advanced space discretization methods are presently under development and they are implemented in commercial crash codes. One instance of recent trial applications in biomechanics of the mesh-less SPH methods is the simulation of coupled structure fluid interaction (FSI) problems, where the fluid is simulated with particles, enclosed in a deformable organ modeled with finite elements. Another emerging method is the coupling of solid codes and flow codes for the simulation of FSI events (LÖHNER [1990]), such as the rupture of the blood-filled aorta in a chest impact scenario.

In many structures, the most important crash simulation finite element is the thin shell element. The most used thin shell element is a bilinear four node quadrilateral element, based on the Mindlin–Reissner plate theory. One of the most efficient thin shell elements was originally developed by BELYTSCHKO and TSAY [1983] and BELYTSCHKO and LIN [1984]. The Mindlin–Reissner plate theory takes the transverse shear deformation of the plate into account and it presumes that lines normal to the plate mid-surface remain straight, but not necessarily normal.

In the classical Kirchhoff–Love plate theory the normal lines to the plate mid-surface remain both, straight and normal, and the influence of the transverse shear deformation is neglected. The implementation of this theory requires slope-compatibility across element edges (C^1-continuity), resulting in a complex finite element formulation. On the other hand, Mindlin–Reissner theory requires only C^0-continuity in the shape functions for assuring complete inter-element deformation compatibility. This greatly simplifies the FE-formulation.

In the case of the Belytschko element, a reduced domain integration technique with one-point quadrature is applied for calculating the nodal forces contributed by the plate and shell elements. This technique avoids membrane-locking, but permits certain zero-energy or kinematic deformation modes of the elements, with which no resisting material stresses are associated. Such zero-energy modes are called hourglass-modes, which, if excited, can lead to numerical instability through uncontrolled spurious oscillations. In order to prevent hourglass modes, a built-in hourglass control algorithm is implemented, as developed by BELYTSCHKO, WONG, LIU and KENNEDY [1984].

It effectively avoids the numerical instability problem associated with one-point quadrature shell elements, by effectively damping out the zero-energy modes.

A.2. Explicit solution method outline

Three dimensional structures. In a three-dimensional solid model there are 3 translational degrees-of-freedom (DOF) per node. In a three-dimensional thin shell numerical model there are 6 degrees of freedom per nodal point, i.e., 3 translations and 3 rotations. In a three-dimensional thin plate numerical model there are 5 local degrees of freedom, i.e., 3 translations and 2 rotations. Crash models are mainly spatial models of elements that can be subdivided into layers with plane stress conditions (plate and thin shell theory).

Lagrangian discretization. This term refers to a choice of independent variables for the problem. In the Lagrangian formulation, each material particle is characterized by its initial coordinates (x_0, y_0, z_0) and its actual coordinates are chosen to be the dependent variables of the problem,

$$x = x(x_0, y_0, z_0, t),$$
$$y = y(x_0, y_0, z_0, t),$$
$$z = z(x_0, y_0, z_0, t).$$

In this formulation, the ordinary differential equations, obtained after spatial discretization, describe the dynamic equilibrium of the material particles which stand for the original continuum. For example,

$$\mathbf{M}\,\mathrm{d}^2\mathbf{x}/\mathrm{d}t^2 + \mathbf{C}\,\mathrm{d}\mathbf{x}/\mathrm{d}t + \mathbf{K}\mathbf{x} = \mathbf{F}_{\mathrm{ext}},$$

are the discretized equations of motion in the linear case, where $\mathbf{M}$ is the mass matrix, $\mathbf{C}$ is the damping matrix, $\mathbf{K}$ is the stiffness matrix and $\mathbf{F}_{\mathrm{ext}}$ is the vector of applied loads.

The acceleration of a particle is equal to the material derivative of the velocity, $\mathbf{v}$, and equal to the partial derivative of the velocity because x_0, y_0, z_0 are constants:

$$\frac{\mathrm{d}\mathbf{v}}{\mathrm{d}t} = \frac{\partial\mathbf{v}}{\partial t} + \frac{\partial\mathbf{v}}{\partial x_0}\frac{\partial x_0}{\partial t} + \frac{\partial\mathbf{v}}{\partial y_0}\frac{\partial y_0}{\partial t} + \frac{\partial\mathbf{v}}{\partial z_0}\frac{\partial z_0}{\partial t} = \frac{\partial\mathbf{v}}{\partial t}$$

The discrete finite element mesh points coincide with material points and have time dependent coordinates. The finite element mesh will thus deform with the material. This will engender distortions of the mesh, which can result in decreasing stable solution time steps and in solution inaccuracies due to excessive element distortions.

Finite elements. Finite element "unstructured" meshes permit to establish numerical models of a physical structure with realistic discretization and representation of the boundary conditions

Only the simplest finite elements are used in the PAM-Crash program because it is believed that fine meshes of simple elements give better results in highly distorting structures than coarse meshes of high-order elements. The most often used finite elements in crash codes are

- 4–8 node solids (1 point and fully integrated);
- 3–4 node bilinear shells (1 point and fully integrated);
- 3–4 node isoparametric membrane elements (fully integrated);
- 2 node beam, bar and joint elements.

The optimization of industrial crash codes has focused on the treatment of the most frequently used shell element.

Explicit integration scheme. A vibrating spring/dashpot-mass system consists of a mass, m, a dashpot with constant c, a spring with constant k, and an external load, $f(t)$, Fig. A.1.

Consider the second order ordinary differential equation which expresses dynamic equilibrium of this system, where x is the displacement, $\dot{x}$ is the velocity c is the damping coefficient, $\ddot{x}$ is the acceleration and m is the mass,

$$m\ddot{x} + c\dot{x} + kx = f(t) \quad \text{(1 DOF)}.$$

The *central difference scheme* considers the following time-axes around a discrete point in time, t_n, Fig. A.2.

The known quantities are the displacement x_n at time t_n, and the velocity $\dot{x}_{n-1/2}$ at the intermediate time $t_{n-1/2}$. The wanted quantities are the displacement x_{n+1} at time t_{n+1}, and the velocity $\dot{x}_{n+1/2}$ at time $t_{n+1/2}$. Dynamic equilibrium at time t_n is expressed

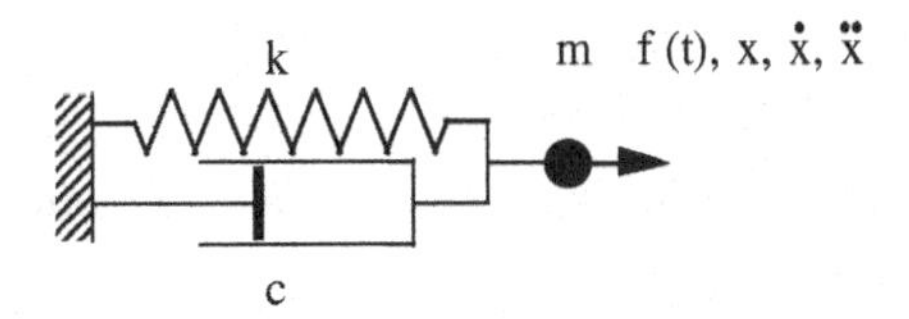

FIG. A.1. Spring/dashpot-mass system.

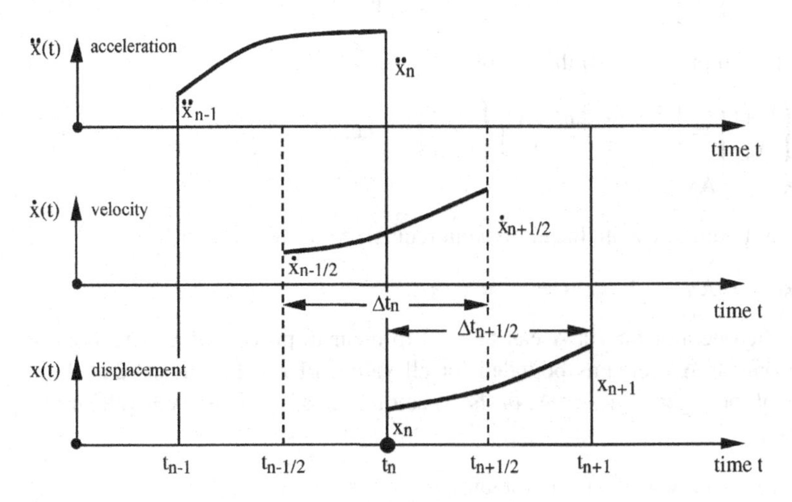

FIG. A.2. Central difference scheme.

as

$$m\ddot{x}_n = f_n - c\dot{x}_n - kx_n.$$

Since all terms on the right hand side are known, one can first solve for the acceleration at time t_n, $\ddot{x}_n$, and then apply the central difference time integration, Fig. A.2, and solve for the unknown quantities, $\dot{x}_{n+1/2}$ at time $t_{n+1/2}$ and x_{n+1} at time t_{n+1} as follows

$$\begin{aligned}
\ddot{x}_n &= m^{-1}\big(f_n - c\dot{x}_n - kx_n\big),\\
\dot{x}_{n+1/2} &= \dot{x}_{n-1/2} + \Delta t_n \ddot{x}_n,\\
x_{n+1} &= x_n + \Delta t_{n+1/2}\dot{x}_{n+1/2}.
\end{aligned}$$

Stable time step. The criterion for solution stability of the central difference explicit time integration can be derived formally as follows (BATHE [1996]). For a 1-DOF free-vibration system the equation of motion becomes

$$\ddot{x}_n + 2\xi\omega\dot{x}_n + \omega^2 x_n = 0 \quad \text{(free vibration)},$$

where $\xi = c/c_{\text{crit}}$ is the damping ratio, $c_{\text{crit}} = \sqrt{2km}$, is the critical damping, $\omega = \sqrt{k/m}$ is the circular frequency of vibration, and k and m are the spring stiffness and vibrating mass, respectively. The central difference scheme dictates, using constant time step, Δt, around time t_n,

$$\begin{aligned}
\dot{x}_{n+1/2} &= (x_{n+1} - x_n)/\Delta t,\\
\dot{x}_{n-1/2} &= (x_n - x_{n-1})/\Delta t,\\
\ddot{x}_n &= (\dot{x}_{n+1/2} - \dot{x}_{n-1/2})/\Delta t = (x_{n+1} - 2x_n + x_{n-1})/\Delta t^2.
\end{aligned}$$

Substitution of these expressions into the above equation of motion yields the following recursive system of equations

$$\begin{Bmatrix} x_{n+1} \\ x_n \end{Bmatrix} = \begin{bmatrix} \frac{2-\omega^2\Delta t^2}{1+\xi\omega\Delta t} & -\frac{1-\xi\omega\Delta t}{1+\xi\omega\Delta t} \\ 1 & 0 \end{bmatrix} \begin{Bmatrix} x_n \\ x_{n-1} \end{Bmatrix}.$$

For zero damping ($\xi = 0$) this becomes

$$\begin{Bmatrix} x_{n+1} \\ x_n \end{Bmatrix} = \begin{bmatrix} 2-\omega^2\Delta t^2 & -1 \\ 1 & 0 \end{bmatrix} \begin{Bmatrix} x_n \\ x_{n-1} \end{Bmatrix}, \quad \text{i.e.,}$$

$$\mathbf{x}_{n+1} = \mathbf{A}\mathbf{x}_n.$$

In general, one can write the equivalent recursive matrix relationship

$$\mathbf{x}_{n+1} = \mathbf{A}\mathbf{x}_n = \mathbf{A}^2\mathbf{x}_{n-1} = \cdots = \mathbf{A}^n\mathbf{x}_1,$$

where the operator matrix $\mathbf{A}$ can be seen to recur in powers of n. This relationship is stable only if $\mathbf{A}^n$ remains bounded for all values of n. This means that the absolute values of the eigenvalues of $\mathbf{A}$, or its "spectral radius", ρ, must be less than or equal to one, i.e.,

$$\rho(\mathbf{A}) = \max|\lambda_i| \leqslant 1 \quad \text{for stability.}$$

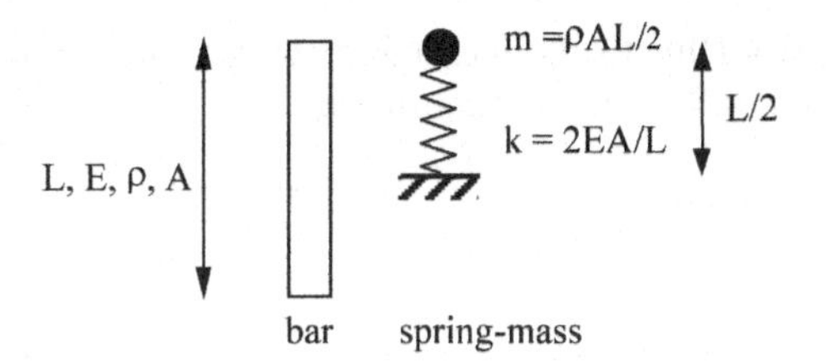

FIG. A.3. Free vibration bar element.

In case of zero damping, the eigenvalues of **A** follow from $\mathbf{Ax} = \lambda\mathbf{x}$, with solutions $\det(\mathbf{A} - \lambda\mathbf{I}) = 0$, where **I** is the 2×2 unit matrix. This leads to $\lambda_{1,2} = a \pm \sqrt{a^2 - 1}$, in which $a = (2 - \omega^2\Delta t^2)/2$. Enforcing $|\lambda_{1,2}| = 1$ one has $\lambda_{1,2}^2 = 1$, from which follows that $a^2 = 1$ satisfies the condition $|\lambda_{1,2}| = 1$ non-trivially. From this equation, the condition for stability of the time step is given by

$$\text{stable } \Delta t \leqslant 2/\omega.$$

For the case of the freely vibrating spring/mass system, stability is enforced if

$$\Delta t \leqslant 2\sqrt{m/k}.$$

Example. Consider the free vibration 1 DOF system, made of a bar element with uniform mass and stiffness distribution, with half length, $L/2$, elastic modulus, E, mass density per unit volume, ρ, and cross section area, A, as shown in Fig. A.3.

The criterion for stability becomes in this case

$$\Delta t_n < 2\sqrt{\frac{m}{k}} = L\sqrt{\frac{\rho}{E}} = \frac{L}{\sqrt{E/\rho}} = \frac{L}{c},$$

where $c = \sqrt{E/\rho}$ is the speed of sound in the bar material. For steel one has roughly $c = 5$ km/s $= 5$ mm/microsecond. This means that for a shell element with dimensions of 5×5 mm the stable time step will be about 1 microsecond, which corresponds to the time it takes of an acoustic signal to travel across the element.

Implicit integration outline. Consider the second order differential equation ($c = 0$)

$$m\ddot{x} + kx = f$$

and a discretized time axis as follows,

Let the known quantities at time t_n be x_n and $\dot{x}_n$. The wanted quantities are then $\dot{x}_{n+1}$ and x_{n+1}. Dynamic equilibrium at time t_{n+1} is expressed as

$$m\ddot{x}_{n+1} + kx_{n+1} = f_{n+1},$$

where x_{n+1} is unknown. Direct solution for $\ddot{x}_{n+1}$ or x_{n+1} is therefore impossible. One can now apply forward differences and substitute

$$\dot{x}_{n+1} = (x_{n+1} - x_n)/\Delta t,$$
$$\ddot{x}_{n+1} = (\dot{x}_{n+1} - \dot{x}_n)/\Delta t$$

and thus:

$$\ddot{x}_{n+1} = (x_{n+1} - 2x_n + x_{n-1})/(\Delta t)^2,$$

which after substitution yields

$$(m/\Delta t^2 + k)x_{n+1} = f_{n+1} - m/\Delta t^2(2x_n - x_{n-1}).$$

This equation can be solved for the unknown displacement at time t_{n+1},

$$x_{n+1} = (m/\Delta t^2 + k)^{-1}(f_{n+1} - m/\Delta t^2(2x_n - x_{n-1})).$$

The other variables are then obtained as

$$\dot{x}_{n+1} = (x_{n+1} - x_n)/\Delta t,$$
$$\ddot{x}_{n+1} = (\dot{x}_{n+1} - \dot{x}_n)/\Delta t.$$

This scheme works independently of the chosen value of Δt. It is said to be "unconditionally stable". It necessitates, however, to perform the solution of $(m/\Delta t^2 + k)^{-1}$, which is in general a costly operation because the stiffness matrix is not diagonal. For large time steps, however, the solution may remain stable, but one may accumulate period elongation and amplitude decay time integration errors in freely vibrating systems. For non-linear equations of motion, the linearized equations of dynamic equilibrium must be solved iteratively to ensure dynamic equilibrium at times t_n. If this is neglected, the solution, may become inaccurate and unstable.

Vectorized and multi-tasked codes. Explicit codes are well suited for vectorized super-computers because elements carry no dependencies, i.e., elements can be treated simultaneously, no K-matrix is stored and no I/O operations are required. This means that the elapsed time is practically equal to the CPU time in a computer run. For the same reasons, explicit codes parallelize well. Most industrial crash codes come with shared and distributed memory parallel architecture.

A.3. Contact treatment outline

Overview on contact algorithms. The successful and efficient treatment of contact events is of prime importance in crash and impact simulation. To identify potential contact surfaces of a given structure, the crash codes allow to define contact interface entities, which are most often given by collections of surface facets, identical to finite element surfaces. Actual contact can be defined a distance t_{contact} (contact thickness) away from the mid-surface defined by the facets represented by shell elements. Most contact algorithms are based on a node-to-segment treatment. After preliminary geometrical proximity searches for potential contact events, the so identified candidate contact segments are fine-checked for mutual penetrations of their nodes and surfaces. Detected penetrations are limited ("penalized") or prevented by the algorithms.

In impact biomechanics several kinds of contact events are of importance. On the one hand, the biomechanical models may contact or collide with external objects (chest-to-seat belt, body-to-airbag, body-to-car interior, etc.). On the other hand, body parts and organs may be in mutual contact (arm-to-chest, chin-to-chest, leg-to-leg, internal organ-to-internal organ, organ-to-wall of body cavity, brain-to-skull, etc.).

Contacts may be sticking ("tied" options), sliding ("slide" options), multiple ("slide-and-void" options), failing (rupture options), and sliding can be with and without friction. Each contact event may require a particular treatment, with or without penalty algorithms. To demonstrate the importance editors of crash codes attribute to contact treatments, Table A.1 lists some often used contact options. The listed rigid wall and nodal constraint options are simpler contact treatments that can serve for detecting collisions with rigid obstacles (rigid walls) or that can be used to constrain specified pairs of nodes to move together (nodal constraints). The interested reader is advised to consult the notes manuals of the commercial crash codes for more detail (see also: HUGHES,

TABLE A.1
Some typical contact options (overview)

Group	Type*	Action
Rigid walls		one-sided moving or fixed boundary condition
Nodal constraints		pairs of nodes are constrained to displace together
Special sliding and tied interfaces	Type 1	sliding without separation
	Type 2	3 DOF or 6 DOF tied with single failure
Internal solid element contact	Type 10	internal solid anti-collapse contact
Safety specific contacts	Type 7	multiply self-impacting contact for airbag
	Type 11	body-to-plane contact (force-deflection)
	Type 12	body-to-body contact (force-deflection)
	Type 21	body-to-multiplane contact (force-deflection or stress-strain)
	Type 37	enhanced self-impacting contact for airbag
Recommended contacts for crash simulation	Type 31	penalty-free node-to-segment contact
	Type 32	penalty or kinematic tied contact with distance and failure
	Type 33	segment-to-segment contact with edge treatment (3D bucket search)
	Type 34	node-to-segment contact with edge treatment (3D bucket search)
	Type 36	self-impacting contact with edge treatment (3D bucket search)
	Type 42	mesh independent spotweld parts
	Type 44	node-to-segment contact with smooth contact surface
	Type 46	edge-to-edge self-impacting contact

*PAM-Crash code.

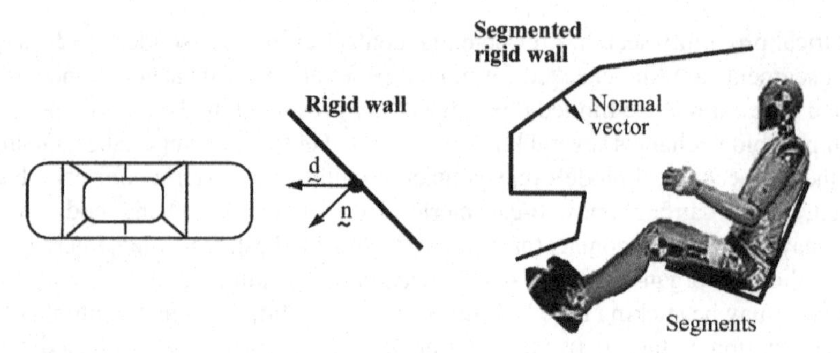

FIG. A.4. Rigid wall contact.

TAYLOR, SACKMAN, CURNIER and KANOKNUKULCHAI [1976], HALLQUIST, GOUDREAU and BENSON [1985] and HAUG, CLINCKEMAILLIE and ABERLENC [1989b]).

In simulations of impact biomechanics all listed options can be applied. Some details are provided next for the most popular rigid wall and penalty contact options. Contact types 11 and 12 are often used in the simple rigid multi-body models (HARB models) and they provide a "soft" penalization of penetrating rigid volumes according to user-defined penetration restoring force-deflection curves.

Rigid walls. Fig. A.4 shows the principle of rigid wall treatment, where a car may collide with an inclined rigid wall or where a car occupant may collide with a simplified model of the car interior, modeled with several rigid wall segments.

The rigid walls are impenetrable and the contacting bodies may slide along the walls with and without specified friction coefficients. The rigid walls may not be able to move, or they can be assigned some specified motion. Penetration of nodes into a rigid wall is prevented by an algorithm that sets the relative normal velocity to zero.

Nodal constraints. Nodal constraints are often used to tie together two non-matching finite element meshes, as it can arise when two internal organs are modeled independently and when the two organs have a common mesh interface. Instead of re-meshing the interfaces, one may simply identify pairs of nodes to move together.

Penalty methods. If penetration of a node into a segment is detected, it will be penalized in most contact algorithms by elastic restoring forces of fictitious (non-)linear springs, which are compressed by the amount of penetration. In dynamic contact events, dashpots can be added to dissipate energy and to prevent from spurious oscillation.

Fig. A.5 shows two potentially contacting surface segments. The algorithm checks if a slave node of the right segment touches the left master segment. Then the master and slave roles are reversed and the algorithm checks if a slave node of the left segment touches the right master segment. Inverting the master-slave roles prevents from contacts to be missed.

Fig. A.6 shows the application of a penalizing contact spring after "penetration" of a slave node into a master segment. If the slave node penetrates a distance beyond

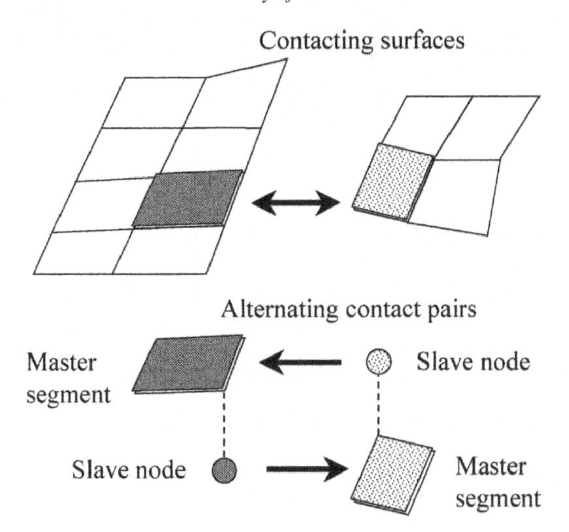

FIG. A.5. Contact interfaces.

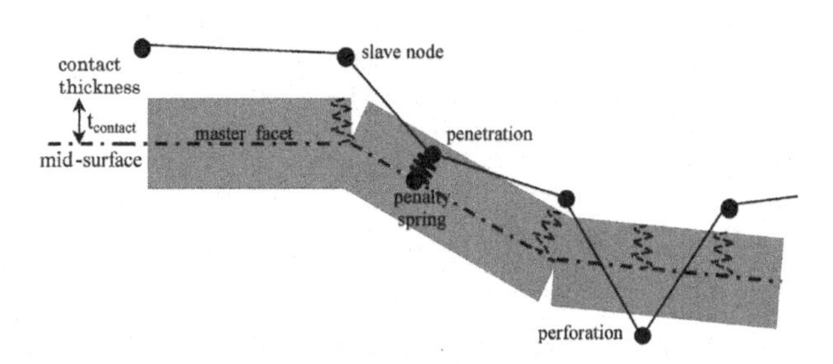

FIG. A.6. Penetration and perforation.

the contact thickness, $t_{contact}$, over the mid-surface of the master segment, then this node will have "perforated" the master segment and it will escape to the other side. Perforations can happen when the impact is too violent to be contained by the penalty springs.

APPENDIX B

Biomaterial Mechanical Properties

The following text and tables are extracted mainly from a report of the HUMOS-1 Project (YANG [1998]):

Yang, J. (1998) Bibliographic Study, Report 3CHA/980529/T1/DB, Chalmers University of Technology, SE-41296 Göteborg, Sweden.

The HUMOS project was funded by the European Commission under the Industrial and Materials Technologies program (Brite–EuRam III). This report has been prepared with the aim to provide input data for human impact models. It contains many references to the abundant literature on biomaterials, and the most important data are summarized hereunder. For simulations in impact biomechanics, the mechanical properties of the musculo–skeletal system are the most needed and known. The static and dynamic mechanical properties of the soft tissues and organs are the least known properties.

The mechanical properties of all living tissues undergo wide scatter through different ages, gender, biological, physical and loading conditions. Most of the listed properties delimit the observed scatter and can serve as guide lines and upper and lower bounds in human model calibration, and in stochastic analyses. Most biomaterials are non-linear, anisotropic, different in tension and compression and they are deformation rate dependent. The majority of the indicated values represent their average linear behavior. The properties of bone and soft tissue materials can depend much on the location in the human body, i.e., vary from member to member and within an individual part. Most properties were obtained from cadaver tests, which can only approximate their behavior in living bodies.

While the mechanical properties of the long bones (femur, tibia, fibula, humerus, ulna, radius) is a well studied subject the in biomechanics, data are much scarcer for pelvic bone, ribs, sternum, clavicle, scapula, short bones of the feet and other parts. For this reason the following data are concentrating on the properties of long bones.

B.1. Bone mechanical properties

Quasi-static properties of bone. Table B.1 contains the average quasi-isotropic properties of the most important skeletal bones, where N is the number of tested specimens, E is the linear isotropic Young's modulus, ν is Poisson's ratio, G is the shear modulus and σ_u is the ultimate stress.

Table B.2 shows the orthotropic stiffness matrix coefficients in directions 1, 2 and 3 of femoral cortical bone from 60 specimens (ASHMAN ET AL. [1984]), where axis 1 is

TABLE B.1
Elastic moduli and ultimate stress for bones (overview)

Authors	Bone	Test (N)	E [Gpa]	ν	G [Gpa]	σ_u [Mpa]
McELHANEY [1966]	general cortical	quasi-static	20 ± 5		3.15 ± 0.25	16 ± 2.5
		strain rate 0.001	15.1			15.0
		strain rate 300	29.2			28.0
	general trabecular	compression	0.001–9.8			0.02–2.5
	vertebral bodies	compression				0.01–1.5
YAMADA [1970] SEDLIN ET AL. [1965] SEDLIN ET AL. [1966] REILLY, BURSTEIN and FRANKEL [1974] REILLY ET AL. [1975]	femur cortical	tensile	14–18			12–15
		compressive	15–19			14–21
		shear	15.5			7–8
		bending	3.28			16–18
		torsion				6.8
	tibia cortical	tensile	18–29			12–17
		compressive	25–35			18–21
		shear				7–8
YAMADA [1970]	femur	tensile	17.6			12.4
		(tensile dry)	(20.2)			(15.1)
	tibia	tensile	18.4			14.3
	fibula	tensile	18.9			14.9
	humerus	tensile	17.1			12.5
	radius	tensile	18.5			15.2
	ulna	tensile	18.8			15.1
	average	tensile	18.3			14.0
QUHAN [1989]	pelvis cortical	bending 63 yr (36)	5.26 ± 2.09			
		bending 23 yr (12)	3.76 ± 1.78			
	pelvis trabecular	bending 63 yr (29)	4.16 ± 2.02			
		bending 23 yr (13)	3.03 ± 1.63			
McELHANEY [1970]	skull sandwich	radial comp.	2.4 ± 1.4	0.19 ± 0.08		740 ± 350
		tangential comp.	5.6 ± 3.0	0.22 ± 0.11		970 ± 360
		tangential tension	5.4 ± 2.9			430 ± 190
GRANIK and STEIN [1973]	ribs	bending	11.5 ± 2.1			11 ± 3
SACRESTE, BRUN-CASSAN, FAYON, TARRIERE, GOT and PATEL [1982]	ribs	bending	6.14 ± 4.26			8.6 ± 5.5

TABLE B.1
(*Continued*)

Authors	Bone	Test (N)	E [Gpa]	ν	G [Gpa]	σ_u [Mpa]
BURGHELE and SCHULLER [1968]	calcaneous	static (10 mm/min) intact bone				2100 [N] (mean force)
		dynamic (500 mm/min) intact bone				2620 [N] (mean force)
		cortical (largest over bone); cut specimens: 2 × 0.5 × 0.5 cm	2890–3200			2.87–3.79 (stress)
	talus	static (10 mm/min) intact bone				446 [N] (mean force)
		dynamic (500 mm/min) intact bone				468 [N] (mean force)

TABLE B.2
Orthotropic stiffness matrix [Gpa] for human femoral cortical bone

E_{11}	E_{22}	E_{33}	G_{12}	G_{13}	G_{23}	ν_{12}	ν_{13}	ν_{23}	ν_{21}	ν_{31}	ν_{32}
12.0	13.4	20.0	4.53	5.61	6.23	0.376	0.222	0.235	0.422	0.371	0.350

TABLE B.3
Transverse isotropic stiffness matrix for human femoral cortical bone

	E_{11}	E_{33}	ν_{31}	ν_{12}	G_{13}
Mean values [GPa]	11.5	17.0	0.46	0.58	3.28
Number of tests (N)	31	170	147	26	166
Standard deviation [%]	15–20% in tension 7–10% in compression		30%	30%	10%

the radial direction, axis 2 is the circumferential (or transverse) direction, axis 3 is the longitudinal direction of the cylindrical bone shaft, E_{ij} are the elastic modules, G_{ij} are the shear modules and ν_{ij} are Poisson's ratios.

Table B.3 contains transverse isotropic stiffness matrix coefficients for human femoral cortical bone ($E_{11} = E_{22}$) from a population over the age spans of 19–80 years (REILLY, BURSTEIN and FRANKEL [1974]), where N is the number of specimens, E_{11} is the elastic modulus for transverse or radial specimens, E_{33} is the elastic modulus for longitudinal specimens, ν_{12} is Poisson's ratio for transverse or radial specimens, ν_{31} is Poisson's ratio for longitudinal specimens and G_{13} is the shear modulus.

Table B.4 contains trabecular orthotropic stiffness properties found in the human proximal tibia from 3 males (ages 52, 55 and 67) (ASHMAN ET AL. [1986]).

TABLE B.4
Trabecular bone properties from human proximal tibias

Module	Average	Standard deviation	Range
E_{11} [MPa]	346.8	218	110–1230
E_{22} [MPa]	457.2	282	140–1750
E_{33} [MPa]	1107.1	634	340–3350
G_{12} [MPa]	98.3	66	30–380
G_{13} [MPa]	132.6	78	35–410
G_{23} [MPa]	165.3	94	45–460
Density [kg/m^3]	263.4	135	130–750

TABLE B.5
Trabecular bone properties from human pelvis

	E_{11}	E_{22}	E_{33}	G_{12}	G_{13}	G_{23}	ν_{12}	ν_{13}	ν_{23}	ν_{21}	ν_{31}	ν_{32}
Average [MPa]	59.8	57.3	43.2	26.0	22.6	22.6	0.18	0.24	0.21	0.17	0.16	0.14
Standard Dev.	44.9	44.6	39.9	19.1	17.1	17.2	0.11	0.14	0.16	0.10	0.07	0.09

Table B.5 lists the trabecular orthotropic stiffness properties of the pelvis bone (DALSTRA ET AL. [1993]).

Tables B.6 and B.7 list further trabecular bone mechanical properties, showing the great variety of these values.

Strain rate dependent properties of bone. Bone is strain rate dependent material. It exhibits stiffer and stronger behavior at high strain rate. Typical values found for cortical are listed in Table B.8.

Age dependent properties of bone. Mechanical properties of human bones change with age as a consequence of changes of density and mineral content. According to BURSTEIN ET AL. [1976], the elastic modulus and tensile strength of human bone decrease slowly after the age of about 45 years. CURREY [1975] showed that the elastic modulus and bending strength both increase with age until the age of about 30 years, but decrease thereafter. The decrease is associated with, and mainly caused by, the increased porosity and demineralization of bone. MCCALDEN ET AL. [1993] found that the change in porosity played a greater role in the reduction in strength than the change in mineral content. Both factors reduce the ability of bones to undergo plastic deformation before fracture starts. Table B.9 shows important trends of the age dependency of the mechanical properties of human bone with age. Table B.10 gives an example for the tensile and compression age dependent mechanical properties of the cortical bone of the femur. Similar trends with age are found in the other bones.

TABLE B.6
Compressive properties of trabecular bone [MPa]

(a) Proximal femur	Test piece	E	σ_u
BROWN [1980]	9.5 mm length	345	–
	5 mm cubes	–	1.2–3.1
CIARELLI ET AL. [1986], CIARELLI ET AL. [1991]	10 mm length	58–2248	2.1–16.2
	8 mm cubes	49–572	–
EVANS ET AL. [1961]	25 × 7.9 mm prisms	–	0.21–14.82
MARTENS ET AL. [1983]	8 mm diameter	1000–9800	0.45–15.6
SCHOENFELD ET AL. [1974]	7.9 mm cubes	20.68–965	–
	4.8 mm diameter	–	0.15–13.5
(b) Distal femur	**Test piece**	E	σ_u
BEHRENS [1974]	5 mm slab	–	2.25–66.2
CIARELLI ET AL. [1986], CIARELLI ET AL. [1991]	8 mm length	58.8–2942	–
	8 mm cubes	7.6–800	19
DUCHEYNE [1977]	5 mm diameter	–	0.98–22.5
(c) Proximal Tibia	**Test piece**	E	σ_u
BEHRENS [1974]	5 mm slab	–	1.8–63.6
	10.3 mm diameter	–	–
CARTER and HAYES [1977]	5 mm length	1.4–79	1.5–45
CIARELLI ET AL. [1986], CIARELLI ET AL. [1991]	8 mm cubes	–	0.52–11
GOLDSTEIN [1983]	7 mm diameter	8–457	1–13
	10 mm length		
HVID ET AL. [1985]	10 mm length	4–430	13.8–116.4
	5 mm slabs		
LINDAHL [1976]	14 × 9 mm		
	Male	34.6	3.9
	Female	23.1	2.2
LINDE [1989]	No constraint	113–853	
WILLIAMS ET AL. [1982]	5–6 mm cubes	10–500	1.5–6.7
(d) Vertebral bodies	**Test piece**	E	σ_u
ASHMAN ET AL. [1986]	5 mm diameter	158–378	–
	10–15 mm length		
BARTLEY [1966]	Lumbar region	–	2.9
GALANTE ET AL. [1970]	7, 10 mm diameter	–	0.39–5.98
LINDAHL [1976]	10 × 9 × 14 mm		
	Male	55.6	4.6
	Female	35.1	2.7
MCELHANEY [1970]	10 mm length	151.7	4.13
ROCKOFF ET AL. [1969]	Lumbar region	–	0.69–6.9
STRUHL ET AL. [1987]	8&6 mm cubes	10–428	0.06–15
WEAVER ET AL. [1966]	10 mm cube	–	0.34–7.72
YAMADA [1970]	40–49 years	88.2	1.86
	60–69 years	68.6	1.37
(e) Pelvic trabecular	**Test piece**	E	σ_u
DALSTRA ET AL. [1993]	6.5 mm cubes	58.9	–

TABLE B.7
Shear properties of trabecular bone [MPa]

	Test piece	G	σ_u
ASHMAN ET AL. [1986]	5 mm diameter 10–15 mm length	58–89	

TABLE B.8
Cortical bone strain rate dependent properties

Cortical bone (MCELHANEY [1966])	Compression (human embalmed)	Compression (fresh cow bone)
Strain rate	from 0.001 to 300	from 0.001 to 300
E [GPa]	from 15.1 to 29.2	from 18.6 to 33
Strength [MPa]	from 150 to 280	from 175 to 280

TABLE B.9
Age dependency of bone properties

(a) *Cortical bone* ultimate tensile strength for the decade age groups (after ZHOU, Q. ET AL.[1996]).
(Reproduced by permission of The Stapp Association.)

Observations:
(1) The absolute strength values obtained by different authors are different, but the trend with age is similar.
(2) The maximum strength is reached around the 30–39 age group.
(3) The ultimate tensile strength of the elderly groups (70–89) drops by 15% to 25% compared to the maximum.

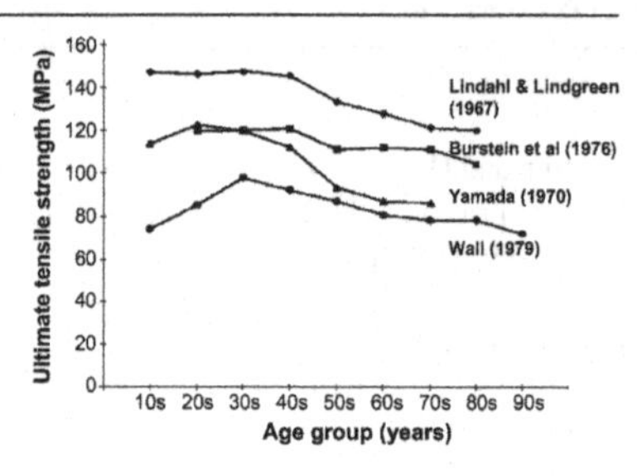

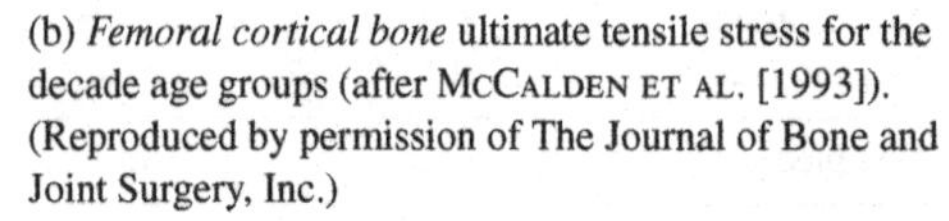

(b) *Femoral cortical bone* ultimate tensile stress for the decade age groups (after MCCALDEN ET AL. [1993]).
(Reproduced by permission of The Journal of Bone and Joint Surgery, Inc.)

The straight line represents the best fit with the use of linear regression analysis (from 253 specimens excised from 47 cadavers).

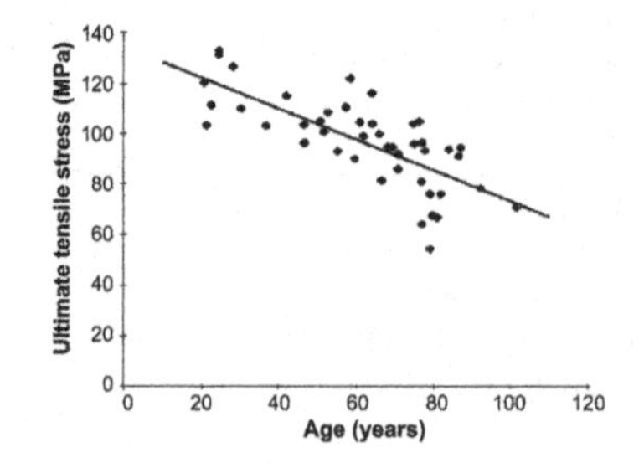

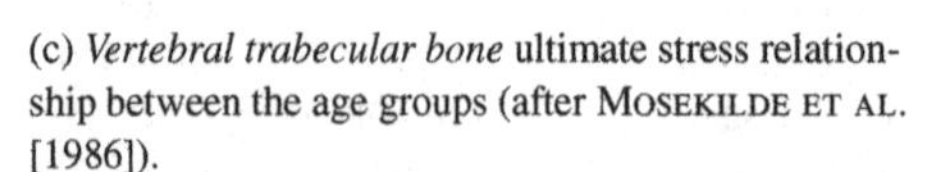

(c) *Vertebral trabecular bone* ultimate stress relationship between the age groups (after MOSEKILDE ET AL. [1986]).
(Reproduced by permission from Elsevier.)

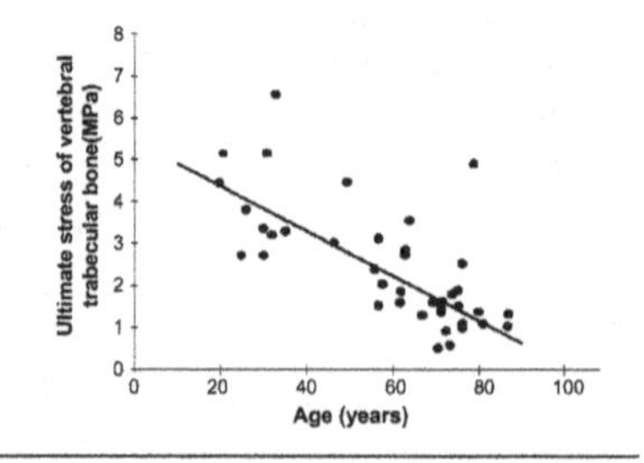

TABLE B.10
Age dependent femur cortical mechanical properties

(a) Tensile properties of femur		Elastic modulus [MPa]	Yield stress [MPa]	Ultimate stress [MPa]
BURSTEIN ET AL. [1976]	20–29 years	17000	120	140
	30–39 years	17600	120	136
	40–49 years	17700	121	139
	50–59 years	16600	111	131
	60–69 years	17100	112	129
	70–79 years	16300	111	129
	80–89 years	15600	104	120
(b) Compressive properties of femur		**Elastic modulus [MPa]**	**Yield stress [MPa]**	**Ultimate stress [MPa]**
BURSTEIN ET AL. [1976]	20–29 years	18100		209
	30–39 years	18600		209
	40–49 years	18700		200
	50–59 years	18200		192
	60–69 years	15900		179
	70–79 years	18000		190
	80–89 years	15400		180
(c) Bending properties of femur		**Elastic modulus [MPa]**	**Yield stress [MPa]**	**Ultimate stress [MPa]**
YAMADA [1970]	20–29 years			151
	30–39 years			174
	40–49 years			174
	50–59 years			162
	60–69 years			154
	70–79 years			139
	80–89 years			139
(d) Torsion properties of femur		**Elastic modulus [MPa]**	**Yield stress [MPa]**	**Ultimate stress [MPa]**
YAMADA [1970]	20–29 years	3430		57
	30–39 years	3430		57
	40–49 years	3140		52.7
	50–59 years	3140		52.7
	60–69 years	2940		48.6
	70–79 years	2940		48.6
	80–89 years	–		48.6

B.2. Ligament mechanical properties

Ligaments are fibrous materials with pronounced non-linear force-displacement response. The mechanical properties of some ligaments are summarized in Table B.11, where v_d is the deflection rate, D is the deflection at rupture, ε is the strain at rupture, E is the elastic modulus. For reasons of practical feasibility, ligaments are most often

TABLE B.11
Some ligament mechanical properties

Specimen type and conditions (source)	v_d [mm/ms]	D [mm]	F [kN]	ε [%]	E [kN/mm^2] (= GPa)	Other properties
(a) Collagen (VIIDIK [1987])						$\sigma_u = 0.045$–0.120 GPa
(b) Typical force-deformation curve for ligament for monotonic forcing (FRANK and SHRIVE [1994]) (Reproduced by permission of the University of Calgary.)	(I) = toe region; (II) = linear region; (III) = region of micro-failure; (IV) = failure region. At the top are schematic representations of fibers going from crimped (I) through recruitment (II) to progressive failure (III and IV)				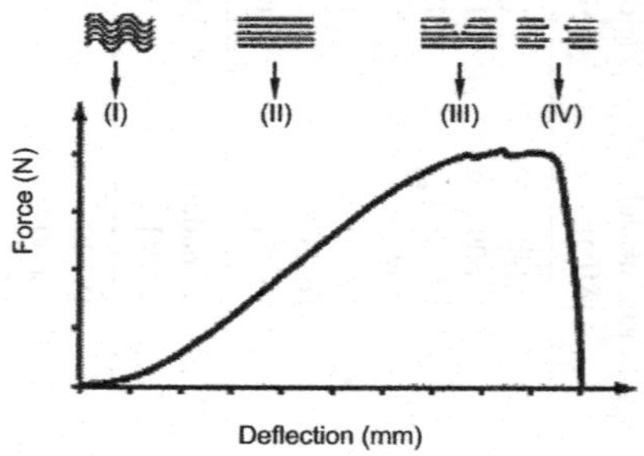	
(c) Human cruciate ligament (NOYES and GROOD [1978]) Non-linear stress-strain behavior (Reproduced by permission of the Journal of Bone and Joint Surgery, Inc.)		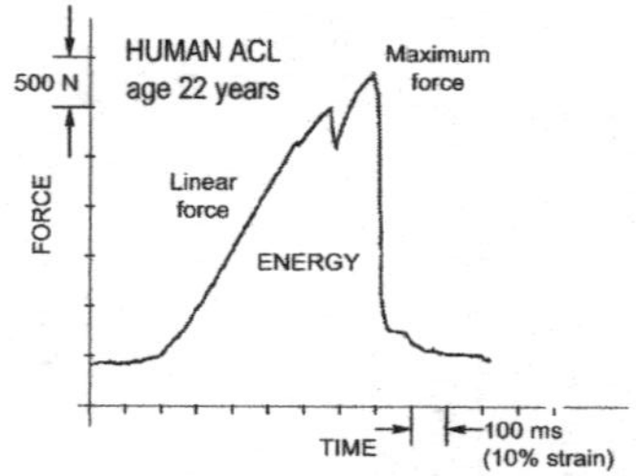				

TABLE B.11
(*Continued*)

Specimen type and conditions (source)	v_d [mm/ms]	D [mm]	F [kN]	ε [%]	E [kN/mm^2] (= GPa)	Other properties
(d) Human calcaneo-fibular and tibio-talar ligaments (BEGEMAN and AEKBOTE [1996]) Static and dynamic isolated bone-ligament-bone test pieces (Reproduced by permission of Prof. Begeman, Wayne State University, Bioengineering Center.)	maximum 0.4 mm/ms with cycle load					
(e) Human ankle/foot ligaments (ATTARIAN, MCCRACKIN, DEVITO, MCELHANEY and GARRETT [1985])						
(i) Anterior tibio-fibular ligament Cyclic load to isolated bone-ligament-bone test piece Effect of strain rate (Reproduced by permission of Foot & Ankle Intl.)	1.50E−04	3	0.016 (m)			
	1.50E−02		0.02 (m)			
	1.00E+00		0.042 (m)			

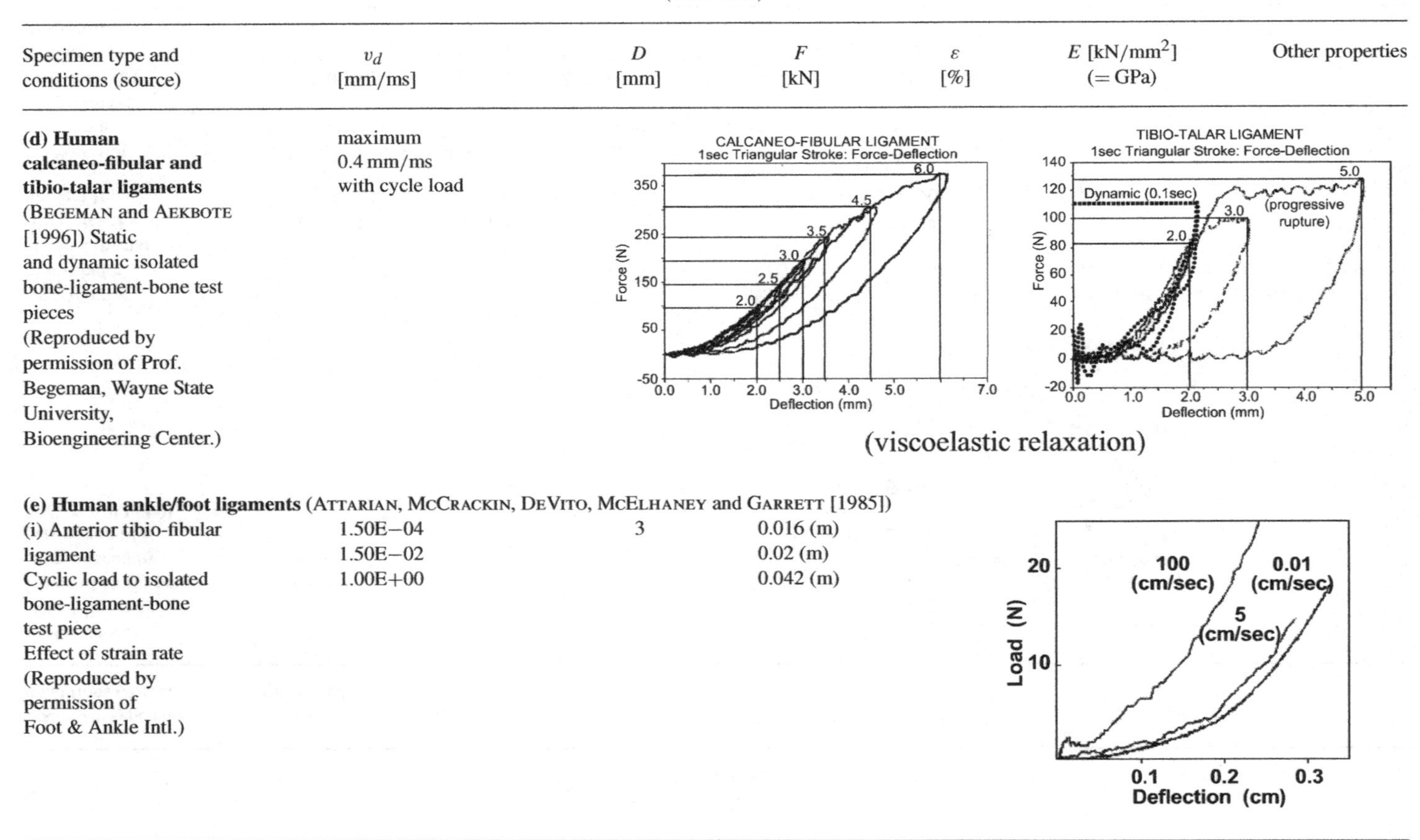

TABLE B.11
(*Continued*)

Specimen type and conditions (source)	v_d [mm/ms]	D [mm]	F [kN]	ε [%]	E [kN/mm^2] (= GPa)	Other properties
(ii) Calcaneo-fibular ligament Effect of strain rate (Reproduced by permission of Foot & Ankle Intl.)	1.50E−04 1.50E−02 1.00E+00	3	0.029 (m) 0.031 (m) 0.087 (m)			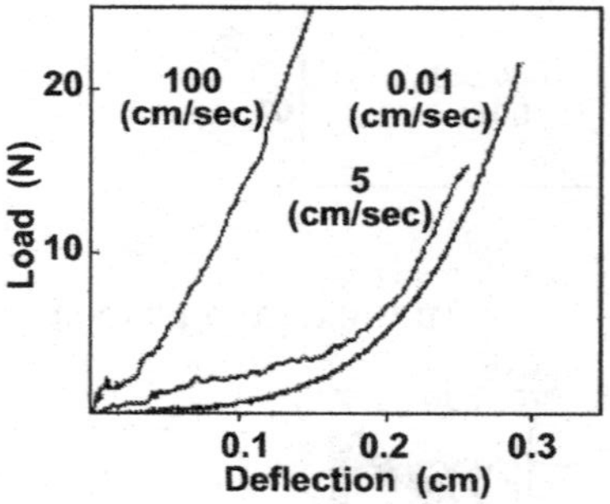 Load (N) 20 10; 100 (cm/sec); 0.01 (cm/sec); 5 (cm/sec); 0.1 0.2 0.3 Deflection (cm)
(iii) Anterior tibio-fibular (12 specimens)	1.01 ± 0.07	5.1 ± 0.5 (f)	0.1389 ±0.0235 (max)	53 ±6 (f)	3.999E−02 ±0.854E−02	maximum load = force required to completely disrupt a ligament (grade III sprain)
(iv) Calcaneo-fibular (16 specimens)	1.06 ± 0.04	6.3 ± 0.5 (f)	0.3457 ±0.0552 (max)	38 ±3 (f)	7.051E−02 ±0.690E−02	
(v) Posterior talo-fibular (4 specimens)	0.82 ± 0.13	13.1 ± 1.6 (f)	0.2612 ±0.0324 (max)	100 ±15 (f)	3.975E−02 ±1.379E−02	
(vi) Tibiotalar ligament (6 specimens)	0.80 ± 0.13	10.5 ± 1.1 (f)	0.7138 ±0.0693 (max)	210 ±23 (f)	1.2282E−01 ±0.2504E−01	

TABLE B.11
(*Continued*)

Specimen type and conditions (source)	v_d [mm/ms]	D [mm]	F [kN]	ε [%]	E [kN/mm^2] (= GPa)	Other properties
(f) Anterior cruciate ligament (WAINWRIGHT, BIGGS and CURREY [1979] (Reproduced by permission of the Princeton University Press.)					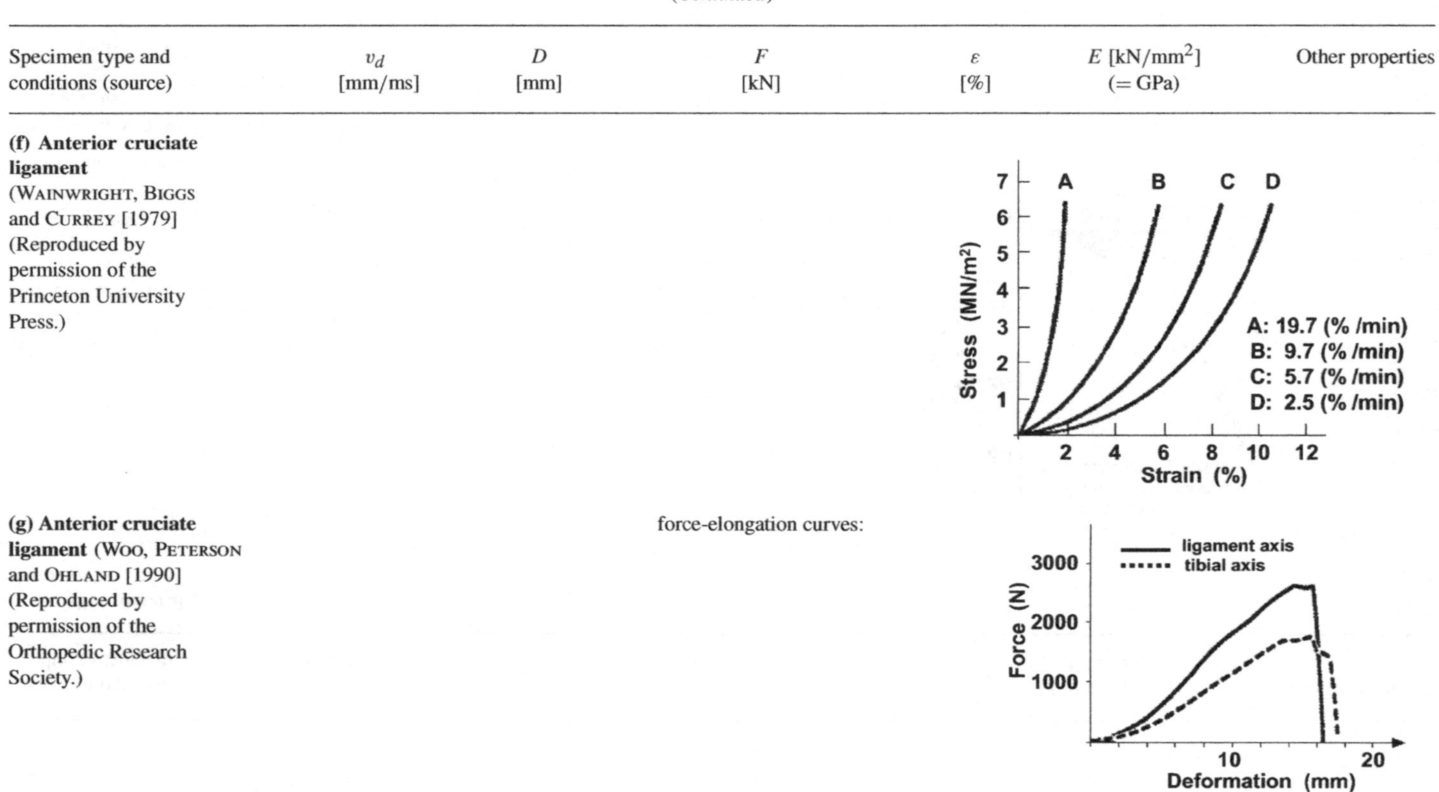	
(g) Anterior cruciate ligament (WOO, PETERSON and OHLAND [1990] (Reproduced by permission of the Orthopedic Research Society.)			force-elongation curves:			

Table B.11
(*Continued*)

Specimen type and conditions (source)	v_d [mm/ms]	D [mm]	F [kN]	ε [%]	E [kN/mm^2] (= GPa)	Other properties
(h) Mediacollateral ligament						
			τ_{max} = 4.0e–04–8.0e–04Gpa			
shear stress-deformation (Shelton, Butler and Feder [1993]) (Reproduced by permission of ASME International.)			force-strain curves:			
(k) Human ACL in tension (Haut [1993])						
(16–26 years)			1.73 ± 0.27 (u)	44.3 ± 8.5 (f)	0.111 ± 0.026	σ_u = 37.8 ± 9.3 MPa (f)
(22–35 years)			2.16 ± 0.175 (u)			
(48–86 years)					0.065 ± 0.024	

TABLE B.11
(*Continued*)

Specimen type and conditions (source)	v_d [mm/ms]	D [mm]	F [kN]	ε [%]	E [kN/mm^2] (= GPa)	Other properties
(l) Spinal ligaments (PANJABI, JORNEUS and GREENSTEIN [1984]) (Reproduced by permission of the Orthopedic Research Society.)			tensile response curves:	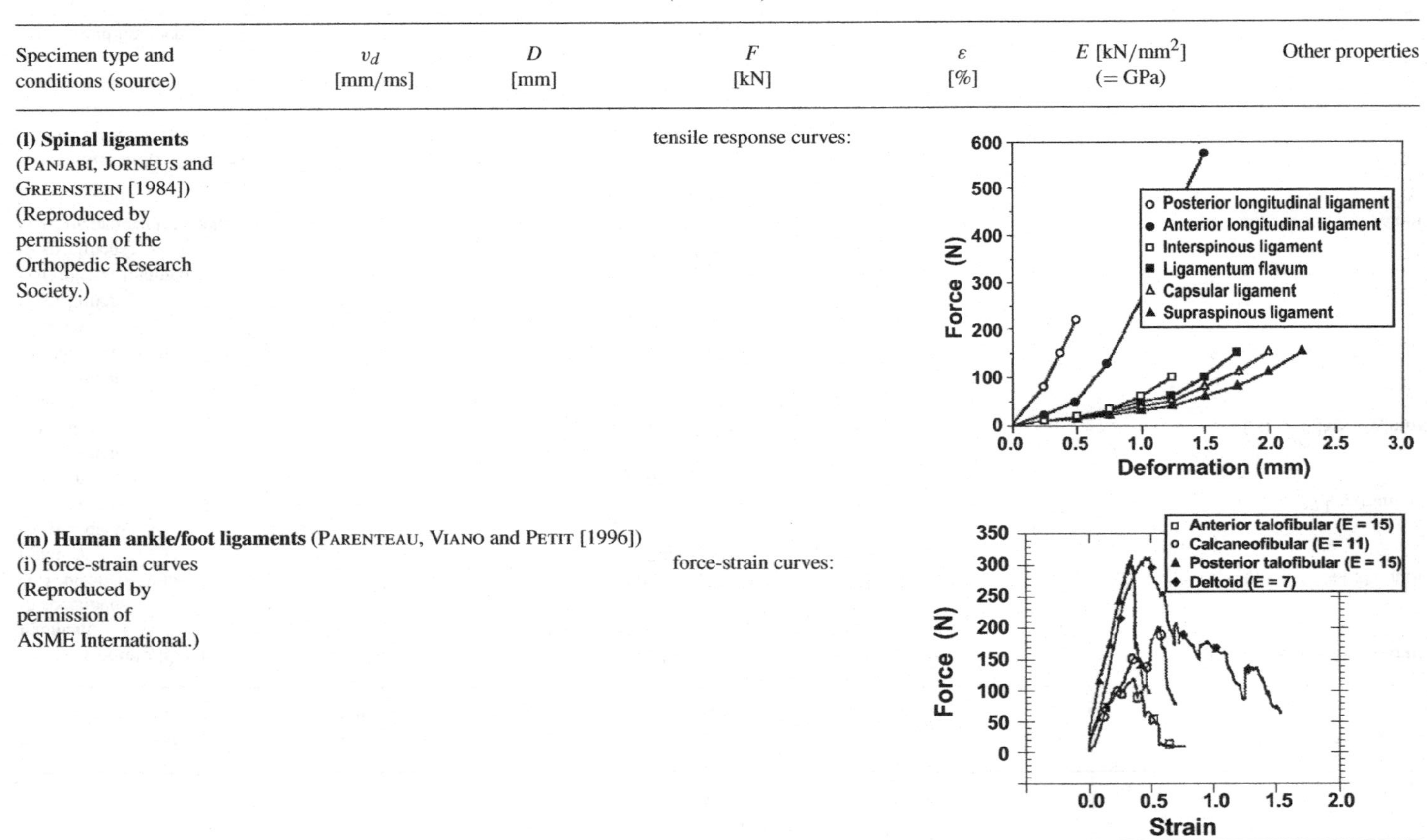		
(m) Human ankle/foot ligaments (PARENTEAU, VIANO and PETIT [1996]) (i) force-strain curves (Reproduced by permission of ASME International.)			force-strain curves:			

TABLE B.11
(*Continued*)

Specimen type and conditions (source)	v_d [mm/ms]	D [mm]	F [kN]	ε [%]	E [kN/mm^2] (= GPa)	Other properties
(ii) anterior talo-fibular ligament (5 specimens)			0.014–0.114 (y) 0.064–0.286 (u)	11–15 24–44		$k = 0.006–0.023$ kN/mm
(iii) calcaneo-fibular ligament (6 specimens)			0.053–0.259 (y) 0.120–0.290 (u)	21–39 30–84		$k = 0.018–0.044$ kN/mm
(iv) posterior talo-fibular ligament (1 specimen)			0.307 (y) 0.307 (u)	35 35		$k = 0.037$ kN/mm
(v) deltoid ligament (6 specimens)			0.119–0.355 (y) 0.239–0.507 (u)	10–50 45–79		$k = 0.026–0.068$ kN/mm
(vi) talo-calcaneal ligament (5 specimens)			0.061–0.117 (y) 0.078–0.156 (u)	13–33 13–33		$k = 0.021–0.031$ kN/mm
(vii) plantar ligament (6 specimens)			0.199–0.511 (y) 0.238–0.506 (u)	8–27 12–32		$k = 0.039–0.495$ kN/mm
(viii) metatarsal ligament (4 specimens)			0.075–0.220 (y) 0.103–0.247 (u)	11–40 0–54		$k = 0.016–0.038$ kN/mm
(n) Human ankle/foot ligaments (NIGG, SKARVAN and FRANK [1990])						
(i) anterior talo-fibular ligament			0.067–0.193	6–60		
(ii) calcaneo-fibular ligament			0.265–0.327	27–81		
(iii) deltoid ligament			0.173–0.315	34–58		

(f) = failure; (u) = ultimate; (m) = mean

tested as bone-ligament-bone test pieces, where the ligament and connected bones are isolated and where the bones are subjected to relative displacements in the direction of the fibers of the attached ligament. In the table, (f) means "failure", (u) means "ultimate" and (m) means "mean".

B.3. Brain mechanical properties

A synthesis of existing bibliography on the biomaterial behavior characterization of the brain is presented in Tables B.12 and B.13. A list of variables is given after Table B.14.

B.4. Joint mechanical properties

The physiological motions of the synovial joints of the human skeleton can be modeled approximately by computationally efficient point-like mechanical joint elements, with six more or less constrained motion degrees of freedom. Motion ranges, stiffness and resistance properties are found in the literature (ex: ROBBINS [1983], ROBBINS, SCHNEIDER, SNYDER, PFLUG and HAFFNER [1983]). Further data are given in Tables B.15 and B.16, in a coordinate system formed by the sagittal, coronal and horizontal planes. The rotation movements of joints are performed about the longitudinal axis of a body segment. The anatomically forbidden joint motions are often penalized by stiff springs, while the natural motions are hardly penalized within the admissible motion ranges and strongly penalized at the limits of these ranges. The corresponding stiffness is also shown in Tables B.15 and B.16. Fig. B.1 shows typical non-linear moment-rotation curves of the hip joint and of the subtalar joint of the ankle. The friction coefficients of the synovial joints of the human skeleton vary between 0.005 and 0.04.

In cases when more detail about the local motion and response under load is required, the joints can be modeled with their true articular surface geometry and their connecting ligaments. Detailed finite element models are also applied in orthopedic analysis. The gain in precision is obtained at a higher computational cost.

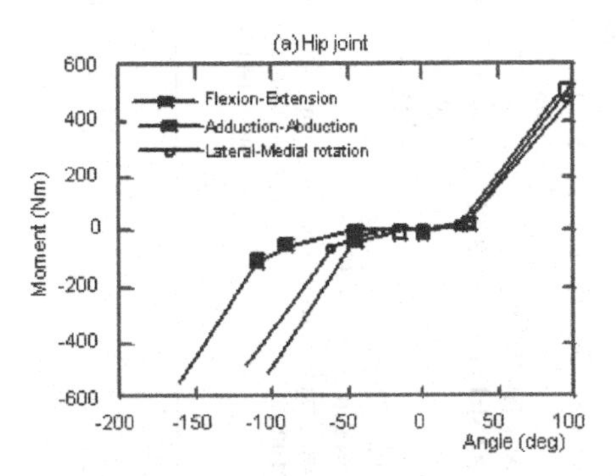

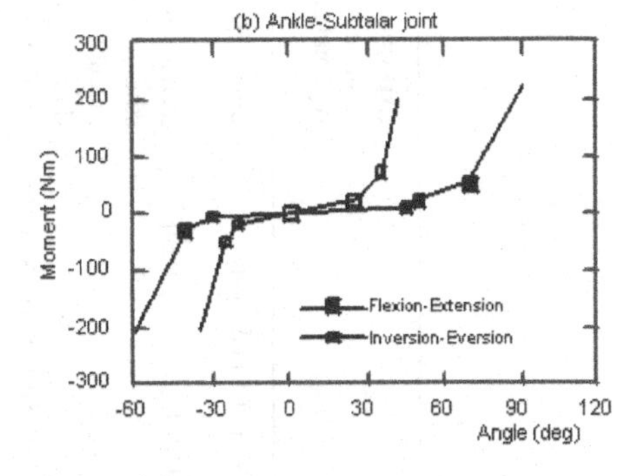

FIG. B.1. Typical moment-angle characteristics of the joints: (a) hip joint, and (b) ankle-subtalar joint (from YANG and LÖVSUND [1997] and PARENTEAU and VIANO [1996]). (Reproduced by permission of Chalmers University of Technology.)

TABLE B.12
Brain material properties

Authors	Experimental configuration type and conditions	Characteristics					
		ρ [kg/mm^3]	E [GPa]	ν	K [GPa]	G [GPa]	Other
DIMASI, MARCUS and EPPINGER [1991]	linear visco elastic				6.895E−02	$G_s = 3.4474E{-}05$ $G_l = 1.723E{-}05$	$\beta = 0.100\ \mathrm{m\,s^{-1}}$
ESTES and MCELHANEY [1970]	incompressible with strain-rate independent bulk modulus (human, monkey)					2.070E+00	
FALLENSTEIN, HULCE and MELVIN [1970]	dynamic complex shear modulus by vibration tests at 10 Hz (human autopsy brain)					$G_1 = 6.00E{-}07$ to 1.10E−06 $G_2 = 3.00E{-}07$ to 6.50E−07	
FIROOZBAKSHK [1975]		1.000E−06				1.920E−04	
GALFORD and MCELHANEY [1970]	dynamic complex tensile modulus by vibration test at 34 Hz (human brain tissue)		$E_1 = 6.67E{-}05$ $E_2 = 2.62E{-}05$				
GOLDSMITH [1972]			7.800E−05		4.320E−10		
LEE, MELVIN and UENO [1987]		1.000E−06		0.49		8.000E−05	
MCELHANEY, MELVIN, ROBERTS and PORTNOY [1973]	dynamic complex shear modulus by vibration tests at 9–10 Hz					$G_1 = 4.30E{-}07$ to 9.50E−07 $G_2 = 3.50E{-}07$ to 6.00E−07	$G2/G1 = 0.72$

TABLE B.12
(*Continued*)

Authors	Experimental configuration type and conditions	Characteristics					
		ρ [kg/mm^3]	E [GPa]	ν	K [GPa]	G [GPa]	Other
MARGULIES and THIBAULT [1989]		1.060E−06	4.000E−05	0.5	2.000E+00	1, 380E−05	
OMMAYA [1967]			8.00E−06 −1.5E−05				
ROSE and GORDON [1974]		1.050E−06			2.100E+00		
RUAN, KHALIL and KING [1991]		1.040E−06	6.670E−05	0.48–0.499	2.190E+00	1.680E−03	
RUAN, KHALIL and KING [1994]	linearly viscoelastic				1.279E−01	$G_s = 5.28E{-}04$ $G_l = 1.68E{-}04$	$\beta = 0.035$ m s^{-1}
SHUGAR [1975]		6.720E−07	1.030E−05	0.5	2.100E+00	3.450E−06	
SHUCK, HAYNES and FOGLE [1970]	dynamic complex shear modulus by vibration tests at 2–400 Hz					$G_1 = 8.30E{-}07$ to 1.38E−04 $G_2 = 3.40E{-}07$ to 8.27E−05	
SHUCK and ADVANI [1972]	dynamic complex shear modulus by vibration tests at 5–350 Hz					$G_1 = 7.60E{-}06$ to 3.39E−05 $G_2 = 2.76E{-}06$ to 8.16E−05	

TABLE B.12
(*Continued*)

Authors	Experimental configuration type and conditions	Characteristics					
		ρ [kg/mm^3]	E [GPa]	ν	K [GPa]	G [GPa]	Other
THIBAULT and MARGULIES [1996]	Age effect on complex shear modulus in custom-designed oscillatory shear testing device, at shear strain 2.5% from 20–200 Hz, 25 °C, 100% humidity only in one location and one direction (neonatal (2–3 days) pigs: curve data)					$G = 7.500E{-}07$ to $1.5938E{-}06$ $G_1 = 6.875E{-}07$ to $1.1875E{-}06$ $G_2 = 1.875E{-}07$ to $1.00E{-}06$	
TROSSEILLE, TARRIERE, LAVASTE, GUILLON and DOMONT [1992]		1.000E−06	2.400E−04	0.49–0.499			
UENO, MELVIN, LI and LIGHTHALL [1995]			2.400E−04	0.49	4.000E−03	8.000E−05	
WARD and THOMSON [1975]		1.040E−06	6.670E−05	0.48			
WARD [1982]			6.500E−04	0.48–0.499			
TURQUIER, KANG, TROSSEILLE, WILLINGER, LAVASTE, TARRIERE and DÖMONT [1996]		1.140E−6	6.750E−04	0.48	5.625E−03	$G_s = 5.28E{-}04$ $G_l = 1.68E{-}04$	$\beta = 0.035\ \mathrm{m\,s^{-1}}$

TABLE B.13
Brain material properties (white vs. gray matter and CSF)

Authors	Experimental configuration type and conditions	Characteristics					
		ρ [kg/mm^3]	E [GPa]	ν	K [GPa]	G [GPa]	Other
White matter:							
TADA and NAGASHIMA [1994]			5.000E−05	0.49			perm. = 1.0E−11 mm^2 poros. = 0.2
ZHOU, C. ET AL. [1996]		1.040E−06		0.4996	3.490E−01	2.680E−04	
ZHOU, C. ET AL. [1996]	SHUCK and ADVANI [1972]				2.190E−400	$G_s = 4.10E-05$ $G_l = 7.6E-06$	$\beta = 0.70\ ms^{-1}$
Gray matter:							
TADA and NAGASHIMA [1994]			5.000E−04	0.49			perm. = 1.0E−14 mm^2 poros. = 0.2
ZHOU, C. ET AL. [1996]		1.040E−06		0.4996	2.190E−01	$G = 1.68E-04$	
ZHOU, C. ET AL. [1996]	SHUCK and ADVANI [1972]				2.190E+00	$G_s = 3.40E-05$ $G_l = 6.3E-06$	$\beta = 0.70\ ms^{-1}$
CSF (cerebro-spinal fluid):							
RUAN, KHALIL and KING [1991]		1.040E−06		0.489	2.190E−02	5.000E−04	
RUAN, KHALIL and KING [1991]		1.000E−06			2.190E+00		
RUAN, KHALIL and KING [1994]		1.040E−06			2.190E−02	5.000E−05	
TADA and NAGASHIMA [1994]			9.930E−10	0.49			perm. = 1.0E−05 mm^2 poros. = 0.99

TABLE B.14
Brain material properties (other tissues)

Tissue & authors	Experimental configuration type and conditions	Characteristics					
		ρ [kg/mm^3]	E [GPa]	ν	K [GPa]	G [GPa]	Other
Cerebellum, brainstem RUAN, KHALIL and KING [1994]		1.040E−06		0.4996	2.19E−01	1.68E−04	
Pia mater ZHOU, Q. ET AL. [1996]		1.133E−06	1.150E−02	0.45			
Bridging veins ZHOU, Q. ET AL. [1996]		1.133E−06	1.100E−04	0.45			
Dura, falx & tentorium ZHOU, Q. ET AL. [1996]		1.133E−06	3.150E−02	0.45			
Dura/Falx MCELHANEY, MELVIN, ROBERTS and PORTNOY [1973]	tensile at frequency: 6.66E−05 msec^{-1} 6.66E−04 msec^{-1} 6.66E−03 msec^{-1}		4.157E−02 4.435E−02 6.069E−02				
MELVIN, MCELHANEY and ROBERTS [1970]			4.1382E−02 to 5.5176E−02				
Membrane RUAN, KHALIL and KING [1993]		1.133E−06	3.150E−02	0.45			
Dura Mater YAMADA [1970]	square 15 mm on each side expansive properties (curve Fig. 198 p. 222) (sample: cerebral rabbit)						$\sigma_u^e = 3.8\text{E}{-}02 \pm 0.0018$ [kg/mm^2] $\sigma_{\lim}^e = 0.2\sigma_u^e$
	rectangular strap 2 cm wide and 2.5 cm long						$T = 1.26$ [kg/mm]
	shearing properties (human adult average)						$\tau_u = 1.98$ [kg/mm^2]

TABLE

List of variables used in Tables B.11 through B.14

ρ = density	E_1 = storage tensile modulus
E = Young's modulus	E_2 = loss tensile modulus
ν = Poisson's ratio	β = decay factor
K = bulk modulus	σ_u^e = ultimate expansive strength
G = shear modulus	σ_{lim}^e = initial expansive strength
G_1 = storage shear modulus	T = shearing breaking load per unit width
G_2 = loss shear modulus	τ_u = ultimate shearing strength
G_e = equivalent shear modulus	perm. = permeability
G_l = long term shear modulus	poros. = porosity
G_s = short term shear modulus	

The shear relaxation behavior is described by the time dependent shear modulus $G(t) = G_l + (G_s - G_l)e^{-\beta t}$, where t is time in a relaxation test.

B.5. Inter-vertebral joint mechanical properties

Fig. B.2 indicates the average range of motion for rotations between the vertebrae, where C2–C7 are the vertebrae of the cerebral spine, T1–T12 of the thoracic spine, L1–L5 of the lumbar spine and S1 indicates the sacrum. The range of rotational motion between C1 (atlas) and C2 (axis) is about 35 degrees and accounts for about 50% of the rotation of the head. The range of motion also differs between individuals, sexes and is strongly age-dependent, decreasing by about 50% from youth to old age. (See Table B.17.)

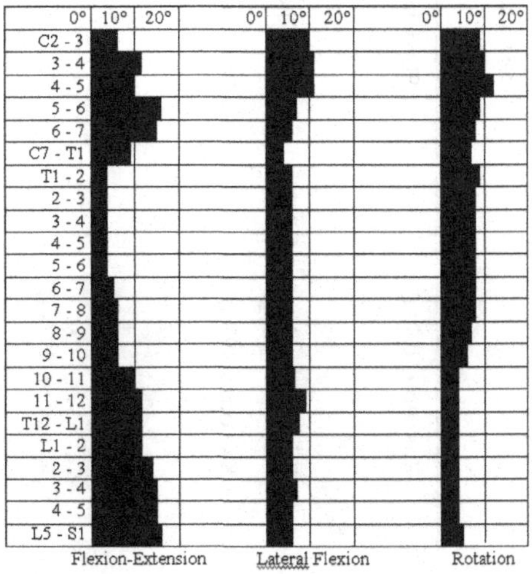

0° 10° 20°
C2 - 3
3 - 4
4 - 5
5 - 6
6 - 7
C7 - T1
T1 - 2
2 - 3
3 - 4
4 - 5
5 - 6
6 - 7
7 - 8
8 - 9
9 - 10
10 - 11
11 - 12
T12 - L1
L1 - 2
2 - 3
3 - 4
4 - 5
L5 - S1
Flexion-Extension
Lateral Flexion
Rotation

FIG. B.2. Inter-vertebral joint range of motion. (Reproduced by permission of Chalmers University of Technology.)

TABLE B.15
Upper limb joint mechanical properties

Joint	Authors	Experimental configuration	Characteristics			
			Physiological feature	Motion range [deg]	Stiffness [Nm/deg]	Mechanical feature
Shoulder	KAPANDJI [1974a]	Physiological motion	Flexion	180	0–0.3	Ball joint
			Extension	45–50	0–0.2	
			Abduction	140	0–0.3	
			Adduction	30–45	–	
	FRANKEL and NORDIN [1980]	Static load	Lateral rotation	80	0.3	
			Medial rotation	95	0.3	
Elbow	KAPANDJI [1974a]	Physiological motion	Flexion	145–160	0–0.2	Pin joint
			Extension	0	–	
			Abduction	0	–	
			Adduction	0	–	
			Lateral rotation	90	0.2	
			Medial rotation	85	0.2	
Wrist	KAPANDJI [1974a]	Physiological motion	Flexion	85		Cardan joint
			Extension	85		
			Abduction	15		
			Adduction	45		

TABLE B.16
Lower limb joint mechanical properties

Joint	Authors	Experimental configuration	Characteristics			
			Physiological feature	Motion range [deg]	Stiffness [Nm/deg]	Mechanical feature
Hip	KAPANDJI [1974b]	Physiological motion	Flexion	90–145	0–2.5	Ball joint
			Extension	20–30	1.2	
	FRANKEL and NORDIN [1980]		Abduction	45	0–1.2	
	MOW and HAYES [1991]	Static load	Adduction	30	0.8	
			Lateral rotation	30	0.6	
			Medial rotation	60	0.6	
Knee	KAPANDJI [1974b]	Physiological motion	Flexion	120–160	0–1.2	Pin joint
			Extension	5–10	2.0	
	FRANKEL and NORDIN [1980]		Abduction	0	–	
			Adduction	0	–	
			Lateral rotation	40	1.0	
			Medial rotation	30	1.0	
Ankle	KAPANDJI [1974b]	Physiological motion	Tibia-Talar:			Pin joint
			Dorsiflexion	20–30	0.5	
			Plantarflexion	30–50	0.3	
	PARENTEAU and VIANO [1996]	Quasi-static load	Subtalar:			Ball joint
			Inversion	15–20	1	
			Eversion	10–15	1.5	
			Abduction	30–40		
			Adduction	25–35		

TABLE B.17
Inter-vertebral joint mechanical properties

Joint	Authors	Experimental configuration	Characteristics			
			Physiological feature	Motion range [deg]	Stiffness [Nm/deg]	Other
Lumbar	KAPANDJI [1974a]	Physiological motion	Flexion	60	1.0–2.1	Tolerance:
	FRANKEL and NORDIN [1980]		Extension	35	0.3–1.8	$M_{flex} = 145–185$ Nm
			Lateral flexion	±20	2.0	
		Static load	Axial rotation	±5	0.9	
Thoracic	KAPANDJI [1974a]	Physiological motion	Flexion	45	1.0–2.1	Tolerance:
	FRANKEL and NORDIN [1980]		Extension	25	0.3–1.8	$M_{flex} = 616$ Nm
			Lateral flexion	±20	2.0	$M_{ext} = 240$ Nm
		Static load	Axial rotation	±35	0.9	
Cervical	KAPANDJI [1974a]	Physiological motion	Flexion	40	1.4	
			Extension	75	2.5	
	WISMANS and SPENNY [1983]	Dynamic load				
		lateral bending	Lateral flexion	±45	0.4–2.2	
		torsion	Axial rotation	±50	0.0–0.5	
	MCELHANEY, DOHERTY, PAVER, MYERS and GRAY [1988]	Dynamic			10.5–67.5 [Nm/rad]	
	MERTZ and PATRIC [1971]	flexion–extension				Tolerance: $M_{flex} = 56.7$ Nm $M_{ext} = 189$ Nm
	BOWMAN, SCHNEIDER, LUSTAK, ANDERSON and THOMAS [1984]	Torsion			2.71–3.74	

TABLE B.18
Articular cartilage and meniscus mechanical properties

Tissue	Authors	Experimental configuration	Characteristics			
			Density [kg/m^3]	Young's modulus [MPa]	Poisson's ratio ν	Other material coefficients
Articular Cartilage	MOW and HAYES [1991]	Tensile Solid matrix		0.3–1.0	0.1–0.4	
	YAMADA [1970]	Compressive		> 1.0		
	VIANO [1986]	Instantaneous		$E = 12$	0.42	$G = 4.1$ MPa $K = 2.5$ MPa
		Asymptotic		$E = 7.1$	0.37	$G = 2.6$ MPa $K = 9.1$ Mpa
	ARMSTRONG, LAI and MOW [1984]		1000	$E = 35$	0.45	
Meniscus	MOW and HAYES [1991]	Tensile		0.1–0.6		

B.6. Articular cartilage and meniscus mechanical properties

Cartilage is known to behave as a biphasic material, where a fluid seeps through a solid porous matrix, which can lead to slow deformation under compressive loads. For the short term behavior in impact studies these time and load dependent properties are not dominant, and only the classical stiffness terms are required. To describe the complete flow and deformational behavior of cartilage and meniscus the biphasic theory was developed by Mow and coworkers. In this theory the solid matrix is linearly elastic and isotropic, the solid matrix and interstitial fluid are intrinsically incompressible and viscous dissipation is due to interstitial fluid flow relative to the solid matrix. (See Table B.18.)

B.7. Inter-vertebral disc mechanical properties

The inter-vertebral discs assure the elastic coherence of the spinal column and they provide a shock absorbing effect. Each disc is composed of the nucleus pulposus, the annulus fibrosus, and a cartilageous end-plate. Between 70 to 90% of the nucleus pulposus by weight is water, and it takes up as much as 40 to 60% of the disc area. The annulus fibrosus is a laminated and hence an anisotropic structure composed of several concentric layers with fibers alternating at plus and minus 30 degree angles of inclination from the horizontal plane. The inner boundary of the annulus fibrosus is attached to the cartilageous end-plate, and the outer surface is directly connected to vertebra body. The discs play dominant role in sustaining the body against compressive load. Under compression, the nucleus pulposus acts like a fluid in a cylinder made of the annulus fibrosus. The discs show greater stiffness for the front/rearward inter-vertebral shearing motion than for the side/side motion, with the annulus fibrosus rather the nucleus pulposus making a major contribution.

The mechanical properties of the inter-vertebral discs are summarized in Table B.19. Table B.20 summarizes age and region dependent tensile properties of the discs and Tables B.21 and B.22 contain the average region dependent compressive and torsional properties (SONODA [1962]). Note that inter-vertebral discs from people in age group 20–39 have the greatest ultimate loads. Discs of females have a breaking moment about five-sixths of that in males. The ultimate torsional strength and angle of twist in whole discs of females are also less than in males.

B.8. Muscle mechanical properties

Skeletal muscles have active and passive properties, Fig. B.3. The active muscle action is usually not of prime importance in car occupant impact simulations, except in low energy collisions, where the activation of the muscles during bracing can modify the injury pattern (example: rear impact/whiplash). In the following tables mostly the passive mechanical behavior of the skeletal muscles is documented.

A quasi-linear visco-elastic model was proposed to model the passive response of skeletal muscle. Within the physiological muscle length, the passive muscle force is usually much lower than the active force. At high elongation, nearing the physiological limits of joint motion, the passive force increases rapidly and reaches the same level and beyond as the maximum active force, while the active force drops to low values. At high stretch, the axial muscle force is therefore dominated by the passive force. At negative stretch velocities the active muscle force drops to almost zero at a given reference velocity, while at positive stretch velocities this force will increase beyond the activation level at zero stretch velocity. The Hill muscle model is often evoked in simulation of the active and passive kinetics of skeletal muscles. This law is described in Appendix C.

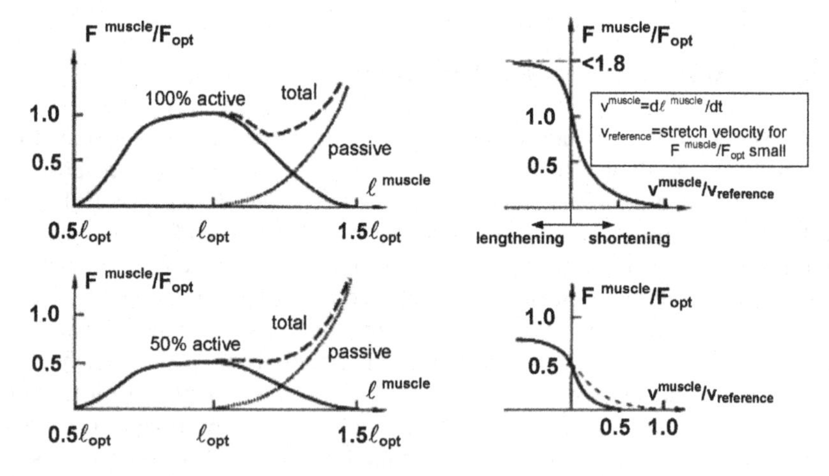

FIG. B.3. Muscle force-length and muscle force-velocity properties (ZAJAC [1989]). (Reproduced by permission of Begell House, Inc.)

TABLE B.19
Inter-vertebral disc mechanical properties

Tissue	Authors	Experimental configuration	Characteristics				
			Density	Young's modulus [MPa]	Poisson ratio ν_{12}	Shear modulus G_{12} [MPa]	Other
Fibers	GALANTE [1967]	Tensile lumbar region		400–500 $E_1 = 500$ $E_2 = 500$	0.3	192	
Ground substance	GOEL, MONROE, GILBERTSON and BRINKMAN [1995]	Compressive lumbar		2–4.2			
	UENO and LIU [1987]	Torsion lumbar		$E_1 = 3$ $E_2 = 3$	0.45	1	
Lamellae	SKAGGS, WEIDENBAUM, IATRIDIS, RATCLIFFE and MOW [1994]	Tensile Posterior of disc Anterior of disc		$E_1 = 70 \pm 42$ $E_1 = 106 \pm 72$			
	KULAK, BELYTSCHKO, SCHULTZ and GALANTE [1976]			$E_1 = 83$ $E_2 = 2.07$	0.45	1.38	
Annulus fibrosus	EBARA, IATRIDIS, SETTON, FOSTER, MOW and WEIDENBAUM [1996]	Tensile lumbar		$E_{circ} = 5–50$			
Annulus fibrosus	SPILKER, JAKOBS and SCHULTZ [1986]	Modeling		$E_{circ} = 33.4$ $E_z = 0.9$	0.5	0.189	
	LIN, LIU, RAY and NIKRAVESH [1978]			$E_{circ} = 22.4$ $E_z = 11.7$	0.45	3.92	

TABLE B.20
Age and region dependent tensile inter-vertebral disc properties

Region	Characteristics								
	20–39 yr			40–79 yr			Adult average		
	F_u [kg]	σ_u [kg/mm^2]	ε_u [%]	F_u [kg]	σ_u [kg/mm^2]	ε_u [%]	F_u [kg]	σ_u [kg/mm^2]	ε_u [%]
Cervical	105 ± 14.5	0.33 ± 0.02	89 ± 4.2	80 ± 8.6	0.29 ± 0.03	71 ± 3.6	88	0.30	77
Upper thoracic	142 ± 16.3	0.24 ± 0.01	55 ± 3.8	106 ± 9.4	0.20 ± 0.02	41 ± 2.1	118	0.21	46
Lower thoracic	291 ± 21.5	0.26 ± 0.02	57 ± 6.3	220 ± 12.8	0.22 ± 0.01	40 ± 2.4	244	0.23	46
Lumbar	394 ± 24.6	0.30 ± 0.01	68 ± 7.1	290 ± 19.5	0.24 ± 0.01	52 ± 6.2	325	0.26	59
Average	233	0.28	67	174	0.24	51	194	0.25	57
Ratio	1	1	1	0.75	0.85	0.76	0.83	0.89	0.85

F_u = breaking load in [kg]
σ_u = ultimate strength in [kg/mm^2]
ε_u = ultimate elongation in [%]

TABLE B.21
Region dependent compressive inter-vertebral disc properties

Region	Characteristics (40 to 59 years of age)		
	Breaking Load [kg]	Ultimate Strength [kg/mm^2]	Ultimate Contraction [%]
Cervical	320	1.08	35.2
Upper thoracic	450	1.02	28.6
Lower thoracic	1150	1.08	31.4
Lumbar	1500	1.12	35.5
Average	(855)	1.08	32.6

TABLE B.22
Region dependent torsional inter-vertebral disc properties

Region	Characteristics (40 to 59 years of age)		
	Breaking Moment [kg cm]	Ultimate Strength [kg/mm^2]	Ultimate Angle of Twist [deg]
Cervical	51	0.48	34
Upper thoracic	84	0.41	26
Middle thoracic	167	0.44	22
Lower thoracic	265	0.45	17
Lumbar	440	0.48	14
Average	201	0.45	23

TABLE B.23
Ultimate tensile strength of skeletal muscles (KATAKE [1961])

Body segment	Muscles	Ultimate tensile strength [MPa]
Trunk	Sternocleidomastoideus	0.19
	Trapezius	0.16
	Pectoralis major	0.13
	Rectus abdominis	0.14
Upper extremity	Biceps brachii	0.17
	Triceps brachii	0.21
	Flexor carpi radialis	0.15
	Brachioradialis	0.18
Lower extremity	Psoas major	0.12
	Sartorius	0.30
	Gracilis	0.20
	Rectus femoris	0.10
	Vastus medialis	0.15
	Adductor longus	0.13
	Semimembranous	0.13
	Gastrocnemius	0.10
	Tibialis anterior	0.22

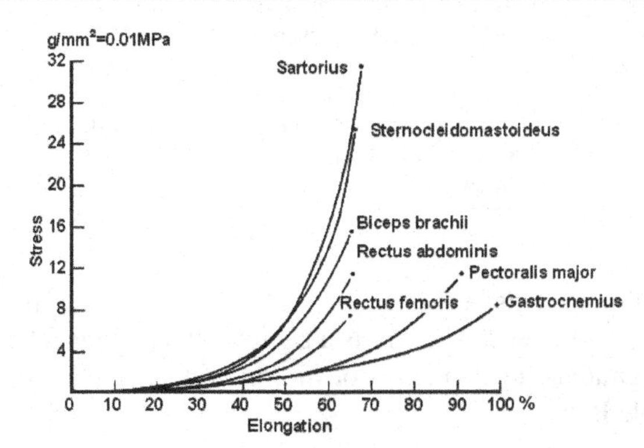

FIG. B.4. Passive stress-strain curves in tension of skeletal muscles for 29 year old persons (YAMADA [1970]). (Reproduced by permission of Lippincott, Williams and Wilkins.)

Table B.23 contains the ultimate tensile strength of a selection of skeletal muscles. The corresponding ultimate passive muscle forces will be obtained by multiplying the strength by the physiological cross section areas of each muscle. Fig. B.4 shows the non-linear passive stress-strain curves of the muscles. Table B.24 contains age differences in passive tensile properties of the rectus abdominis muscle. There is no significant sexual difference in the ultimate strength of skeletal muscles. (See Table B.25.)

B.9. Tendon mechanical properties

Tendons connect muscles to the skeletal bones. There is almost no age effect on the ultimate elongation, δ_u, a moderate age effect on the elastic modulus, E, and a marked

TABLE B.24
Age differences in passive tensile properties of rectus abdominis muscle (KATAKE [1961])

	Age group							Adult average
	10–19 yr	20–29 yr	30–39 yr	40–49 yr	50–59 yr	60–69 yr	70–79 yr	
	Ultimate tensile strength [g/mm^2 = 0.01 Mpa]							
	19 ± 1.2	15 ± 0.6	13 ± 1.0	11 ± 0.6	10 ± 0.5	9 ± 0.3	9 ± 0.3	11
ratio	1.00	0.79	0.68	0.58	0.53	0.47	0.47	
	Ultimate percentage elongation							
	65 ± 1.2	64 ± 1.1	62 ± 0.7	61 ± 0.9	61 ± 1.5	58 ± 1.8	58 ± 1.8	61
ratio	1.00	0.98	0.95	0.94	0.94	0.89	0.89	

TABLE B.25
Summary of studies on mechanical properties of muscles

Authors	Experimental configuration	Characteristics	
		σ_u [MPa]	Other
YAMADA [1970]	Experiments on various animal and human specimens	0.1–0.32	
WINTERS and STARK [1985], WINTERS and STARK [1988]		0.5–1.0	
SCHNECK [1992]	C = damping coefficient K = stiffness of whole muscle	0.2–1.0	$C = 10$–1000 Ns/m $K = 32.5$–250 kN/m

age effect on the ultimate tensile strength, σ_u. The ultimate tensile strength can exceed the insertion strength of a tendon. There is a significant strain rate effect on the elastic modulus. The longitudinal tendon strips of the supraspinatus muscle are not of equal strength. (See Table B.26.)

B.10. Skin mechanical properties

Skin is a non-linear elastic material and its response is orthotropic. In human models, skin should be modeled as the enveloping membrane of flesh and fatty tissue. Skin assures the stability of the female breasts, which consists of fatty tissue, enveloped by resistant skin. The behavior of skin is best modeled as a material with perpendicular layers of fibers, each described with non-linear stress-elongation curves (see Figs. B.5–B.7). Some mechanical properties are listed in Table B.27.

B.11. Internal organ mechanical properties

Rather little information is available on the mechanical properties of the internal organs. Table B.28 contains some preliminary data on the lungs, liver and spleen. The data on

TABLE B.26
Mechanical properties of tendons

Type of tissue	Authors	Experimental configuration type and conditions	E [GPa]	ν Poissons ratio	σ_u [MPa] Ultimate strength	δ_u Ultimate elongation
	MOW and HAYES [1991]		1.2–1.8	Incomp.		
Achilles tendon	LEWIS and SHAW [1997a], LEWIS and SHAW [1997b]	Tension (fiber direction) 0.1sec^{-1} and $1\ \text{sec}^{-1}$ embalmed, age 36–100	2.00 ± 0.99	0.4		
Patellar tendon	JOHNSON, TRAMAGLINI, LEVINE, ONO, CHOI and WOO [1994]	Tension (fiber direction) fresh frozen, non-irradiated				
		(i) age 29–50	0.66 ± 0.266		64.7 ± 15	0.14 ± 0.06
		(ii) age 64–93	0.504 ± 0.222		53.6 ± 10	0.15 ± 0.05
Human tendon	VOIGT, BOJSEN-MOLLER, SIMONSEN and DYHRE-POULSEN [1995]	Tension (fiber direction)	1.2		50	0.06 (yield point)
Patellar tendon	WOO, JOHNSON and SMITH [1993]	Tension (fiber direction) Stress–Relaxation	0.58			
Finger flexor	PRING, AMIS and COOMBS [1985]	Tension (fiber direction)				0.13
Supraspinatus	ITOI, BERGLUND, GRABOWSKI, SCHULTZ, GROWNEY, MORREY and AN [1995]	Tension (fiber direction) fresh				
		(i) anterior			16.5 ± 7.1	
		(ii) middle			6.0 ± 2.6	
		(iii) posterior			4.1 ± 1.3	

the liver and spleen are estimations used in a project on virtual abdominal surgery, DAN and MILCENT [2002].

Fig. B.8 shows the non-linear force-displacement curve of an entire human liver under compression between two parallel plates under quasi-static loading, DAN [1995]. The tested liver was pressurized in the sense that the natural in- and outflow of body fluids was maintained artificially with a circulating substitute fluid. The liver as a whole

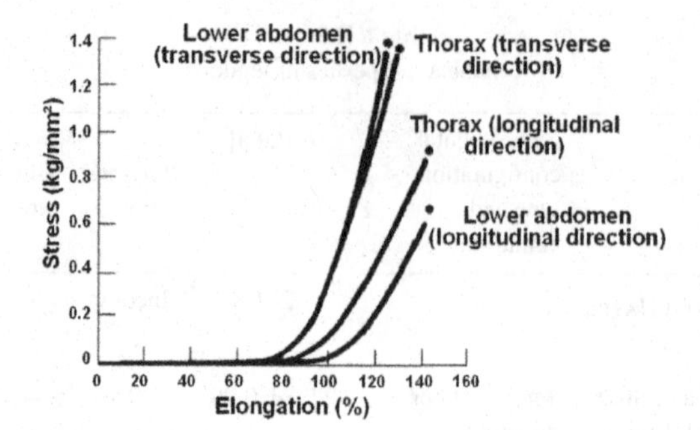

FIG. B.5. Stress-strain curves in tension of the skin of persons 20 to 29 years of age (YAMADA [1970]). (Reproduced by permission of Lippincott, Williams and Wilkins.)

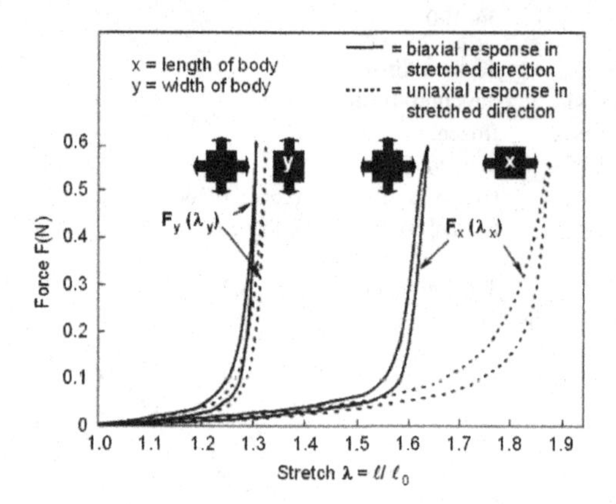

FIG. B.6. Force-stretch relation of rabbit skin at stretch rate 0.2 mm/s (LANIR and FUNG [1974]). (Reproduced by permission of the Journal of Biomechanics.)

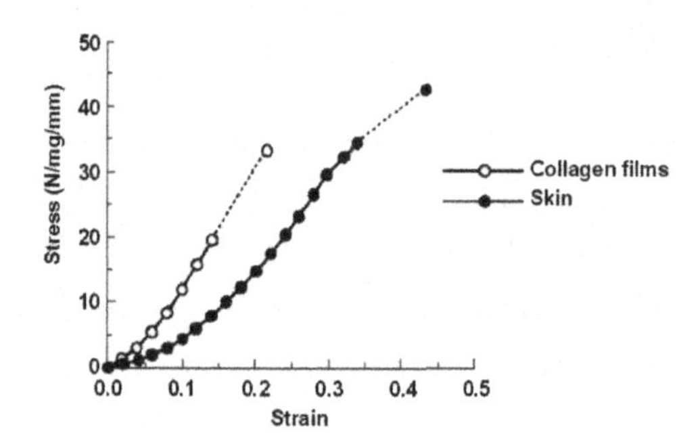

FIG. B.7. Stress-strain relation of rat skin (OXLUND and ANDREASSON [1980]). (Reproduced by permission of Blackwell Publishing Ltd.)

body was then loaded. It resists the applied compression load with the combined action of the stored circulating fluid and the resistance of its bulk material.

TABLE B.27
Mechanical properties of skin

Tissue	Authors	Experimental Configuration type and conditions	E [Gpa]	Tensile breaking load per unit width [kg/mm^2]	σ_u [UTS] [kg/mm^2]	δ_{max} Ultimate percentage elongation
Skin of the calf	MANSCHOT and BRAKKEE [1986a], MANSCHOT and BRAKKEE [1986b]	Tension Slow rates in Vivo				
		(i) along tibial axis	Non-linear			
		(ii) across tibial axis	Non-linear			
Skin	YAMADA [1970]	Tension human skin				
		(i) 30–49 years	Non-linear	0.2 to 3.4	0.29 to 1.47	–
		(ii) 10–29 years	Non-linear	–	–	56 to 144
		(iii) Adult average	Non-linear	0.2 to 3.1	0.26 to 1.32	43 to 111

TABLE B.28
Mechanical properties of some organs

Tissue	Authors	Experimental configuration type and conditions	Density ρ [kg/m^3]	Modulus E [Mpa]	Poisson Ratio ν	Shear mod. G [Mpa]
Lungs	HOPPIN, LEE and DAWSON [1975]	Compression (3-D loading)		$E/Pt^* = 4$	0.3	
	HAJJI, WILSON and LAI-FOOK [1979]					$G/07Pt = 1$ to 1.5
Liver	DAN [1999], DAN and MILCENT [2002]	Small test cylinders of parenchyma	1158	0.3E–4 to 5E–4	0.4	
		Venous vessels	1168	0.158	0.49	
		Glisson capsule	1168	1.0	0.49	
Spleen	CARTER [1999]	(Indentation tests for the study of force feedback in virtual surgery)		(much smaller)		

$^*Pt$ = the transpulmonary pressure

Fig. B.9 represents static compression results of small test cylinders of pure liver parenchyma (diameter 12 mm by 15 mm initial height), from which the interstitial fluid can escape, DAN [1999]. It indicates a non-linear distribution of the elastic modulus,

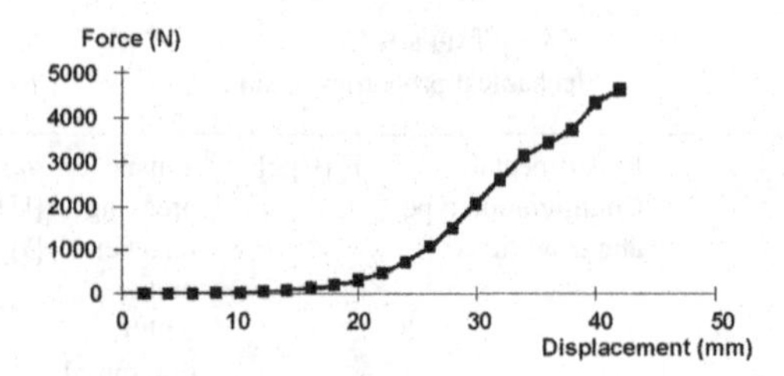

FIG. B.8. Whole human liver force vs. displacement curves (DAN [1995]). (Private communication by the author.)

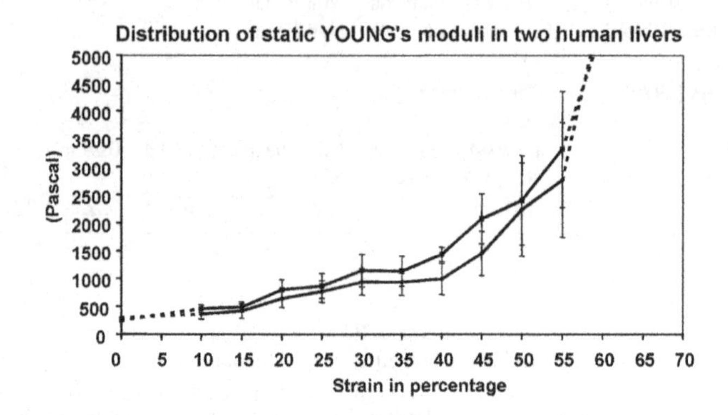

FIG. B.9. Liver parenchyma elastic modulus vs. displacement curves (DAN [1999]). (Private communication by the author.)

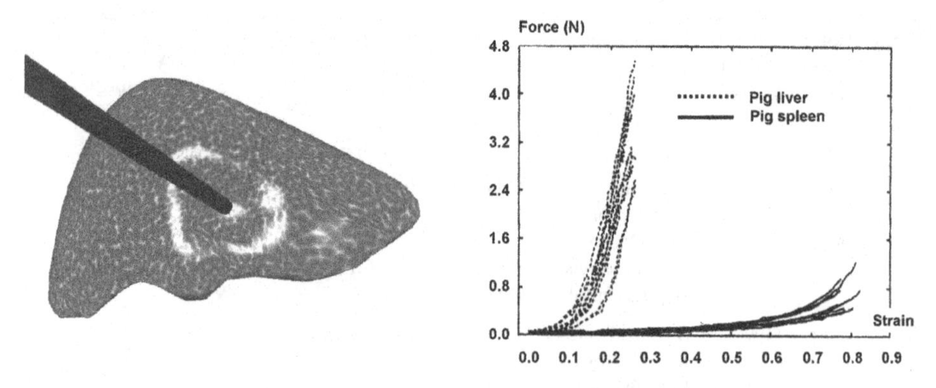

FIG. B.10. Indentation tests on the liver and spleen of pigs (CARTER [1999]). (Reproduced by permission of Fiona Carter of the University of Dundee.)

and the response of the so isolated material is much weaker than the response of the pressurized complete organ.

Fig. B.10 shows results of indentation tests on pig livers and spleens, CARTER [1999], which were performed in projects on the study of haptic force feedback in virtual surgery. The picture shows that the spleen is much less resistant than the liver.

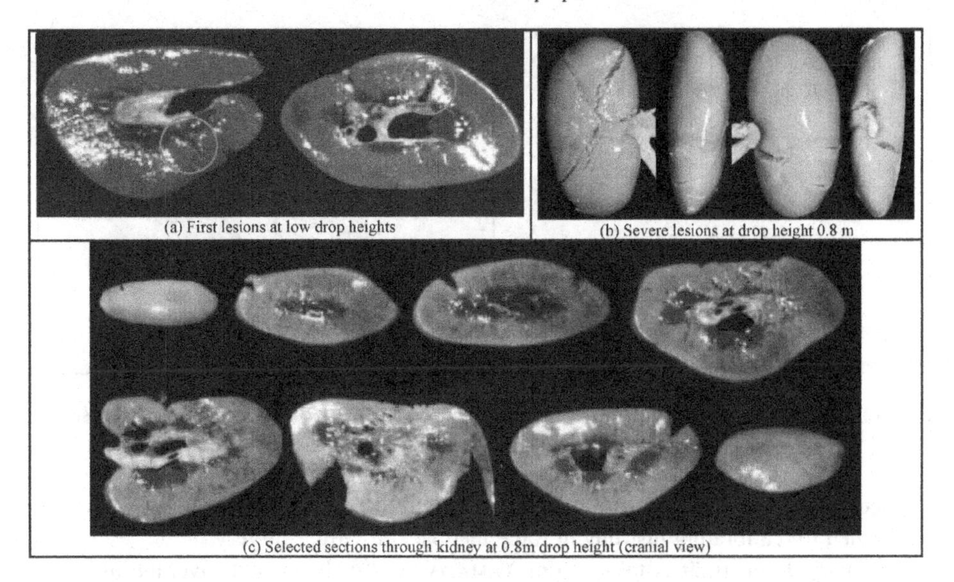

(a) First lesions at low drop heights

(b) Severe lesions at drop height 0.8 m

(c) Selected sections through kidney at 0.8m drop height (cranial view)

FIG. B.11. Typical drop test lesions in porcine kidneys (BSCHLEIPFER [2002]). (Reproduced by permission of Dr. med. Thomas Bschleipfer.)

FAZEKAS, KOSA, JOBBA and MESZARO [1971a], FAZEKAS, KOSA, JOBBA and MESZARO [1971b], FAZEKAS, KOSA, JOBBA and MESZARO [1972] (in German) published data on the compression resistance of the whole human cadaver liver, kidneys and spleen, respectively. They found that the liver showed superficial ruptures at a compressive stress of 169 kPa and multiple ruptures at 320 kPa. The first superficial ruptures of the spleen occurred at a compressive *stress* of 44 kPa and of the kidneys at a *load* of 60.2 ± 28.2 daN, the latter of which showed multiple ruptures at loads of 109.44 ± 51.4 daN.

BAUDER [1985] (in German) investigated the compressive resistance of the isolated human liver with blunt drop weight impact tests. The tests showed for 3–4 m/s impact velocities mean compressive loads of the organ of 175.6 ± 39.2 daN, which were associated with mean compressive deformations of 29.5 ± 3.5 mm. The observed injuries were contusions, superficial ruptures and crushing of the livers. The thickness of the organs and the portion of connective tissue were important parameters for the severity of the injuries.

BSCHLEIPFER [2002] investigated the lesions inflicted on isolated porcine kidneys under blunt drop test impacts, from heights of 0.1 to 1.0 m, with a cylindrical impactor (⌀10 cm) of mass 1.45 kg (1.4 to 14.2 Joule), with and without ligatured urethers. Typical lesions found in the tests are shown in Fig. B.11, for the first lesions at low load, (a), for severe lesions at 0.8 m drop height, views, (b), and sections through the organ, (c).

Fig. B.12, finally, gives an overview on the nonlinear visco-elastic modulus response (Pa) of live porcine livers, based on in vivo semi-infinite elastic body indentation with a vibrating cylindrical indenter, as a function of vibration frequency (Hz) and median

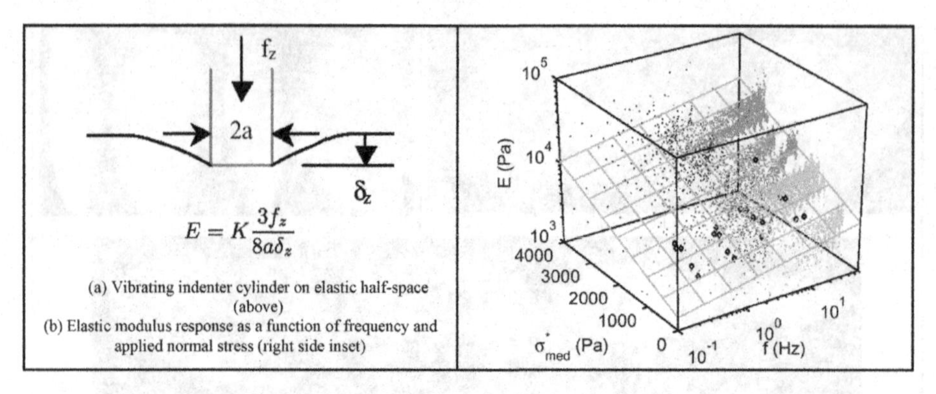

FIG. B.12. In vivo elastic modulus measurements of live porcine livers (OTTENSMEYER and SALISBURY [2001]). (Reproduced by permission of Springer Verlag.)

applied normal stress (Pa) (after OTTENSMEYER and SALISBURY [2001]). This work was performed for calibrating instrument force feedback in virtual surgery.

Summary Table B.29 is taken from YAMADA [1970]. It contains average ultimate strength (stress) and percentage elongation (strain) for a variety of human tissues and organs.

Data about many internal organs of humans are still missing. Table B.30 lists some such data found from animals (YAMADA [1970]).

Further data on liver and kidneys of rhesus monkeys are reported by MELVIN, STALNAKER and ROBERTS [1973]. These data are used by MILLER [2000] in modeling constitutive relationships of the abdominal organs.

B.12. Quasi-linear visco-elastic solids

Many biological tissues exhibit visco-elastic behavior. A simple such law is described next. The deviatoric response of a linear viscoelastic solid material is governed by Zener's model, which can be considered as a Maxwell spring-dashpot model in parallel with a spring, Fig. B.13.

The "slow motion" response to small strain rates, $d\varepsilon/dt$, is governed by the long term shear modulus, $G_\infty (\equiv G_l)$, while the instantaneous response to a step loading, $H(t)$, is according to the long term modulus $G_0 (\equiv G_s)$.

The elastic behavior of this material is described by the deformation rate dependent shear modulus, G, and by the constant bulk modulus, K.

$$\text{Shear modulus} \quad G = \frac{E}{2(1+\nu)} \quad \text{where } E = \text{Young's Modulus}$$

and

$$\text{Bulk modulus} \quad K = \frac{E}{3(1-2\nu)} \quad \nu = \text{Poisson's Ratio}.$$

TABLE B.29
Average adult human mechanical tissue properties (YAMADA [1970])

Tissue	σ_u [kg/mm^2]	δ_{max} [%]	Tissue	σ_u [kg/mm^2]	δ_{max} [%]
Hair	19.7	40	Ureter (L)	0.18	36
Compact bone (femur)	10.9	1.4	Mixed arterial tissue (L)	0.17	87
Chorda tendinea	6.4	33	Venous tissue (L)	0.17	89
Tendinous tissue (calcaneal)	5.4	9	Umbilical cord (mature fetus)	0.15	59
Nail	1.8	14	Mixed arterial tissue (T)	0.14	69
Fascia	1.4	16	Muscular arterial tissue (L)	0.14	102
Nerve (secondary fiber bundle)	1.3	18	Spinal dura mater (T)	0.13	34
Fibrocartilage (annulus fibrosus) (L)	1.3	14	Spongy bone (vertebra)	0.12	0.6
Skin (thorax, neck)	1.3	90	Coronary artery (L)	0.11	64
Spinal dura mater (L)	1.1	21	Renal calyx (L)	0.11	35
Skin (abdomen, back, foot, arm)	0.97	90	Elastic arterial tissue (T)	0.10	82
Skin (leg, hand)	0.74	90	Muscular arterial tissue (T)	0.10	75
Sclera (E)	0.69	17	Cardiac valve (R)	0.094	17
Fibrocartilage (annulus fibrosus) (T)	0.53	12	Elastic arterial tissue (L)	0.08	80
Sclera (M)	0.48	17	Large intestine (L)	0.069	117
Skin (face, head, genitals)	0.38	69	Esophagus (L)	0.06	73
Vertebra	0.35	0.8	Stomach (L)	0.056	93
Cornea	0.35	15	Small intestine (L)	0.056	43
Auricle	0.34	26	Small intestine (T)	0.053	89
Elastic cartilage (auricle)	0.31	26	Renal calyx (T)	0.048	48
Thyroid cartilage (L)	0.30	15	Large intestine (T)	0.045	137
Venous tissue (T)	0.30	66	Ureter (T)	0.045	89
Hyaline cartilage (costal)	0.29	18	Stomach (T)	0.044	127
Intervertebral disc	0.28	57	Tracheal membranous wall (T)	0.036	81
Cardiac valve (C)	0.25	13	Urinary bladder	0.024	126
Tracheal cartilage	0.24	18	Papillary muscle tissue	0.023	30
Amnion (normal labor)	0.24	42	Esophagus (T)	0.018	124
Renal fibrous capsule	0.23	29	Skeletal muscle tissue (rectus abdominis)	0.011	61
Tracheal membrane wall (L)	0.22	61	Cardiac muscle tissue	0.011	64
Tracheal intercartlagnious membrane	0.19	138	Renal parynchyma	0.005	52

(C) = circumferentially, (E) = equatorially, (L) = longitudinally, (M) = meridionally, (R) = radially, (T) = transversely.

TABLE B.30
Mechanical properties of some internal organs of animals (YAMADA [1970])

Tissue	Animal	σ_u [kg/mm^2]	δ_{max} [%]
Liver parenchyma	Rabbits	0.0024	46
Gall bladder	Rabbits	0.21	53
Uterus	Rabbits	0.018	150
Cerebral dura mater	Rabbits	0.038	

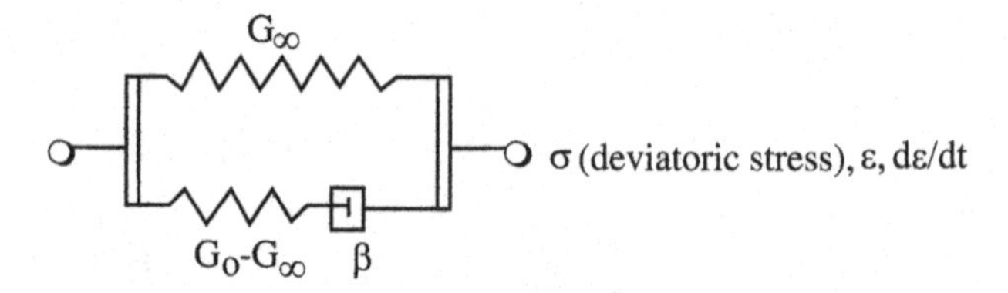

FIG. B.13. Zener type model.

The shear relaxation behavior is given by the response to a step function, and is described by the shear relaxation modulus

$$G(t) = G_\infty + (G_0 - G_\infty)e^{-\beta t}\big(\equiv G_l + (G_s - G_l)e^{-\beta t}\big).$$

Time t is the current simulation time, but relaxation starts only when the material experiences a deviatoric strain. The decay constant, β, has the unit of $(\text{time})^{-1}$, which must be consistent with the chosen time units.

The deviatoric stress rate, s_{ij}, depends on the shear relaxation modulus as follows

$$s_{ij} = 2\int_{\tau=0}^{t} G(t-\tau)D'_{ij}(\tau)\,\mathrm{d}\tau,$$

where D'_{ij} is the deviatoric velocity strain tensor. The above convolution expression for the deviatoric stress rate may be understood as follows: the deviatoric strain is approximated by a piecewise constant function. The material responds to each step function $H(\tau)$ following the relaxation law. This material model does not describe volumetric viscous effects, as might be present in the compression of foams.

B.13. Further references on biomaterials

The literature on biomaterials and related subjects is abundant. Some references are indicated in this appendix. While many of the indicated references deal with the experimental evaluation of biomaterial properties, others deal with the aspects of their modeling, the use of these materials in biomechanical models and the characterization of trauma and injury.

Many further references for biomaterials could be cited by separate topics on bones, ligaments, brain, joints, spine and inter-vertebral discs, muscles, tendons, skin and organs. These references are not mentioned explicitly in this appendix, but constitute further valuable sources of historical and actual information.

APPENDIX C

The Hill Muscle Model

Skeletal muscles. The Hill muscle model is one of the simplest phenomenological engineering models of the active and passive biomechanical behavior of skeletal muscles (HILL [1970]). Its simplest implementation is with bar finite elements. More involved implementations can be in 2D and 3D composite finite elements, where the composite fibers are assigned the properties of Hill-type muscle models.

Fig. C.1 gives an overview on the anatomical detail of skeletal muscles.

Active voluntary muscle contraction can be considered a material behavior that has no parallel in conventional engineering material models. Whereas the passive mechanical impact behavior of biological tissues can often be approximated using standard engineering material models, active muscle behavior clearly distinguishes living and non-living materials. For this reason it is interesting to briefly outline the standard Hill muscle model.

Standard Hill muscle model. This model, its implementation into a crash code and its application is described by WITTEK and KAJZER [1995], WITTEK and KAJZER [1997]; WITTEK, HAUG and KAJZER [1999]; WITTEK, KAJZER and HAUG [1999]; WITTEK, ONO and KAJZER [1999]; WITTEK, ONO, KAJZER, ÖRTENGREN and INAMI [2001]. Authors KAJZER, ZHOU, KHALIL and KING [1996] describe the application of modeling of ligaments and muscles under transient loads.

Fig. C.2 summarizes the Hill muscle bar model.

Inset (a) of Fig. C.2 is an overview of the types of skeletal muscles (WIRHED [1985]). Inset (b) shows the schematics of the Hill model for a fusiform tendon-muscle-tendon assembly. Inset (c) depicts the (normalized) active muscle component force versus length diagrams, F_{CE}/F_{max}. Inset (d) contains the active component force versus (normalized) stretch velocity diagram, $F_V(V/V_{max})$, and inset (e) shows the active component activation versus time function, $N_a(t)$.

Inset (b) of Fig. C.2 shows a simple mechanical model of a fusiform muscle with the contractile sub-element, (CE), the parallel elastic sub-element, (PE) and the parallel dashpot sub-element, (DE), of its active "muscle" element, and the nonlinear spring sub-element, (SE), and dashpot sub-element, (DSE), of its two "tendon" elements, switched in series. The tendons are not discussed here because as passive materials their mechanical response can be approximated with standard engineering visco-elastic-damaging material models. Their action can be modeled by arranging serial bars together with the central muscle bar element.

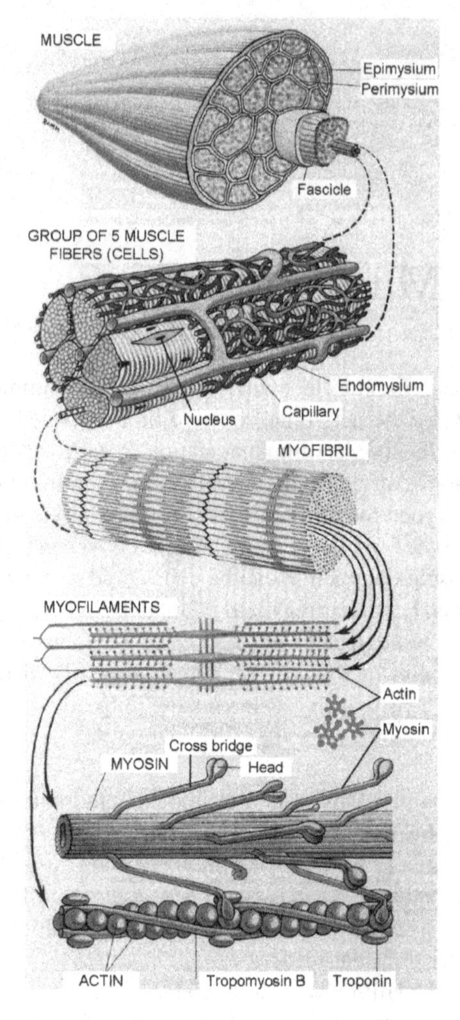

FIG. C.1. Skeletal muscle structure. (Reproduced by permission of the Longman Group UK Ltd.)

For the central "muscle" bar element, the total muscle force consists in an active and a passive component,

$$F_{\text{muscle}} = F_{\text{active muscle}} + F_{\text{passive muscle}} = F_{\text{CE}} + (F_{\text{PE}} + F_{\text{DE}}). \tag{C.1}$$

The normalized "active branch" of the muscle force, $F_{\text{CE}}/F_{\max}$ acting in the contractile sub-element, CE, is modeled by

$$F_{\text{CE}}/F_{\max} = N_{\text{a}}(t)F_V(V/V_{\max})F_L(L/L_{\text{opt}}) = F_{\text{active muscle}}/F_{\max}. \tag{C.2}$$

In this expression $F_{\max} = \sigma A_{\text{phys}}$ is the maximum muscle force at 100% voluntary muscle activation, with the maximum active muscle stress $\sigma \cong 0.001$ Gpa, which is fairly intrinsic to all skeletal muscles, and A_{phys} = physiological cross section area

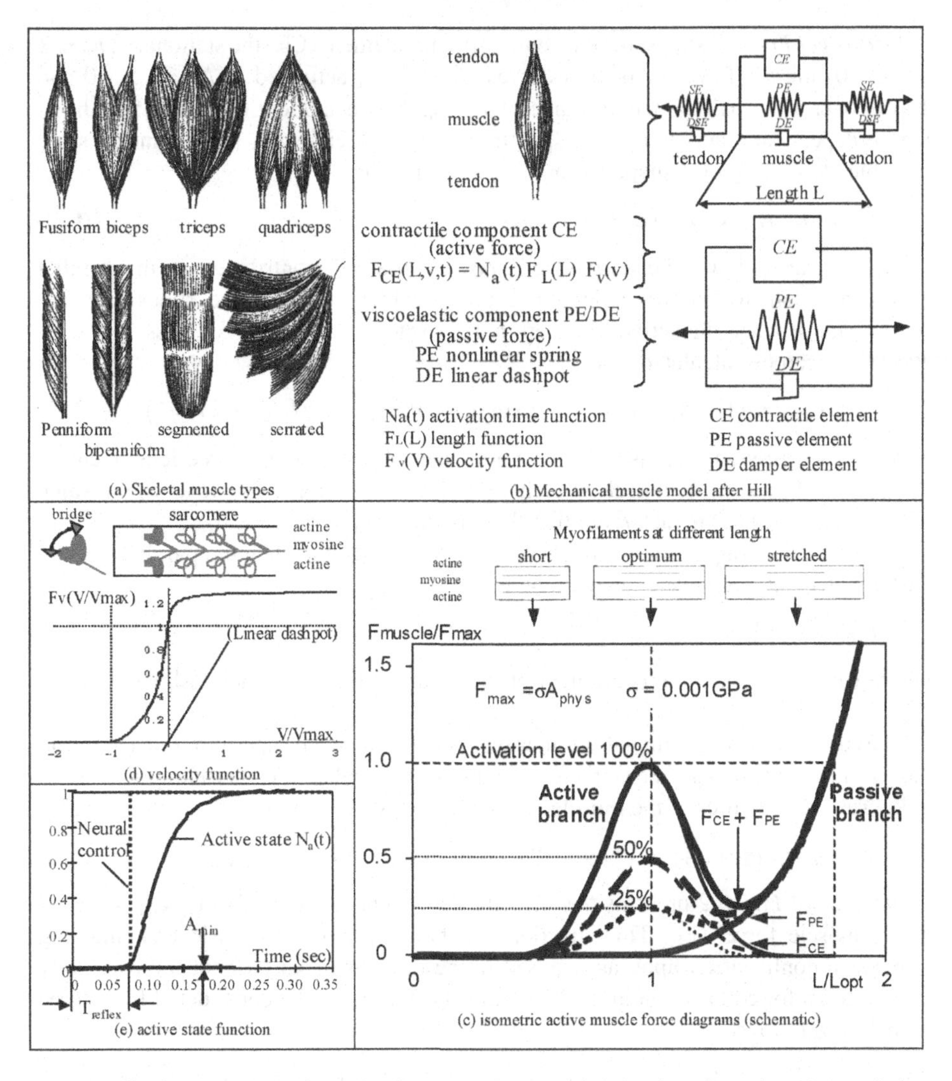

FIG. C.2. Hill's model of skeletal muscles. (Insets (a), (b) anatomical drawings: Reproduced by permission of Rolf Wirhed, WIRHED [1985].)

of the muscle; $N_a(t)$ is the neurological muscle activation state versus time function (voluntary and reflexes); $F_V(V/V_{max})$ is a muscle stretch velocity dependent function, where V_{max} is a reference muscle stretch velocity; $V = dL/dt$ is the muscle length rate of change or stretch velocity; $F_L(L/L_{opt})$ is a muscle length dependent shape function, where L is the current length and L_{opt} is the optimal length of the skeletal muscle "at rest", at which the voluntary muscle force can reach its peaks. The optimal muscle length is sometimes attributed to the freely floating position of a dormant astronaut.

Inset (c) of Fig. C.2 shows for the contractile sub-element (CE) the stationary ($F_v = 1$ at $V = 0$) muscle force-elongation curves, F_{CE}/F_{max}, activated at $N_a = 25$, 50 and 100%, over a normalized length range of L/L_{opt} between about 0.5 and 1.5 with the length dependency curve, $F_L(L/L_{opt})$ (thin lines). Other length shape functions are possible, depending on a shape factor, C_{sh}, according to

$$F_L(L/L_{opt}) = \exp\bigl(-(L/L_{opt} - 1)/C_{sh}\bigr). \tag{C.3}$$

Inset (c) also shows the normalized "passive branch" contributed as the parallel elastic sub-element (PE) force-length response, which is due to the cohesive material resistance when the active fibers are not activated. The normalized passive forces, F_{PE}/F_{max}, can be calculated from

$$F_{PE}/F_{max} = \bigl(1/(\exp(C_{PE}) - 1)\bigr)\exp\bigl((C_{PE}/PE_{max})((L/L_{fib}) - 1) - 1\bigr). \tag{C.4}$$

In this expression C_{PE} is a shape parameter of the passive force-length curve, $PE_{max} = L/L_{fib}$ at is the muscle stretch when the passive force F_{PE} reaches the value of F_{max} and L_{fib} is a characteristic fiber length, often set to $L_{fib} = L_{opt}$.

If the muscle elongates at a stretch velocity of $V = dL/dt$, then the parallel passive dashpot element (DE) responds with the force

$$F_{DE} = C_{DE} V, \tag{C.5}$$

where C_{DE} is the damping coefficient of the assumed linear parallel dashpot element, DE.

The curves drawn with thick lines in inset (c) schematize the superimposed active and passive muscle forces under isometric conditions, i.e., when the shortening or lengthening stretch velocities of the muscle are small or zero, $V \approx 0$,

$$F_{muscle} = F_{CE} + F_{PE} + (F_{DE} = 0) = F_{active\ muscle} + F_{passive\ muscle}. \tag{C.6}$$

Inset (d) of Fig. C.2 presents the velocity dependent function, $F_V(V/V_{max})$, of the active muscle force F_{CE}. This function can be interpreted in classical engineering terms as a nonlinear dashpot, as opposed to the familiar linear dashpot force-velocity curve shown for comparison in the diagram. The curves can be constructed from three branches as follows.

$$F_V(V/V_{max}) = \begin{cases} 0 & \text{for } v = V/V_{max} \leqslant -1, \\ (1+v)/(1-v/C_{short}) & \text{for } -1 < v \leqslant 0, \\ (1+vC_{mvl}/C_{leng})/(1+v/C_{leng}) & \text{for } v > 0, \end{cases} \tag{C.7}$$

where C_{short} is a shape parameter for the non-zero curve segment at shortening stretch velocities, C_{leng} is a shape parameter for the curve segment at lengthening stretch velocities, C_{mvl} is the asymptotic value of the curve for large positive stretch velocities, $v = V/V_{max} \ll 1$.

The physiological origin of the stretch velocity dependency seems to stem from the actions of the so-called cross-bridges inside the actine-myosine components of the sarcomere cells of the active muscle fibers, see the bottom zooms of Fig. C.1 and insets (c) and (d) of Fig. C.2. At zero stretch velocity, $V = 0$, the muscle can afford an isometric active muscle force, when the cross bridges of the recruited muscle fibers continually connect, flex forward and disconnect the telescoping actine and myosin muscle fiber components. The number of recruited fibers can be modeled with the percentage level of the (normalized) activation function, $0 \leqslant N_a(t) \leqslant 1$. The continuous process of connection, flexion and disconnection creates a forward motion of the myosin fibers into the actine tubes, which counteracts the backward slipping motion due to the constant pull of the section force. This action can be compared roughly to the action of rowers in a boat in still waters, who must keep rowing on the spot in order to create a steady pull on a rope that retains their boat in place. If the retaining rope is cut, the boat will move forward, which corresponds to the shortening of the unconstrained muscle if the external force vanishes. The image of the tied rowers also helps understanding how physiological energy must be spent in order to keep a muscle at the same length under active tension, i.e., when the muscle does no external mechanical work. Although the tied down boat does not move, the rowers will fatigue and eventually stop rowing. Furthermore, it is easy to understand why in inset (c) the isometric active force-length curves, $F_L(L/L_{opt})$, are not constant with the muscle length. If the muscle is longer than optimal, $L > L_{opt}$, then the overlap of the myosin and actine components in a sarcomere decreases in length, and less connecting cross bridges are available to create the active muscle force. On the other hand, if the muscle has shortened, $L < L_{opt}$, the efficiency of the cross bridge action decreases because of the hindrance created by the shortening.

The shape of the velocity dependent function, $F_V(V/V_{max})$, in inset (d) of Fig. C.2 is discussed next. If to a muscle at a given instantaneous length, L, and under a given active force, F, a positive stretch velocity is imposed, $V/V_{max} > 0$, the connected bridges tend to be pulled in fiber direction and the force output at the same voluntary activation level increases by the factor $F_V(V/V_{max}) > 1$. At negative stretch velocities the connecting bridges do not re-connect fast enough to make up for the negative length rate of change of the muscle fibers, and the force output falls drastically and reaches the value of zero at negative stretch velocity $V = -V_{max}$. This again might be compared to rowers in a boat who are more efficient when rowing downstream than upstream. When the face stream velocity becomes equal to the rowing velocity, the action of the rowers will no longer produce any force on the retaining rope.

Inset (e) of Fig. C.2 shows the active muscle state function $N_a(t)$. In a simplified approach, this function depends on a muscle neurological reflex time, T_{reflex}, which is the time that elapses between, say, the onset of an impact event where a defensive muscle action should ideally start and the time when the activation actually starts. After the reflex time has elapsed, a neuro-control flag, $u(t)$, is set equal to one, and the muscle activation process sets in,

$$u(t) = \begin{cases} 0 & \text{for } t \leqslant T_{reflex}, \\ 1 & \text{for } t > T_{reflex}. \end{cases} \tag{C.8}$$

Reflex times for skeletal muscles are known to range from about 25 to 100 milliseconds (ms) and the reflex time is set to 80 ms in inset (e). After the reflex time has passed, the muscle force must be activated. This physiological process takes a certain time, and the maximum muscle force occurs at about 250 ms in inset (e). More details about skeletal muscle excitation and activation can be found in the literature cited in the given references.

The muscle activation function of the Hill model can be calculated from the differential equations

$$\begin{aligned}\big(\mathrm{d}N_e(t)/\mathrm{d}t\big) &= \big(u(t) - N_e(t)\big)/T_{ne},\\ \big(\mathrm{d}N_a(t)/\mathrm{d}t\big) &= \big(N_e(t) - N_a(t)\big)/T_a,\end{aligned} \tag{C.9}$$

where $N_e(t)$ is the neuro-muscular excitation function, $N_a(t)$ is the muscle force excitation function, $0.02 \leqslant T_{ne} \leqslant 0.05$ s and $0.005 \leqslant T_a \leqslant 0.02$ s are time constants. For the assumed binary form of the neural control flag $u(t)$, there exists an analytical solution of the form

$$Na(t) = \begin{cases} A_{init} & \text{for } t \leqslant T_{reflex},\\ 1 + A_a/B_a + A_{ne}/B_{ne} & \text{for } t > T_{reflex},\end{cases} \tag{C.10}$$

where

$$\begin{aligned}
A_a &= (A_{init} - 1)(T_a - T_{ne}) - T_{ne},\\
B_a &= (T_a - T_{ne})\exp\big((t - T_{reflex})/T_a\big),\\
A_{ne} &= T_{ne},\\
B_{ne} &= (T_a - T_{ne})\exp\big((t - T_{reflex})/T_{ne}\big),\\
A_{init} &= A_{min} = 0.005,\\
T_{ne} &= C_1 + C_2 m C_{slow},\\
T_a &= B_1 + B_2 m (C_{slow})^2,\\
C_1 &= 0.025\ \text{s},\\
C_2 &= 0.01\ \text{s},\\
B_1 &= 0.005\ \text{s},\\
B_2 &= 0.0005\ \text{s},\\
m &= \text{muscle mass in grams},\\
C_{slow} &= \text{fraction of slow muscle fibers}.
\end{aligned}$$

Applications. The mechanical behavior of the skeletal muscles in the directions of their fibers can be modeled to first order accuracy by Hill-type muscle bars. Each Hill-type muscle bar element is characterized by the physiological cross section area of the muscle, cut perpendicular to the fibers, and by the muscle fiber stretch and stretch velocity dependent active and passive mechanical properties of the Hill muscle model, described above. The bars cannot, in general, transmit compressive forces.

Fig. C.3 shows a couple of postures and the maximum sustainable limit loads as calculated from the muscled skeleton model, compared to values found in the literature (BOUISSET and MATON [1995], p. 135). The calculated values follow from the application of the optimization process described in Chapter II, Section 6, where the applied loads were incremented until the process could no longer find a solution for the given

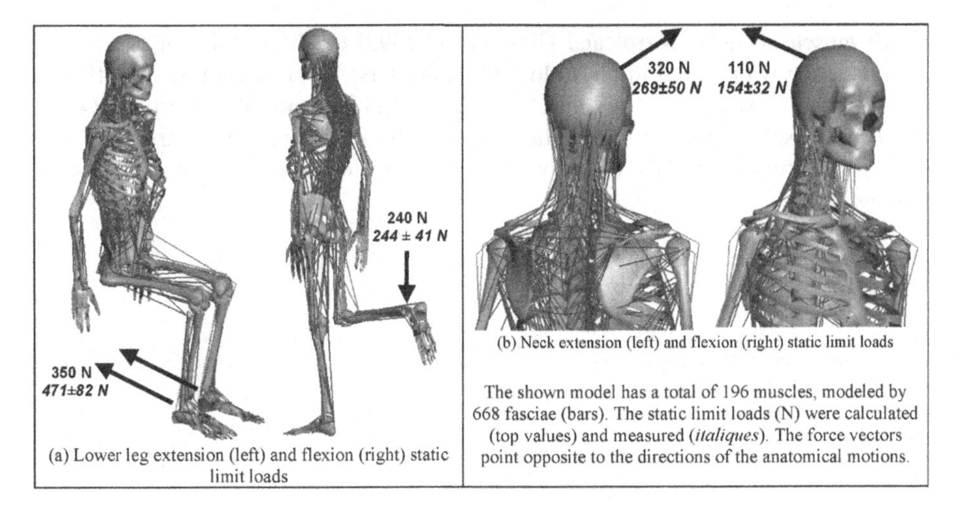

240 N
244 ± 41 N
350 N
471±82 N
320 N
269±50 N
110 N
154±32 N
(a) Lower leg extension (left) and flexion (right) static limit loads
(b) Neck extension (left) and flexion (right) static limit loads
The shown model has a total of 196 muscles, modeled by 668 fasciae (bars). The static limit loads (N) were calculated (top values) and measured (*italiques*). The force vectors point opposite to the directions of the anatomical motions.

FIG. C.3. Validation of the limit load muscle force calculation (BENES [2002]).

posture, i.e., the respective limit loads were attained. The shown model has a total of 196 anatomical muscle groups, modeled by 668 fasciae (bars). The colors of the bars range between red (100% activation) and blue (no activation). The comparison with the reference values is considered fair in view of the sensitivity of the results to the joint geometry, the lever arms of the muscle bars, the muscle section area and trajectory, the true strategy of activation, the chosen objective function (physiological muscle energy/work) in the optimization process and the uncertainty of the experimental conditions, such as the exact anatomy of the volunteers, the exact posture and the point of load application (BENES [2002]).

References on muscle materials. The following references deal with the structure and the modeling of skeletal muscle: BAHLER, FALES and ZIERLER [1968] on the dynamic properties of skeletal muscle; COLE, BOGERT, HERZOG and GERRITSEN [1996] on modeling of forces in stretched muscles; CRAWFORD and JAMES [1980] on the design of muscles; HARRY, WARD, HEGLUD, MORGAN and MCMAHON [1990] on cross bridge action; HERZOG [1994] on the biomechanics of the musculo-skeletal system; HAWKINS and BEY [1994], HAWKINS and BEY [1997] on muscle-tendon mechanics and mechanical properties; HILL [1970] on experiments in muscle mechanics; KIRSCH, BOSKOV and RYMER [1994] on stiffness of moving cat muscles; KRYLOW and SANDEROCK [1996] on dynamic force response of muscles under excentric contraction; MA and ZAHALAK [1991] on a distribution-moment model of energetics in skeletal muscle; MORGAN [1990] on the behavior of muscle under active lengthening; MYERS, VAN EE, CAMACHO, WOOLLEY and BEST [1995] on the structural properties of mammalian skeletal muscle in the neck; RACK and WESTBURY [1969] on the effect of length rate on tension in muscles; SCHNECK [1992] on the mechanics of muscle; VANCE, SOLOMONOV, BARATTA, ZEMBO and D'AMBROSIA [1994] on the comparison of two muscle models; Wank and Gutewort [1993] on the simulation of muscular contraction with regard to physiological parameters; Winters and Stark [1985], Winters and Stark [1988] on muscle modeling and mechanical properties; Zajac [1989] on muscle and tendon properties and models; Zuurbier, Everard, van Der Wees and Huijing [1994] on the force-length characteristics in active and passive muscles.

APPENDIX D

Airbag Models

The following paragraphs are based mainly on Pam-Safe documentation of ESI Software.

Airbags. Airbags are considered volumes of ideal gas that are enclosed by a flexible envelope, Fig. D.1. Their physics requires fluid-structure interaction (FSI) simulation. In simple airbag models, pressure and temperature of the gas are assumed to be distributed uniformly throughout the airbag volume. More complex approaches (not described here) model the enclosed gas and the envelope separately, using fluid-solid and multi-physics formulations. The gas can be confined by a bag made of flexible (visco-)elastic industrial fabric or of any other material, such as the walls of hollow organs.

Airbag models can simulate gas inflow and outflow through orifices, and leakage of gas through the fabric of the envelope can be defined. The input data for the gas are atmospheric pressure p, temperature T, constants γ and R, where R is the perfect gas constant from $pV = nRT$, where n is the number of moles in the volume. In the described simple airbag model, the gas is confined in a single chamber and it obeys the thermodynamic equation of a perfect gas. Solution of that equation at each explicit structural solution time step yields a pressure load to be applied to the inside of the envelope. Solution of the equations of motion of the pressurized envelope yields a volume change to be applied to the enclosed gas. This process is repeated over the duration of the simulation.

Airbag gas model. The volume of enclosed gas is subject to the equation of state $pV^{\gamma} =$ constant, where p is the pressure, V is the volume and $\gamma = c_p/c_v$ is the specific heat ratio for a perfect gas under adiabatic conditions, and where c_p and c_v are the specific heat of the gas at constant pressure and at constant volume, respectively. The gas constant is defined as

$$R = c_p - c_v = R_u/W,$$

where R_u is the universal gas constant related to moles and W is the molecular weight of the gas.

For nitrogen gas, N_2, one has in SI-units (m, kg, s) and with the molecular weight $W(N_2) = 0.028014$ kg/mole

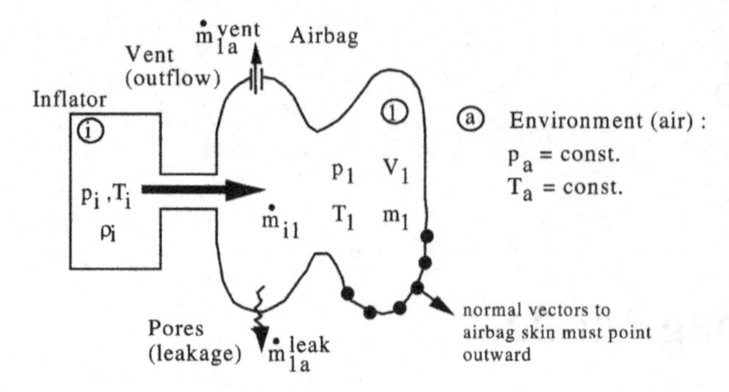

FIG. D.1. Basic airbag model (schematic) (ESI Software).

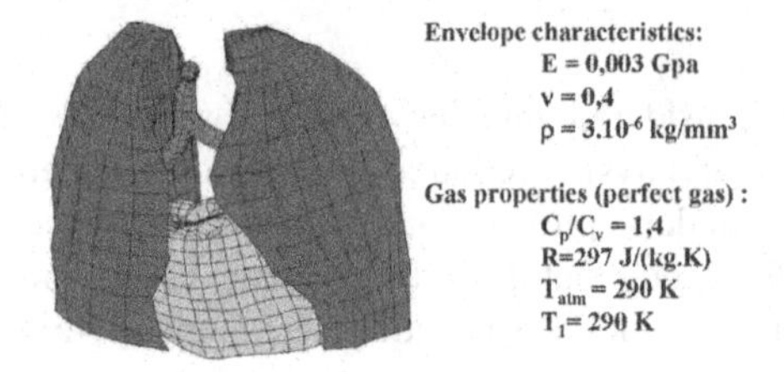

FIG. D.2. Basic airbag model for the lungs (ESI software).

$$c_p = 1038\ \text{J/(kgK)}, \qquad c'_p = 29.08\ \text{J/(mole K)},$$
$$c_v = 741\ \text{J/(kgK)}, \qquad c'_v = 20.76\ \text{J/(mole K)},$$
$$R = 297\ \text{J/(kgK)}, \qquad R_u = 8.32\ \text{J/(mole K)},$$

where J = Joules = Nm (Newton meters), K = °Kelvin and c'_p and c'_v are the molar heat capacities.

The gauge pressure is the differential pressure between the airbag pressure and the atmospheric pressure, Fig. D.1,

$$p_{\text{gauge}} = p_1 - p_a.$$

The specific heat ratio, γ, is defined as

$$\gamma = c_p/c_v = c_p/(c_p - R),$$

which for nitrogen gas (N_2) is equal to 1.4. The meaning of the remaining parameters is illustrated on Fig. D.1.

Bio-bag models. Bio-bag models are derived from airbag models to model hollow organs. Hollow organs have flexible walls and are filled with quasi-incompressible material (blood, body fluids, food), but also (partly or fully) with compressible material (gas in stomach or lungs). Airbag gas models can be used to approximate quasi-incompressible fluid-filled hollow organs by re-setting their input data to model a linear

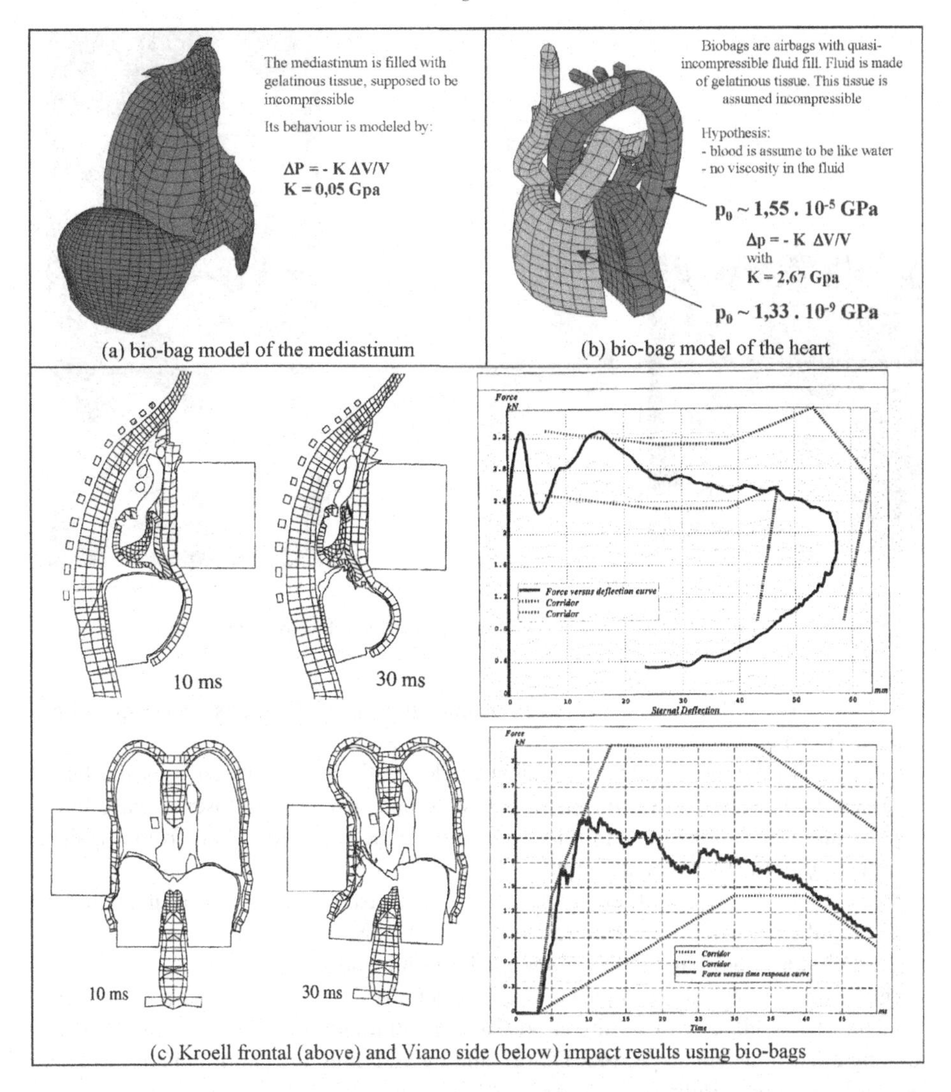
The mediastinum is filled with gelatinous tissue, supposed to be incompressible
Its behaviour is modeled by:
ΔP = - K ΔV/V
K = 0,05 Gpa
Biobags are airbags with quasi-incompressible fluid fill. Fluid is made of gelatinous tissue. This tissue is assumed incompressible
Hypothesis:
- blood is assume to be like water
- no viscosity in the fluid
P0 ~ 1,55 . 10-5 GPa
Δp = - K ΔV/V
with
K = 2,67 Gpa
P0 ~ 1,33 . 10-9 GPa
10 ms
30 ms
Force versus deflection curve
Corridor
Corridor
Sternal Deflection
Force versus time response curve
Time

(a) bio-bag model of the mediastinum

(b) bio-bag model of the heart

(c) Kroell frontal (above) and Viano side (below) impact results using bio-bags

FIG. D.3. Bio-bag models of the mediastinum and heart (ESI software).

pressure-volume equation of the type

$$p = p_0 + K(\rho/\rho_0 - 1),$$

where K is the desired bulk modulus for modeling quasi-incompressibility, p_0 is the initial pressure and ρ and ρ_0 are the mass density in the compressed and uncompressed gas, respectively. This can be achieved approximately by setting c_p to a large value (isothermal conditions) with $\gamma = c_p/c_v$ close to 1.

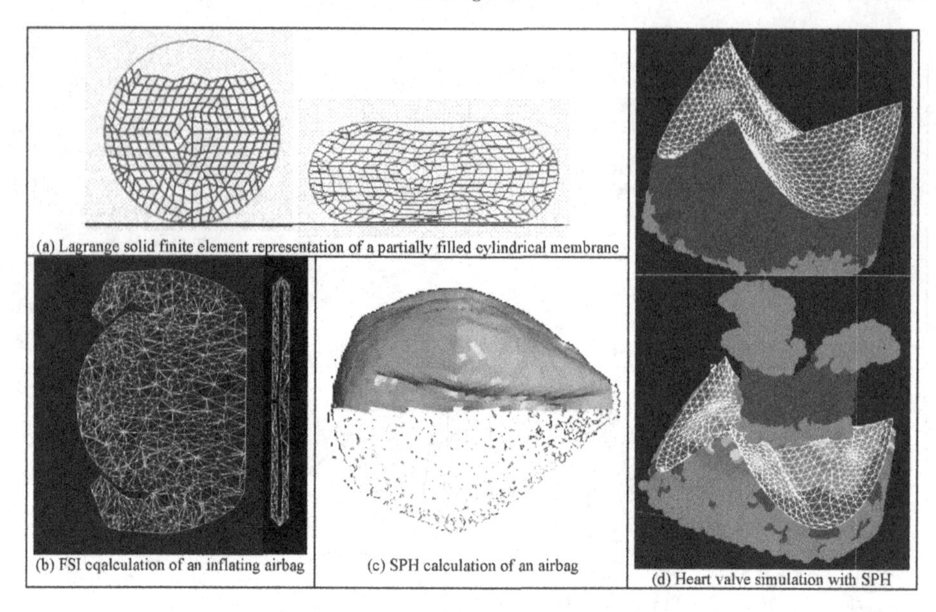

(a) Lagrange solid finite element representation of a partially filled cylindrical membrane
(b) FSI cqalculation of an inflating airbag
(c) SPH calculation of an airbag
(d) Heart valve simulation with SPH

FIG. D.4. Alternative airbag modeling techniques (ESI software).

Examples. The following examples demonstrate the use of "bio-bags" in the modeling of the hollow internal organs.

Fig. D.2 is an approximate mechanical model of the lungs, which uses the basic airbag model with perfect gas properties. A fictitious envelope encloses a space filled with air, and vent holes can be provided to simulate the expulsion of air from a violent compression of the thorax in an impact.

This model is relatively efficient when the lungs must not be simulated in detail, but when only their resistance to compression of the rib cage is of interest.

Fig. D.3 shows bio-bag models of the mediastinum and heart.

The mediastinum is the complex space between the thoracic organs and vessels and it is considered filled with an incompressible gelatinous fluid. It can be modeled by defining a fictitious envelope, Fig. D.3(a), and by assigning the conditions of bio-bags for quasi-incompressibility.

Similarly, the heart can be modeled by a bio-bag with a high degree of incompressibility, Fig. D.3(b). For capturing the effect of expelled blood during a violent chest impact, the modeled heart chambers can be provided with outflow vents. The thorax models can so be calibrated to well represent the results from Kroell frontal and Viano side pendulum impact tests, Fig. D.3(c).

Other fluid-structure interaction modeling techniques. While the described biobag models are efficient but approximate models, the volume enclosed by a hollow organ can be modeled with more precision with *Lagrangian solid finite elements*, which undergo the equation of state given by a fluid, Fig. D.4, inset (a). This model is possible when the fluid transport is small, i.e., the distortion of the Lagrange mesh is limited.

Coupled fluid-structure interaction (FSI) models (LÖHNER [1990]), as shown for an airbag in inset (b), will be the most accurate (and the most expensive) representations of hollow organs. This technique is indicated if the detailed interaction of the moving fluids with the confining wall is of interest (example: aorta rupture). In that case a fluid code and a structure code are coupled, where the fluid code provides the wall pressure loads to the structure code and where the structure code provides the wall positions and velocities to the fluid code at each common solution time step. Finally, the fluid or gas can also be represented by *SPH or FPM "particle" techniques* (MONAGHAN and GINGOLD [1983], MONAGHAN [1988]), insets (c), (d). While in FSI techniques the fluid domain is meshed, inset (b), particle methods do not require a domain mesh. This is particularly convenient when the fluid domain connectivity changes, as in the example of the heart valve, inset (d).

APPENDIX E

Interactions between Parts

Contact simulation. The numerical treatment of contact with various types of contact options was mentioned in Appendix A. Effective treatment of contact is not only of prime importance for modeling impact biomechanics, but certain contact algorithms serve also in assembling the complex geometries of the parts of the human body. Some examples, taken from the HUMOS model (ESI version), explain how different modeling strategies can provide viable solutions to the complex mechanical interactions between organs and parts of the human body.

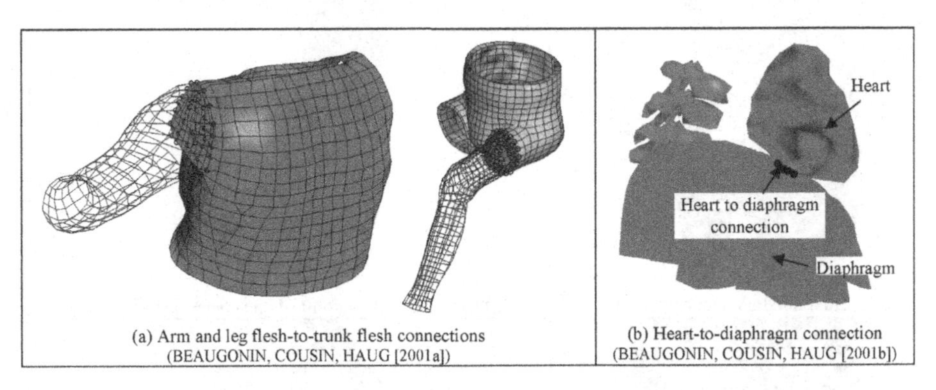

Heart
Heart to diaphragm connection
Diaphragm
(a) Arm and leg flesh-to-trunk flesh connections (BEAUGONIN, COUSIN, HAUG [2001a])
(b) Heart-to-diaphragm connection (BEAUGONIN, COUSIN, HAUG [2001b])

FIG. E.1. Connections between parts via contact (HUMOS-ESI model).

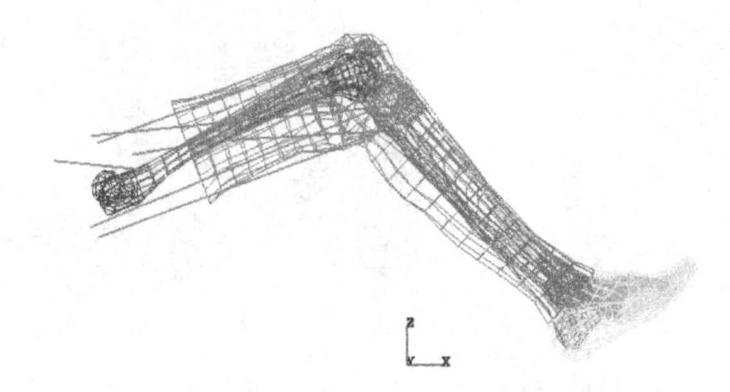

FIG. E.2. Non-matching flesh and bone meshes connected by tied contacts (HUMOS-ESI model).

Mesh merging. Apart from simulating dynamic collisions between moving parts, special contact options are frequently used within models of articulations, to delimit adjacent organs and to conveniently join differently meshed parts, Fig. E.1 (BEAUGONIN, COUSIN and HAUG [2001a], BEAUGONIN, COUSIN and HAUG [2001b]).

In another example, Fig. E.2 shows the independent meshes of the leg bones and the surrounding flesh, BEAUGONIN and HAUG [2001].

This figure demonstrates how different constituents of body parts, such as flesh and bone, are linked with connective membranes and tissue, and are often meshed independently for convenience. Both material constituents are connected by a tied contact option (type 32 in Table A.1, Appendix A).

Example: Lower extremity. To control the interaction between the different components involved in the lower limb segment of the HUMOS model, several types of sliding interfaces have been defined in its PAM-Crash version, Fig. E.3 (BEAUGONIN and HAUG [2001]).

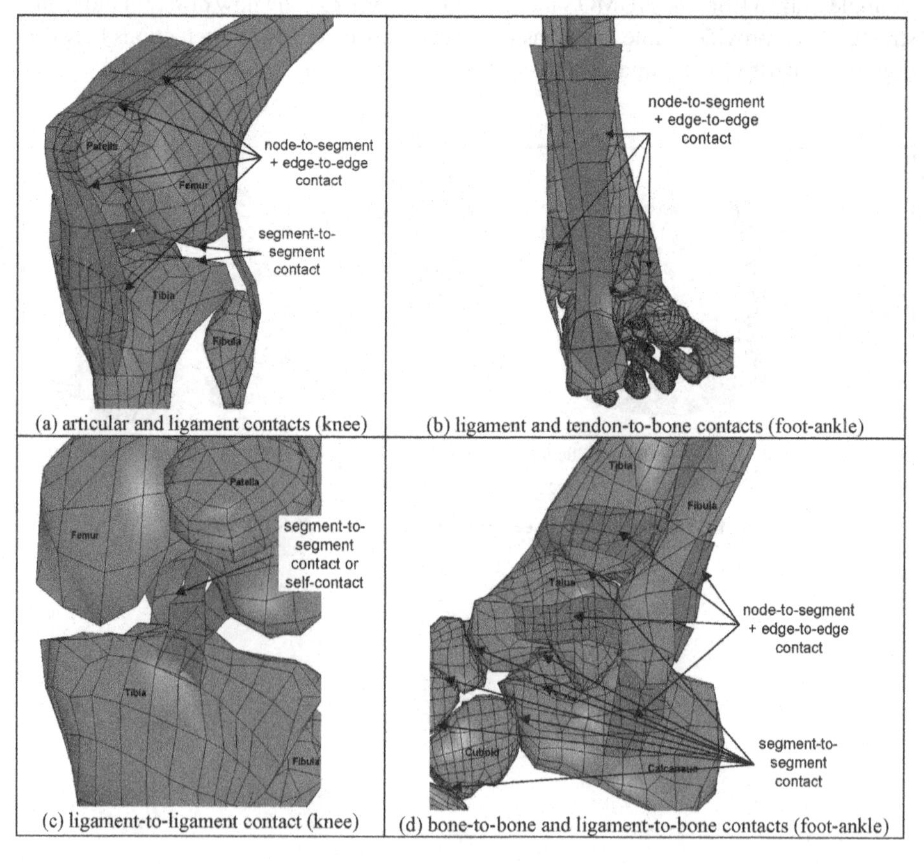

FIG. E.3. Articular, ligamant/tendon-to-bone and self-contacts in the lower extremity (HUMOS-ESI model) (BEAUGONIN and HAUG [2001]).

(i) The sliding interfaces between bones near their articulations are defined with a segment-to-segment contact (type 33 in Table A.1, Appendix A).
(ii) The sliding interfaces between bone and ligament or tendon, and between bone and skin are defined with a segment-to-segment or a node-to-segment contact (type 34 in Table A.1, Appendix A). If necessary, an edge-to-edge contact (type 46 in Table A.1, Appendix A) is added to avoid the penetration between the components.
(iii) The interaction between ligaments is controlled by a segment-to-segment contact or a self-contact (type 36 in Table A.1, Appendix A). The sliding interface between ligament/tendon and skin is defined by a node-to-segment contact.

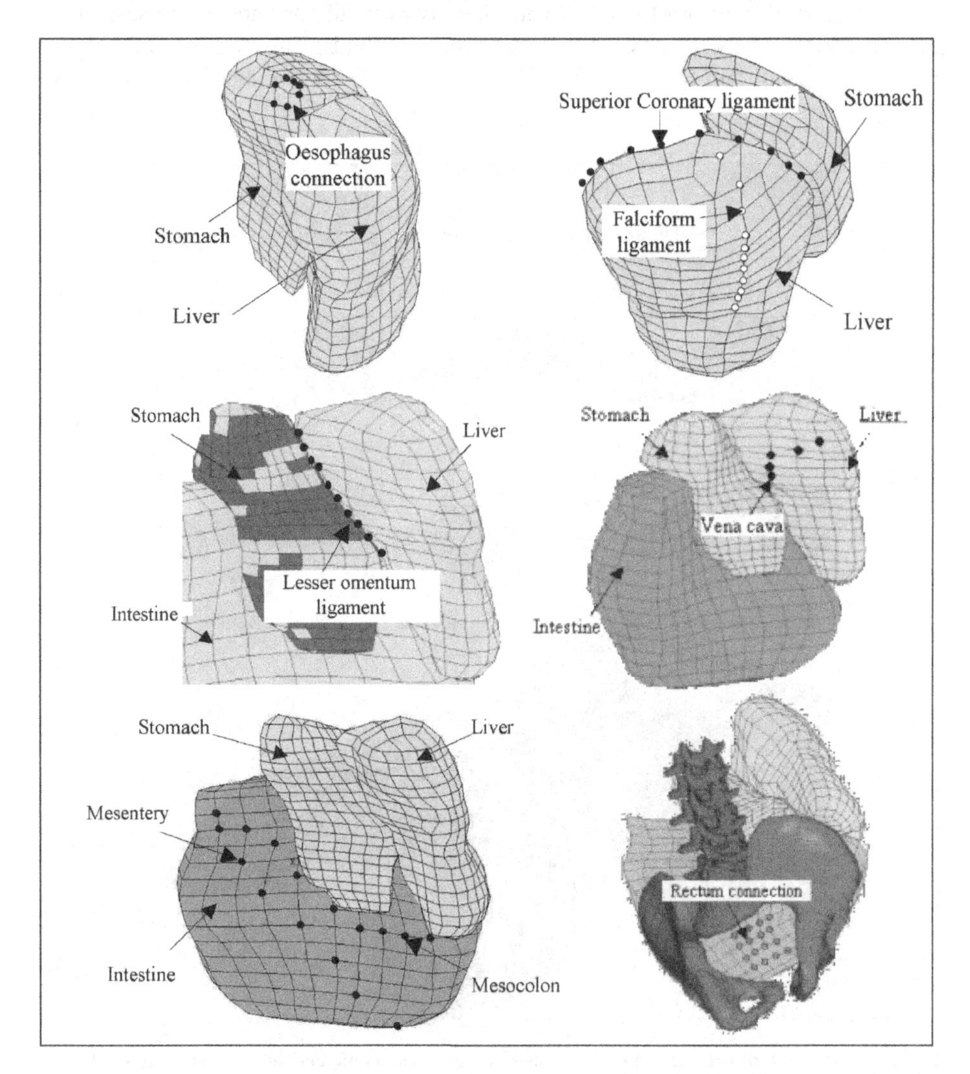

FIG. E.4. Attachments between abdominal organs (HUMOS-ESI model).

Example: Abdominal organs. The abdominal organs are tied together and are loosely tethered to the abdominal walls (peritoneum) by ligaments or folds of the peritoneum. These features can be modeled with tied contact options in the ESI HUMOS model. The tied contact options permit no relative motion between the tied parts, but can be allowed to break when certain contact force limits are exceeded. This option can approximate the rupture of ligaments and other tissue connections. Some of the connections are shown as examples in Fig. E.4 (after BEAUGONIN, ALLAIN and HAUG [2001]).

The abdominal organs, as well as the brain and the thoracic organs, interact with neighbors and cavity walls under considerable relative sliding motions. This can be modeled with the standard sliding contact options.

The organs fill their host cavities with literally no voids or gaps in the sense that the space between the organs is filled with tissue or fluids. This can be modeled approximately with "bio-bags", as shown in Appendix D for the mediastinum, where

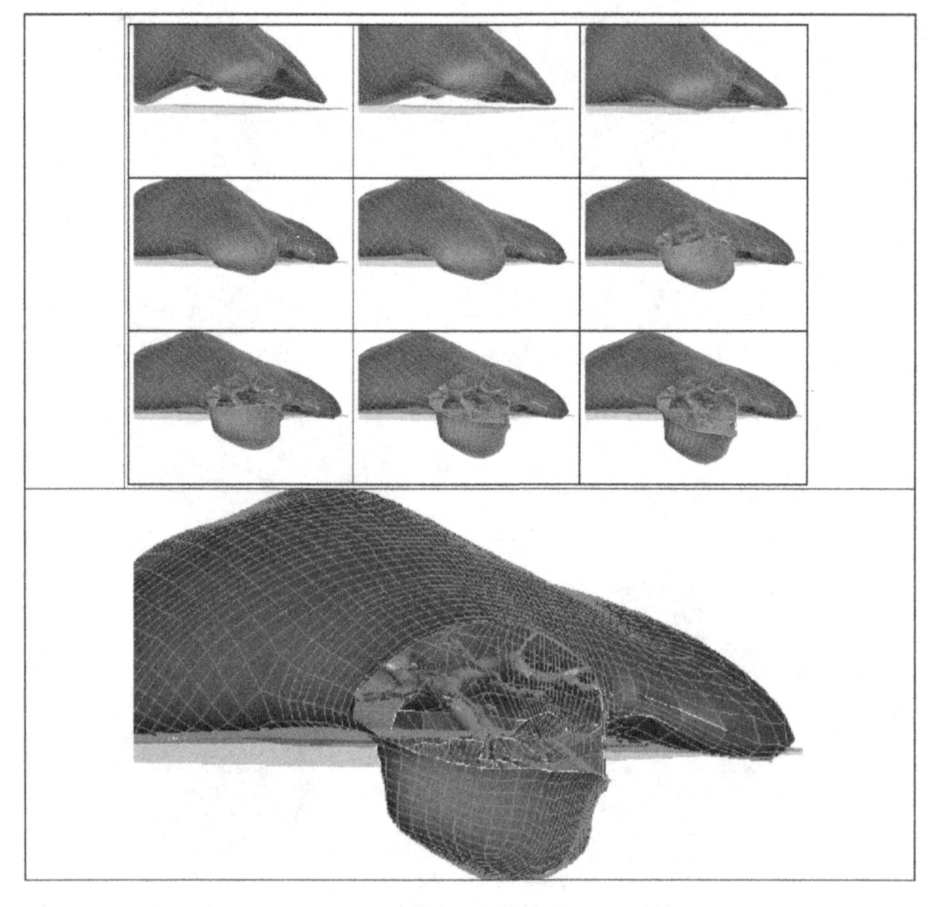

FIG. E.5. Simulation of cut liver with tied meshes between vena porta and parenchyma (CAESARE-ESI model: DAN and MILCENT [2002]).

the organ-to-organ contact can be modeled with slide-and-void contacts between the organs, while the incompressibility of the liquid-filled space between them is assured by the bio-bag feature (Appendix D). The same feature was applied in Chapter III, Section 8 (Fig. 8.8), for the modeling of the Cavanaugh bar impact test on the abdomen.

Example: Liver. For practical reasons (convenience, meshing freedom, mesh size limitations, etc.) vessels inside internal organs are often modeled apart from the bulk matter of the organs. Automatic mesh merging techniques, or tied contact options, can then be applied to tie the non-congruent meshes between the surface of the meshed vessels (often: shells) and the organ bulk matter (solids). Fig. E.5 shows a model of a human liver with the internal arborescence of a systems of vessels (vena porta) exposed through a simulated progressive cut into the parenchyma (from the CAESARE Project DAN and MILCENT [2002]). Glisson's capsule around the liver is modeled with thin membranes. The cut-in-progress was simulated with assigning almost zero resistance to the sectioned elements. In order to expose the incision, the cut portion of the liver is not supported by the horizontal support plate and it deflects through the action of gravity. It is still connected by its uncut portion with the main portion of the organ. The tougher vessels become visible by assigning transparency to the surrounding parenchyma solid elements, which appear only through their exposed surface grids (blue color).

References

ABEL, J.M., GENNARELLI, T.A., SEGAWA, H. (1978). Incidence and severity of cerebral concussion in the rhesus monkey following sagittal plane angular acceleration. In: *Proc. 22nd Stapp Car Crash Conference*, pp. 35–53. Paper No. 780886.

AL-BSHARAT, A.S., HARDY, W.N., YANG, K.H., KHALIL, T.B., TASHMAN, S., KING, A.I. (1999). Brain-skull relative displacement magnitude due to blunt head impact: New experimental data and model. In: *Proc. 43rd Stapp Car Crash Conference*, pp. 101–160. Paper No. 99SC22.

ALLAIN, J.C. (1998). Etude et calibration d'un modèle numérique de thorax, Internal Report, ESI Software S.A., 99, rue des Solets, BP 80112, 94513 Rungis Cedex, France.

ALLEN, B.L., FERGUSON, R.L., LEHMANN, T.R., O'BRIEN, R.P. (1982). A mechanistic classification of closed indirect fractures and dislocations of the lower cervical spine. *Spine* **7** (1), 1–27.

ALLSOP, D.L. (1993). Skull and facial bone trauma: Experimental aspects. In: Nahum, A.M., Melvin, J.W. (eds.), *Accidental Injury – Biomechanics and Prevention* (Springer, Berlin), pp. 247–267 (Chapter 11).

ARMSTRONG, C.G., LAI, W.M., MOW, V.C. (1984). An analysis of the unconfined compression of articular cartilage. *J. Biomech. Engrg.* **106**, 165–173.

ASHMAN, R.B., et al. (1984). A continuous wave technique for the measurement of the elastic properties of cortical bone. *J. Biomech.*, 349.

ASHMAN, R.B., et al. (1986). Ultrasonic technique for the measurement of the structural elastic modulus of cancellous bone. *Trans. Orthopedic Res. Soc.*, 43.

ATTARIAN, D.E., MCCRACKIN, H.J., DEVITO, D.P., MCELHANEY, J.H., GARRETT, W.E. (1985). Biomechanical characteristics of human ankle ligaments. *Foot & Ankle* **6** (2).

BACH, J.M., HULL, M.L., PATTERSON, H.A. (1997). Direct measurement of strain in the postero-lateral bundle of the anterior cruciate ligament. *J. Biomech.* **3** (3), 281–283.

BAHLER, A.S., FALES, J.T., ZIERLER, K.L. (1968). The dynamic properties of mammalian skeletal muscle. *J. Gen. Physiol.* **51**, 369–384.

BANDAK, F.A. (1996). Biomechanics of impact traumatic brain injury. In: *Proceedings of the NATO ASI on Crashworthiness of Transportation Systems, Structural Impact and Occupant Safety, Troia, Portugal*, pp. 213–253.

BANDAK, F.A., EPPINGER, R.H. (1994). A three-dimensional finite element analysis of the human brain under combined translational and rotational acceleration. In: *Proc. 38th Stapp Car Crash Conference*, pp. 145–163. Paper No. 942215.

BANDAK, F.A., TANNOUS, R.E., ZHANG, A.X., DIMASI, F., MASIELLO, P., EPPINGER, R.H. (2001). SIMon: A simulated injury monitor, application to head injury assessment. In: *17th International Technical Conference on the Enhanced Safety of Vehicles, Amsterdam, Holland.*

BANDAK, F.A., TANNOUS, R.E., ZHANG, A.X., TORIDIS, T.G., EPPINGER, R.H. (1996). Use of finite element analysis and dummy test measurements for the assessment of crash injury traumatic brain injury. In: *Advisory Group for Aerospace Research and Development, Mescalero, New Mexico*, pp. 10–10.13.

BARTLEY, M.H., et al. (1966). The relationship of bone strength and bone quantity in health, disease and aging. *J. Gerontol.* **21**, 517.

BATHE, J. (1996). *Finite Element Procedures* (Prentice-Hall, Englewood Cliffs, NY).

BAUDER, B. (1985). The dynamic load tolerance of the human liver (Die dynamisch mechanische Belastbarkeit der menschlichen Leber). PhD Dissertation, Medicine, University of Heidelberg.

BAUDRIT, P., HAMON, J., SONG, E., ROBIN, S., LE COZ, J.-Y. (1999). Comparative studies of dummy and human body models in frontal and lateral impact conditions. In: *Proc. 43rd Stapp Car Crash Conference*, pp. 55–75. Paper No. 99SC05.

BEAUGONIN, M., ALLAIN, J.C., HAUG, E. (2001). Pam–Crash modeling and validation: Pelvis–Abdomen segment. HUMOS Report 5ESI/010131/T3/DA, ESI Software S.A., 99, rue des Solets, BP 80112, 94513 Rungis Cedex, France.

BEAUGONIN, M., COUSIN, G., HAUG, E. (2001a). Pam–Crash: Whole model validation. HUMOS Report 6ESI/010131/T5/DA, ESI Software S.A., 99, rue des Solets, BP 80112, 94513 Rungis Cedex, France.

BEAUGONIN, M., COUSIN, G., HAUG, E. (2001b). Pam–Crash modeling and validation: Thorax-shoulder segment. HUMOS Report 6ESI/010131/T2/DA, ESI Software S.A., 99, rue des Solets, BP 80112, 94513 Rungis Cedex, France.

BEAUGONIN, M., HAUG, E. (2001). Pam–Crash modeling and validation: Lower limb segment. HUMOS Report 5ESI/010131/T5/DA, ESI Software S.A., 99, rue des Solets, BP 80112, 94513 Rungis Cedex, France.

BEAUGONIN, M., HAUG, E., CESARI, D. (1996). A numerical model of the human ankle/foot under impact loading in inversion and eversion. In: *Proc. 40th Stapp Car Crash Conference Proceedings, Albuquerque, NM*, pp. 239–249. Paper No. 962428.

BEAUGONIN, M., HAUG, E., CESARI, D. (1997). Improvements of numerical ankle/foot model: Modeling of deformable bone. In: *Proc. 41st Stapp Car Crash Conference*, pp. 225–237. Paper No. 973331.

BEAUGONIN, M., HAUG, E., HYNCIK, L. (1998). Robby2 – alpha-version. Internal Report, ESI Software S.A., 99, rue des Solets, BP 80112, 94513 Rungis Cedex, France.

BEAUGONIN, M., HAUG, E., MUNCK, G., CESARI, D.A. (1995). Preliminary numerical model of the human ankle under impact loading. In: *Pelvic and Lower Extremity Injuries (PLEI) Conference, Washington, DC, USA*.

BEAUGONIN, M., HAUG, E., MUNCK, G., CESARI, D. (1996). The influence of some critical parameters on the simulation of the dynamic human ankle dorsiflexion response. In: *15th ESV Conference, 96-S10-W-31, Melbourne*.

BEDEWI, P.G., MIYAMOTO, N., DIGGES, K.H., BEDEWI, N.E. (1998). Human femur impact and injury analysis utilizing finite element modeling and real-world case study data. In: *PUCA '98 Proceedings*, Nihon ESI, ESI Software S.A., 99, rue des Solets, BP 80112, 94513 Rungis Cedex, France, vol. 1, pp. 311–320.

BEGEMAN, P., AEKBOTE, K. (1996). Axial load strength and some ligament properties of the ankle joint. In: *Injury Prevention Through Biomechanics Symposium Proceedings, Hutzel Hospital, Wayne State University, Detroit, Michigan, USA*, pp. 125–135.

BEGEMAN, P., BALAKRISHNAN, P., LEVINE, R., KING, A. (1992). Human ankle response in dorsiflexion. In: *Injury Prevention Through Biomechanics Symposium Proceedings, Wayne State University*.

BEGEMAN, P., BALAKRISHNAN, P., LEVINE, R., KING, A. (1993). Dynamic human ankle response to inversion and eversion. In: *Proc. 37th Stapp Car Crash Conference Proceedings*, pp. 83–93. Paper No. 933115.

BEGEMAN, P.C., KING, A.I., PRASAD, P. (1973). Spinal loads resulting from $-gx$ acceleration. In: *Proc. 17th Stapp Car Crash Conference, Warrendale, PA*. Paper No. 730977.

BEGEMAN, P., KOPACZ, J.M. (1991). Biomechanics of human ankle impact in dorsiflexion. In: *Injury Prevention Through Biomechanics Symposium Proceedings, Wayne State University*.

BEHRENS, J.C., et al. (1974). Variation in strength and structure of cancellous bone at the knee. *J. Biomech.* **7**, 201–207.

BEILLAS, P., BEGEMAN, P.C., YANG, K.H., KING, A., ARNOUX, P.-J., KANG, H.-S., KAYVANTASH, K., BRUNET, C., CAVALLERO, C., PRASAD, P. (2001). Lower limb: Advanced FE model and new experimental data. *Stapp Car Crash J.* **45**, 469–494. Paper No. 2001-22-0022.

BEILLAS, P., LAVASTE, F., NICOLOPOULOS, D., KAYVANTASH, K., YANG, K.H., ROBIN, S. (1999). Foot and ankle finite element modeling using CT-scan data. In: *Proc. 43rd Stapp Car Crash Conference*, pp. 1–14. Paper No. 99SC11.

BELYTSCHKO, T., LIN, J.I. (1984). Explicit algorithms for the nonlinear dynamics of shells. *Comput. Methods Appl. Mech. Engrg.* **42**, 225–251.

BELYTSCHKO, T., KULAK, R.F., SCHULTZ, A.B., GALANTE, J.O. (1972). Numerical stress analysis of intervertebral disk. In: *ASME (Biomechanical & Human Factors Division), Winter Annual Meeting, NY*.

BELYTSCHKO, T., KULAK, R.F., SCHULTZ, A.B., GALANTE, J.O. (1974). Finite element stress analysis of an intervertebral disc. *J. Biomech.* **7**, 277–285.

BELYTSCHKO, T., TSAY, C.S. (1983). A stabilization procedure for the quadrilateral plate element with one-point quadrature. *Internat. J. Numer. Methods Engrg.* **19**, 405–419.

BELYTSCHKO, T., WONG, J.S., LIU, W.K., KENNEDY, J.M. (1984). Hourglass control in linear and nonlinear problems. *Comput. Methods Appl. Mech. Engrg.* **43**, 251–276.

BENES, K. (2002). ROBBY 2 Model with Muscles, Internal Report, ESI Software S.A., 99, rue des Solets, BP 80112, 94513 Rungis Cedex, France.

BLACK, J., HASTINGS, G. (eds.) (1998). *Handbook of Biomaterial Properties, Part I* (Chapman & Hall, London).

BOKDUK, N., YOGANANDAN, N. (2001). Biomechanics of the cervical spine, Part 3: Minor injuries. *Clinical Biomech.* **16**, 267–275.

BOUISSET, S., MATON, B. (1995). *Muscles, Posture et Mouvement* (Hermann, Paris).

BOUQUET, R., RAMET, M., BERMOND, F., CESARI, D. (1994). Thoracic and human pelvis response to impact. In: *Proceedings of the 14th International Technical Conference on Enhanced Safety of Vehicles (ESV)*, pp. 100–109.

BOWMAN, B.M., SCHNEIDER, L.W., LUSTAK, L.S., ANDERSON, W.R., THOMAS, D.J. (1984). Simulation analysis of head and neck dynamic response. In: *Proc. 28th Stapp Car Crash Conference*, pp. 173–205. Paper No. 841668.

BROWN, T.D., et al. (1980). Mechanical property distribution in the cancellous bone of the human proximal femur. *Acta Orthop. Scand.* **51**, 429–437.

BSCHLEIPFER, Th. (2002). Das experimentelle stumpfe Nierentrauma: Biomechanik, Traumaverhalten und bildgebende Diagnostik. M.D. Dissertation, Faculty of Medicine, University Ulm, Germany.

BURGHELE, N., SCHULLER, K. (1968). Die Festigkeit der Knochen Kalkaneus und Astragallus. Aus der Klinik fur Orthopädie des Bukarester Unfallkrankenhauses und dem Lehrstuhl fur Festigkeit und Materialprufungswesen der Technischen Hochschule, Bukarest, Rumänien (in German).

BURSTEIN, A.H., et al. (1976). Aging of bone tissue: Mechanical properties. *J. Bone Joint Surgery A* **58**, 82.

CARTER, D.R., HAYES, W.C. (1977). The compressive behavior of bone as a two-phase porous structure. *J. Bone Joint Surgery A* **59**, 954–962.

CARTER, F. (1999). Work at the University of Dundee. Private communication to INRIA, Sophia-Antipolis.

CAVANAUGH, J.M. (1993). The biomechanics of thoracic trauma. In: Nahum, A.M., Melvin, J.W. (eds.), *Accidental Injury – Biomechanics and Prevention* (Springer, Berlin), pp. 362–390 (Chapter 15).

CAVANAUGH, J.M., NYQUIST, G.W., GOLDBERG, S.J., KING, A.I. (1986). Lower abdominal tolerance and response. In: *Proc. 30th Stapp Car Crash Conference*, pp. 41–63. Paper No. 861878.

CAVANAUGH, J.M., WALILKO, T.J., MALHOTRA, A., ZHU, Y., KING, A.I. (1990). Biomechanical response and injury tolerance of the pelvis in twelve sled side impact tests. In: *Proc. 34th Stapp Car Crash Conference*, pp. 23–38. Paper No. 902307.

CESARI, D., BERMOND, F., BOUQUET, R., RAMET, M. (1994). Virtual predictive testing of biomechanical effects of impacts on the human leg. In: *Proceedings of PAM '94*, ESI Software S.A., 99, rue des Solets, BP 80112, 94513 Rungis Cedex, France, pp. 177–188.

CESARI, D., BOUQUET, R. (1990). Behavior of human surrogates thorax under belt loading. In: *Proc. 34th Stapp Car Crash Conference*, pp. 73–81. Paper No. 902310.

CESARI, D., BOUQUET, R. (1994). Comparison of Hybrid III and human cadaver thorax deformation loaded by a thoracic belt. In: *Proc. 38th Stapp Car Crash Conference*, pp. 65–76. Paper No. 942209.

CHANCE, G.O. (1948). Note on a type of flexion fracture of the spine. *Br. J. Radiol.* **21**, 452–453.

CHANDLER, R.F. (1993). Development of crash injury protection in civil aviation. In: Nahum, A.M., Melvin, J.W. (eds.), *Accidental Injury – Biomechanics and Prevention* (Springer, Berlin), pp. 151–185 (Chapter 6).

CHAPON, A., VERRIEST, J.P., DEDOYAN, J., TRAUCHESSEC, R., ARTRU, R. (1983). Research on brain vulnerability from real accidents, ISO Document No. ISO/TC22SC12/GT6/N139.

CHAZAL, J., TANGUY, A., BOURGES, M. (1985). Biomechanical properties of spinal ligaments and a historical study of the supraspinal ligament in traction. *J. Biomech.* **18** (3), 167–176.

CHOI, H.-Y. (2001). Numerical human head model for traumatic injury assessment. *KSME Internat. J.* **15** (7), 995–1001.

Choi, H.-Y., Eom, H.-W. (1998). Finite element modeling of human cervical spine. *Hongik J. Sci. Technol. (Special volume)*.

Choi, H.-Y., Lee, I.-H. (1999a). Finite element modeling of human thorax for occupant safety simulation. In: *10th Proceedings of PUCA*, Japan; ESI Software S.A., 99, rue des Solets, BP 80112, 94513 Rungis Cedex, France, pp. 469–478.

Choi, H.-Y., Lee, I.-H. (1999b). Advanced finite element modeling of the human body for occupant safety simulation. In: *EuroPam 1999*, ESI Software S.A., 99, rue des Solets, BP 80112, 94513 Rungis Cedex, France.

Choi, H.-Y., Lee, I.-H., Eom, H.-W., Lee, T.-H. (1999). A study on the whiplash injury due to the low velocity rear-end collision. In: *3rd World Congress of Biomechanics, Japan*, p. 522.

Choi, H.-Y., Lee, I.-H, Haug, E. (2001a). Finite element modeling of human upper extremity for occupant safety simulation. In: *JSAE Spring Conference, No. 34-01*, 20015358.

Choi, H.-Y., Lee, I.-H., Haug, E. (2001b). Finite element modeling of human head–neck complex for crashworthiness simulation. In: *First MIT Conference Proceedings*.

Ciarelli, M.J., et al. (1986). Experimental determination of the orthogonal mechanical properties, density, and distribution of human trabecular bone from the major metaphyseal regions utilizing material testing and computed tomography. *Trans. Orthop. Res. Soc.*, 42.

Ciarelli, M.J., et al. (1991). Evaluation of orthogonal mechanical properties and density of human trabecular bone from the major metaphyseal regions with materials testing and computed tomography. *J. Orth. Res.* **9**, 674–682.

Claessens, M. (1997). *Finite Element Modelling of the Human Head under Impact Conditions*. PhDThesis. Eindhoven University. ISBN 90-386-0369-X.

Claessens, M., Sauren, F., Wismans, J. (1997). Modeling the human head under impact conditions: A parametric study. In: *Proc. 41st Stapp Car Crash Conference*, pp. 315–328. Paper No. 973338.

Clemente, C.D. (1981). *Anatomy (A Regional Atlas of the Human Body)*, second ed. (Urban & Schwarzenberg, Baltimore–Munich).

Cohen, D.S. (1987). The safety problem for passengers in frontal impacts: Analysis of accidents, laboratory and model simulation data. In: *11th ESV International Technical Conference on Experimental Safety Vehicles, Washington, DC*.

Cole, G.K., van den Bogert, A.J., Herzog, W., Gerritsen, K.G.M. (1996). Modelling of force prediction in skeletal muscle undergoing stretch. *J. Biomech.* **29** (8), 1091–1104.

Compton, C.P. (1993). The use of public crash data in biomechanical research. In: Nahum, A.M., Melvin, J.W. (eds.), *Accidental Injury – Biomechanics and Prevention* (Springer, Berlin), pp. 49–65 (Chapter 3).

Cooper, P.R. (1982a). Post-traumatic intracranial mass lesions. In: Cooper, P.R. (ed.), *Head Injury* (Williams and Wilkins, Baltimore/London), pp. 185–232.

Cooper, P.R. (1982b). Skull fracture and traumatic cerebrospinal fluid fistulas. In: Cooper, P.R. (ed.), *Head Injury* (Williams and Wilkins, Baltimore/London), pp. 65–82.

Cooper, G.J., Pearce, B.P., Stainer, M.C., Maynard, R.L. (1982). The biomechanical response of the thorax to non-penetrating trauma with particular reference to cardiac injuries. *J. Trauma* **22** (12), 994–1008.

Costa-Paz, M., Ranalletta, M., Makino, A., Ayerza, M., Muscolo, L. (2002). Displaced patella fracture after cruciate ligament reconstruction with patellar ligament graft, *SICOT Online Report E006 February 2002*, http://www.sicot.org/.

Crandall, J.R., Duma, S.M., Bass, C.R., Pilkey, W.D., Kuppa, S.M., Khaewpong, N., Eppinger, R. (1999). Thoracic response and trauma in airbag deployment tests with out-of-position small female surrogates. *J. Crash Prevention Injury Control* **1** (2), 101–102.

Crawford, G.N.C., James, N.T. (1980). The design of muscles. In: Owen, R., Goodfellow, J., Bullough, P. (eds.), *Scientific Foundations of Orthopaedics and Traumatology* (William Heinemann, London), pp. 67–74.

Currey, J.D. (1975). The effects of strain rate, reconstruction and mineral content on some mechanical properties of bovine bone. *J. Biomech.*, 81.

Cusick, J.F., Yoganandan, N. (2002). Biomechanics of the cervical spine 4: Major injuries. *Clinical Biomech.* **17**, 1–20.

Dalstra, M., Huiskes, R. (1995). Load transfer across the pelvic bone. *J. Biomech.* **6** (2), 715–724.

Dalstra, M., et al. (1993). Mechanical and textural properties of pelvic trabecular bone. *J. Biomech.* **26** (4–5), 523–535.

Dan, D. (1995). Elaboration d'un capteur de pression. Application au foie humain. Projet de fin d'étude D.E.S.S., Collaboration entre L'Université Paris 7 et le LAB Laboratoire d'Accidentologie, de Biomécanique et d'Etude du Comportement Humain de PSA Peugeot Citroën–Renault.

Dan, D. (1999). Caractérisation mécanique du foie humain en situation de choc. PhD Thesis, University Paris 7 – Denis Diderot.

Dan, D., Milcent, G. (2002). Caesare – chirurgie abdominale et simulation a retour d'effort, French Ministry of Education, Research and Technology Project, Final Technical Report RE/02.1600/A, Contract FSP9E2045, ESI France, 99 rue des Solets, 94513 Rungis Cedex, France.

Davidsson, J., Flogard, A., Lövsund, P., Svensson, M.Y. (1999). BioRID P3—Design and performance compared to Hybrid III and Volunteers in Rear Impacts of $\Delta V = 7$ km/h. In: *Proc. 43rd Stapp Car Crash Conference*, pp. 253–265. Paper No. 99SC16.

Deng, Y.C. (1985). Human head/neck/upper-torso model response to dynamic loading, PhD Thesis, University of California.

Deng, Y.C., Goldsmith, W. (1987). Response of a human head/neck/upper-torso replica to dynamic loading – II. Analytical/numerical model. *J. Biomech.* **20**, 487–497.

Digimation/Viewpoint Catalog (2002). External and Skeletal Anatomy, 2002 Edition.

DiMasi, F., Marcus, J., Eppinger, R. (1991). 3-D anatomic brain model for relating cortical strains to automobile crash loading. In: *13th ESV International Technical Conference on Experimental Safety Vehicles*. Paper No. 91-S8-O-11.

Doherty, B.J., Esses, S.L., Heggeness, M.H. (1992). A biomechanical study of odontoid fractures and fracture fixation. In: *Cervical Spine Research Society*.

Donnelly, B.R., Medige, J. (1997). Shear properties of human brain tissue. *J. Biomech. Engrg.* **119**, 423–432.

Dostal, W.F. (1981). A three-dimensional biomechanical model of hip musculature. *J. Biomech.* **14** (11), 803–812.

Ducheyne, P., et al. (1977). The mechanical behavior of intracondylar cancellous bone of the femur at different loading rates. *J. Biomech.* **10**, 747–762.

Dvorak, J., Hayek, J., Zehnuder, R. (1987). CT-functional diagnostics of rotary instability of upper cervical spine. Part II: An evaluation on healthy adults and patients with suspected instability. *J. Spine* **12** (8), 726.

Dvorak, J., Panjabi, M.M., Froehlich, D., et al. (1988). Functional radiographic diagnosis of the cervical spine: Flexion/extension. *J. Spine* **13** (7), 748.

Ebara, S., Iatridis, J.C., Setton, L.A., Foster, R.J., Mow, V.C., Weidenbaum, M. (1996). Tensile properties of nondegenerate human lumbar anulus fibrosus. *J. Spine* **21** (4), 452–461.

Engin, A.E. (1979). Passive resistance torques about long bone axes of major human joints. *Aviation, Space and Environmental Medicine* **50** (10), 1052–1057.

Engin, A.E. (1980). On the biomechanics of the shoulder complex. *J. Biomech.* **13-7**, 575–590.

Engin, A.E. (1983). Dynamic modelling of human articulating joints. *Math. Modelling* **4**, 117–141.

Engin, A.E. (1984). On the damping properties of the shoulder complex. *J. Biomech. Engrg.* **106**, 360–363.

Engin, A.E., Chen, S.M. (1988a). On the biomechanics of the human hip complex in vivo – I. Kinematics for determination of the maximal voluntary hip complex sinus. *J. Biomech.* **21** (10), 785–796.

Engin, A.E., Chen, S.M. (1988b). On the biomechanics of the human hip complex in vivo – II. Passive resistive properties beyond the hip complex sinus. *J. Biomech.* **21** (10), 797–806.

Engin, A.E., Chen, S.M. (1989). A statistical investigation of the in vivo biomechanical properties of the human shoulder complex. *Math. Comput. Modelling* **12** (12), 1569–1582.

Engin, A.E., Peindl, R.D. (1987). On the biomechanics of human shoulder complex – I. Kinematics for determination of the shoulder complex sinus. *J. Biomech.* **20** (2), 103–117.

Engin, A.E., T'Mer, S.T. (1989). Three-dimensional kinematic modelling of the human shoulder complex – Part I: Physical model and determination of joint sinus cones. *J. Biomech. Engrg.* **111**, 107–112.

Eppinger, R.H. (1976). Prediction of thoracic injury using measurable experimental parameters. In: *6th ESV International Technical Conference on Experimental Safety Vehicles*, pp. 770–779.

EPPINGER, R.H. (1993). Occupant restraint systems. In: Nahum, A.M., Melvin, J.W. (eds.), *Accidental Injury – Biomechanics and Prevention* (Springer, Berlin), pp. 186–197 (Chapter 8).

EPPINGER, R.H., MARCUS, J.H., MORGAN, R.M. (1984). Development of dummy and injury index for NHTSA's thoracic side impact protection research program. In: *Government/Industry Meeting and Exposition, Washington, DC*. SAE 840885.

ESTES, M.S., MCELHANEY, J.H. (1970). Response of brain tissue to compressive loading. ASME Paper No. 70-BHF-13.

EVANS, F.G., et al. (1961). Regional differences in some physical properties of human spongy bone. In: EVANS, F.G. (ed.), *Biomechanical Studies of the Musculo–Skeletal System* (CC Thomas, Springfield, IL), pp. 49–67.

EWING, C., THOMAS, D., LUSTICK, L., MUZZY, W. III, WILLEMS, G., MAJEWSKI, P. (1976). The effect of duration, rate of onset, and peak sled acceleration on the dynamic response of the human head and neck. In: *Proc. 20th Stapp Car Crash Conference*, pp. 3–41. Paper No. 760800.

EWING, C., THOMAS, D., LUSTICK, L., MUZZY, W. III, WILLEMS, G., MAJEWSKI, P. (1978). Effect of initial position on the human head and neck response to $+Y$ impact acceleration. In: *Proc. 22nd Stapp Car Crash Conference*, pp. 103–138. Paper No. 780888.

FALLENSTEIN, G.T., HULCE, V.D., MELVIN, J.W. (1970). Dynamic mechanical properties of human brain tissue. *J. Biomech.* **2**, 217–226.

FAZEKAS, I.G., KOSA, F., JOBBA, G., MESZARO, E. (1971a). Die Druckfestigkeit der Menschlichen Leber mit besonderer Hinsicht auf Verkehrsunfälle. *Z. Rechtsmedizin* **68**, 207–224.

FAZEKAS, I.G., KOSA, F., JOBBA, G., MESZARO, E. (1971b). Experimentelle Untersuchungen über die Druckfestigkeit der Menschlichen Niere. *Zacchia* **46**, 294–301.

FAZEKAS, I.G., KOSA, F., JOBBA, G., MESZARO, E. (1972). Beiträge zur Druckfestigkeit dser Menschlichen Milz bei stumpfen Gewalteinwirkungen. *Arch. Kriminol.* **149**, 158–174.

FIROOZBAKSHK (1975). A model of brain shear under impulsive torsional loads. *J. Biomech.* **8**, 65–73.

FOGRASCHER, K. (1998). Development of a simulation model for the head protection system ITS and integration of the component model into the full structural vehicle model, *EuroPam 1998*, ESI Software S.A., 99, rue des Solets, BP 80112, 94513 Rungis Cedex, France.

FOREMAN, S.M., CROFT, A.C. (1995). *Whiplash Injuries: The cervical Acceleration/Deceleration Syndrome*, second ed. (Williams & Wilkins).

FRANK, C.B., SHRIVE, N.G. (1994). Biomechanics of the musculo–skeletal System – 2.3: Ligament. In: Nigg, B.M., Herzog, W. (eds.), *Biomechanics of the Musculo–Skeletal System* (University of Calgary, Alberta, Canada), pp. 106–132.

FRANKEL, V.H., NORDIN, M. (1980). *Basic Biomechanics of the Skeletal System* (Lea & Febiger, Philadelphia, PA).

FUKUBAYASHI, T., KUROSAWA, H. (1980). The contact area and pressure distribution pattern of the knee. *J. Acta Orthop. Scand.* **51**, 871–879.

FUNG, Y.C. (1993a). *Biomechanics – Mechanical Properties of Living Tissues* (Springer, Berlin).

FUNG, Y.C. (1993b). The application of biomechanics to the understanding of injury and healing. In: Nahum, A.M., Melvin, J.W. (eds.), *Accidental Injury – Biomechanics and Prevention* (Springer, Berlin), pp. 1–11 (Chapter 1).

FUNG, Y.C., YEN, M.R. (1984). Experimental investigation of lung injury mechanisms, Topical Report, US Army Medical Research and Development Command. Contract No. DAMD 17-82-C-2062.

FURUKAWA, K., FURUSU, K., MIKI, K. (2002). A development of child FEM model – Part I: Skeletal model of six-year-old child. In: *Proc. 2002 JSME Annual Congress, No. 02-1*, pp. 89–90 (in Japanese).

GADD, C.W. (1961). Criteria for injury potential. In: *Impact Acceleration Stress Symposium*, National Research Council Publication, No. 977 (National Academy of Sciences, Washington, DC), pp. 141–144.

GADD, C.W. (1966). Use of a weighted impulse criterion for estimating injury hazard. In: *Proc. 10th Stapp Car Crash Conference*, pp. 164–174. Paper No. 660793.

GALANTE, J.O. (1967). Tensile properties of human lumbar anulus fibrosus. *Acta Orthop Scand (Suppl.)* **100**.

GALANTE, J.O., et al. (1970). Physical properties of trabecular bone. *Calcif. Tissue Res.* **5**, 236–246.

GALFORD, J.E., MCELHANEY, J.H. (1970). A visco-elastic study of scalp, brain and dura. *J. Biomech.* **3**, 211–221.

GENNARELLI, T.A. (1980). Analysis of head injury severity by AIS-80. In: *24th Annual Conference of the American Association of Automotive Medicine* (AAAM, Morton Grove, IL), pp. 147–155.

GENNARELLI, T.A., THIBAULT, L.E. (1982). Biomechanics of acute subdural hematoma. *J. Trauma* **22** (8), 680–686.

GENNARELLI, T.A., THIBAULT, L.E., ADAMS, J.H., GRAHAM, D.I., THOMPSON, C.J., MARCINCIN, R.P. (1982). Diffuse axonal injury and traumatic coma in the primate. *Ann. Neuron.* **12**, 564–574.

GENNARELLI, T.A., THIBAULT, L.E., TOMEI, G., WISER, R., GRAHAM, D., ADAMS, J. (1987). Directional dependence of axonal brain injury due to centroidal and non-centroidal acceleration. In: *Proc. 31st Stapp Car Crash Conference*, pp. 49–53. Paper No. 872197.

GOEL, V.K., GOYAL, S., CLARK, C., NISHIYAMA, K., NYE, T. (1985). Kinematics of the whole lumbar spine effect of dissectomy. *J. Spine* **10** (6).

GOEL, V.K., MONROE, B.T., GILBERTSON, L.G., BRINKMAN, P. (1995). Interlaminar shear stresses and laminae separation in a disc: finite element analysis of the L3–L4 motion segment subjected to axial compressive loads. *Spine* **20** (6), 689–698.

GOLDSMITH (1972). Biomechanics of Head Injury. In: Fung, Y.C., Perrone, N., Anliker, M. (eds.), *Biomechanics: It's Foundation and Objectives* (Prentice-Hall, Englewood Cliffs, NJ).

GOLDSTEIN, S., FRANKENBURG, E., KUHN, J. (1993). Biomechanics of bone. In: Nahum, A.M., Melvin, J.W. (eds.), *Accidental Injury – Biomechanics and Prevention* (Springer, Berlin), pp. 198–223 (Chapter 9).

GOLDSTEIN, S.A., et al. (1983). The mechanical properties of human tibia trabecular bone as a function of metaphyseal location. *J. Biomech.* **16**, 965–969.

GRANIK, G., STEIN, I. (1973). Human ribs: static testing as a promising medical application. *J. Biomech.* **6**, 237–240.

GRAY'S ANATOMY (1989). Williams, P.L., Warwick, R., Dyson, M., Bannister, L. (eds.), Thirty-seventh ed., Churchill Livington, ISBN 0 443 02588 6, after Figure 5.8A, p. 554.

GURDJIAN, E.S., LISSNER, H.R. (1944). Mechanism of head injury as studied by the cathode ray oscilloscope, Preliminary report. *J. Neurosurgery* **1**, 393–399.

GURDJIAN, E.S., ROBERTS, V.L., THOMAS, L.M. (1966). Tolerance curves of acceleration and intracranial pressure and protective index in experimental head injury. *J. Trauma* **6**, 600–604.

GURDJIAN, E.S., WEBSTER, J.E., LISSNER, H.R. (1955). Observations of the mechanism of brain concussion, contusion, and laceration. *Surgery Gynecol. Obstet.* **101**, 680–690.

HAJJI, M.A., WILSON, T.A., LAI-FOOK, S.J. (1979). Improved measurements of shear modulus and pleural membrane tension of the lung. *J. Appl. Physiol.* **47**, 175–181.

HALLQUIST, J.O., GOUDREAU, G.L., BENSON, D.J. (1985). Sliding surfaces with contact–impact in large-scale Lagrangian computations problems. *Comput. Methods Appl. Mech. Engrg.* **51**, 107–137.

HARDY, W.N. (1993). Instrumentation in experimental design. In: Nahum, A.M., Melvin, J.W. (eds.), *Accidental Injury – Biomechanics and Prevention* (Springer, Berlin), pp. 12–48 (Chapter 2).

HARDY, W.N., FOSTER, C.D., MASON, M.J., YANG, K.H., KING, A.I., TASHMAN, S. (2001). Investigation of head injury mechanisms using neutral density technology and high-speed biplanar X-ray. *Stapp Car Crash Conference J.* **45**, 337–368. Paper No. 2001-22-0016.

HARRY, J.D., WARD, A.W., HEGLUD, N.C., MORGAN, D.L., MCMAHON, T.A. (1990). Cross-bridge cycling theories cannot explain high-speed lengthening behavior in frog muscle. *Biophys. J.* **57**, 201–208.

HAUG, E. (1995). Biomechanical models in vehicle accident simulation. In: *PAM – User's Conference in Asia, PUCA '95*, Shin-Yokohama, ESI Software S.A., 99, rue des Solets, BP 80112, 94513 Rungis Cedex, France, pp. 233–256.

HAUG, E., BEAUGONIN, M., TRAMECON, A., HYNCIK, L. (1999). Current status of articulated and deformable human models for impact and occupant safety simulation at ESI group. In: *European Conference on Computational Mechanics, EEVC'99*, August 31 – September 3, 1999, Munic.

HAUG, E., CLINCKEMAILLIE, J.C., ABERLENC, F. (1989a). Computational mechanics in crashworthiness analysis. In: *Post Symposium Short Course of the 2nd Symposium of Plasticity, Nagoya, Japan.*

HAUG, E., CLINCKEMAILLIE, J.C., ABERLENC, F. (1989b). Contact–impact problems for crash. In: *Post Symposium Short Course of the 2nd Symposium of Plasticity, Nagoya, Japan.*

HAUG, E., LASRY, D., GROENENBOOM, P., MUNCK, G., ROGER, J., SCHLOSSER, J., RÜCKERT, J. (1993). Finite element models of dummies and biomechanical applications using PAM-CRASH™. In: *PAM – User's*

Conference in Asia, PUCA '93, Shin-Yokohama, ESI Software S.A., 99, rue des Solets, BP 80112, 94513 Rungis Cedex, France, pp. 123–158.

HAUG, E., TRAMECON, A., ALLAIN, J.C., CHOI, H.-Y. (2001). Modeling of ergonomics and muscular comfort. *KSME Internat. J.* **15** (7), 982–994.

HAUG, E., ULRICH, D. (1989). The PAM–CRASH code as an efficient tool for crashworthiness simulation and design. In: *Second European Cars/Trucks Simulation Symposium, Schliersee (Munich), AZIMUTH* (Springer, Berlin).

HAUT, R.C. (1993). Biomechanics of soft tissues. In: Nahum, A.M., Melvin, J.W. (eds.), *Accidental Injury – Biomechanics and Prevention* (Springer, Berlin), pp. 224–246 (Chapter 10).

HAWKINS, D., BEY, M. (1994). A comprehensive approach for studying muscle-tendon mechanics. *J. Biomech. Engrg.* **116**, 51.

HAWKINS, D., BEY, M. (1997). Muscle and tendon force-length properties and their interactions in vivo. *J. Biomech.* **30** (1), 63–70.

HAYASHI, S., CHOI, H.-Y., LEVINE, R.S., YANG, K.H., KING, A.I. (1996). Experimental and analytical study of a frontal knee impact. In: *Proc. 40th Stapp Car Crash Conference*, pp. 161–173. Paper No. 962423.

HERZOG, W. (1994). Muscle. In: Nigg, B.M., Herzog, W. (eds.), *Biomechanics of the Musculo–Skeletal System* (Wiley, New York), pp. 154–187.

HILL, A.V. (1970). *First and Last Experiments in Muscle Mechanics* (Cambridge).

HIRSCH, C. (1955). The reaction of intervertebral discs to compression force. *J. Bone Joint Surgery A* **37** (6).

HOLBOURN, A.H.S. (1943). Mechanics of head injury. *Lancet* **2**, 438–441.

HOPPIN, F.G., LEE, G.C., DAWSON, S.V. (1975). Properties for lung parenchyma in distortion. *J. Appl. Physiol.* **39**, 742–751.

HORST, M.J. VAN DER, THUNNISSEN, J.G.M., HAPPEE, R., HAASTER VAN, R.M.H.P., WISMANS, J.S.H.M. (1997). The influence of muscle activity on head–neck response during impact. In: *Proc. 41st Stapp Car Crash Conference*, pp. 487–507. Paper No. 973346.

HUANG, Y., KING, A.I., CAVANAUGH, J.M. (1994a). A MADYMO model of near-side human occupants in side impacts. *J. Biomech. Engrg.* **116**, 228–235.

HUANG, Y., KING, A.I., CAVANAUGH, J.M. (1994b). Finite element modelling of gross motion of human cadavers in side impact. In: *Proc. 38th Stapp Car Crash Conference*, pp. 35–53. Paper No. 942207.

HUELKE, D.F., NUSHOLTZ, G.S., KAIKER, P.S. (1986). Use of quadruped models in thoraco-abdominal biomechanics research. *J. Biomech.* **19** (12), 969–977.

HUGHES, T.J.R., PISTER, J.S., TAYLOR, R.L. (1979). Implicit–explicit finite elements in non-linear transient analysis. *Comput. Methods Appl. Mech. Engrg.* **17/18**, 159–182.

HUGHES, T.J.R., TAYLOR, R.L., SACKMAN, J.L., CURNIER, A., KANOKNUKULCHAI, W. (1976). A finite element method for a class of contact–impact problems. *Comput. Methods Appl. Mech. Engrg.* **8**, 249–276.

HVID, I., et al. (1985). Trabecular bone strength patterns at the proximal tibial epiphysis. *J. Orthop. Res.* **3**, 464–472.

HYNCIK, L. (1997). Human articulated rigid body model (ROBBY1). ESI Group Internal Report, ESI Software S.A., 99, rue des Solets, BP 80112, 94513 Rungis Cedex, France.

HYNCIK, L. (1999a). Human shoulder model. In: *15th Conference Computational Mechanics*, pp. 117–124. ISBN 80-7082-542-1 (in Czech).

HYNCIK, L. (1999b). Human inner organs model. In: *15th Conference Computational Mechanics*, pp. 125–132. ISBN 80-7082-542-1.

HYNCIK, L. (2000). Biomechanical human model. In: *8th Conference Biomechanics of Man*, pp. 52–55. ISBN 80-244-0193-2.

HYNCIK, L. (2001a). Rigid body based human model for crash test purposes. *Engrg. Mech.* **5**, 337–342.

HYNCIK, L. (2001b). Biomechanical model of human inner organs and tissues. In: *17th Conference Computational Mechanics*, pp. 113–120. ISBN 80-7082-780-7.

HYNCIK, L. (2002a). Multi-body human model. In: *Preprint of seminar Virtual Nonlinear Multibody Systems*. In: Preprint of NATO Advanced Study Institute **1**, pp. 89–94.

HYNCIK, L. (2002b). Deformable human abdomen model for crash test purposes. In: *4th Conference Applied Mechanics*, pp. 159–164. ISBN 80-248-0079-9.

HYNCIK, L. (2002c). Biomechanical model of abdominal inner organs and tissues for crash test purposes. PhD Thesis, Department of Mechanics of the Faculty of Applied Sciences of the University of West Bohemia in Pilsen.

IRWIN, A.L. (1994), Analysis and CAL3D model of the shoulder and thorax response of seven cadavers subjected to lateral impacts. PhD Thesis, Wayne State University, 1994.

IRWIN, A., MERTZ, H.J. (1997). Biomechanical basis for the CRABI and Hybrid III child dummies. In: *Proc. 41st Stapp Car Crash Conference*, pp. 261–272. Paper No. 973317.

ITOI, E., BERGLUND, L.J., GRABOWSKI, J.J., SCHULTZ, F.M., GROWNEY, E.S., MORREY, B.F., AN, K.N. (1995). Tensile properties of the supraspinatus tendon. *J. Orthop. Res.* **13** (4), 578–584.

IWAMOTO, M., KISANUKI, Y., WATANABE, I., FURUSU, K., MIKI, K., HASEGAWA, J. (2002). Development of a finite element model of the total human model for safety (THUMS) and application to injury reconstruction. In: *2002 International IRCOBI Conference*, pp. 31–42.

JAGER, M. DE (1996). Mathematical head–neck models for acceleration impact. Ph.D. Thesis, Eindhoven University.

JAGER, M. DE, SAUREN, A., THUNNISSEN, J., WISMANS, J. (1994). A three-dimensional neck model: Validation for frontal and lateral impact. In: *38th Stapp Car Crash Conference*, pp. 93–109. Paper No. 942211.

JOHNSON, G.A., TRAMAGLINI, D.M., LEVINE, R.E., ONO, K., CHOI, N.Y., WOO, S.L. (1994). Tensile and viscoelastic properties of human patellar tendon. *J. Orthop. Res.* **12** (6), 796–803.

KAJZER, J. (1991). Impact biomechanics of knee injuries. Doctoral Thesis, Dept. of Injury Prevention, Chalmers University of Technology.

KAJZER, J., SCHROEDER, G., ISHIKAWA, H., MATSUI, Y., BOSCH, U. (1997). Shearing and bending effects at the knee joint at high speed lateral loading. In: *Proc. 41st Stapp Car Crash Conference*, pp. 151–165. Paper No. 973326.

KAJZER, J., ZHOU, C., KHALIL, T.B., KING, A.I. (1996). Modelling of ligaments and muscles under transient loads: application of PAM–CRASH material models. In: *PAM User's Conference in Asia PUCA '96*, Nihon ESI, ESI Software S.A., 99, rue des Solets BP 80112, 94513 Rungis Cedex, France, pp. 223–231.

KALLIERIS, D., SCHMIDT, G. (1990). Neck response and injury assessment using cadavers and the US-SID for far-side lateral impacts of rear seat occupants with inboard-anchored shoulder belts. In: *Proc. 34th Stapp Car Crash Conference*, pp. 93–99. Paper No. 902313.

KANG, H.S., WILLINGER, R., DIAW, B.M., CHINN, B. (1997). Validation of a 3D anatomic human head model and replication of head impact in motorcycle accident by finite element modeling. In: *Proc. 41*st *Stapp Car Crash Conference*, pp. 329–338. Paper No. 973339.

KAPANDJI, I.A. (1974a). *The Physiology of Joints, vol. 1: Upper Limb* (Churchill Livingstone).

KAPANDJI, I.A. (1974b). *The Physiology of Joints, vol. 2: Lower Limb* (Churchill Livingstone).

KAPANDJI, I.A. (1974c). *The Physiology of Joints, vol. 3: The Trunk and the Vertebral Column* (Churchill Livingstone).

KATAKE, K. (1961). The strength for tension and bursting of human fasciae. *J. Kyoto Pref. Med. Univ.* **69**, 484–488.

KAZARIAN, L.E. (1982). Injuries to the human spinal column: Biomechanics and injury classification. *Exerc. Sport Sci. Rev.* **9**, 297–352.

KAZARIAN, L.E., BEERS, K., HERNANDEZ, J. (1979). Spinal injuries in the F/FB-111 crew escape system. *Aviat. Space Environ. Med.* **50**, 948–957.

KEITHEL, L.M. (1972). Deformation of the thoracolumbar intervertebral joints in response to external loads. *J. Bone Joint Surgery A* **54** (3).

KIMPARA, H., IWAMOTO, M., MIKI, K. (2002). Development of a small female FEM model. In: *Proc. JSAE Spring Congress*, No. 59-02, Paper No. 20025242, pp. 1–4 (in Japanese).

KING, A.I. (1993). Injury to the thoraco-lumbar spine and pelvis. In: Nahum, A.M., Melvin, J.W. (eds.), *Accidental Injury – Biomechanics and Prevention* (Springer, Berlin), pp. 429–459 (Chapter 17).

KIRSCH, R.F., BOSKOV, D., RYMER, W.Z. (1994). Muscle stiffness during transient and continuous movements of cat muscle: Perturbation characteristics and physiological relevance. *IEEE Trans. Biomedical Engrg.* **41** (8), 758–770.

KISIELEWICZ, T.K., ANDOH, K. (1994). Critical issues in biomechanical tests and simulations of impact injury. In: *PAM – User's Conference in Asia, PUCA '94*, Shin-Yokohama, ESI Software S.A., 99, rue des Solets,

BP 80112, 94513 Rungis Cedex, France, pp. 209–222, and: *World Congress on Computational Mechanics, WCCM '94*, Makuhari, Japan, August 1994.

KLEINBERGER, M., SUN, E., EPPINGER, R.H., KUPPA, S., SAUL, R. (1998). Development of improved injury criteria for the assessment of advanced automotive restraint systems. NHTSA report.

KRESS, T.A., SNIDER, J.N., PORTA, D.J., FULLER, P.M., WASSERMAN, J.F., TUCKER, G.V. (1993). Human femur response to impact loading. In: *Int. IRCOBI Conf. on the Biomechanics of Trauma*, Eindhoven, the Netherlands. IRCOBI Secretariat, Bron, France, pp. 93–104.

KROELL, C.K. (1971). Thoracic response to blunt frontal loading. In: Backaitis, S.H. (ed.), *Biomechanics of Impact Injury and Injury Tolerances of the Thorax-Shoulder Complex* (Society of Automotive Engineers), pp. 51–80. Paper No. PT-45.

KROELL, C.K., ALLEN, S.D., WARNER, C.Y., PERL, T.R. (1986). Interrelationship of velocity and chest compression in blunt thoracic impact to swine II. In: *Proc. 30th Stapp Car Crash Conference*, pp. 99–121. Paper No. 861881.

KROELL, C.K., SCHNEIDER, D.C., NAHUM, A.M. (1971). Impact tolerance and response of the human thorax. In: *Proc. 15th Stapp Car Crash Conference*, pp. 84–134. Paper No. 710851.

KROELL, C.K., SCHNEIDER, D.C., NAHUM, A.M. (1974). Impact tolerance and response of the human thorax II. In: *Proc. 18th Stapp Car Crash Conference*, pp. 383–457. Paper No. 741187.

KROONENBERG, A. VAN DEN, THUNNISSEN, J., WISMANS, J. (1997). A human model for low severity rear-impacts. In: *IRCOBI Conference*.

KRYLOW, A.M., SANDEROCK, T.G. (1996). Dynamic force responses of muscle involving eccentric contraction. *J. Biomech.* **30** (1), 27–33.

KULAK, R.F., BELYTSCHKO, T.B., SCHULTZ, A.B., GALANTE, J.O. (1976). Nonlinear behavior of the human intervertebral disc under axial load. *J. Biomech.* **9**.

LANIR, Y., FUNG, Y.C. (1974). Two-dimensional mechanical properties of rabbit skin – II. Experiment results. *J. Biomech.* **7**, 171–182.

LASKY, I.I., SIEGEL, A.W., NAHUM, A.M. (1968). Automotive cardio-thoracic injuries: A medical-engineering analysis. In: *Automotive Engineering Congress, Detroit, MI*. Paper No. 680052.

LEE, S-H., CHOI, H-Y. (2000). Finite element modeling of human head–neck complex for crashworthiness simulation. *Hongik J. Sci. Technol.* **4**, 1–15.

LEE, M.C., HAUT, R.C. (1989). Insensitivity of tensile failure properties of human bridging veins to strain rate: implications in biomechanics of subdural hematoma. *J. Biomech.* **22**, 537–542.

LEE, M.C., MELVIN, J.W., UENO, K. (1987). Finite element analysis of traumatic subdural hematoma. In: *Proc. 31st Stapp Car Crash Conference*, pp. 67–77. Paper No. 872201.

LEE, J.B., YANG, K.H. (2001). Development of a finite element model of the human abdomen. In: *Proc. 45th Stapp Car Crash Conference*, pp. 1–22. Paper No. 2001-22-0004.

LESTINA, D.C., KUHLMANN, T.P., KEATS, T.E., MAXWELL ALLEY, R. (1992). Mechanism of fracture in ankle and foot injuries to drivers in motor vehicles. In: *Proc. 36th Stapp Car Crash Conference*, pp. 59–68. Paper No. 922515.

LEVINE, R. (1993). Injury to the extremities. In: Nahum, A.M., Melvin, J.W. (eds.), *Accidental Injury – Prevention* (Springer, Berlin), pp. 460–491 (Chapter 18).

LEWIS, G., SHAW, K.M. (1997a). Modeling the tensile behavior of human achilles tendon. *Biomed. Mater. Engrg.* **7** (4), 231–244.

LEWIS, G., SHAW, K.M. (1997b). Tensile properties of human tendo achillis: Effect of donor age and strain rate. *J. Foot Ankle Surgery* **36** (6), 435–445.

LIN, H.S., LIU, Y.K., RAY, G., NIKRAVESH, P. (1978). Mechanical response of the lumbar intervertebral joint under physiological (complex) loading. *J. Bone Joint Surgery A* **60** (1), 41–55.

LINDAHL, O. (1976). Mechanical properties of dried defatted spongy bone. *Acta Orthop. Scand.* **47**, 11–19.

LINDE, F., et al. (1989). Energy absorptive properties of human trabecular bone specimens during axial compression. *J. Orth. Res.* **7**, 432–439.

LISSNER, H.R., LEBOW, M., EVANS, F.G. (1960). Experimental studies on the relation between acceleration and intracranial pressure changes in man. *Surgery Gynecol. Obstet.* **111**, 329–338.

LIZEE, E., ROBIN, S., SONG, E., BERTHOLON, N., LECOZ, J.Y., BESNAULT, B., LAVASTE, F. (1998). Development of a 3D finite element model of the human body. In: *Proc. 42nd Stapp Car Crash Conference*, pp. 1–23. Paper No. 983152.

LIZEE, E., SONG, E., et al. (1998). Finite element model of the human thorax validated in frontal, oblique and lateral impacts: A tool to evaluate new restraint systems. In: *Proceedings of the 1998 International IRCOBI Conference*.

LÖHNER, R. (1990). Three-dimensional fluid-structure Interaction using a finite element solver and adaptive re-meshing. *Comput. Syst. Engrg.* **1** (2–4), 257–272.

LOWENHIELM, P. (1974). Dynamic properties of parasagittal bridging veins. *Z. Rechtsmedizin* **74**, 55–62.

LUNDBERG, A., GOLDIE, I., KALIN, B., SELVIK, G. (1989). Kinematics of the ankle/foot complex: Plantarflexion and dorsiflexion. *Foot & Ankle* **9** (4), 194–200.

LUNDBERG, A., SVENSSON, O., BYLUND, C., GOLDIE, I., SELVIK, G. (1989). Kinematics of the ankle/foot complex Part 2: Pronation and supination. *Foot & Ankle* **9** (5), 248–253.

MA, S., ZAHALAK, G.I. (1991). A distribution-moment model of energetics in skeletal muscle. *J. Biomech.* **24** (1), 21–35.

MA, D., OBERGEFELL, L.A., RIZER, L.A. (1995). Development of human articulating joint model parameters for crash dynamics simulations. In: *Proc. 39th Stapp Car Crash Conference*, pp. 239–250. Paper No. 952726.

MAENO, T., HASEGAWA, J. (2001). Development of a finite element model of the total human model for safety (THUMS) and application to car-pedestrian impacts. In: *17th International ESV Conference*. Paper No. 494.

MAKHSOUS, M., HÖGFORS, C., SIEMIEN'SKI, A., PETERSON, B. (1999). Total shoulder and relative muscle strength in the scapular plane. *J. Biomech.* **32**, 1213–1220.

MANSCHOT, J.F., BRAKKEE, A.J. (1986a). The measurement and modelling of the mechanical properties of human skin in vivo – I. The measurement. *J. Biomech.* **19** (7), 511–515.

MANSCHOT, J.F., BRAKKEE, A.J. (1986b). The measurement and modelling of the mechanical properties of human skin in vivo – II. The model. *J. Biomech.* **19** (7), 517–521.

MARGULIES, S.S., THIBAULT, L.E. (1989). An analytic model of traumatic diffuse brain injury. *J. Biomech. Engrg.* **111**, 241–249.

MARGULIES, S.S., THIBAULT, L.E. (1992). A proposed tolerance criterion for diffuse axonal injury in man. *J. Biomech.* **25** (8), 917–923.

MARGULIES, S.S., THIBAULT, L.E., GENNARELLI, T.A. (1990). Physical model simulations of brain injury in the primate. *J. Biomech.* **23**, 823–836.

MARKOLF, K.L. (1972). Deformation of the thoracolumbar intervertebral joints in response to external loads. *J. Bone Joint Surgery A* **54** (3).

MARKOLF, K.L., MORRIS, J.M. (1974). The structural components of the intervertebral disc. *J. Bone Joint Surgery A* **56** (4).

MARTENS, M., et al. (1983). The mechanical characteristics of cancellous bone at the upper femoral region. *J. Biomech.* **16**, 971–983.

MARTIN, J., THOMPSON, G. (1986). Achilles tendon rupture. *CORR* **210**, 216–218.

MARTINI, F.H., TIMMONS, M.J., TALLITSCH, B. (2003). Human Anatomy, fourth ed. (Pearson Education, Inc., publishing as Benjamin Cummings, Upper Saddle River, NJ), ISBN 0-13-061569-2.

MASSON, C., CESARI, D., BASILE, F., BEAUGONIN, M., TRAMECON, A., ALLAIN, J.C., HAUG, E. (1999). Quasi static ankle/foot complex behavior: Experimental tests and numerical simulations. In: *IRCOBI Conference, Spain*.

MAUREL, W. (1998). 3D modelling of the human upper limb including the biomechanics of joints, muscles and soft tissues, Thèse N° 1906, Ecole Polytechnique Federale de Lausanne.

MCCALDEN, R.W., et al. (1993). Age-related in the tensile properties of cortical bone. *J. Bone Joint Surgery A* **75**, 1193–1205.

MCCLURE, P., SIEGLER, S., NOBILINI, R. (1998). Three-dimensional flexibility characteristics of the human cervical spine in vivo. *Spine* **23** (2), 216–223.

MCELHANEY, J.H. (1966). Dynamic response of bone and muscle tissue. *J. Appl. Physiol.*, 1231.

MCELHANEY, J.H., DOHERTY, B.J., PAVER, J.G., MYERS, B.S., GRAY, L. (1988). Combined bending and axial loading responses of the human cervical spine. In: *Proc. 32nd Stapp Car Crash Conference*, pp. 21–28. Paper No. 881709.

MCELHANEY, J.H., MELVIN, J.W., ROBERTS, V.L., PORTNOY, H.D. (1973). Dynamic characteristics of the tissues of the head. In: Kennedi, R.M. (ed.), *Perspectives in Biomedical Engineering* (MacMillan, London), pp. 215–222.

McElhaney, J.H., Myers, B.S. (1993). Biomechanical aspects of cervical trauma. In: Nahum, A.M., Melvin, J.W. (eds.), *Accidental Injury – Biomechanics and Prevention* (Springer, Berlin), pp. 311–361 (Chapter 14).

McElhaney, J.H., Paver, J.G., McCrackin, H.J., Maxwell, G.M. (1983). Cervical spine compression responses. In: *Proc. 27th Stapp Car Crash Conference*, pp. 163–177. Paper No. 831615.

McElhaney, D.A., et al. (1970). Mechanical properties of cranial bone. *J. Biomech.* **3**, 495–511.

Meany, D.F., Smith, D., Ross, T., Gennarelli, T.A. (1993). Diffuse axonal injury in the miniature pig: Biomechanical development and injury threshold. *ASME Crashworthiness and Occupant Protection Systems* **25**, 169–175.

Melvin, J.W., Lighthall, J.W., Ueno, K. (1993). Brain injury biomechanics. In: Nahum, A.M., Melvin, J.W. (eds.), *Accidental Injury – Biomechanics and Prevention* (Springer, Berlin), pp. 268–291 (Chapter 12).

Melvin, J.W., McElhaney, J.H., Roberts, V.L. (1970). Development of mechanical model of human head – determination of tissue properties and synthetic substitute materials. In: *Proc. 14th Stapp Car Crash Conference*, pp. 221–240. Paper No. 700903.

Melvin, J.W., Stalnaker, R.L., Roberts, V.L. (1973). Impact injury mechanisms in abdominal organs. In: *Proc. 17th Stapp Car Crash Conference Proceedings*, pp. 115–126. Paper No. 730968.

Mertz, H.J., Patric, L.M. (1971). Strength and response of human neck. In: *Proc. 15th Stapp Car Crash Conference*, pp. 207–255. Paper No. 710855.

Mertz, H.J. (1984). A procedure for normalizing impact response data. SAE Paper 840884, Washington, DC.

Mertz, H.J. (1993). Anthropomorphic test devices. In: Nahum, A.M., Melvin, J.W. (eds.), *Accidental Injury – Biomechanics and Prevention* (Springer, Berlin), pp. 66–84 (Chapter 4).

Miller, K. (2000). Constitutive modeling of abdominal organs. *J. Biomech.* **33**, 367–373.

Miller, K., Chinzei, K. (1997). Constitutive modelling of brain tissue: Experiment and theory. *J. Biomech.* **30**, 1115–1121.

Miller, J.A.A., Schultz, A.B., Warwick, D.N., Spencer, D.L. (1986). Mechanical properties for lumbar spine motion segments under large loads. *J. Biomech.* **19** (1), 79–84.

Moffatt, C.A., Advani, S.H., Lin, C.J. (1971). Analytical end experimental investigations of human spine flexure. *Amer. Soc. Mech. Engrg.* **71-WA/BHF-7**.

Momersteeg, T.J.A., Blaskevoort, L., Huiskes, R., Kooloos, J.G., Kauer, J.M.G. (1996). Characterization of the mechanical behavior of human knee ligaments: A numerical–experimental approach. *J. Biomech.* **29** (2), 151–160.

Monaghan, J.J. (1988). An introduction into SPH. *Comput. Phys. Comm.* **48**, 89–96.

Monaghan, J.J., Gingold, R.A. (1983). Shock simulation by the particle method SPH. *Comput. Phys.* **52**, 374–398.

Morgan, D.L. (1990). New insights into the behavior of muscle during active lengthening. *Biophys. J.* **57**, 209–221.

Morgan, R.M., Eppinger, R.H., Hennessey, B. (1991). Ankle joint injury mechanism for adults in frontal automotive impact. In: *Proc. 35th Stapp Car Crash Conference Proceedings*, pp. 189–198. Paper No. 912902.

Moroney, S.P., Schultz, A.B., Miller, J.A.A. (1988). Analysis and measurement of neck load. *J. Orthopaedic Res.* **6**, 713–720.

Moroney, S.P., Schultz, A.B., Miller, J.A.A., Andersson, G.B.J. (1988). Load-displacement properties of lower cervical spine motion segments. *J. Biomech.* **21** (9), 769–779.

Mosekilde, L., et al. (1986). Normal vertebral body size and compressive strength: Relations to age and to vertebral and iliac trabecular bone compressive strength. *Bone* **7**, 207–212.

Mow, V.C., Hayes, W.C. (1991). *Basic Orthopaedic Biomechanics* (Raven Press, New York).

Myers, B.S., McElhaney, J.H., Doherty, B.J., Paver, J.G., Gray, L. (1991). The role of torsion in cervical spinal injury. *Spine* **16** (8), 870–874.

Myers, B.S., McElhaney, J.H., Richardson, W.J., Nightingale, R., Doherty, B.J. (1991). The influence of end condition on human cervical spine injury mechanisms. In: *Proc. 35th Stapp Car Crash Conference*, pp. 391–400. Paper No. 912915.

MYERS, B.S., VAN EE, C.A., CAMACHO, D.L.A., WOOLLEY, C.T., BEST, T.M. (1995). On the structural properties of mammalian skeletal muscle and its relevance to human cervical impact dynamics. In: *Proc. 39th Stapp Car Crash Conference*, pp. 203–214. Paper No. 952723.

MYKLEBUST, J.B., PINTAR, F. (1988). Tensile strength of spinal ligaments. *J. Spine* **13** (5).

NAGASAKA, K., IWAMOTO, M., MIZUNO, K., MIKI, K., HASEGAWA, J. (2002). Pedestrian injury analysis using the human FE model (Part I, development and validation of the lower extremity model and application to damage evaluation of knee ligaments). In: *Proc. JSME 14th Bioengineering Conference*, pp. 141–142 (in Japanese).

NAHUM, M., GADD, C.W., SCHNEIDER, D.C., KROELL, C.K. (1970). Deflection of the human thorax under sternal impact. In: *Proceedings of the International Automobile Safety Conference, Detroit*. SAE 700400.

NAHUM, A.M., MELVIN, J.W. (eds.) (1993). *Accidental Injury: Biomechanics and Prevention* (Springer, New York).

NAHUM, M., SCHNEIDER, D.C., KROELL, C.K. (1975). Cadaver skeletal response to blunt thoracic impact. In: *Proc. 19th Stapp Car Crash Conference*, pp. 259–293. Paper No. 751150.

NAHUM, A.M., SIEGEL, A.W., HIGHT, P.V., BROOKS, S.H. (1968). Lower extremity injuries of front seat occupants. In: *Proc. 11th Stapp Car Crash Conference Proceedings*. Paper No. 680483.

NAHUM, A.M., SMITH, R., WARD, C.C. (1977). Intracranial pressure dynamics during head impact. In: *Proc. 21st Stapp Car Crash Conference*, pp. 339–366. Paper No. 770922.

NEUMANN, P., KELLER, T.S., EKSTROM, L., PERRY, L., HANSSON, T.H., SPENGLER, D.M. (1992). Mechanical properties of the human lumbar anterior longitudinal ligament. *J. Biomech.* **25** (10), 1185–1194.

NEWMAN, J.A. (1993). Biomechanics of head trauma: Head protection. In: Nahum, A.M., Melvin, J.W. (eds.), *Accidental Injury – Biomechanics and Prevention* (Springer, Berlin), pp. 292–310 (Chapter 13).

NHTSA/CIREN (1997). Crash test results from William Lehman Injury Research Center at the Ryder Trauma Center. In: *First Annual CIREN Conference, October 20, 1997*, http://www-nrd.nhtsa.dot.gov/departments/nrd-50/ciren/ciren1.html.

NICOLL, E.A. (1949). Fractures of the dorso-lumbar spine. *J. Bone Joint Surgery B* **31**, 376–393.

NIGG, B.M., SKARVAN, G., FRANK, C.B. (1990). Elongation and forces of ankle ligaments in a physiological range of motion. *Foot & Ankle* **11** (1), 30–40.

NIGHTINGALE, R.W., WINKELSTEIN, B.A., KNAUB, K.E., RICHARDSON, W.J., LUCK, J.F., MYERS, B.S. (2002). Comparative strengths and structural properties of the upper and lower cervical spine in flexion and extension. *J. Biomech.* **35**, 725–732.

NITSCHE, S., HAUG, E., KISIELEWICZ, L.T. (1996). Validation of a finite element model of the human neck. In: *PUCA '96*, Nihon ESI, Japan, ESI Software S.A., 99, rue des Solets, BP 80112, 94513 Rungis Cedex, France, pp. 203–220.

NORDIN, M., FRANKEL, V.H. (1989). *Basic Biomechanics of the Musculoskeletal System*, second ed. (Lea & Febiger, Philadelphia, PA).

NOVOTNY, J.E. (1993). Spinal biomechanics. *J. Biomech. Engrg.* **115**.

NOYES, F.R., GROOD, E.S. (1978). The strength of the anterior cruciate ligaments in humans and rhesus monkey. *J. Bone Joint Surgery A* **58** (68), 1074–1082.

NUSHOLTZ, G.S., WILEY, B., GLASCOE, L.G. (1995). Cavitation/boundary effects in a simple head impact model. *Aviation Space and Environmental Medicine* **66** (7), 661–667.

NYQUIST, G.W., CHENG, R., EL-BOHY, A.A.R., KING, A.I. (1985). Tibia bending: Strength and response. In: *Proc. 29th STAPP Car Crash Conference*, pp. 99–112. Paper No. 851728.

OMMAYA, A.K. (1967). Mechanical properties of tissues of the nervous system. *J. Biomech.* **1**, 127–138.

OMMAYA, A.K., GENNARELLI, T.A. (1974). Cerebral concussion and traumatic unconsciousness: Correlation of experimental and clinical observations on blunt head injuries. *Brain* **97**, 633–654.

OMMAYA, A.K., HIRSCH, A.E. (1971). Tolerances for cerebral concussion from head impact and whiplash in primates. *J. Biomech.* **4**, 13–31.

OMMAYA, A.K., HIRSCH, A.E., FLAMM, E.S., MAHONE, R.H. (1966). Cerebral concussion in the monkey: An experimental model. *Science* **153**, 211–212.

ONO, K., KANEOKA, K., SUN, E.A., TAKHOUNTS, E.G., EPPINGER, R.H. (2001). Biomechanical response of human cervical spine to direct loading of the head. In: *IRCOBI Conference*.

ONO, K., KANEOKA, K., WITTEK, A., KAJZER, J. (1997). Cervical injury mechanism based on the analysis of human cervical vertebral motion and head-neck-torso kinematics during low speed rear impacts. In: *Proc. 41st Stapp Car Crash Conference*, pp. 339–356. Paper No. 973340.

ONO, K., KIKUCHI, A., NAKAMURA, M., KOBAYASHI, H., NAKAMURA, N. (1980). Human head tolerance to sagittal impact reliable estimation deduced from experimental head injury using subhuman primates and human cadaver skulls. In: *Proc. 24th Stapp Car Crash Conference*, pp. 101–160. Paper No. 801303.

ONO, K., et al. (1999). Relationship between localized spine deformation and cervical vertebral motions for low speed rear impacts using human volunteers. In: *IRCOBI Conference, Spain*, pp. 149–164.

OSHITA, F., OMORI, K., NAKAHIRA, Y., MIKI, K. (2002). Development of a finite element model of the human body. In: *Proc. 7th International LS-DYNA Users Conference, Detroit*, pp. 3-37–3-48.

OTTE, D., VON RHEINBABEN, H., ZWIPP, H. (1992). Biomechanics of injuries to the foot and ankle joint of car drivers and improvements for an optimal car floor development. In: *Proc. 36th Stapp Car Crash Conference Proceedings*, pp. 43–58. Paper No. 922514.

OTTENSMEYER, M.P., SALISBURY, J. (2001). In vivo data acquisition instrument for solid organ mechanical property measurement. In: Niessen, W., Viergever, M. (eds.), *MICCAI 2001*. In: Lecture Notes in Comput. Sci. **2208** (Springer, Berlin), pp. 975–982.

OXLUND, H., ANDREASSON, T.T. (1980). The role of hyaluronic acid, collagen and elastin in the mechanical properties of connective tissues. *J. Anat.* **131**, 611–620.

PALANIAPPAN, P. JR., WIPASURAMONTON, P., BEGEMAN, P., TANAVDE, A.S., ZHU, F.A. (1999). Three dimensional finite element model of the human arm. In: *Proc. 43rd Stapp Car Crash Conference*, pp. 351–363. Paper No. 99SC25.

PANJABI, M.M., BRAND, R.A. (1976). Mechanical properties of the human thoracic spine as shown by three-dimensional load displacement curves. *J. Bone Joint Surgery A* **58** (5).

PANJABI, M.M., CRISCO, J.J., VASAVADA, A., ODA, T., CHOLEWICKI, J., NIBU, K., SHIN, E. (2001). Mechanical properties of the human cervical spine as shown by three-dimensional load-displacement curves. *Spine* **26** (24), 2692–2700.

PANJABI, M.M., DVORAK, J., DURANCEAU, J., et al. (1988). Three-dimensional movements of the upper cervical spine. *J. Spine* **13** (7), 726.

PANJABI, M., JORNEUS, L., GREENSTEIN, G. (1984). Physical properties of lumbar spine ligaments. *Trans. Orthop. Res. Soc.* **9**, 112.

PARENTEAU, C.S., VIANO, D.C. (1996). Kinematics study of the ankle-subtalar joints. *J. Biomech. Engrg.*, Ph.D. Thesis work.

PARENTEAU, C.S., VIANO, D.C., PETIT, P.Y. (1996). Biomechanical properties of ankle-subtalar joints in quasi-static loading to failure. *J. Biomech. Engrg.*

PATHRIA, M.N., RESNIK, D. (1993). Radiologic analysis of trauma. In: Nahum, A.M., Melvin, J.W (eds.), *Accidental Injury – Biomechanics and Prevention* (Springer, Berlin), pp. 85–101 (Chapter 5).

PATTIMORE, D., WARD, E., THOMAS, P., BRADFORD, M. (1991). The nature and cause of lower limb injuries in car crashes. In: *Proc. 35th Stapp Car Crash Conference*, pp. 177–188. Paper No. 912901.

PEINDL, R.D., ENGIN, A.E. (1987). On the biomechanics of human shoulder complex – II. Passive resistive properties beyond the shoulder complex sinus. *J. Biomech.* **20** (2), 118–134.

PENN, R.D., CLASEN, R.A. (1982). Traumatic brain swelling and edema. In: Cooper, P.R. (ed.), *Head Injury* (Williams and Wilkins, Baltimore), pp. 233–256.

PENNING, L. (1979). Normal movements of the cervical spine. *Amer. J. Roentgenol.*, 130–317.

PENNING, L., WILMARK, J.T. (1987). Rotation of the cervical spine. *J. Spine* **12** (8), 732.

PIKE, J.A. (1990). Automotive Safety (Society of Automotive Engineers). ISBN 1-56091-007-0.

PINTAR, F.A., YOGANANDAN, N., EPPINGER, R.H. (1998). Response and tolerance of the human forearm to impact loading. In: *Proc. 42nd Stapp Car Crash Conference*, pp. 1–8. Paper No. 983149.

PLANK, G.R., KLEINBERGER, M., EPPINGER, R.H. (1994). Finite element modeling and analysis of thorax/restraint system interaction. In: *The 14th ESV International Technical Conference on the Enhanced Safety of Vehicles* (Munich, Germany).

POPE, M.E., KROELL, C.K., VIANO, D.C., WARNER, C.Y., ALLEN, S.D. (1979). Postural influences on thoracic Impact. In: *Proc. 23rd Stapp Car Crash Conference*, pp. 765–795. Paper No. 791028.

PORTIER, L., TROSSEILLE, X., LE COZ, J.-Y., LAVASTE, F., COLTAT, J.-C. (1993). Lower leg injuries in real-world frontal accidents. In: *28th International IRCOBI Conference on the Biomechanics of Trauma*, pp. 57–78.

PRADAS, M.M., CALLEJA, R.D. (1990). Nonlinear viscoelastic behaviour of the flexor tendon of the human hand. *J. Biomech.* **23** (8), 773–781.

PRASAD, P. (1990). Comparative evaluation of the dynamic responses of the Hybrid II and the Hybrid III dummies. In: *Proc. 34th Stapp Car Crash Conference*, pp. 175–183. Paper No. 902318.

PRASAD, P., CHOU, C.C. (1993). A review of mathematical occupant simulation models. In: Nahum, A.M., Melvin, J.W. (eds.), *Accidental Injury – Biomechanics and Prevention* (Springer, Berlin), pp. 102–150 (Chapter 6).

PRASAD, P., KING, A.I. (1994). An experimentally validated dynamic model of the spine. *J. Appl. Mech.*, 546–550.

PRING, D.J., AMIS, A.A., COOMBS, R.R. (1985). The mechanical properties of human flexor tendons in relation to artificial tendons. *J. Hand Surgery [Br]* **10** (3), 331–336.

PUTZ, R., PABST, R. (2000). *Sobotta – Atlas der Anatomie des Menschen*, twenty-first ed. (Urban&Fischer, München, Jena).

QUHAN, et al. (1989). Comparison of trabecular and cortical tissue moduli from human iliac crests. *J. Orthop. Res.* **7**, 876–884.

RACK, P.M.H., WESTBURY, D.R. (1969). The effect of length and stimulus rate on tension in the isometric cat soleus muscle. *J. Physiol.* **204**, 443–460.

REILLY, D.T., BURSTEIN, A.H., FRANKEL, V.H. (1974). The elastic modulus for bone. *J. Biomech.* **7**, 271–275.

REILLY, D.T., et al. (1975). The elastic and ultimate properties of compact bone tissue. *J. Biomech.* **8**, 395–405.

RENAUDIN, F., GUILLEMOT, H., PÉCHEUX, C., LESAGE, F., LAVASTE, F., SKALLI, W. (1993). A 3D finite element model of the pelvis in side impact. In: *Proc. 37th Stapp Car Crash Conference*, pp. 241–252. Paper No. 933130.

RIETBERGEN, B. VAN (1996). Mechanical behavior and adaptation of trabecular bone in relation to bone morphology. PhD Thesis Nijmegen University. ISBN 90-9010006-7.

RIETBERGEN, B. VAN, MÜLLER, R., ULRICH, D., RÜEGSEGGER, P., HUISKES, R. (1998). Tissue stresses and strain in trabeculae of a canine proximal femur can be quantified from computer reconstructions. *J. Biomech.*

ROAF, R. (1960). A study of the mechanics of spinal injury. *J. Bone Joint Surgery B* **42**, 810–823.

ROBBINS, D.H. (1983). *Anthropometry of Motor Vehicle Occupants, vol. 2: Mid-sized Male, vol. 3: Small Female and Large Male* (UMTRI-83-53-2, University of Michigan).

ROBBINS, D.H., SCHNEIDER, L.W., SNYDER, R.G., PFLUG, M., HAFFNER, M. (1983). Seated posture of vehicle occupants. In: *Proc. 32nd Stapp Car Crash Conference*, pp. 199–224. Paper No. 831617.

ROBBY1 (1997). *User's Guides*, ESI Software S.A., 99, rue des Solets, BP 80112, 94513 Rungis Cedex, France.

ROBBY2 (1998). *User's Guides*, ESI Software S.A., 99, rue des Solets, BP 80112, 94513 Rungis Cedex, France.

ROBIN, S. (1999). Validation data base, Report PSA/990331/T0/DA, LAB PSA Peugeot Citroën – Renault, 132 Rue des Suisses, F-92000 Nanterre.

ROBIN, S. (2001). HUMOS: Human model for safety – A joint effort towards the development of refined human-like car occupant models. In: *17th ESV Conference*. Paper Number 297.

ROBINA (1998). Internal draft report, ESI Software S.A., 99, rue des Solets, BP 80112, 94513 Rungis Cedex, France.

ROCKOFF, S.D., et al. (1969). The relative contribution of trabecular and cortical bone to the strength of human lumbar vertebrae. *Calc. Tiss. Res.* **3**, 163.

ROHEN, J.W., YOKOCHI, C. (1983). In: *Color Atlas of Anatomy* (IGAKU–SHOIN Medical Publishers), p. 349.

ROSE, J.L., GORDON, S.L., MOSKOWITZ, G. (1974). Dynamic photoelastic model analysis of impact to the human skull. *J. Biomech.* **7**, 193–199.

ROUHANA, S.W. (1993). Biomechanics of abdominal trauma. In: Nahum, A.M., Melvin, J.W. (eds.), *Accidental Injury – Biomechanics and Prevention* (Springer, Berlin), pp. 391–428 (Chapter 16).

RUAN, J.S., KHALIL, T.B., KING, A.I. (1991). Human head dynamic response to side impact by finite element modeling. *J. Biomech. Engrg.* **113**, 267–283. PhD Dissertation, Wayne State University.

RUAN, J.S., KHALIL, T.B., KING, A.I. (1993). Finite element modeling of direct head impact. In: *Proc. 37th Stapp Car Crash Conference*, pp. 69–81. Paper No. 933114.

RUAN, J.S., KHALIL, T.B., KING, A.I. (1994). Dynamic response of the human head to impact by three-dimensional finite element analysis. *ASME J. Biomech. Engrg.* **116**, 44–50.

RUAN, J.S., PRASAD, P. (1994). Head injury assessment in frontal impacts by mathematical modeling. In: *Proc. 38th Stapp Car Crash Conference*, pp. 111–121. Paper No. 942212.

RUDOLF, C., FELLHAUER, A., SCHAUB, S., MARCA, C., BEAUGONIN, M. (2002). OOP Simulation with a 5th Percentile Deformable Human Model. In: *EuroPam 2002*, ESI Software S.A., 99, rue des Solets, BP 80112, 94513 Rungis Cedex, France.

SAE ENGINEERING AID 23 (1986). User's Manual for the 50th Percentile HYBRID III Test Dummy.

SANCES JR, A., et al. (1982). Head and spine injuries. In: *AGARD Conference on Injury Mechanisms, Prevention and Cost, Köln, Germany*, pp. 13-1–13-33.

SACRESTE, J., BRUN-CASSAN, F., FAYON, A., TARRIERE, C., GOT, C., PATEL, A. (1982). Proposal for a thorax tolerance level in side impacts based on 62 tests performed with cadavers having known bone condition. In: *Proc. 26th Stapp Car Crash Conference*, pp. 155–171. Paper No. 821157.

SCHMIDT, G., KALLIERIS, D., BARZ, J., MATTERN, R., SCHULZ, F., SCHÜLER, F. (1978). Biomechanics – Determination of the mechanical loadability limits of the occupants of a motor vehicle. Final report at the end of project (31-12-1978) – Research project, No. 3906, Institute of Forensic Medicine, Heidelberg.

SCHNECK, D.J. (1992). *Mechanics of Muscle* (New York University Press, New York).

SCHNEIDER, L.W., KING, A.I., BEEBE, M.S. (1990). Design requirements and specifications, Thorax abdomen development task. Interim report: Trauma assessment device development program. Report No. DOT-HS-807-511.

SCHOENFELD, C.M., et al. (1974). Mechanical properties of human cancellous bone in the femoral head. *Med. Biol. Engrg.* **12**, 313–317.

SCOTT, W.E. (1981). Epidemiology of head and neck trauma in victims of motor vehicle accidents. Head and Neck Criteria. In: Ommaya, A.K. (ed.), *A Consensus Workshop* (US Department of Transportation, National Highway Traffic Safety Administration, Washington, DC), pp. 3–6.

SEDLIN, E.D., et al. (1965). A rheological model for cortical bone. In: *A Study of the Physical Properties of Human Femoral Samples*. In: Acta Orthop Scand. Supplementum **83**.

SEDLIN, E.D., et al. (1966). Factors affecting the determination of the physical properties of femoral cortical bone. *Acta Orthop. Scand.* **37**, 29–48.

SEIREG, A., ARVIKAR, R. (1989). *Biomechanical Analysis of the Musculoskeletal Structure for Medicine and Sports* (Hemisphere, Washington, DC).

SHAH, C.S., YANG, K.H., HARDY, W.N., WANG, H.K., KING, A.I. (2001). Development of a computer model to predict aortic rupture due to impact loading. *Stapp Car Crash J.* **45**, 161–182.

SHELTON, F.E., BUTLER, D.L., FEDER, S.M. (1993). Shear stress transmission in the patellar tendon and mediacollateral ligament. *Adv. Bioengrg. BED ASME* **26**, 271.

SHUCK, L.Z., ADVANI, S.H. (1972). Rheological response of human brain tissue in shear. *ASME J. Basic Engrg.*, 905–911.

SHUCK, L.Z., HAYNES, R.R., FOGLE, J.L. (1970). Determination of viscoelastic properties of human brain tissue, ASME Paper N°70-BHF-12.

SHUGAR, T.A. (1975). Transient structural response of the linear skull-brain system. In: *Proc. 19th Stapp Car Crash Conference*, pp. 581–614. Paper No. 751161.

SKAGGS, D.L., WEIDENBAUM, M., IATRIDIS, J.C., RATCLIFFE, A., MOW, V.C. (1994). Regional variations in tensile properties and biomechanical composition of the human lumbar annulus fibrosus. *Spine* **19**, 1310–1319.

SONNERUP, L. (1972). A semi-experimental stress analysis of the human intervertebral disk in compression. *Experimental Mech.*

SONODA, T. (1962). Studies on the strength for compression, tension and torsion of the human vertebral column. *J. Kyoto Pref. Med. Univ.* **71**, 659–702.

Spilker, R.L., Jakobs, D.M., Schultz, A.B. (1986). Material constants for a finite element model of the intervertebral disc with a fiber composite annulus. *J. Biomech. Engrg.* **108**, 1–11.

Spitzer, V.M., Whitlock, D.G. (1998). *Atlas of the Visible Human Male* (Jones & Bartlett Publishers), pp. 6, 7.

Stalnaker, R.L., McElhaney, J.H., Roberts, V.L., Trollope, L.L. (1973). Human torso response to blunt trauma. In: *Proceedings of the Symposium: Human Impact Response – Measurement and Simulation, General Motors Research laboratories, New York* (Plenum Press, London), pp. 181–199.

Stalnaker, R.L., Mohan, D. (1974). Human chest impact protection criteria. In: *Proc. 3rd International Conference on Occupant Protection* (Society of Automotive Engineers, New York), pp. 384–393.

States, J.D. (1986). Adult occupant injuries of the lower limb. In: *Biomechanics and Medical Aspects of Lower Limb Injuries*. SAE 861927.

Stockier, R.M., Epstein, J.A., Epstein, B.S. (1969). Seat belt trauma to the lumbar spine: An unusual manifestation of the seat belt syndrome. *J. Trauma* **9**, 508–513.

Struhl, S., et al. (1987). The distribution of mechanical properties of trabecular bone within vertebral bodies and iliac crest: Correlation with computed tomography density. *Trans. Orthop. Res. Soc.*, 262.

Stürtz, G. (1980). Biomechanical data of children. In: *Proc. 24th Stapp Car Crash Conference*, pp. 513–559. Paper No. 801313.

Svensson, M.Y., Lövsund, P. (1992). A dummy for rear-end collisions – development and validation of a new dummy-neck. In: *Proceedings of IRCOBI Conference, Verona, Italy.*

Tada, Y., Nagashima, T. (1994). Modeling and simulation of brain lesions by the finite element method. *IEEE Engineering in Medicine and Biology* **13** (4).

Tarriere, C. (1981). Investigation of brain injuries using the CT scanner. In: Ommaya, A.K. (ed.), *Head and Neck Injury Criteria: A Consensus Work-Shop* (US Department of Transportation, National Highway Traffic Safety Administration, Washington, DC), pp. 39–49.

Tennyson, S.A., King, A.I. (1976a). A biodynamic model of the spinal column. In: *Mathematical Modeling Biodynamic Response to Impact, SAE Special Publication*. SAE 760711.

Tennyson, S.A., King, A.I. (1976b). Electromyographic signals of the spinal musculature during +Gz accelerations. In: *1976 Meeting of the International Society for the Study of the Lumbar Spine, Bermuda.*

Thibault, K.L., Margulies, S.S. (1996). Material properties of the developing porcine brain. In: *1996 International IRCOBI Conference on the Biomechanics of Impact, Dublin, Ireland.*

Thunnissen, J., Wismans, J., Ewing, C.L., Thomas, D.J. (1995). Human volunteer head–neck response in frontal flexion: A new analysis. In: *Proc. 39th Stapp Car Crash Conference*, pp. 439–460. Paper No. 952721.

T'mer, S.T., Engin, A.E. (1989). Three-dimensional kinematic modelling of the human shoulder complex – Part II: Mathematical modelling and solution via optimization. *J. Biomech. Engrg.* **111**, 113–121.

Torg, J.S. (ed.) (1982). *Athletic Injuries to the Head, Neck and Face* (Lea & Febiger, Philadelphia, PA).

Torg, J.S., Pavlov, H. (1991). Axial load "teardrop" fracture. In: *Athletic Injuries to the Head, Neck and Face*, second ed. (Lea & Febiger, Philadelphia, PA).

Trosseille, X., Tarriere, C., Lavaste, F., Guillon, F., Domont, A. (1992). Development of an F.E.M. of the human head according to a specific test protocol. In: *Proc. 36th Stapp Car Crash Conference*, pp. 235–253. Paper No. 922527.

Turquier, F., Kang, H.S., Trosseille, X., Willinger, R., Lavaste, F., Tarriere, C., Dömont, A. (1996). Validation study of a 3D finite element head model against experimental data. In: *Proc. 40th Stapp Car Crash Conference*, pp. 283–293. Paper No. 962431.

Ueno, K., Liu, Y.K. (1987). A three-dimensional nonlinear finite element model of lumbar intervertebral joint in torsion. *J. Biomech. Engrg.* **109**, 200–209.

Ueno, K., Melvin, J.W., Li, L., Lighthall, J.W. (1995). Development of tissue level brain injury criteria by finite element analysis. *J. Neurotrauma* **12** (4).

Ueno, K., Melvin, J.W., Lundquist, E., Lee, M.C. (1989). Two-dimensional finite element analysis of human brain impact responses: Application of a scaling law. In: *Crashworthiness and Occupant Protection in Transportation Systems*. In: AMD **106** (The American Society of Mechanical Engineers, New York), pp. 123–124.

ULRICH, D. (1998). Evaluation of the mechanical properties of bone with consideration of its microarchitecture. Dissertation for the degree of doctor of the technical sciences of the Swiss Federal Institute of Technology, Zürich, Switzerland.

VANCE, T.L., SOLOMONOV, M., BARATTA, R., ZEMBO, M., D'AMBROSIA, R.D. (1994). Comparison of isometric and load moving length-tension models of two bicompartmental muscles. *IEEE Trans. Biomed. Engrg.* **41** (8), 771–781.

VERRIEST, J.-P., CHAPON, A. (1994). Validity of thoracic injury criteria based on the number of rib fractures. In: Backaitis, S.H. (ed.), *Biomechanics of Impact Injury and Injury Tolerances of the Thorax-Shoulder Complex* (Society of Automotive Engineers), pp. 719–727. SAE PT-45.

VIANO, D.C. (1986). Biomechanics of bone and tissue: A review of material properties and failure characteristics. In: *Symposium on Biomechanics and Medical Aspects of Lower Limb Injuries* (San Diego, California), pp. 33–63. SAE 861923.

VIANO, D.C. (1989). Biomechanical responses and injuries in blunt lateral impact. In: *Proc. 33rd Stapp Car Crash Conference*, pp. 113–142. Paper No. 892432.

VIANO, D.C., LAU, V.K. (1983). Role of impact velocity and chest compression in thoracic injury. *Aviat. Space Environ. Med.* **54** (1), 16–21.

VIIDIK, A. (1987). Properties of tendons and ligaments. In: Skalak, R., Chien, S. (eds.), *Handbook of Bioengineering* (McGraw-Hill, New York), pp. 6.1–6.19.

VIRGIN, W.J., LUDHIANA, P. (1951). Experimental investigations into the physical properties of the intervertebral disc. *J. Bone Joint Surgery B* **33** (4).

VISIBLE HUMAN PROJECT (1994) (public release of male dataset on CD Rom), National Library of Medicine. Visible Human Database. 8600 Rockville Pike, Bethseda, MD 20894.

VOIGT, M., BOJSEN-MOLLER, F., SIMONSEN, E.B., DYHRE-POULSEN, P. (1995). The influence of tendon Youngs modulus, dimensions and instantaneous moment arms on the efficiency of human movement. *J. Biomech.* **28** (3), 281–291.

VOO, L., KUMARESAN, S., PINTAR, F.A., YOGANANDAN, N., SANCES JR, A. (1996). Finite element models of the human head. *Medical Biological Engrg. Comput.* **34**, 375–381.

WAINWRIGHT, S.A., BIGGS, W.D., CURREY, J.D. (1979). *Mechanical Design in Organisms – Pliant Materials* (Edward Arnold, London), pp. 110–143.

WALKE, A.E., KOLLROS, J.J., CASE, T.J. (1944). The physiological basis of concussion. *J. Neurosurg.* **1**, 103–116.

WANK, V., GUTEWORT, W. (1993). Modelling and simulation of muscular contractions with regard to physiological parameters. In: *Proc. of the XIV International Society of Biomechanics Congress, Paris* (International Society of Biomechanics), pp. 1450–1451.

WARD, C.C. (1982). Finite element models of the head and their use in brain injury research. In: *Proc. 26*th *Stapp Car Crash Conference*, pp. 71–85. Paper No. 821154.

WARD, C.C., THOMSON, R.B. (1975). The development of a detailed finite element brain model. In: *Proc. 19th Stapp Car Crash Conference*, pp. 641–674. Paper No. 751163.

WATANABE, I., ISHIHARA, T., FURUSU, K., KATO, C., MIKI, K. (2001). Basic research of spleen FEM model for impact analysis. In: *Proc. of 2001 JSME Annual Congress*, pp. 115–116 (in Japanese).

WEAVER, J.K., et al. (1966). Cancellous bone: Its strength and changes with aging and an evaluation of some methods for measuring its mineral content. *J. Bone Joint Surgery A* **48**, 289–299.

WIRHED, R. (1985). *Anatomie et Science du Geste Sportif* (Edition VIGOT). ISBN 2-7114-0944-9.

WHITE, A.A., PANJABI, M.M. (1978). *Biomechanics of the Spine* (Lippincott, Philadelphia, PA).

WHITE, A.A., PANJABI, M.M. (1990). Clinical biomechanics of the spine. *J. Biomech.*, Lippincott Company, second ed.

WILLIAMS, J.L., et al. (1982). Properties and an anisotropic model of cancellous bone from the proximal tibia epiphysis. *J. Biomech. Engrg.* **104**, 50–56.

WILLINGER, R., KANG, H.S., DIAW, B. (1999). Three-dimensional human finite-element model validation against two experimental impacts. *Ann. Biomedical Engrg.* **27**, 403–410.

WILSON-MACDONALD, J., WILLIAMSON, D.M. (1988). Severe ligamentous injury of the ankle with ruptured tendo achillis and fractured neck talus. *J. Trauma* **28**, 872–874.

WINTERS, J.M., STARK, L. (1985). Analysis of fundamental human movement patterns through the use of in-depth antagonistic muscle models. *IEEE Trans. Biomedical Engrg.* **12**, 826–839.

WINTERS, J.M., STARK, L. (1988). Estimated mechanical properties of synergistic muscles involved in movements of a variety of human joints. *J. Biomech.* **21**, 1027–1041.

WISMANS, J.S., SPENNY, C. (1983). Performance requirements of mechanical necks in lateral flexion. In: *Proc. 27th Stapp Car Crash Conference*, pp. 56–65. Paper No. 831613.

WISMANS, J., VAN OORSCHOT, H., WOLTRING, H.J. (1986). Omni-directional human head–neck response. In: *Proc. 30th Stapp Car Crash Conference*, pp. 313–331. Paper No. 861893.

WISMANS, J.S.H.M., et al. (1994). *Injury Biomechanics* (Eindhoven University of Technology), p. 49.

WITTEK, A., HAUG, E., KAJZER, J. (1999). Hill-type muscle model for analysis of muscle tension on the human body response in a car collision using an explicit finite element code. *JSME Internat. J. Ser. A: Solid Material Mech.* **43**, 8–18.

WITTEK, A., KAJZER, J. (1995). A review and analysis of mathematical models of muscle for application in the modeling of musculoskeletal system response to dynamic load. In: Högfors, C., Andreasson, G. (eds.), *Proceedings of the 9th Biomechanics Seminar* (Center for Biomechanics Chalmers University of Technology and Göteborg University, Göteborg), pp. 192–216.

WITTEK, A., KAJZER, J. (1997). Modeling of muscle influence on the kinematics of the head–neck complex in impacts. *Mem. School Engrg. Nagoya University* **49**, 155–205.

WITTEK, A., KAJZER, J., HAUG, E. (1999). Finite element modeling of the muscle effects on kinematic responses of the head–neck complex in frontal impact at high speed. *Japan Soc. Automotive Engrg. Rev.*

WITTEK, A., ONO, K., KAJZER, J. (1999). Finite element modeling of kinematics and dynamic response of cervical spine in low-speed rear-end collisions: mechanical aspects of facet joint injury. *J. Crash Prevention Injury Control.*

WITTEK, A., ONO, K., KAJZER, J.,ÖRTENGREN, R., INAMI, S. (2001). Analysis and comparison of reflex times and electro-myograms of cervical muscles under impact loading obtained using surface and fine-wire electrodes. *IEEE Trans. Biomedical Engrg.* **48** (2), 143–153.

WOO, S.L.Y., JOHNSON, G.A., SMITH, B.A. (1993). Mathematical modelling of ligaments and tendons. *J. Biomech. Engrg.* **115**, 468–473.

WOO, S.L.Y., PETERSON, R.H., OHLAND, K.J., et al. (1990). The effects of strain rate on the properties of the medial collateral ligament. Skeletally Immature and Mature Rabbits: A Biomechanical and Histological Study. *J. Orthop. Res.* **8**, 712–721.

WYKOWSKI, E., SINNHUBER, R., APPEL, H. (1998). Finite element model of human lower extremities in a frontal impact. In: *IRCOBI Conference.*

YAMADA, H. (1970). In: EVANS, F.G. (ed.), *Strength of Biological Materials* (Williams &Wilkins, Baltimore, MD).

YANG, J. (1998). Bibliographic study. Report 3CHA/980529/T1/DB, Chalmers University of Technology, SE-41296 Göteborg, Sweden.

YANG, J., LÖVSUND, P. (1997). Development and validation of a human body mathematical model for simulation of car-pedestrian collisions, in: *IRCOBI Conference*, Hannover.

YANG, K.H., ZHU, F., LUAN, F., ZHAO, L., BEGEMAN, P. (1998). Development of a finite element model of the human neck. In: *Proc. 42nd Stapp Car Crash Conference*, pp. 1–11. Paper No. 983157.

YOGANANDAN, N., HAFFNER, M., MALMAN, D.J., NICHOLS, H., PINTAR, F.A., JENTZEN, J., WEINSHEL, S.S., LARSON, S.J., SANCES JR., A. (1989). Epidemiology and injury biomechanics of motor vehicle related trauma to the human spine. In: *Proc. 33rd Stapp Car Crash Conference*, pp. 22–242. Paper No. 892438.

YOGANANDAN, N., PINTAR, F.A. (1998). Biomechanics of human thoracic ribs. *J. Biomech. Engrg.* **120**, 100–104.

YOGANANDAN, N., SANCES JR., A., PINTAR, F. (1989b). Biomechanical evaluation of the axial compressive responses of the human cadaveric and manikin necks. *J. Biomech. Engrg.* **111**, 250–255.

YOGANANDAN, N., SRIRANGAM, K., PINTAR, F.A. (2001). Biomechanics of the cervical spine Part 2. Cervical spine soft tissue responses and biomechanical modeling. *Clinical Biomech.* **16**, 1–27.

YOO, W.-H., CHOI, H.-Y. (1999). Finite element modeling of human lower extremity. *Hongik J. Sci. Technol.* **3**, 187–212.

ZAJAC, F.E. (1989). Muscle and tendon: Properties, models, scaling and applications to biomechanics and motor control. *Critical Rev. Biomedical Engrg.* **17**, 359–411.

ZEIDLER, F., STÜRTZ, G., BURG, H., RAU, H. (1981). Injury mechanisms in head-on collisions involving glance-off. In: *Proc. 25th Stapp Car Crash Conference*, pp. 825–860. Paper No. 811025.

ZINK, L. (1997). Simulation von Unfällen mit Airbagauslösung für den Out-of-Position Fall. Diploma Thesis prepared at TRW Germany, University of Stuttgart, IVK.

ZHANG, L., BAE, J., HARDY, W.N., MONSON, K.L., MANLEY, G.T., GOLDSMITH, W., YANG, K.H., KING, A. (2002). Computational study of the contribution of the vasculature on the dynamic response of the brain. *Stapp Car Crash J.* **46**, 145–165. Paper No. 2002-22-0008.

ZHANG, L., YANG, K.H., DWARAMPUDI, R., OMORI, K., LI, T., CHANG, K., HARDY, W.N., KHALIL, T.B., KING, A.I. (2001). Recent advances in brain injury research: A new human head model development and validation. In: *Proc. 45th Stapp Car Crash Conference*, pp. 369–394. Paper No. 2001-22-0017.

ZHOU, C., KHALIL, T.B., KING, A.I. (1996). Viscoelastic response of the human brain to sagittal and lateral rotational acceleration by finite element analysis. In: *International IRCOBI Conference on the Biomechanics of Impact, Dublin, Ireland*.

ZHOU, Q., ROUHANA, S.W., MELVIN, J.W. (1996). Age effects on thoracic injury tolerance. In: *Proc. 40th Stapp Car Crash Conference*, pp. 137–148. Paper No. 962421.

ZUURBIER, C.J., EVERARD, P., VAN DER WEES, A.J., HUIJING, P.A. (1994). Length-force characteristics of the aponeurosis in the passive and active muscle condition and in the isolated condition. *J. Biomech.* **27** (4), 445–453.

CHAPTER III

Soft Tissue Modeling for Surgery Simulation

Hervé Delingette, Nicholas Ayache

INRIA Sophia–Antipolis, 2004, route des Lucioles,
BP 93, 06902 Sophia–Antipolis, France
E-mail addresses: Herve.Delingette@sophia.inria.fr (H. Delingette),
Nicholas.Ayache@inria.fr (N. Ayache)

Foreword

In this chapter, we address a specific issue belonging to the field of biomechanics – modeling living tissue deformation with real-time constraints. This issue was raised by the emergence, in the middle of the 1990s, of a very specific application – the simulation of surgical procedures. This new concept of surgery simulation was in large part advocated by the American Department of Defense (SATAVA [1994]), for which this concept was a key part of their vision of the future of emergency medicine.

Since then, the concept of having surgeons being trained on simulators (just like pilots on flight simulators) has been refined. First, the development of commercial simulators has proved that there was a demand for products that help to optimize the learning curve of surgeons.[1] Second, the emergence of medical robotics and more precisely of minimally invasive surgery robots, has reinforced the need for simulating surgical procedures, since these robots require a very specific hand–eye coordination. Third, there is a large consensus among the medical community that current simulators are not realistic enough to provide advanced gesture training. In particular, the modeling of living tissue, and their ability to deform under the contact of an instrument is one of the important aspect of simulators that should be improved.

[1] This curve represents the number of incidents occurring during the performance of surgery as a function of time. This curve is generally monotonically decreasing under the effect of training and usually reaches an asymptotic value after a certain number N of real interventions. The objective of the simulators is to reduce this number N as much as possible.

Essential Computational Modeling for the Human Body
Special Volume (N. Ayache, Guest Editor) of
HANDBOOK OF NUMERICAL ANALYSIS, VOL. XII
P.G. Ciarlet (Editor)

ISSN 1570-8659
DOI 10.1016/B978-0-444-53755-3.00003-2

In this chapter, we present different algorithms for modeling soft tissue deformation in the context of surgery simulation. These algorithms make radical simplifications about tissue material property, tissue visco-elasticity and tissue anatomy. The first section of this chapter describes the principles and the components of a surgical simulator. In particular, we insist on the different constraints of soft tissue models in this application, the most important being the real-time computation constraint. In Section 2, we present the process of building a patient-specific hepatic surgery simulator from a set of medical images. The different stages of computation leading to the creation of a volumetric tetrahedral mesh from a medical image are especially emphasized. In Section 3, we detail the five main hypotheses that are made in the proposed soft tissue models. Furthermore, we recall the main equations of isotropic and transversally anisotropic linear elasticity in continuum mechanics. The discretization of these equations are presented in Section 4 based on finite element modeling. Because we rely on the simple linear tetrahedron element, we provide closed form expressions of local and global stiffness matrices. After describing the types of boundary conditions existing in surgery simulation, we derive the static and dynamic equilibrium equations in their matrix form. In Section 5, a first model of soft tissue is proposed. It is based on the off-line inversion of the stiffness matrix and can be computed very efficiently as long as no topology change is required. In such case, in Section 6, a second soft tissue model allows to perform cutting and tearing but with less efficiency as the previous model. A combination of the two previous models, called "hybrid model" is also presented in this section. Finally, in Section 7, we introduce an extension of the second soft tissue model that implements large displacement elasticity.

1. General issues in surgery simulation

1.1. Surgical simulators

1.1.1. Medical impact of surgical simulators

Surgery simulation aims at reproducing the visual and haptic senses experienced by a surgeon during a surgical procedure, through the use of computer and robotics systems. The medical interest of this technology is linked with the development of minimally invasive techniques especially video-surgery (endoscopy, laparoscopy, ...). More precisely, laparoscopy consists in performing surgery by introducing different surgical instruments in the patient abdomen through one centimeter-wide incisions. The surgeon can see the abdominal anatomy with great clarity by watching a high resolution monitor connected to an endoscope introduced inside the patient abdomen. This technique bears several advantages over traditional open surgery. On one hand, it decreases the trauma entailed by the surgical procedure on the patient body. This allows to decrease the patient stay in hospitals and therefore decreases the cost of health care. On the other hand, it reduces the morbidity as demonstrated by the Hunter and Sackier study (BERCI, HUNTER and SACKIER [1994]).

However, if these minimally invasive techniques are clearly beneficial to the patients, they also bring new constraints on the surgical practice. First, they significantly degrade the surgeon access to the patient body. In laparoscopy, for instance, the surgical procedure is made more complex by the limited number of degrees of freedom of each surgical instrument. Indeed, they must go through fixed points where the incisions in the patient's abdomen were done. Furthermore, because the surgeon cannot see his hand on the monitor, this technique requires a specific hand–eye coordination. Therefore, an important training phase is required before a surgeon acquires the skills necessary to adequately perform minimally invasive surgery (corresponding to a plateau in the learning curve).

Currently, surgeons are trained to perform minimally invasive surgery by using mechanical simulators or living animals. The former method is based on "endotrainers" representing an abdominal cavity inside which are placed plastic objects representing human organs. These systems are sufficient for acquiring basic surgical skills but are not realistic enough to represent fully the complexity of the human anatomy and physiology (respiratory motion, bleeding, ...). The latter training method consists in practicing simple or complex surgical procedures on living animals (often pigs for abdominal surgery). This method has two limitations. First, the similarity between the human and animal anatomy is limited and therefore certain procedures cannot be precisely simulated with this technique. Also, the evolution of the ethical code in most countries may forbid the use of animals for this specific training, as it is already the case in several European and North American countries.

Because of the limitations of current training methods, there is a large interest in developing video-surgery simulation software for providing efficient and quantitative gesture training systems (AYACHE and DELINGETTE [2003]). Indeed, such systems should bring a greater flexibility by providing scenarios including different types of pathologies. Furthermore, thanks to the development of medical image reconstruction algorithms, surgery simulation allows surgeons to verify and optimize the surgical strategy of a procedure on a given patient.

1.1.2. Classification of surgical simulators

SATAVA [1996] et al. proposed to classify surgical simulators into three categories (see Fig. 1.1). The first generation simulators are solely based on anatomical information, in particular on the geometry of the anatomical structures included in the simulator. In these simulators, the user can virtually navigate inside the human body but has a limited interaction with the modeled organs. Currently, several first generation surgical simulators are available including commercial products linked to medical imaging systems (CT or MRI scanners) that are focusing on virtual endoscopy (colonoscopy, tracheoscopy, ...). In general, they are used as a complementary examination tools establishing a diagnosis (for instance, when using virtual endoscopy) or as a surgical planning tool before performing surgery.

In addition to geometrical information, second generation simulators describe the physical properties of the human body. For instance, the modeling of soft tissue biomechanical properties enables the simulation of basic surgical gestures such as cutting or suturing. Currently, several prototypes of second generation simulators are being developed including the simulation of cholecystectomy (COVER, EZQUERRA and O'BRIEN [1993], KUHN, KÜHNAPFEL, KRUMM and NEISIUS [1996]), of arthroscopy of

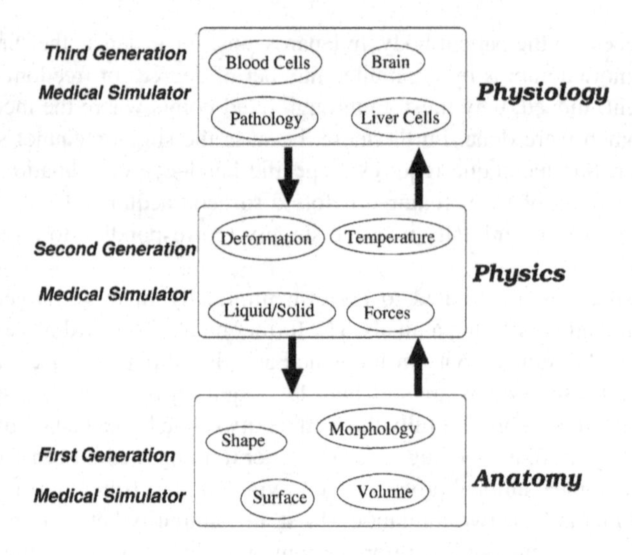

FIG. 1.1. The different generations of medical simulators.

the knee (GIBSON, SAMOSKY, MOR, FYOCK, GRIMSON, KANADE, KIKINIS, LAUER and MCKENZIE [1997]) and of gynecological surgery (SZEKELY, BAIJKA and BRECHBUHLER [1999]). Section 2 will shortly describe the hepatic surgery simulator being developed at INRIA.

Third generation of surgical simulators provides an anatomical, physical and physiological description of the human body. There are very few simulators including these three levels of modeling, essentially because of the difficulty to realistically describe the coupling between physiology and physics. A good example of an attempt in this direction is given by the work of KAYE, PRIMIANO and METAXAS [1997] who modeled the mechanical cardiopulmonary interactions. Another important example is the study of the contraction of the right and left ventricles of the heart under the propagation of the action potential which is being carried out by the group of Prof. McCulloch (this work is published in this book) but also by the INRIA ICEMA group (SERMESANT, COUDIÈRE, DELINGETTE and AYACHE [2002], SERMESANT, FARIS, EVANS, MCVEIGH, COUDIÈRE, DELINGETTE and AYACHE [2003]). Finally, it should be noted that a comprehensive effort for creating computational physiological models has been recently launched in the international Physiome Project (BASSINGTHWAIGHTE [2000]).

1.2. Simulator architecture

In this section, we present the basic components of simulators for surgical gesture training and especially in the context of minimally invasive therapy. For the acquisition of basic skills, it is necessary to simulate the behavior of "living" tissues and therefore to develop a second generation surgical simulator. However, it raises important technical and scientific issues. The different components of these simulators are summarized in Fig. 1.2.

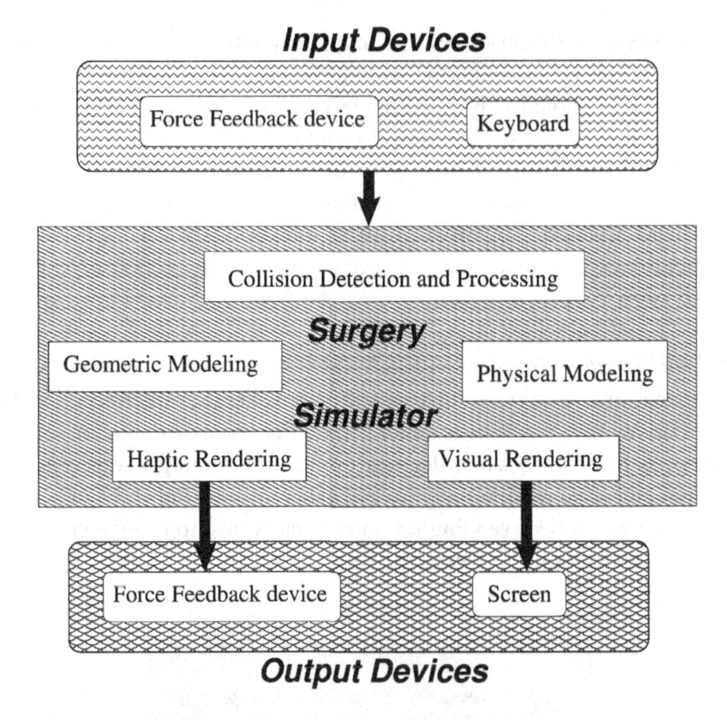

FIG. 1.2. The different components of a second generation surgery simulator.

The input devices in such simulators usually consist of one or several mechanical systems that drive the motion of virtual surgical tools or of virtual endoscopes. In fact, as input devices they do not need to be motorized and they are usually equipped with simple optical encoders or position trackers. A keyboard and electronic mouse are also useful to modify the scenario of the simulation.

The core of a simulator consists of several modules. For instance, a first module provides the enabling tools for the creation of geometric models from medical images (see Section 1.2.1). Another module, detailed in Section 1.3, computes the deformation of soft tissues under the action of virtual instruments. These interactions between virtual instruments and virtual organs, performed in a separate module, mainly consists in detecting collisions followed with modeling contact forces and displacements (see Section 1.2.2).

Finally, a surgical simulator must provide an advanced user interface that includes visual and force feedback (respectively presented in Sections 1.2.3 and 1.2.4). Last but not least, it is necessary to rely on advanced software engineering methodology to make these different modules communicate within the same framework: some of these implementation issues are introduced in Section 1.2.5.

1.2.1. Geometric modeling

In general, the extraction of tridimensional geometric models of anatomical structures is based on medical imagery: CT scanner images, MRI images, cryogenic images, 3D ultrasound images,.... Because medical image resolution and contrast have greatly improved over the past few years, the tridimensional reconstruction of certain structures have become possible by using computerized tools. For instance, the availability in 1995 of the "Visible Human" dataset provided by the National Library of Medicine has allowed the creation of a complete geometric human model (ACKERMAN [1998]). However, the automatic delineation of structures from medical images is still considered an unsolved problem. Therefore, a lot of human interaction is usually required for reconstructing the human anatomy. DUNCAN and AYACHE [2000], AYACHE [2003], provide a survey on the past and current research effort in medical image analysis.

1.2.2. Interaction with a virtual instrument

A key component of a surgery simulation software is the user interface. The hardware interface that drives the virtual instrument essentially consists in one or several force-feedback devices having the same degrees of freedom and appearance as the actual surgical instruments used in minimally invasive therapy (see Fig. 1.3). In general, these systems are force-controlled, sending the instrument's position to the simulation software and receiving reaction force vectors.

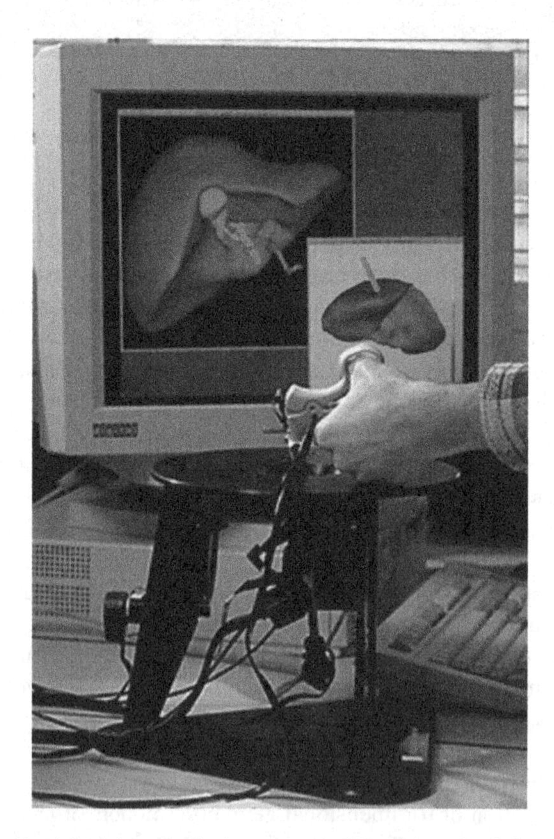

FIG. 1.3. A force feedback system suited for surgery simulation.

Once the position of the virtual instrument is known, it is necessary to detect possible collisions with other instruments or surrounding anatomical structures. In this case, it is particularly difficult to obtain a computationally efficient collision detection algorithm because the geometry of objects may change at each iteration. Therefore, algorithms based on pre-computed data structures (such as the approach proposed in GOTTSCHALK, LIN and MANOCHA [1996]) are not appropriate. LOMBARDO, CANI and NEYRET [1999] proposed an original collision detection method based on the OpenGL graphics library which is especially well-suited for elongated instruments shaped like those used in laparoscopic surgery. Although this technique cannot be used for the detection of self-collisions, several algorithms have been proposed recently (TESCHNER, HEIDELBERGER, MULLER, POMERANETS and GROSS [2003], KNOTT and PAI [2003]) to tackle this complex task.

When a collision is detected, a set of geometrical or physical constraints are applied on soft tissue models. However, modeling the physics of contacts can lead to complex algorithms and therefore purely geometric approaches are often preferred.

1.2.3. Visual feedback

A surgery simulator must provide a realistic visual representation of the surgical procedure. Visual feedback is especially important in video-surgery because it helps the surgeon to acquire a tridimensional perception of his environment. In particular, the effects of shading, shadows and textures are important clues that must be reproduced in a simulator.

The quality of visual feedback is directly related to the availability and performance of graphics accelerators. In the past few years, the market of graphics cards has evolved in three directions: improved price-performance ratio, increased geometric transformation and rasterization performance and the emergence of programmable pixel rendering. Combined with the development of more efficient computer graphics algorithms, we can foresee that realistic visual feedback for surgery simulation could be achieved in the next few years if this additional graphics rendering is focused on the three-dimensional clues used by surgeons to understand the surgical field.

1.2.4. Haptic feedback

Haptic display serves at least two purposes in a surgical simulator: kinesthetic and cognitive. First, it provides the sensation of movement to the user and therefore it significantly enhances surgical performance. Second, it helps to distinguish between tissues by testing their mechanical properties.

However, the addition of a haptic display in a simulation system increases by a large factor its complexity and the required computational power (MARK, RANDOLPH, FINCH, VAN VERTH and TAYLOR II [1996]): it leads to an increase by a factor 10 of the required bandwidth, synchronization between visual and haptic displays, force computation, ... Only a few papers have assessed the importance of haptic feedback in surgery (MARCUS [1996]). In general, it is accepted that the combination of visual and haptic displays is optimal for surgery training or pre-planning.

In video-surgery, the surgical instruments slide inside a trocard and are constrained to enter the abdomen through a fixed point. This entails substantial friction, specifically for laparoscopy where airtightness must be enforced. The friction of the instruments inside trocards perturbes the forces sensed by the end-user. Despite those perturbations, it appears that it is still necessary to provide force-feedback for realistic user immersion.

1.2.5. Implementation of a simulator

Most of the difficulties encountered when implementing a surgical simulator originate from the trade-off that must be found between real-time interaction and the necessary surgical realism of a simulator.

The first constraint indicates that a minimum bandwidth between the computer and the interface devices must be available in order to provide a satisfactory visual and haptic feedback. If this bandwidth is too small, the user cannot properly interact with the simulator and it becomes useless for surgery gesture training. However, the "real-time" constraint can be interpreted in different ways. Most of the time, it implies that the mean update rate is high enough to allow a suitable interaction. However, it is possible that during the simulation, some events (such as the collision with a new structure) may increase the computational load of the simulation engine. This may result in a lack of synchronicity between the user gesture and the feedback the user gets from the simulator. When the computation time is too irregular, the user may even not be able to use the simulator properly. In order to guarantee a good user interaction, it is necessary to use a dedicated "real-time" software that supervises all tasks executed on the simulator.

The second constraint is related to the targeted application of a simulator: training surgeons to new gestures or procedures. To reach this goal, the user must "believe" that the simulator environment corresponds to a real procedure. The level of realism of a simulator is therefore very dependent on the type of surgical procedures and is also connected with physio-psychological parameters. In any case, increasing the realism of a simulator requires an increase of computational time which is contradictory with the constraint of real-time interaction.

The main difficulty in implementing a simulator is to optimize its credibility, given an amount of graphics and computational resources. Therefore, an analysis of the training scenario should be performed to find the most important elements that contribute to the realism of the simulation.

1.3. Constraints of soft tissue models

In the scope of a surgical simulator, it is not possible to model the biomechanical complexity of living soft tissue. Instead, authors have resorted to simplified models to decrease the implementation complexity and to optimize computational efficiency. A survey on soft tissue modeling can be found in DELINGETTE [1998].

Before presenting the main features of our approach (available in Section 3.1), we list three constraints that should be taken into account when specifying a soft tissue model for surgery simulation.

1.3.1. Visualization constraints

To obtain high quality visual rendering, two techniques are traditionally used: surface and volume rendering. A comparison between these two rendering techniques for surgery simulation is described in SOFERMAN, BLYTHE and JOHN [1998]. Surface rendering is by far the most commonly used technique, and uses basic polygonal elements (triangles, quads, ...) to achieve the rendering of an entire scene. A rule of thumb in surface rendering states that the quality of rendering is proportional to the number of polygonal elements. Unfortunately, the screen refresh rate of a graphics display is inversely proportional to the number of elements.

Therefore, an important concern arises when specifying a soft tissue model: is it compatible with high quality visual rendering? For some models, it is clearly not the case. For instance, the chain–mail algorithm (GIBSON, SAMOSKY, MOR, FYOCK, GRIMSON, KANADE, KIKINIS, LAUER and MCKENZIE [1997]) represents soft tissue with the help of cubic lattices that are allowed to move slightly with respect to their neighbors. For this representation, as well as for particle-based representations (FRANCE, ANGELIDIS, MESEURE, CANI, LENOIR, FAURE and CHAILLOU [2002], DESBRUN and GASCUEL [1995]) and multigrid representations (DEBUNNE, DESBRUN, CANI and BARR [2001]), authors use a two-layer strategy: a volumetric soft tissue model is combined with a surface model dedicated to visual rendering. These two models are often coupled with a linear relationship based on barycentric coordinates: once the shape of a soft tissue model is modified, the surface model is updated in an efficient manner. Similarly, the collision detection is performed on the surface model, but contact forces and displacements are imposed on the volumetric model. However, this approach has two limitations. First, the modeling of contact between a virtual tool and a soft tissue model is usually not satisfactory because the mapping between surface and volumetric model is complex (though mapping from volumetric to surface models is often trivial). Second, this approach makes the modeling of tissue cutting very complex where the surface and volumetric topology is altered.

For soft tissue models based on tetrahedral or hexahedral meshes, the problem of high quality visual rendering is posed in a different manner since the shell of these meshes (made of triangular or quadrangular elements) can be used directly for rendering. However, in general, coarse volumetric meshes are used in order to achieve real-time performances (see next section). Therefore, it is often required to compensate the poor geometrical quality by using specific computer graphics algorithms such as subdivision surfaces (ZORIN, SCHROEDER and SWELDENS [1996]), using avatars (DECORET, SCHAUFLER, SILLION and DORSEY [1999]) or by replacing elements with texture (SILLION, DRETTAKIS and BODELET [1997]). In the case of the hepatic surgery simulator, we have used the PN triangles algorithms (VLACHOS, PETERS, BOYD and MITCHELL [2001]) in order to provide a smooth visual rendering of the liver. The idea behind PN triangles is to subdivide each triangle and its normal vector into subtriangles in order to produce a smoother looking surface (see Fig. 1.4 for an example).

1.3.2. Real time deformation constraints

A surgical simulator is an example of a virtual reality system. To succeed in training surgeons, a simulator must provide an advanced user interface that leads to the immersion of surgeons into the virtual surgical field. To reach this level of interaction, three basic rules must be formulated:

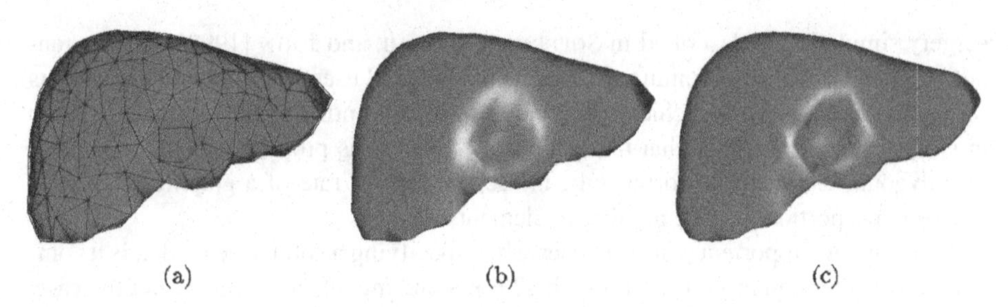

FIG. 1.4. Display of a liver being resected: (a) display of the triangles corresponding to the shell of the liver tetrahedral mesh; (b) surface rendering based on Gouraud shading without PN triangles; (c) surface rendering based on PN triangles with two levels of subdivisions.

Rule 1. *Minimum bandwidth for visual and haptic feedback.* An acceptable bandwidth for visual display is in the range of 20–60 Hz while the acceptable bandwidth for haptic display is on the range of 300–1000 Hz (300 Hz is the free-hand gesture frequency). In fact, this notion of minimal bandwidth depends on the nature of the scene to be displayed: for objects moving slowly on the screen, an update rate of 20 Hz is sufficient. Similarly, a frequency of 300 Hz may be enough to render the contact with very soft objects.

Rule 2. *Low latency.* Latency measures the time between measurements performed by the sensor (for instance, the position of the surgical instrument) and action (visual or haptic display). Latency is critical for user immersion. The hardware configuration of the system can greatly influence latency since communication between elements may be responsible for additional delays. In Fig. 1.5, the architecture of the simulation system used at INRIA (COTIN, DELINGETTE, CLEMENT, TASSETTI, MARESCAUX and AYACHE [1996]) in 1996 is presented. It is composed of one haptic display, a personal computer and a graphics workstation. There are several causes contributing to latency: communication delays between the haptic display and the PC, communication between the PC and the graphics workstation, the delay caused by the graphics display, the computation time for collision detection, force feedback and deformation. Since some of the communication links between elements are asynchronous, the total latency is not the sum of those delays but it is important to reduce them to their minimum values. The latency depends greatly on hardware, specifically on computation and graphics performance.

Rule 3. *Realistic motion of soft tissue.* It is important that the dynamic behavior of a deformable tissue is correctly simulated. To assess the visco-elastic behavior of a material, one can measure the speed at which an object returns to its rest position after it is perturbed. Soft tissue undergoes a damped motion whereas stiff objects react almost instantaneously to any perturbation. At the limit, very stiff objects can be considered to have a quasi-static motion, implying that static equilibrium is reached at each time-step (see Section 5 for a discussion about quasi-static motion).

In terms of soft tissue modeling, two parameters are important for real-time deformation constraints. The first parameter is the *update frequency* f_{u} which controls the rate at which the shape of a soft tissue model is modified. If we write X_t as the position of

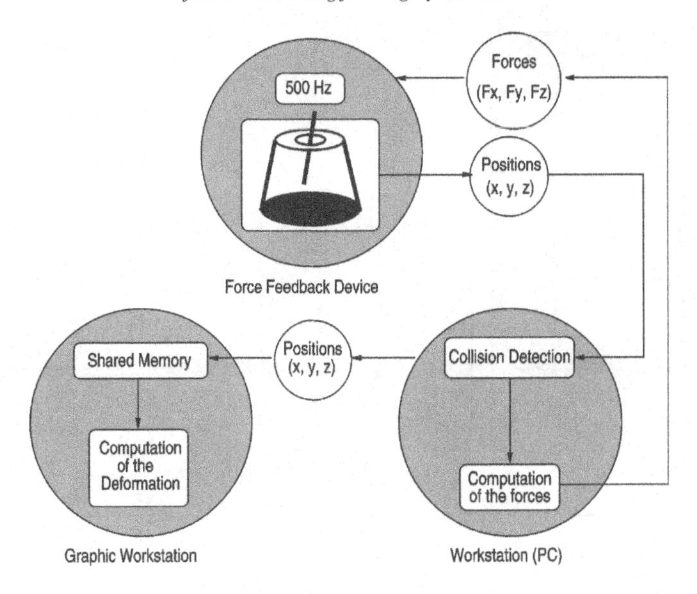

FIG. 1.5. Architecture of a simulator composed of a personal computer driving an haptic device and a graphics workstation.

the tissue model at iteration t, the *computation time* $T_c = 1/f_u$ is the time needed to compute the new position X_{t+1}. The second parameter is the *relaxation time* $T_{\text{relaxation}}$ which is the time needed for a material to return to its rest position once it has been perturbed.

To reach the required bandwidth for haptic and visual rendering (Rule 1) it is necessary that the computation time T_c is bounded by a constant $T_{\text{interaction}}$ that depends on the architecture of the simulator. For instance, in Fig. 1.6 we display three different software architectures for handling soft tissue deformation, visual and haptic feedback.

In a first architecture (Fig. 1.6(a)), all three tasks are performed sequentially, one after the other. The advantage of this approach lies in its simplicity of implementation. However, it has two drawbacks. The main problem is that the computation time T_c must be short enough to follow the minimum frequency for haptic feedback: 300 to 1000 Hz. This implies that $T_{\text{interaction}} \approx 2$ ms which is a very high requirement for a soft tissue model of reasonable size. In fact, to the best of our knowledge, only methods based on pre-computation of the static solution such as the one proposed in Section 5 can comply with this constraint. The second problem with this approach is that a delay in any of the three tasks perturbs the other tasks. For instance, when the user performs tissue cutting, an additional task is needed to update the mesh topology which translates into a delay in the visual and haptic rendering.

The second architecture shown in Fig. 1.6(b) is the most commonly used in today's surgical simulators: the haptic rendering is performed in a different process or different thread than the visual rendering and soft tissue modeling tasks. Its purpose is to sharply decrease the real-time constraint on the soft tissue computation from haptic frequency (≈ 500 Hz, $T_{\text{interaction}} \approx 2$ ms) to visual frequency (≈ 25 Hz, $T_{\text{interaction}} \approx 40$ ms). In

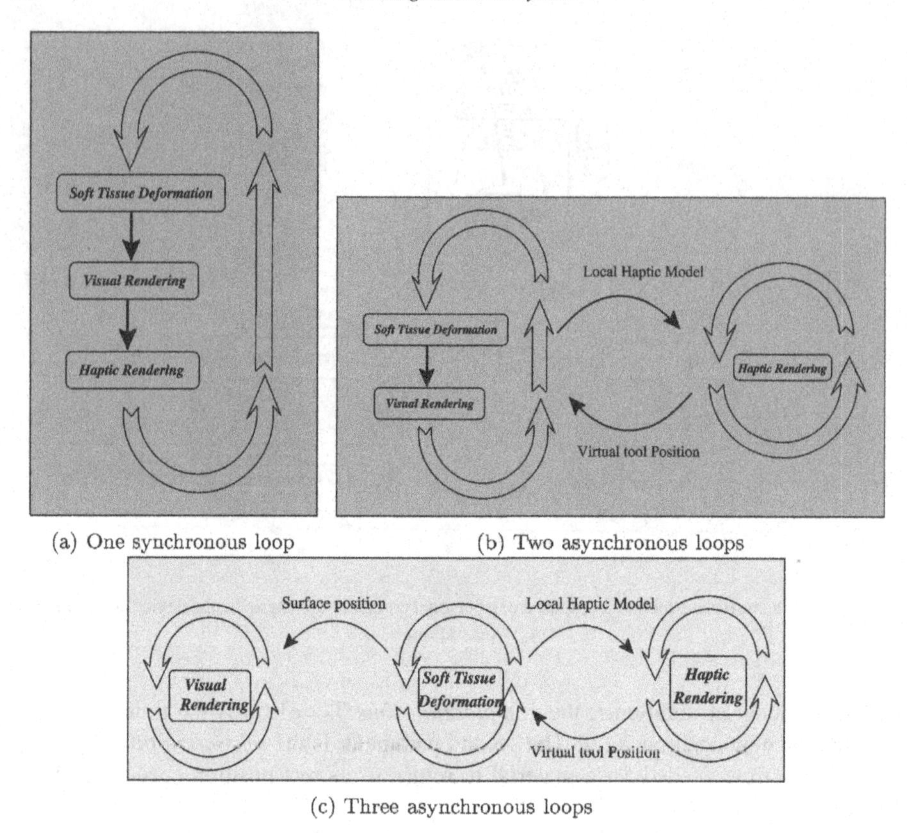

(a) One synchronous loop (b) Two asynchronous loops

(c) Three asynchronous loops

FIG. 1.6. Different software architecture for handling visual rendering, haptic rendering and soft tissue modeling.

order to keep a satisfactory force feedback, a separate thread or process, running at haptic frequency, computes the force intensity for the haptic device based on a simplified local model. This local model, that may consist of a sphere (SERRANO and LAUGIER [2001]) or a plane (FOREST, DELINGETTE and AYACHE [2002a]) is updated by the soft tissue modeling loop while the position of the virtual surgical tool, necessary to compute its contact with soft tissue, is updated by the haptic rendering process and sent to the process. This asynchronous communication between haptic and visual rendering gives satisfactory results when some temporal smoothing is performed during the computation of force intensity. The main drawback of this approach is that it increases software complexity compared to the previous architecture. However, since only little information must be shared between the two processes, it has been adopted in several simulators, including the current version of the INRIA hepatic surgery simulator.

In the third architecture described in Fig. 1.6(c), the visual and haptic rendering tasks are performed in separate processes or threads in order to remove the latency caused by graphics hardware. Furthermore, this architecture makes the computation of soft tissue deformation more efficient (decrease of T_c) when compared to previous solutions. However, it has little effect on the maximum computation time per iteration $T_{\text{interaction}} \approx 40$ ms since the geometric model still requires to be updated at 25 Hz for visual rendering. This approach is more difficult to implement because the amount of information to transmit to the visual rendering task is quite large. Furthermore, a change in mesh topology during simulation requires to modify the data structure of the computer graphics algorithm. An example of such architecture is provided in BIELSER and GROSS [2000].

To summarize, we can state that a soft tissue model in a surgical simulator must essentially meet two constraints: computation time T_c per iteration less than a constant $T_{\text{interaction}}$, and relaxation time $T_{\text{relaxation}}$ defined by the visco-elastic behavior of the material.

1.3.3. Tissue cutting and suturing

The ability to cut and suture tissue is of primary importance for designing a surgery simulation system. The impact of such operations in terms of tissue modeling is considerable since it implies changing tissue topology over time. The cost of such a topological change depends largely on the chosen geometric representation but also on the numerical method that is adopted to compute tissue deformation (see discussion in previous section).

In addition, the tissue behavior must be adapted at locations where cutting or suturing occurs. Little is known about the stress/strain relationship occurring during and after cutting. The basic assumption that is made is that the physical properties of tissue are only modified locally. However, in practice, cutting can modify the boundary conditions significantly between tissue and the surrounding organs which implies considerable change with respect to their ability to deform.

Finally, when cutting volumetric or surface models, it is very likely that the new geometric and physical representation of tissue leads to self-intersections. The detection of self-intersections is very computationally expensive and, therefore repulsive force between neighboring vertices are sometimes added to prevent self-intersections.

1.4. Computational methods for soft tissue modeling

Several computational methods can be employed for modeling the deformation of soft tissue. We simplify the taxonomy of these methods by proposing the three classes of algorithms most commonly used:

- *Direct methods.* This category contains all methods that solve the static or dynamic equilibrium equation at each iteration (quasi-static motion). To reach such performance, some kind of pre-computation is performed. The algorithm presented in Section 5 is a direct method as well as the algorithm described in DEBUNNE, DESBRUN, CANI and BARR [2001], RADETZKY [1998].
- *Explicit iterative methods.* With iterative methods, the deformation is computed as the limit (in finite time) of a converging series that have been initialized. The closer the initial value is from the solution the faster the convergence. Iterative methods can be implemented based on implicit or explicit schemes. With

TABLE 1.1
Comparison between the three soft tissue models: direct methods (pre-computed quasi-static model), explicit iterative schemes (tensor–mass and spring–mass models) and implicit iterative schemes (Houbolt or Newmark methods)

	Direct methods	Explicit iterative scheme	Implicit iterative scheme
Computation time	low	low	high
Relaxation time	low	high	low
Cutting simulation	very difficult	possible	difficult

explicit schemes, the next position of the tissue model X_{t+1} is obtained from the application of internal forces estimated at iteration t. These methods encompass the most common algorithms found in the literature for modeling soft tissue deformation, including spring–mass model (KUHNAPFEL, AKMAK and MAA [2000]), tensor–mass models (COTIN, DELINGETTE and AYACHE [2000]) (presented in Section 6), the "chain–mail" algorithm (GIBSON, SAMOSKY, MOR, FYOCK, GRIMSON, KANADE, KIKINIS, LAUER and MCKENZIE [1997]) and others (BRO-NIELSEN [1998]).

- *Semi-implicit iterative methods.* With implicit or semi-implicit schemes, the next position of the tissue model X_{t+1} is obtained from the application of internal forces estimated at iteration $t + 1$. Therefore, a linear system of equations needs to be solved entirely or partially (BARAFF and WITKIN [1998]).

In Table 1.1, we present the general features of these three types of numerical methods with respect to the constraints enumerated in Section 1.3. More precisely, the time of computation, the relaxation time (inversely proportional to the speed of convergence towards the rest position) and the ability to support any change of mesh topology during the simulation of cutting or suturing is estimated qualitatively for each method.

Direct methods can support high frequency update f_{u} and may have a low relaxation time to model stiff material, but they cannot simulate tissue cutting since they rely on the precomputation of some parameters.

On the other hand, explicit iterative methods are well-suited for the simulation of cutting, but these method often suffer from a high relaxation time, which makes their dynamic behavior somewhat unrealistic (jelly-like behavior). This high relaxation time originates from a lack of synchronicity, where the time step Δt used in the discretization of the explicit scheme, is much smaller than the computation time T_{c}. To obtain satisfactory results, it is often required to use a mesh with a small number of nodes (typically less than 1000 vertices on a standard PC).

Finally, with implicit iterative methods, the time step Δt can be increased by an order of magnitude compared to the explicit case. This allows to obtain much better dynamical behavior, but, on the other hand, they suffer from a higher computation time than explicit methods since a (sparse) linear system of equations needs to be solved at each iteration. Again, to achieve real-time performance, these methods are limited to meshes with a small number of vertices.

2. The INRIA hepatic surgery simulator

2.1. *Objectives*

In the sequel we use the hepatic surgery simulator developed at INRIA in the Epidaure project[2] as a case study to illustrate the different algorithms and the practical issues involved when building soft tissue models.

The INRIA hepatic surgery simulator was initiated in 1995 as a part of the European project MASTER in collaboration with the IRCAD research center[3] which hosts the European Institute of Tele-Surgery (EITS). The motivations that led us to propose the development of an hepatic surgery simulator were twofold.

First, hepatic pathologies are among the major causes of death worldwide. For instance, hepatocellular carcinoma (HCC) is a primary liver cell cancer and it accounts for most of cancer tumors. It causes the death of 1 250 000 people mainly in Asia and Africa. Furthermore, hepatic metastases (secondary tumorous cells) are mainly caused by colorectal cancers (in 30 to 50% of cases) and patients have little chance to survive hepatic carcinoma without any therapy (0 to 3% of survival for a 5 year period with an average survival period of 10 months).

The second motivation is related to the nature of hepatic resection surgery. Indeed, this surgical procedure involves many generic surgical gestures (large displacement motion, grasping, cutting, suturing) that can be useful in the simulation of different procedures. Also, because of the presence of hepatic parenchyma, the tissue models must be of volumetric nature which departs significantly from previously developed simulators simulating hollow organs like the gall bladder. Finally, tissue being a soft material allows to employ low-end force feedback systems for simulating contact forces between surgical tools and hepatic tissue.

This work has greatly benefited from the INRIA incentive action AISIM[4] which gathered different INRIA teams working in the fields of medical image analysis (Epidaure), robotics (Sharp) (Boux de Casson and Laugier [1999]), computer graphics (Imagis) (Debunne, Desbrun, Cani and Barr [2001]) and numerical analysis (Sinus, Macs) (Vidrascu, Delingette and Ayache [2001]).

2.2. *Liver anatomy*

The liver is the largest gland (average length of about 28 cm, average height of about 16 cm and average greatest thickness of about 8 cm) in the human body. It has numerous physiological functions: to filter, metabolize, recycle, detoxify, produce, store and destroy. It is located in the right hypochondriac and epigastric regions (see Fig. 2.1). The liver has a fibrous coat, the so-called Glisson's capsule. Its rheological behavior is quite different from the glandular parenchyma. Five vessel types run through the liver parenchyma: biliary and lymphatic ducts on one hand, blood vessels (internal portal

[2]Description of the Epidaure project is provided at http://www.inria.fr/epidaure/.

[3]Institut de Recherche Centre le Cancer de l'Appareil Digestif, 1, place de l'Hôpital, 67091 Strasbourg cedex, France, http://www.ircad.com/, funded by Prof. J. Marescaux.

[4] http://www-sop.inria.fr/epidaure/AISIM/.

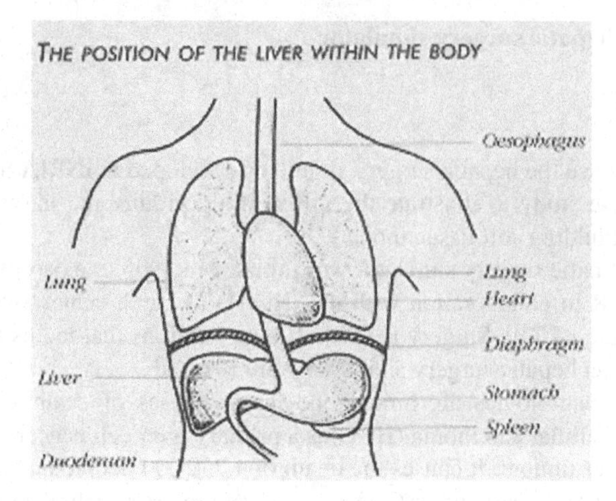

FIG. 2.1. Description of the liver anatomy with its neighboring structures (source Children's Liver Disease Foundation).

supply, hepatic arterial tree and collecting venous network) on the other hand. The portal vein, which conveys blood from the digestive tract to be detoxified and metabolized, is deep to the proper hepatic artery and common bile duct. This hepatic triad runs to the liver; it enters the liver via the hilum. This region is thus supposed to be wholly stable.

2.3. *Creation of an anatomical model of the liver*

In order to produce a model of the liver with anatomical details, the Visible Human dataset (ACKERMAN [1998]) provided by the *National Library of Medicine* was used. This dataset consists of axial MRI images of the head and neck and longitudinal sections of the rest of the body. The CT data consists of axial scans of the entire body taken at 1 mm intervals. The axial anatomical images are scanned pictures of cryogenic slices of the body. They are 24-bit color images whose size is 2048×1216 pixels. These anatomical slices are also at 1 mm interval and are registered with the CT axial images. There are 1878 cross-sections for each modality.

To extract the shape of the liver from this dataset, we used the anatomical slices (cf. Fig. 2.2), which give a better contrast between the liver and the surrounding organs. The dataset corresponding to the liver is reduced to about 180 slices. After contrast enhancement, we apply an edge detection algorithm to extract the contours of the image, and then using a simple thresholding technique, we retain only those with higher-strength contours are considered for further processing. Next, we use semi-automatic deformable contour (KASS, WITKIN and TERZOPOULOS [1988], DELINGETTE and MONTAGNAT [2001]) to extract a smooth two-dimensional boundary of each liver slice. These contours are finally transformed into a set of two-dimensional binary images (cf. Fig. 2.2). The slices generated are then stacked to form a tridimensional binary image (MONTAGNAT and DELINGETTE [1998]) (cf. Fig. 2.3).

In order to capture the shape of the external surface of the liver, one could use a sub-voxel triangulation provided by the marching-cubes algorithm (LORENSEN and CLINE

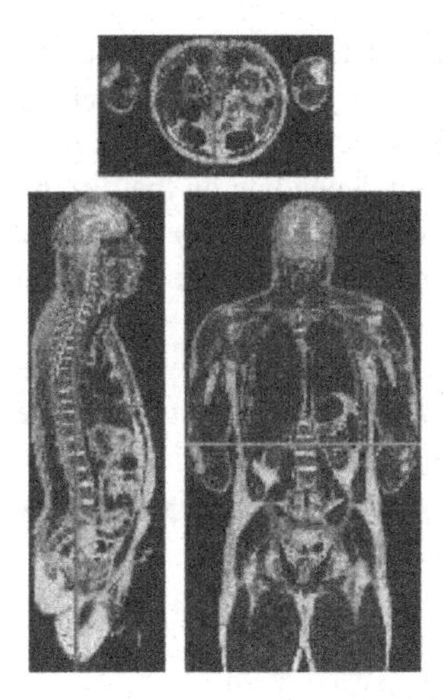

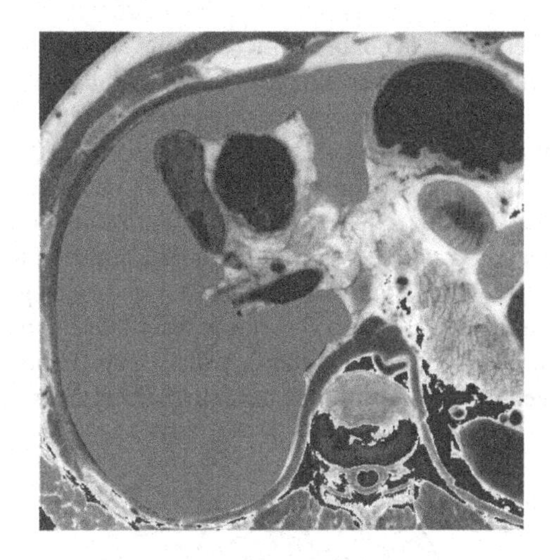

FIG. 2.2. Slice-by-slice segmentation of the liver. The initial data (left) is a high resolution photography of an anatomical slice of the abdomen. The binary image (right) corresponds to the segmented liver cross-section.

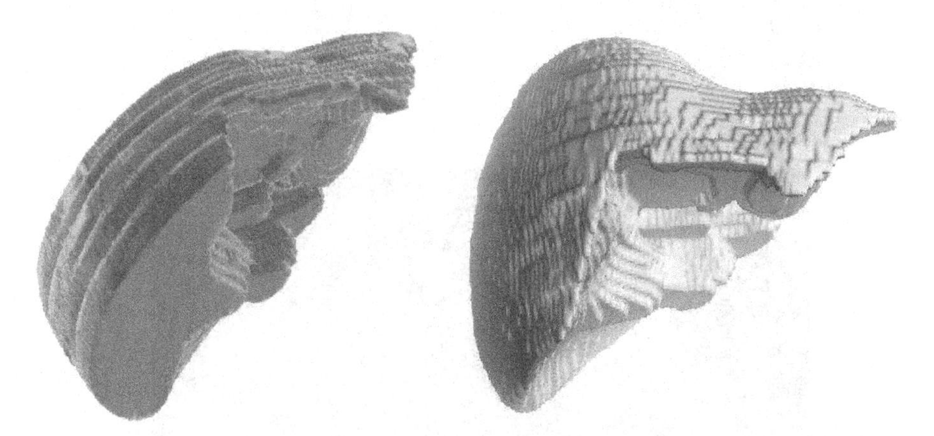

FIG. 2.3. After segmentation, the binary images are stacked (left) to give a 3D binary image. We see the step-effect on the shape of the liver (right) when extracted using the marching-cubes algorithm.

[1987]), however the number of triangles generated is too large for further processing. Moreover, a smoothing of the surface is necessary to avoid staircase effects (see Fig. 2.3). A possible solution consists in decimating an iso-surface model by using a mesh simplification tool (SCHROEDER, ZARGE and LORENSEN [1992]). However, for more flexibility, in both the segmentation and simplification processes, liver reconstruction was performed using *simplex meshes*.

Simplex meshes are an original representation of tridimensional objects developed by DELINGETTE [1999]. A simplex mesh is a deformable discrete surface mesh that is well-suited for generating geometric models from volumetric data. A simplex mesh can be deformed under the action of regularizing and external forces. Additional properties like a constant connectivity between vertices and a duality with triangulations have been defined. Moreover, simplex meshes are adaptive, for example, by concentrating vertices in areas of high curvature (thereby achieving an optimal shape description for a given number of vertices). The mesh may be refined or decimated depending on the distance of the vertices from the dataset. The decimation can also be interactively controlled. Fig. 2.4 shows the effect of the mesh adaptation and where the vertices are nicely concentrated at highly high curvature locations of the liver.

By integrating simplex meshes in the segmentation process, we have obtained smoothed triangulated surfaces, very close to an iso-surface extraction, but with fewer faces to represent the shape of the organs. In our example, the liver model has been created by fitting a simplex mesh to the tridimensional binary image previously described.

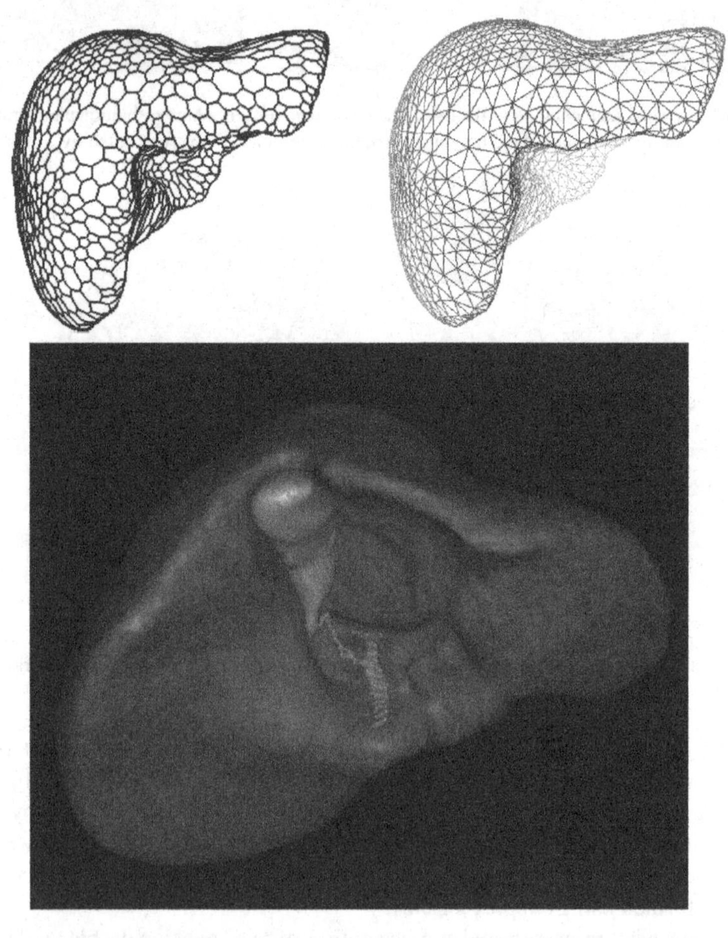

FIG. 2.4. Different representations of the geometric liver model. The simplex mesh (MONTAGNAT and DELINGETTE [1998]) fitting the data (top left) with a concentration of vertices in areas of high curvature, the triangulated dual surface (top right) and a texture-mapped model with anatomical details (gall bladder and ducts) from an endoscopic viewpoint (bottom).

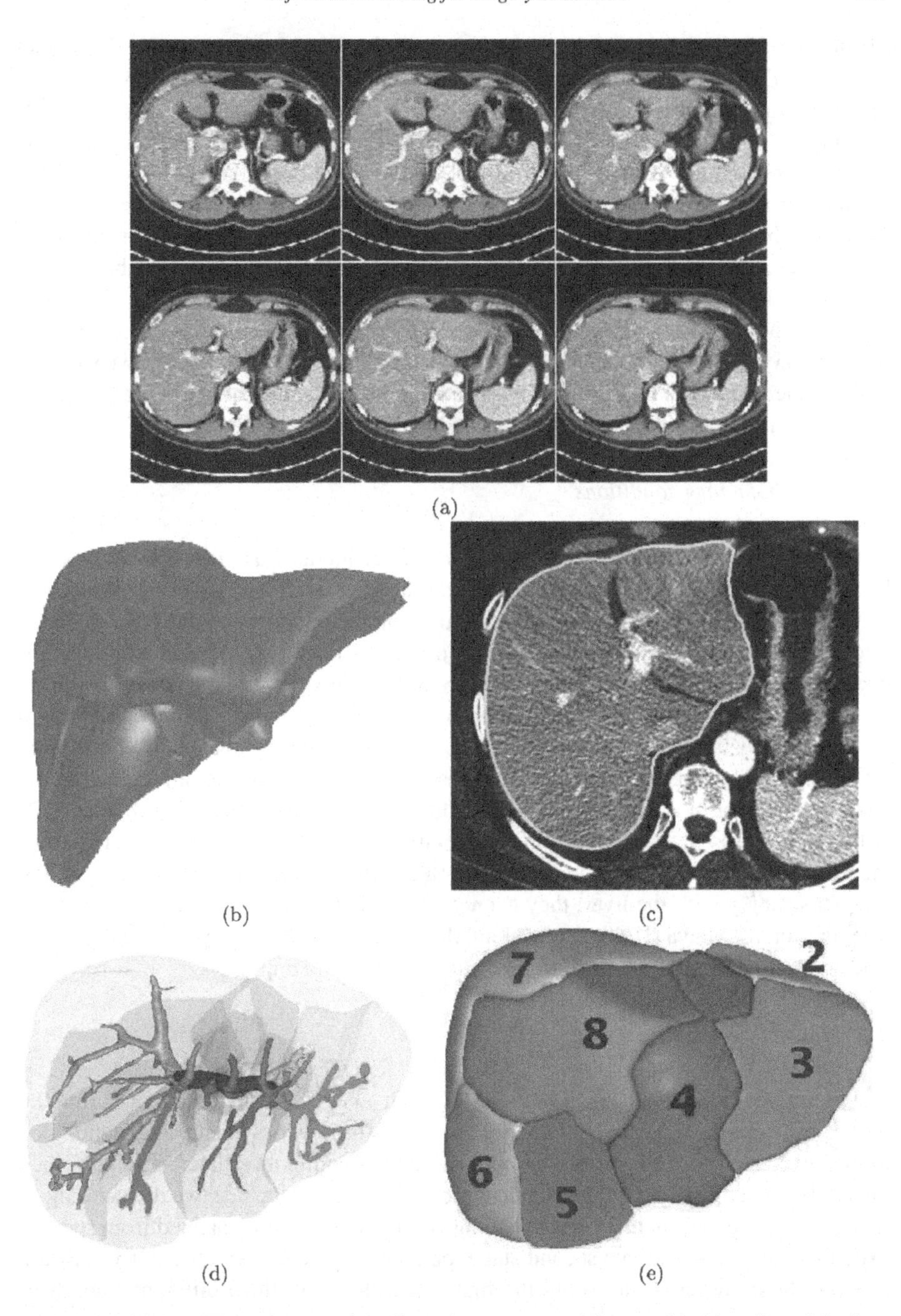

FIG. 2.5. (a) Original CT-scan images of the liver; (b) reconstructed liver model; (c) outline of the liver surface model in a CT-scan image; (d) segmentation of the portal vein (SOLER, DELINGETTE, MALANDAIN, MONTAGNAT, AYACHE, CLÉMENT, KOEHL, DOURTHE, MUTTER and MARESCAUX [2000]); (e) reconstruction of the eight anatomical segments (Couinaud segmentation).

Thanks to the adaptation and decimation properties of the simplex meshes, this model is composed of only 14 000 triangles, whereas the marching-cubes algorithm produced 94 000 triangles (cf. Figs. 2.3 and 2.4).

Although this approach is very useful for building a "generic" liver model, it is essential to integrate "patient-based" models in the simulator. In the framework of this research project, MONTAGNAT and DELINGETTE [1998] have developed a method for extracting liver models from CT scan images. The principle of this algorithm is to deform a generic simplex mesh (for instance, the one extracted from the Visible Human dataset) such that its surface coincides with the liver boundary in the image. The work of SOLER, MALANDAIN and DELINGETTE [1998], SOLER, DELINGETTE, MALANDAIN, MONTAGNAT, AYACHE, CLÉMENT, KOEHL, DOURTHE, MUTTER and MARESCAUX [2000] has extended this work by additionally extracting the main bifurcations of the portal and hepatic veins but also the hepatic lesions and gall-bladder (see Fig. 2.5).

2.4. *Liver boundary conditions*

In the scope of the AISIM project, a reference liver model was created by VIDRASCU, DELINGETTE and AYACHE [2001]. They define the liver environment (VIDRASCU, DELINGETTE and AYACHE [2001]) in order to set up the boundary conditions associated to computational models. The right liver extremity is thick and rounded while the left one is thin and flattened. Both extremities are not submitted to specific loads. The anterior border is thin, sharp and free. The posterior border is connected to the diaphragm by the coronary ligament. The upper surface, covered by the peritoneum, is divided into 2 parts by the suspensory ligament. However, this ligament does not affect the biomechanical behavior of the liver. The lower surface is connected with the gall-blader (GB) within the GB fossa, the stomach, the duodenum, the right kidney and the right part of the transverse colon. These organs are in contact with the liver surface, but they do not interact strongly with the liver; they cannot be considered as being supporting organs. The inferior vena cava (IVC) travels along the posterior surface, very often in a groove. The connection implies another strong fitting condition (clamp).

2.5. *Material characteristics*

The literature on the mechanical property of the liver is relatively poor, but during the past four years, there has been a renewed attention on soft tissue characterization due to the development of new robotics tools and new imaging modalities. The published materials concerning liver biomechanical properties usually include two distinct stages. In a first stage, experimental curves relating strain and stress are obtained from specific experimental setups and in a second stage, parameters of a known constitutive law are fitted to these curves. Concerning the first stage, there are three different sources of rheological data:

- *ex-vivo testing* where a sample of a liver is positioned inside a testing rig,
- *in-vivo testing* where a specific force and position sensing device is introduced inside the abdomen to perform indentation,

TABLE 2.1
List of published articles providing some quantitative data about the biomechanical properties of the liver

Authors	Experimental technique	Liver origin	Young modulus (kPa)
YAMASHITA and KUBOTA [1994]	image-based	human	not available
BROWN, ROSEN, KIM, CHANG, SINANAN and HANNAFORD [2003]	*in-vivo*	porcine liver	≈ 80
CARTER [1998]	*in-vivo*	human liver	≈ 170
DAN [1999]	*ex-vivo*	porcine liver	≈ 10
LIU and BILSTON [2002], LIU and BILSTON [2000]	*ex-vivo*	bovine liver	not available
NAVA, MAZZA, KLEINERMANN, AVIS and MCCLURE [2003]	*in-vivo*	porcine liver	≈ 90
MILLER [2000]	*in-vivo*	porcine liver	not available
SAKUMA, NISHIMURA, KONG CHUI, KOBAYASHI, INADA, CHEN and HISADA [2003]	*ex-vivo*	bovine liver	not available

- *image-based elastometry* where an imaging modality like ultrasound (YAMASHITA and KUBOTA [1994]), Magnetic Resonance Elastometry (MANDUCA, MUTHUPILLAI, ROSSMAN, GREENLEAF and EHMAN [1996]) or CT-scan (O'MAHONY, WILLIAMS and KATZ [1999], HODGSKINSON and CURREY [1992]) provides relevant information to assess the Young modulus of living materials.

A non-comprehensive list of articles describing the liver material characteristics is provided in Table 2.1. From this wide variety of studies, it is difficult to pick one particular constitutive model since each of experimental setup has its advantages and drawbacks. For instance, the rich perfusion of the liver affects deeply its rheology (the liver receives one fifth of the total blood flow at any time) and therefore it is still an open question whether *ex-vivo* experiments can assess the property of living liver tissue, even when specific care is taken to prevent the swelling or drying of the tissue. Conversely, data obtained from *in-vivo* experiments should also be considered with caution because the response may be location-dependent (linked to specific boundary conditions or non-homogeneity of the material) and the influence of the loading tool caliper on the deformation may not be well-understood. Furthermore, both the respiratory and circulatory motions may affect *in-vivo* data.

Furthermore, little is known about the variability of liver characteristics between species (does a porcine liver behave like a human liver?) but also between patients. For instance, studies from NAVA, MAZZA, KLEINERMANN, AVIS and MCCLURE [2003] suggest that a 20% difference in stiffness between normal and diseased livers whereas BROWN, ROSEN, KIM, CHANG, SINANAN and HANNAFORD [2003] show significant differences between *in-vivo* pig livers and *ex-vivo* cow livers.

Another important source of uncertainty in those measurements is the strain state of the liver during indentation. Indeed, as pointed out by BROWN, ROSEN, KIM, CHANG, SINANAN and HANNAFORD [2003], most researchers precondition their liver samples by applying several cycles of indentation in order to have more consistent estimates of stiffness and hysteresis. However, during surgery, (rightfully) surgeons do not pre-condition living tissues which may imply that only measurements obtained *in-vivo* and *in-situ* through modified surgical instruments (like those developed in CARTER [1998], BROWN, ROSEN, KIM, CHANG, SINANAN and HANNAFORD [2003], NAVA, MAZZA, KLEINERMANN, AVIS and MCCLURE [2003]) are relevant for modeling soft tissue in a surgical simulator.

Finally, the rheology of the liver is not only influenced by its perfusion, but also by the Glisson's capsule. For instance, CARTER [1998] et al. have showed that the stiffness of cylindrical samples of liver parenchyma with part of Glisson's capsule is twice the one without Glisson's capsule, using similar rheological tests (CARTER [1998]).

To conclude, more experimental studies are needed to reach a good understanding of the liver biomechanical properties. Methods based on *in-vivo* and *in-situ* indentations seem to be the most promising ones for building realistic soft tissue models in surgery simulation. All studies demonstrate that the liver is a strongly visco-elastic material, while LIU and BILSTON [2002] suggest that the liver can be considered as linear elastic for strains smaller than 0.2%.

Fortunately, in many surgical simulators, the boundary conditions governing the deformation of soft tissues, consist of imposed displacements only. In such case, the computation of soft tissue deformation requires to solve a homogeneous system of equations $\mathbf{FU} = \mathbf{0}$ which is not sensitive to the absolute value of stiffness materials but to the relative stiffness between materials (GLADILIN [2002]). Hopefully, we can expect that the relative stiffness between the liver and its neighboring organs is less variable and easier to assess, for instance, through medical imagery.

3. Linear elastic models for surgery simulation

3.1. Main features of our approach

In the next sections, we propose three different soft tissue models that are well-suited for the simulation of surgery and which are compatible with the constraints described in Section 1.3. These models bear many common features that are listed below:

- volumetric structures;
- continuum mechanics based deformation;
- finite element modeling;
- linear tetrahedron finite element;
- strong approximation in dynamical modeling.

We explain the motivations of such characteristics in the next sections.

3.1.1. Using volumetric models

We can classify the geometry of anatomical structures depending on their "idealized" dimensionality, even though they all consist of an assembly of tridimensional cells. For instance, at a coarse scale, a blood vessel can be thought as a one-dimensional structure (QUARTERONI, TUVERI and VENEZIANI [2000]) whereas the gall-blader can be represented as a two-dimensional structure (KUHNAPFEL, AKMAK and MAA [2000]) (a closed surface filled with bile). Similarly, the behavior of most parenchymatous organs such as the brain, lungs, liver or kidneys are intrinsically volumetric. But one should notice that at a fine enough scale, all anatomical structures can be considered as volumetric.

In surgical simulators, it is frequent to rely on such dimensionality simplification in order to speed-up computation: tubular surfaces, such as the colon, are modeled as a deformable spline (FRANCE, LENOIR, MESEURE and CHAILLOU [2002]) whereas deformable volumetric structures, such as the liver, are represented with their surrounding surface envelope (KUHNAPFEL, AKMAK and MAA [2000]).

However, such artifices cannot be used in a hepatic resection simulator when the removal of hepatic parenchyma is performed.

3.1.2. Using continuum mechanics

We have chosen to rely on the theory of continuum mechanics to govern the deformation of our volumetric soft tissue models. Other alternative representations exist such as spring–mass models (KUHNAPFEL, AKMAK and MAA [2000]), chain–mail (GIBSON, SAMOSKY, MOR, FYOCK, GRIMSON, KANADE, KIKINIS, LAUER and MCKENZIE [1997]) or long element models (COSTA and BALANIUK [2001]). Spring–mass models correspond to small deformation one-dimensional elastic elements (see Section 6.1.7 for an extended comparison) but are no longer valid for two- or three-dimensional elasticity. These models are especially popular in computer graphics since they are easy to implement and are based on straightforward point mechanics. The chain–mail (GIBSON, SAMOSKY, MOR, FYOCK, GRIMSON, KANADE, KIKINIS, LAUER and MCKENZIE [1997]) is an original quasi-static deformable model based on a hexahedral mesh which is well-suited for stiff material but does not allow any topological change. Long element models (COSTA and BALANIUK [2001]) correspond to valid tridimensional cylindrical elastic models but are used to approximate the deformation of general volumetric shapes.

We chose to base our soft tissue models on continuum mechanics since it offers a well-studied and validated framework for modeling the deformation of volumetric objects unlike the methods cited above. Furthermore, it offers the following advantages:

- *Scalability*: when modifying the mesh topology (refinement or cutting for instance), the behavior of the mesh is guaranteed to evolve continuously.
- *Physical parameter identification*: the elastic parameters of a biomaterial (Young modulus, for instance) can be estimated from various methods (incremental rheological experiments, elastography or solving inverse problems). Parameter identification for spring–mass models is known to be more difficult and requires stochastic optimization (genetic algorithms (LOUCHET, PROVOT and CROCHEMORE [1995]) or simulated annealing (DEUSSEN, KOBBELT and TUCKE [1995])).

3.1.3. Using finite element modeling

Finite Element Modeling (FEM) is certainly the most popular technique for the computation of structure deformation based on the elasticity theory. Furthermore, it is well-formalized and understood and there exists many software implementations although none of them deals with real-time deformation. Nonetheless, there exists alternative approaches such as Boundary Element Modeling (BEM) or the Finite Difference Method (FDM).

The BEM is well-suited for the simulation of linear elastic isotropic and homogeneous materials (for which there exists a Green function) and is indeed a good alternative to FEM when the mesh topology is not modified. In fact, BEM has the important advantage over FEM of not requiring the construction of a volumetric mesh. A more thorough discussion is provided in Section 5.3.2 but this approach is not well-suited when cutting is simulated.

The FDM is well-suited when the domain is discretized over a structured grid in which case partial derivatives can be easily discretized. They often lead to the same equation as FEM when specific finite elements (based on linear interpolation) are employed (BATHE [1982]). On unstructured meshes such as tetrahedral meshes, some extensions of the finite difference method have been proposed (DEBUNNE, DESBRUN, CANI and BARR [2001]) also leading to a similar equation as FEM (see discussion in Section 4.4). With non-linear elasticity however, FEM (PICINBONO, DELINGETTE and AYACHE [2003]) and FDM (DEBUNNE, DESBRUN, CANI and BARR [2001]) differ significantly and no formal proof has been given that the FDM converges towards the right solution as the mesh resolution increases.

3.1.4. Using linear tetrahedron finite element

For all finite element models described in the remainder, a simple finite element is used: a 4-node tetrahedron with linear interpolation ($P1$). The Linear Tetrahedron (LT) is known to be a poor element (in terms of convergence) compared to the Linear Hexahedron (LH) for static linear and non-linear elastic analysis (BENZLEY, PERRY, CLARK, MERKLEY and SJAARDEMA [1995]). Also this paper shows that LH performs better than the Quadratic Tetrahedron (10 nodes) even in a static linear elastic analysis.

The motivation for using tetrahedra rather than hexahedra clearly comes from a geometrical point of view. Indeed, meshing most anatomical structures with hexahedra is known to be a difficult task especially for structures having highly curved or circumvoluted parts such as the liver or the brain parenchyma (Fig. 3.1). To obtain a smooth surface envelope, it is then necessary to employ many hexahedra where a smaller number of tetrahedra would suffice. Furthermore, there exist several efficient commercial and academic software (SIMAIL, OWEN [2000]) to fill automatically a closed triangulated surface with tetrahedra of high shape quality (PARTHASARATHY, GRAICHEN and HATHAWAY [1993]). A second motivation for using tetrahedra rather than hexahedra is related to the simulation of cutting soft tissue that involves removing and remeshing of local elements. With hexahedral meshes, it is not possible to simulate general surface of cut without resorting to add new element types (such as prismatic elements). Such multi-element models (BATHE [1982]) would make the matrix assembly and local remeshing algorithms more complex to manage.

Regarding the choice of the interpolation function (linear versus quadratic), our choice has been mainly governed by computational issues. Given that a minimum number of tetrahedra is necessary to get a realistic visual rendering of a structure, the QT element involves one additional node per edge compared to the LT element which on a

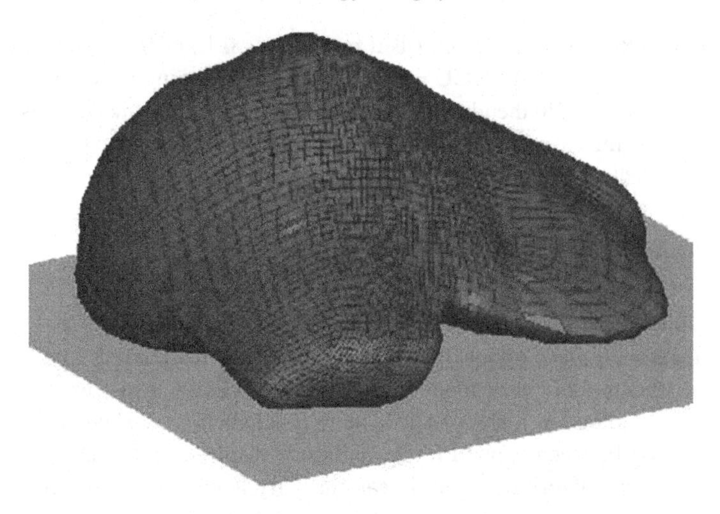

FIG. 3.1. Example of liver meshed with hexahedra (courtesy of ESI SA).

typical tetrahedral mesh implies at least a sixfold increase of the number of nodes. Furthermore, we believe that the loss of accuracy in the deformation computation entailed by the use of LT elements remains small compared to the large uncertainty on the physical parameter values (Young modulus, ...) existing for most soft tissues.

Finally, by using linear elements, the computation of local stiffness matrices can be done explicitly (analytically) even for non-linear elasticity. Also, the gradient of the displacement field which is constant inside each element (constant strain) has a simple geometric interpretation using area vectors (see Section 4.2). A significant speed-up is therefore obtained when compared to higher order elements that require numerical integration methods such as Gauss quadrature for estimating stiffness matrices.

3.1.5. Using large approximations of dynamic behavior

Despite the development of new *in vivo* rheological equipment (KAUER, VUSKOVIC, DUAL, SZÉKELY and BAJKA [2001]), the dynamical behavior is only known quantitatively for a few anatomical structures: skin, muscle, myocardium, ... The viscoelastic properties of liver tissue have been studied by LIU and BILSTON [2000] but for most organs, constitutive laws of dynamics and their parameters must be hypothesized and validated qualitatively.

In a surgical simulator, the boundary conditions caused by the contact with surgical instruments can change between two iterations. Given that surgeons typically move their instruments at low speed (typically a few millimiters per second) and making the hypothesis that the mass of these instruments is the same or smaller than the mass of anatomical structures, we chose to neglect the dynamics of soft tissue models in two different ways. For a first class of models described in Section 5, we solve the static problem $\mathbf{F} = \mathbf{KU}$ (where $\mathbf{F}$ is the force vector, $\mathbf{K}$ the stiffness matrix and $\mathbf{U}$ the displacement vector) at each iteration thus leading to a quasi-static approximation.

For a second class of models, described in Sections 6.1 and 7, we solve the Newtonian equation of motion $\mathbf{M}\ddot{\mathbf{U}} + \mathbf{C}\dot{\mathbf{U}} = -\mathbf{K}\mathbf{U}$ with the following hypotheses: the mass matrix $\mathbf{M}$ is proportional to the identity matrix while the damping matrix $\mathbf{C}$ is diagonal. Furthermore, in some cases, the computational time T_c is longer than the time step Δt which creates a lack of synchronicity in the simulation.

3.2. *Tridimensional linear elasticity*

The fast computation of soft tissue deformation in a surgical simulator requires that some hypotheses are made about the nature of the tissue material. A first hypothesis, which leads to the two soft tissue models described in Sections 5 and 6, assumes that soft tissue can be considered as linear elastic. The rationale behind this hypothesis is clear: the linear relation between applied forces and node displacements leads to very computationally efficient algorithms. But, linear elasticity is not only a convenient mathematical model for deformable structures: it is also a quite realistic hypothesis. Indeed, all hyperelastic materials can be approximated by linear elastic materials when small displacements (and therefore small deformations) are applied (FUNG [1993], MAUREL, WU, MAGNENAT THALMANN and THALMANN [1998]). It is often admitted as reasonable to consider that a material is linear elastic when observed displacements are less than 5% of the typical object size. In the case of hepatic tissue, a recent publication (LIU and BILSTON [2000]) indicates that the linear domain is only valid for strain less than 0.2%.

Whether this constraint on the amount of displacement is valid or not in a surgical simulator depends both on the anatomical structure and the type of surgery. For instance, when simulating the removal of the gall bladder (cholelysectomy), the liver undertakes small displacements but it is not the case when simulating hepatic resection or any other surgical procedure that requires a large motion of the left lobe.

When large displacements are applied to a linear elastic material, the approximation of hyperelasticity is no longer valid and large errors in the computation of deformation and reaction forces can be perceived both visually and haptically. Section 7.1 describes the shortcomings of linear elasticity in such cases.

To summarize the general equations of linear elastic materials, we proceed in four steps. In Section 3.2.1, we provide some general definitions whereas Sections 3.2.3 and 3.2.4 give the main equations of isotropic and transversally anisotropic material. Finally, the principle of virtual work is formulated in Section 3.2.5.

3.2.1. *Definition of infinitesimal strain*

We consider a three-dimensional body defined in a tridimensional Euclidian space $\mathbb{R}^3$. We describe the geometry of this body in its rest position $\mathcal{M}_{\text{rest}}$ by using material coordinates $\mathbf{X} = (x, y, z)^{\mathrm{T}}$ defined over the volume of space Ω occupied by $\mathcal{M}_{\text{rest}}$.

This body is deformed under the application of boundary conditions: these may be either geometric boundary conditions (also called essential boundary conditions (BATHE [1982])) or natural boundary conditions, i.e., prescribed boundary forces.

We note $\mathcal{M}_{\text{def}}$ the body in its deformed state and $\boldsymbol{\Phi}(x, y, z)$, the deformation function that associates to each material point $\mathbf{X}$ located in the body at its rest position, its new

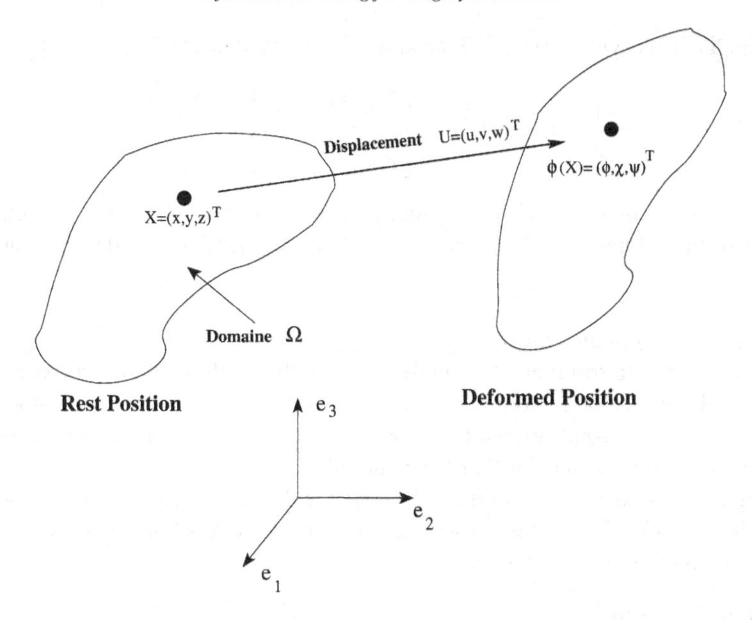

FIG. 3.2. Definition of deformation and displacement between rest and deformed positions.

position $\boldsymbol{\Phi}(\mathbf{X})$ after the body has been deformed

$$\boldsymbol{\Phi}: \Omega \subset \mathbb{R}^3 \mapsto \boldsymbol{\Phi}(\Omega), \quad \mathbf{X} \rightarrow \boldsymbol{\Phi}(\mathbf{X}) = \begin{cases} \phi(x,y,z), \\ \chi(x,y,z), \\ \psi(x,y,z). \end{cases}$$

The displacement vector field $\mathbf{U}$ is defined as the variation between the deformed position and the rest position (see Fig. 3.2):

$$\mathbf{U}(\mathbf{X}) : \Omega \mapsto \mathbb{R}^3, \quad \mathbf{X} \rightarrow \mathbf{U} = \begin{cases} u(x,y,z), \\ v(x,y,z), \\ w(x,y,z). \end{cases}$$

The observed deformation can be characterized and quantified through the analysis of the spatial derivatives of the deformation function $\boldsymbol{\Phi}(\mathbf{X})$. More precisely, the right Cauchy–Green strain tensor $\mathbf{C}(\mathbf{X})$ which is a symmetric 3×3 matrix (therefore, has 3 real eigenvalues) is simply computed from the deformation gradient

$$\mathbf{C}(\mathbf{X}) = \nabla\boldsymbol{\Phi}^{\mathrm{T}}\nabla\boldsymbol{\Phi}. \tag{3.1}$$

The Green–Lagrange strain tensor $\mathbf{E}(\mathbf{X})$, derived from the right Cauchy–Green strain tensor, allows to analyze the deformation after rigid body motion has been removed:

$$\mathbf{E}(\mathbf{X}) = \frac{1}{2}(\mathbf{C} - \mathbf{I}_3) = \frac{1}{2}\left(\nabla\mathbf{U} + \nabla\mathbf{U}^{\mathrm{T}} + \nabla\mathbf{U}^{\mathrm{T}}\nabla\mathbf{U}\right), \tag{3.2}$$

where $\mathbf{I}_3$ is the 3×3 identity matrix.

In the linear elasticity framework, applied displacements are considered as infinitesimal and the Green–Lagrange strain tensor $\mathbf{E}(\mathbf{X})$ is linearized into the infinitesimal strain tensor $\mathbf{E}_\mathrm{L}(\mathbf{X})$. This symmetric 3×3 tensor is simply written as

$$\mathbf{E}_\mathrm{L}(\mathbf{X}) = [e_{ij}] = \frac{1}{2}(\nabla\mathbf{U} + \nabla\mathbf{U}^\mathrm{T}) = \begin{bmatrix} e_{xx} & e_{xy} & e_{xz} \\ e_{xy} & e_{yy} & e_{yz} \\ e_{xz} & e_{yz} & e_{zz} \end{bmatrix}. \tag{3.3}$$

The diagonal elements e_{ii} of the symmetric matrix describe the relative stretch in the direction of the reference frame, whereas off-diagonal elements e_{ij} describe shearing quantities.

3.2.2. Definition of infinitesimal stress

The deformation of a tridimensional body is caused by applying external forces: these forces may be either body forces $\mathbf{F}^\mathrm{B}$ (such as gravity forces) or surface forces $\mathbf{F}^\mathrm{S}$ (applied pressure) or concentrated forces $\mathbf{F}^\mathrm{P}$. As a reaction to external forces, internal forces are created inside the elastic body material.

Through Cauchy theorem (CIARLET [1987]), it is demonstrated that for each volume element inside the deformed body, the force per unit area $\mathbf{t}(\mathbf{X}, \mathbf{n})$ at a point $\mathbf{X}$ and along the normal direction $\mathbf{n}$ is written as

$$\mathbf{t}(\mathbf{X}, \mathbf{n}) = \mathbf{T}(\mathbf{X})\mathbf{n},$$

where $\mathbf{T}(\mathbf{X})$ is the Cauchy stress tensor. The Cauchy stress tensor is a 3×3 symmetric tensor and can be written as

$$\boldsymbol{\Sigma}(\mathbf{X}) = [\sigma_{ij}] = \begin{bmatrix} \sigma_{xx} & \sigma_{xy} & \sigma_{xz} \\ \sigma_{xy} & \sigma_{yy} & \sigma_{yz} \\ \sigma_{xz} & \sigma_{yz} & \sigma_{zz} \end{bmatrix}.$$

The Cauchy stress $\boldsymbol{\Sigma}$ and infinitesimal strain $\mathbf{E}_\mathrm{L}$ are conjugated variables (BATHE [1982]) which implies the following relations:

$$\sigma_{ij} = \frac{\partial W}{\partial e_{ij}}, \qquad e_{ij} = \frac{\partial W}{\partial \sigma_{ij}}, \tag{3.4}$$

where $W(\mathbf{X})$ is the amount of elastic energy per unit volume.

3.2.3. Isotropic linear elastic materials

For an isotropic linear elastic material, the elastic energy $W(\mathbf{X})$ is a quadratic function of the first two invariants of the infinitesimal strain tensor (CIARLET [1987]):

$$W(\mathbf{X}) = \frac{\lambda}{2}(\mathrm{tr}\,\mathbf{E}_\mathrm{L})^2 + \mu\,\mathrm{tr}\,\mathbf{E}_\mathrm{L}^2, \tag{3.5}$$

where λ and μ are the two Lamé coefficients characterizing the material stiffness. These two parameters are simple functions of Young modulus E and Poisson coefficients ν, which belong to the material's physical properties:

$$\lambda = \frac{E\nu}{(1+\nu)(1-2\nu)}, \qquad \mu = \frac{E}{2(1+\nu)},$$
$$E = \frac{\mu(3\lambda + 2\mu)}{\lambda + \mu}, \qquad \nu = \frac{\lambda}{2(\lambda + \mu)}.$$

Through Eq. (3.4), we can derive the linear relationship, known as Hooke's law, between the stress and the infinitesimal strain tensors,

$$\boldsymbol{\Sigma} = \lambda(\operatorname{tr} \mathbf{E}_{\mathrm{L}})\mathbf{I}_3 + 2\mu \mathbf{E}_{\mathrm{L}}. \tag{3.6}$$

Note that the elastic energy can be written simply as a function of the linearized strain and stress tensors,

$$W = \frac{1}{2} \operatorname{tr}(\mathbf{E}_{\mathrm{L}} \boldsymbol{\Sigma}).$$

3.2.4. Transversally anisotropic linear elastic materials

Most anatomical structures like muscles, ligaments or blood vessels are strongly anisotropic material. This anisotropy is caused by the presence of different fibers (collagen, muscle, ...) that are wrapped together within the same tissue. For instance, anisotropic materials have been successfully used to model the deformation of the eye (KAISS and LE TALLEC [1996]), of the heart (HUMPHREY and YIN [1987], HUMPHREY, STRUMPF and YIN [1990], PAPADEMETRIS, SHI, DIONE, SINUSAS, CONSTABLE and DUNCAN [1999]) or the knee ligaments (WEISS, GARDINER and QUAPP [1995], PUSO and WEISS [1998]). In the scope of our hepatic surgery simulator, we have added an anisotropic behavior where the first branches of the portal vein are located inside the hepatic parenchyma.

We have chosen to focus only transversally anisotropic material only where there is one direction $\mathbf{a}_0$ along which the material stiffness differs from the stiffness in the orthogonal plane. Indeed, one major obstacle when modeling anisotropic material is to get reliable data from rheological experiments regarding the directions of anisotropy and the Young modulus in all directions. With transversal anisotropy, it is sufficient to provide a single direction $\mathbf{a}_0$ and an additional pair of Lamé coefficients $\lambda^{\mathbf{a}_0}$ and $\mu^{\mathbf{a}_0}$ (see Fig. 3.3).

The theoretical description of elastic energy of transversally anisotropic material is largely based on the work of SPENCER [1972], SPENCER [1984] and FUNG [1993]. For the sake of clarity, we introduce the notion of direction invariant and the concept of anisotropic stretching and shearing.

We decompose the elastic energy of a transversally anisotropic material as the sum of the isotropic energy, provided by Eq. (3.5) and by a corrective term ΔW_{Ani} which only depends on the variation of Lamé coefficients:

$$\Delta\lambda = \lambda^{\mathbf{a}_0} - \lambda, \qquad \Delta\mu = \mu^{\mathbf{a}_0} - \mu,$$
$$W_{\mathrm{Transv.Ani}}(\mathbf{X}) = W(\mathbf{X}) + \Delta W_{\mathrm{Ani}}(\mathbf{X}, \Delta\lambda, \Delta\mu).$$

Without loss of generality, we can assume that $\mathbf{a}_0$ coincides with the z direction of the reference frame. The isotropic elastic energy can then be written as a function of the stretch e_{zz} and shear (e_{xz}, e_{yz}) in the direction $\mathbf{a}_0$:

$$W(\mathbf{X}) = \left(\frac{\lambda}{2} + \mu\right)\left(e_{xx}^2 + e_{yy}^2 + e_{zz}^2\right) + \lambda(e_{xx}e_{yy} + e_{xx}e_{zz} + e_{yy}e_{zz})$$
$$+ 2\mu\left(e_{xy}^2 + e_{yz}^2 + e_{xz}^2\right).$$

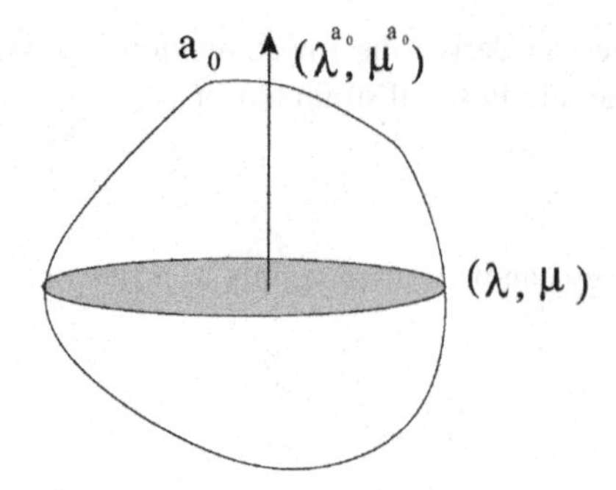

FIG. 3.3. Definition of Lamé coefficients along the direction $\mathbf{a}_0$ are $\lambda^{\mathbf{a}_0}$ and $\mu^{\mathbf{a}_0}$.

The purpose of the corrective term ΔW_{Ani} is to modify the isotropic Lamé coefficients in the direction of anisotropy:

$$\Delta W_{\mathrm{Ani}}(\mathbf{X}) = \left(-\frac{\Delta\lambda}{2} + \Delta\mu\right) e_{zz}^2 + \Delta\lambda e_{zz}(\operatorname{tr}\mathbf{E}_{\mathrm{L}}) + 2\Delta\mu\left(e_{yz}^2 + e_{xz}^2\right).$$

The equation above can be written using the two parameters I_4 and I_5 which characterize the strain tensor in the direction $\mathbf{a}_0$ (PICINBONO, LOMBARDO, DELINGETTE and AYACHE [2002]):

$$I_4 = \mathbf{a}_0^{\mathrm{T}}\mathbf{E}_{\mathrm{L}}\mathbf{a}_0, \tag{3.7}$$

$$I_5 = \mathbf{a}_0^{\mathrm{T}}\mathbf{E}_{\mathrm{L}}^2\mathbf{a}_0. \tag{3.8}$$

The first parameter I_4 is simply the amount of stretch in the direction $\mathbf{a}_0$ whereas the total amount of shearing $e_{xz}^2 + e_{yz}^2$ is given by $I_5 - I_4^2$. With these notations, the corrective term can be written as

$$\Delta W_{\mathrm{Ani}}(\mathbf{X}) = \Delta\lambda I_4 \operatorname{tr}\mathbf{E}_{\mathrm{L}} + 2\Delta\mu I_5 - \left(\frac{\Delta\lambda}{2} + \Delta\mu\right) I_4^2. \tag{3.9}$$

PICINBONO, LOMBARDO, DELINGETTE and AYACHE [2002] proposed to decompose the anisotropic term $\Delta W_{\mathrm{Ani}}(\mathbf{X})$ into a stretching and shearing part:

$$\Delta W_{\mathrm{Ani}}(\mathbf{X}) = \Delta W_{\mathrm{Str.Ani}} + \Delta W_{\mathrm{Sh.Ani}},$$

$$\Delta W_{\mathrm{St.Ani}} = \left(-\frac{\Delta\lambda}{2} + \Delta\mu\right) I_4^2 + \Delta\lambda I_4 \operatorname{tr}\mathbf{E}_{\mathrm{L}},$$

$$\Delta W_{\mathrm{Sh.Ani}} = 2\Delta\mu\left(I_5 - I_4^2\right).$$

In Fig. 3.4, the distinction between stretching and shearing effects of a transversally anisotropic material is pictured by applying a force F_1 and F_2 on a cylinder respectively along and orthogonal to the direction.

3.2.5. *Principle of virtual work*

The equilibrium equation of a deformed body is derived through the *principle of virtual displacements*. This principle states that for any compatible virtual displacement $\hat{\mathbf{u}}(\mathbf{X})$

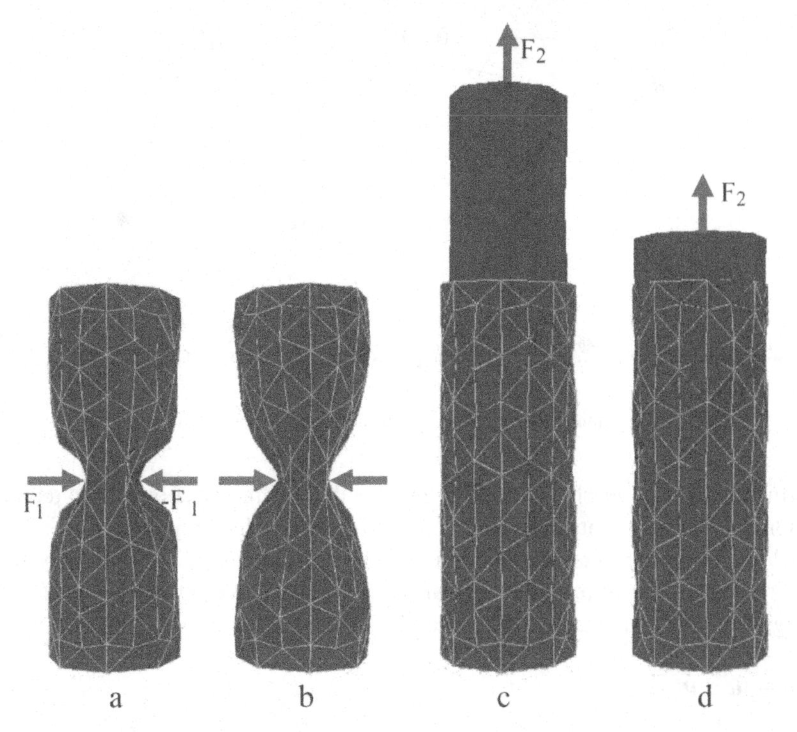

FIG. 3.4. Comparison between isotropic (a and c) and anisotropic (b and d) cylinders (PICINBONO, DELINGETTE and AYACHE [2003]). The same horizontal (respectively vertical) loads F_1 (respectively F_2) are applied in the two leftmost (respectively rightmost) figures.

applied on a body $\mathcal{M}_{\text{def}}$, the total internal virtual work is equal to the total external work. The total internal work is given by the integral of elastic energy over the body volume whereas the external work is created by the application of body and surface forces:

$$\int_{\Omega} \widehat{W}(\mathbf{X})\,\mathrm{d}V = \int_{\Omega} \hat{\mathbf{u}}^{\mathrm{T}}\mathbf{f}^{\mathrm{B}}\,\mathrm{d}V + \int_{\partial\Omega} \hat{\mathbf{u}}^{\mathrm{T}}\mathbf{f}^{\mathrm{S}}\,\mathrm{d}S \tag{3.10}$$

where $\mathbf{f}^{\mathrm{B}}$ and $\mathbf{f}^{\mathrm{S}}$ are the applied body and surface forces. Note that in Eq. (3.10), the virtual displacement field $\hat{\mathbf{u}}(\mathbf{X})$ is supposed to be compatible with the geometric boundary constraints (imposed displacements). Furthermore, this relation is only valid for small virtual displacements such that the linearized strain hypothesis still holds.

4. Finite element modeling

4.1. Linear tetrahedron element

As justified in Section 3.1.4, the computation of soft tissue deformation is based on the finite element method. Anatomical structures of interest are spatially discretized into a

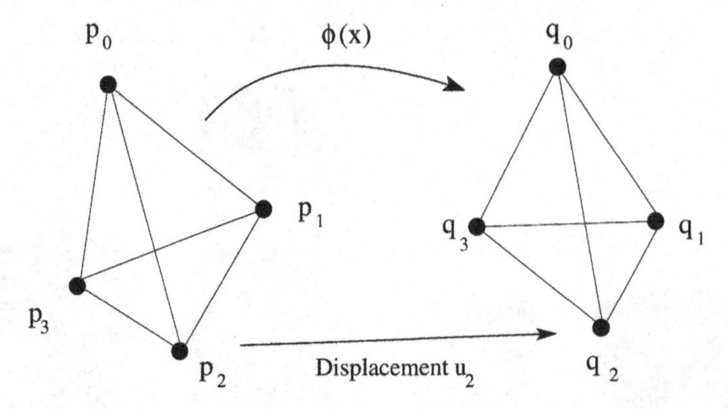

FIG. 4.1. Notations for the position and displacement vectors of a tetrahedron.

conformal tetrahedral mesh. Conformity implies that the intersection of two tetrahedra of that mesh is either empty or consists of a vertex or an edge or a triangle.

Let $\mathcal{M}_{\text{rest}}$ be a conformal tetrahedral mesh at its rest position. The initial position of each vertex is written as $\mathbf{p}_i$ while its position in the deformed position is written as $\mathbf{q}_i$ (see Fig. 4.1). The displacement at each node is then defined as

$$\mathbf{u}_i = \mathbf{q}_i - \mathbf{p}_i.$$

We use a linear tetrahedron finite element, denoted in the literature as P_1. This amounts to assuming a C^0 continuity of the displacement vector across the domain and equivalently assuming constant strain inside each tetrahedron (since the gradient matrix is constant inside each tetrahedron).

More precisely, if $\mathcal{T}$ is a tetrahedron defined by its four vertices $\mathbf{p}_j$, $j = 0, \ldots, 3$, in their *rest position*, then the displacement vector at a given point $\mathbf{X} = (x, y, z) \subset \mathcal{T}$ is defined as

$$\mathbf{u}(\mathbf{X}) = \sum_{j=0}^{3} h_j(\mathbf{X})\mathbf{u}_j,$$

where $h_j(\mathbf{X})$, $j = 0, \ldots, 3$, are the shape functions that correspond to the linear interpolation inside tetrahedron $\mathcal{T}$. These shape functions $h_j(\mathbf{X})$ correspond to the barycentric coordinates of point $\mathbf{X}$ with respect to vertices $\mathbf{p}_i$. The analytical expression of these shape functions is obtained from the linear relation

$$\begin{bmatrix} x \\ y \\ z \\ 1 \end{bmatrix} = \begin{bmatrix} \mathbf{p}_0^x & \mathbf{p}_1^x & \mathbf{p}_2^x & \mathbf{p}_3^x \\ \mathbf{p}_0^y & \mathbf{p}_1^y & \mathbf{p}_2^y & \mathbf{p}_3^y \\ \mathbf{p}_0^z & \mathbf{p}_1^z & \mathbf{p}_2^z & \mathbf{p}_3^z \\ 1 & 1 & 1 & 1 \end{bmatrix} \begin{bmatrix} h_0 \\ h_1 \\ h_2 \\ h_3 \end{bmatrix} = \mathbf{PH},$$

where $\mathbf{p}_i = (\mathbf{p}_i^x, \mathbf{p}_i^y, \mathbf{p}_i^z)^{\mathrm{T}}$ are the coordinates of each tetrahedron vertex. The matrix $\mathbf{P}$ completely encapsulates the shape of the tetrahedron $\mathcal{T}$ at its rest position. Since its determinant $|\mathbf{P}| = 6V(\mathcal{T})$ is the volume of $\mathcal{T}$, for non-degenerate tetrahedra $\mathbf{P}$ can be

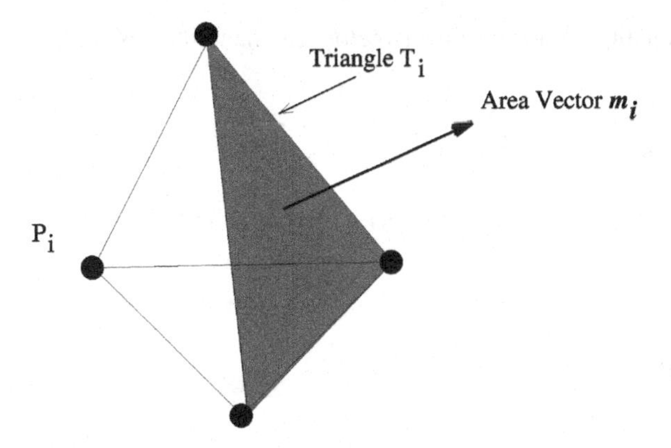

FIG. 4.2. Definition of area vector $\mathbf{m}_i$ on the triangle T_i opposite to vertex $\mathbf{p}_i$ in tetrahedron $\mathcal{T}$.

inverted,

$$\left[\mathbf{P}^{-1}\right] = \frac{-1}{6V(\mathcal{T})}\begin{bmatrix} \mathbf{m}_0^x & \mathbf{m}_0^y & \mathbf{m}_0^z & -V_0 \\ \mathbf{m}_1^x & \mathbf{m}_1^y & \mathbf{m}_1^z & -V_1 \\ \mathbf{m}_2^x & \mathbf{m}_2^y & \mathbf{m}_2^z & -V_2 \\ \mathbf{m}_3^x & \mathbf{m}_3^y & \mathbf{m}_3^z & -V_3 \end{bmatrix},$$

where:

- $\mathbf{m}_i = (m_i^x, m_i^y, m_i^z)^{\mathrm{T}}$ is the ith area vector opposite to vertex $\mathbf{p}_i$ (see description below),
- $V_i = (-1)^{i+1}|\mathbf{p}_{i+1}, \mathbf{p}_{i+2}, \mathbf{p}_{i+3}|$ can be interpreted[5] geometrically as 6 times the volume of the tetrahedron made by the origin $\mathbf{o}$ and vertices $\mathbf{p}_{i+1}$, $\mathbf{p}_{i+2}$ and $\mathbf{p}_{i+3}$.

  To simplify notations, the index $i+k$ should be understood as $(i+k)$ mod 4.

Area vectors $\mathbf{m}_i$ have a very simple interpretation: they are directed along the outer normal direction of the triangle T_i opposite to $\mathbf{p}_i$ and their norm is equal to twice the area of that triangle (see Fig. 4.2). More precisely, they can be computed as

$$\mathbf{m}_i = (-1)^{i+1}(\mathbf{p}_{i+1} \times \mathbf{p}_{i+2} + \mathbf{p}_{i+2} \times \mathbf{p}_{i+3} + \mathbf{p}_{i+3} \times \mathbf{p}_{i+1}), \tag{4.1}$$

where $\mathbf{p}_{i+1} \times \mathbf{p}_{i+2}$ is the cross product between the two vectors $\mathbf{p}_{i+1}$ and $\mathbf{p}_{i+2}$.

Because they are computed from the inverse of matrix $\mathbf{P}$, these area vectors also capture the shape of $\mathcal{T}$ completely, and thus play a key role when computing the stiffness matrix $\mathbf{K}$. Further properties of area vectors are described in Section 4.2.

The shape functions $h_i(\mathbf{X})$ can then be written as

$$h_i(\mathbf{X}) = -\frac{\mathbf{m}_i \cdot \mathbf{X} - V_i}{6V(\mathcal{T})}, \tag{4.2}$$

where $\mathbf{m}_i \cdot \mathbf{X}$ is the dot product between the two vectors $\mathbf{m}_i$ and $\mathbf{X}$.

If we note that elementary volumes V_i can be expressed as

$$V_i = \mathbf{m}_i \cdot \mathbf{p}_{i+1},$$

[5] $|\mathbf{a}, \mathbf{b}, \mathbf{c}|$ is the triple product of vectors $\mathbf{a}$, $\mathbf{b}$ and $\mathbf{c}$.

then the interpolation of displacement vectors can be written as

$$\mathbf{u}(\mathbf{X}) = -\sum_{i=0}^{3} \frac{\mathbf{m}_i \cdot (\mathbf{X} - \mathbf{p}_{i+1})}{6V(\mathcal{T})} \mathbf{u}_i. \tag{4.3}$$

Finally, the interpolation matrix $\mathbf{H}(\mathbf{X})$ widely used in the finite element literature is defined as

$$\mathbf{u}(\mathbf{X}) = \mathbf{H}(\mathbf{X}) \begin{bmatrix} \mathbf{u}_0 \\ \mathbf{u}_1 \\ \mathbf{u}_2 \\ \mathbf{u}_3 \end{bmatrix},$$

$$\mathbf{H}(\mathbf{X}) = \begin{bmatrix} h_0 & 0 & 0 & h_1 & 0 & 0 & h_2 & 0 & 0 & h_3 & 0 & 0 \\ 0 & h_0 & 0 & 0 & h_1 & 0 & 0 & h_2 & 0 & 0 & h_3 & 0 \\ 0 & 0 & h_0 & 0 & 0 & h_1 & 0 & 0 & h_2 & 0 & 0 & h_3 \end{bmatrix}.$$

4.2. *Properties of area vectors*

Area vectors have a major significance with respect to the geometry of a tetrahedron for instance through the law of cosine. To write essential geometric relations, we need to introduce the following quantities:

- *normal vector* $\mathbf{n}_i$ of triangle T_i defined as the normalized area vector, $\mathbf{n}_i = \mathbf{m}_i / \|\mathbf{m}_i\|$. The normal vector is pointing outward if the tetrahedron $\mathcal{T}$ is positively oriented, i.e., if its volume $V(\mathcal{T})$ is positive;
- *dihedral angle* $\theta_{i,j}$ existing between triangle T_i and T_j and therefore between their normal vectors $\mathbf{n}_i$ and $\mathbf{n}_j$;
- *triangle area* A_i, area of triangle T_i;
- *edge length* $l_{i,j}$ is the length between vertex $\mathbf{p}_i$ and $\mathbf{p}_j$ (see Fig. 4.3);
- *foot height* f_i is the height of vertex $\mathbf{p}_i$ above triangle T_i (see Fig. 4.3).

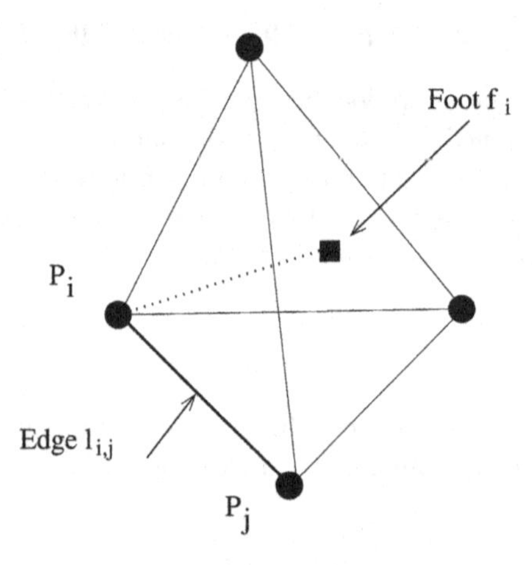

FIG. 4.3. Definition of foot height f_i and edge length $l_{i,j}$ in tetrahedron $\mathcal{T}$.

The definition of area vectors gives the relation

$$\mathbf{m}_i = 2A_i\mathbf{n}_i.$$

Noting that the tetrahedron volume is simply related to the foot height and area, we get

$$\mathbf{m}_i = \frac{2V(\mathcal{T})}{3f_i}\mathbf{n}_i.$$

From the relations $[\mathbf{P}^{-1}][\mathbf{P}] = \mathbf{I}_3$ and $[\mathbf{P}][\mathbf{P}^{-1}] = \mathbf{I}_3$, the following relations are obtained:

$$\sum_i \mathbf{m}_i = 0, \tag{4.4}$$

$$\sum_i \mathbf{m}_i \cdot \mathbf{p}_i = -18V(\mathcal{T}), \tag{4.5}$$

$$\sum_i \mathbf{m}_i \cdot \mathbf{p}_{i+1} = 6V(\mathcal{T}), \qquad (\mathbf{p}_{i+1} - \mathbf{p}_i) \cdot \mathbf{m}_i = 6V(\mathcal{T}),$$

$$\sum_{i \neq j,\, i<j} \mathbf{m}_i \cdot \mathbf{m}_j l_{i,j}^2 = 108V(\mathcal{T})^2, \qquad |\mathbf{m}_i, \mathbf{m}_{i+1}, \mathbf{m}_{i+2}| = (-1)^{i+1} 36V(\mathcal{T})^2.$$

The most important result is that all area vectors sum to zero. A result of this property is the *law of cosine*:

$$A_0^2 = A_1^2 + A_2^2 + A_3^2 - 2A_1A_2\cos\theta_{1,2} - 2A_1A_3\cos\theta_{1,3} - 2A_2A_3\cos\theta_{2,3}. \tag{4.6}$$

In fact, area vectors sum to zero for any closed triangulated surface. Indeed, through Green's formulae (BRONSHTEIN and SEMENDYAYEV [1985]), the sum of area vectors can be interpreted as the total flow of a constant field across a closed surface.

4.3. Computation of stiffness matrix: isotropic case

We use a displacement based finite element method which is equivalent to the classical Ritz analysis (BATHE [1982]). On a single tetrahedron, the (linear) isotropic elastic energy is equal to

$$W(\mathcal{T}) = \int_{\mathcal{T}} \left(\frac{\lambda}{2} (\operatorname{tr} \mathbf{E}_{\mathrm{L}})^2 + \mu \operatorname{tr} \mathbf{E}_{\mathrm{L}}^2 \right) \mathrm{d}V.$$

The gradient of the displacement $\nabla\mathbf{u}(\mathbf{X})$ is constant inside $\mathcal{T}$,

$$\nabla\mathbf{u}(\mathbf{X}) = -\sum_{i=0}^{3} \nabla \frac{\mathbf{m}_i \cdot (\mathbf{X} - \mathbf{p}_{i+1})}{6V(\mathcal{T})}\mathbf{u}_i = -\sum_{i=0}^{3} \frac{1}{6V(\mathcal{T})}\mathbf{m}_i \otimes \mathbf{u}_i,$$

where $\mathbf{m}_i \otimes \mathbf{u}_i = \mathbf{m}_i\mathbf{u}_i^{\mathrm{T}}$ is the tensor product of the two vectors,

$$\mathbf{m}_i \otimes \mathbf{u}_i = \begin{bmatrix} \mathbf{m}_i^x\mathbf{u}_i^x & \mathbf{m}_i^x\mathbf{u}_i^y & \mathbf{m}_i^x\mathbf{u}_i^z \\ \mathbf{m}_i^y\mathbf{u}_i^x & \mathbf{m}_i^y\mathbf{u}_i^y & \mathbf{m}_i^y\mathbf{u}_i^z \\ \mathbf{m}_i^z\mathbf{u}_i^x & \mathbf{m}_i^z\mathbf{u}_i^y & \mathbf{m}_i^z\mathbf{u}_i^z \end{bmatrix}.$$

The infinitesimal strain tensor $\mathbf{E}_\mathrm{L}$ is also constant inside $\mathcal{T}$:

$$\mathbf{E}_\mathrm{L} = \frac{1}{2}\left(\nabla\mathbf{u} + \nabla\mathbf{u}^\mathrm{T}\right) = \frac{-1}{12V(\mathcal{T})}\sum_{i=0}^{3}(\mathbf{m}_i \otimes \mathbf{u}_i + \mathbf{u}_i \otimes \mathbf{m}_i).$$

The first invariant $(\operatorname{tr}\mathbf{E}_\mathrm{L})^2$ is simply

$$(\operatorname{tr}\mathbf{E}_\mathrm{L})^2 = \frac{1}{144V(\mathcal{T})^2}\left(\sum_{i}^{3}\mathbf{m}_i \cdot \mathbf{u}_i\right)^2 = \frac{1}{144V(\mathcal{T})^2}\sum_{i,j=0}^{3}\mathbf{u}_i^\mathrm{T}[\mathbf{m}_i \otimes \mathbf{m}_j]\mathbf{u}_j.$$

The second invariant is slightly more complex to obtain:

$$\begin{aligned}\operatorname{tr}\mathbf{E}_\mathrm{L}^2 &= \frac{1}{144V(\mathcal{T})^2}\operatorname{tr}\Bigg(\sum_{i,j=0}^{3}(\mathbf{m}_i \otimes \mathbf{u}_j)(\mathbf{u}_i \cdot \mathbf{m}_j) + (\mathbf{u}_i \otimes \mathbf{m}_j)(\mathbf{m}_i \cdot \mathbf{u}_j) \\ &\qquad + (\mathbf{m}_i \otimes \mathbf{m}_j)(\mathbf{u}_i \cdot \mathbf{u}_j) + (\mathbf{u}_i \otimes \mathbf{u}_j)(\mathbf{m}_i \cdot \mathbf{m}_j)\Bigg) \\ &= \frac{1}{72V(\mathcal{T})^2}\sum_{i,j=0}^{3}\mathbf{u}_i^\mathrm{T}\left[(\mathbf{m}_j \otimes \mathbf{m}_i) + (\mathbf{m}_j \cdot \mathbf{m}_i)\mathbf{I}_3\right]\mathbf{u}_j.\end{aligned}$$

Finally, the linear elastic energy is a quadratic function of the displacement and is written as

$$W(\mathcal{T}) = \frac{1}{72V(\mathcal{T})}\sum_{i,j=0}^{3}\mathbf{u}_i^\mathrm{T}\left[\lambda(\mathbf{m}_i \otimes \mathbf{m}_j) + \mu(\mathbf{m}_j \otimes \mathbf{m}_i) + \mu(\mathbf{m}_i \cdot \mathbf{m}_j)\mathbf{I}_3\right]\mathbf{u}_j,$$

$$W(\mathcal{T}) = \frac{1}{2}\sum_{i,j=0}^{3}\mathbf{u}_i^\mathrm{T}[\mathbf{B}_{ij}^{\mathcal{T}}]\mathbf{u}_j, \tag{4.7}$$

where $[\mathbf{B}_{ij}^{\mathcal{T}}]$ is the 3×3 stiffness matrix of tetrahedron $\mathcal{T}$ between vertices i and j. Noting that $[\mathbf{m}_i \otimes \mathbf{m}_j]\mathbf{a} = \mathbf{m}_i(\mathbf{m}_j \cdot \mathbf{a})$, we can write the local elastic energy in terms of dot products,

$$\begin{aligned}W(\mathcal{T}) = \frac{1}{72V(\mathcal{T})}\sum_{i,j=0}^{3}\big(&\lambda(\mathbf{u}_i \cdot \mathbf{m}_i)(\mathbf{m}_j \cdot \mathbf{u}_j) + \mu(\mathbf{u}_i \cdot \mathbf{m}_j)(\mathbf{m}_i \cdot \mathbf{u}_j) \\ &+ \mu(\mathbf{m}_i \cdot \mathbf{m}_j)(\mathbf{u}_i \cdot \mathbf{u}_j)\big).\end{aligned}$$

Since $\mathbf{m}_i \otimes \mathbf{m}_j = (\mathbf{m}_j \otimes \mathbf{m}_i)^\mathrm{T}$, it is clear that local tensors are symmetric matrices: $[\mathbf{B}_{ij}^{\mathcal{T}}] = [\mathbf{B}_{ji}^{\mathcal{T}}]^\mathrm{T}$. Therefore, there are only 10 distinct local stiffness matrices with *four vertex matrices* $[\mathbf{B}_{ii}^{\mathcal{T}}]$ and six *edge matrices* $[\mathbf{B}_{ij}^{\mathcal{T}}]$, $i \neq j$.

4.3.1. Local vertex stiffness matrix

Vertex stiffness matrices take the simple form with normal vector $\mathbf{n}_i$:

$$[\mathbf{B}_{ii}^{\mathcal{T}}] = \frac{A_i^2}{9V(\mathcal{T})}((\lambda + \mu)(\mathbf{n}_i \otimes \mathbf{n}_i) + \mu\mathbf{I}_3). \tag{4.8}$$

These matrices have eigenvalues $(\lambda + 2\mu, \mu, \mu)$, $\mathbf{n}_i$ being the first eigenvector, and two directions orthogonal to $\mathbf{n}_i$ being the two other eigenvectors.

4.3.2. Local edge stiffness matrix

The stiffness matrix between vertex i and j is

$$[\mathbf{B}_{ij}^{\mathcal{T}}] = \frac{1}{36V(\mathcal{T})}\big(\lambda(\mathbf{m}_i \otimes \mathbf{m}_j) + \mu(\mathbf{m}_j \otimes \mathbf{m}_i) + \mu(\mathbf{m}_i \cdot \mathbf{m}_j)\mathbf{I}_3\big). \tag{4.9}$$

This matrix has the edge direction $(\mathbf{p}_j - \mathbf{p}_i)/\|\mathbf{p}_j - \mathbf{p}_i\|$ as first eigenvector associated with eigenvalue $(\mu(\mathbf{m}_i \cdot \mathbf{m}_j))/(36V(\mathcal{T}))$. The existence of the other two eigenvectors depends on the sign of the following matrix determinant:

$$\begin{vmatrix} (\lambda+\mu)(\mathbf{m}_i \cdot \mathbf{m}_j) & \mu\|\mathbf{m}_i\|^2 \\ \lambda\|\mathbf{m}_j\|^2 & 2\mu(\mathbf{m}_i \cdot \mathbf{m}_j) \end{vmatrix} = \lambda\mu A_i^2 A_v^2\left(2\left(1+\frac{\lambda}{\mu}\right)\cos^2\theta_{i,j} - 1\right).$$

4.3.3. Global stiffness matrix

The elastic energy of the whole deformed body is then computed by summing Eq. (4.7) over all tetrahedra. This total energy $W(\mathcal{M}_{\text{def}})$ may be written with the displacement vector $\mathbf{U}$, gathering all displacement vectors $\mathbf{u}_i$, and a global stiffness matrix $\mathbf{K}$:

$$W(\mathcal{M}_{\text{def}}) = \frac{1}{2}\mathbf{U}^{\mathrm{T}}\mathbf{K}\mathbf{U}. \tag{4.10}$$

This stiffness matrix $\mathbf{K}$ is built by assembling local stiffness matrices $[\mathbf{B}_{ij}^{\mathcal{T}}]$. Because these local matrices are symmetric with respect to the swap of indices $[\mathbf{B}_{ij}^{\mathcal{T}}] = [\mathbf{B}_{ji}^{\mathcal{T}}]$, the global stiffness matrix $\mathbf{K}$ is symmetric.

4.3.4. Global vertex stiffness matrix

The 3×3 submatrix $[\mathbf{K}_{i,j}]$ associated with vertices i (row index) and j (column index) is computed as the sum of local stiffness matrices for all tetrahedra containing both vertices. The set of tetrahedra adjacent to a given vertex (respectively edge) is called the *shell* $\mathcal{S}(i)$ of this vertex (respectively edge). In particular, for diagonal submatrices, we get

$$[\mathbf{K}_{i,i}] = \sum_{\mathcal{T}\in\mathcal{S}(i)} \frac{1}{36V(\mathcal{T})}\Big((\lambda_{\mathcal{T}} + \mu_{\mathcal{T}})(\mathbf{m}_i \otimes \mathbf{m}_i) + \mu_{\mathcal{T}} A_i^2 \mathbf{I}_3\Big).$$

In fact, we can provide a rather simple interpretation of this matrix expression. Its first term can be seen as the inertial matrix (second order moment) of area vectors $\mathbf{m}_i$ weighted by $(\lambda_{\mathcal{T}} + \mu_{\mathcal{T}})/(72V(\mathcal{T}))$ (see Fig. 4.4). Indeed, on a manifold tetrahedral mesh (but not on all conformal tetrahedral meshes), the shell of an interior vertex is homeomorphic to a sphere and it can be easily proved that the sum of its area vectors is null (Minkowsky's sum). Therefore $\mathbf{m}_i \otimes \mathbf{m}_i$ represents the local contribution to an inertia matrix. If vertex $\mathbf{p}_i$ is surrounded by semi-regular tetrahedra, then the matrix of inertia is proportional to identity. Note also that because it is weighted by the inverse of the tetrahedron's volume, it is very sensitive to the disparity in tetrahedra size. The second part is simply the sum of the second Lamé coefficient weighted with the inverse of the tetrahedron volume.

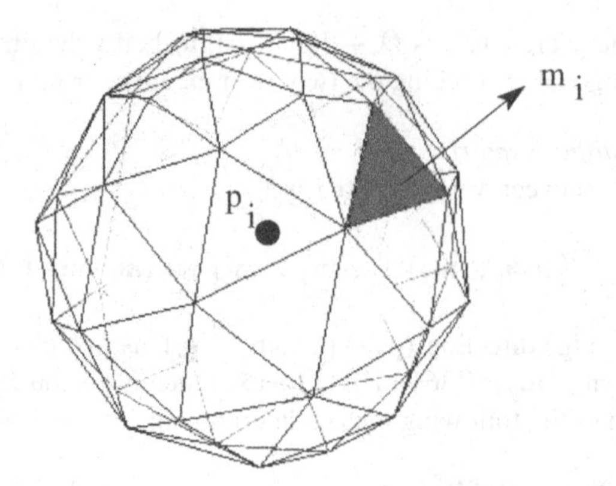

FIG. 4.4. Shell of a vertex $\mathbf{p}_i$ in a tetrahedral mesh: only the opposite triangle is drawn. For each triangle the area vector $\mathbf{m}_i$ is pointing outward. The sum of these area vectors is null and their weighted matrix of inertia determines the global stiffness matrix at $\mathbf{p}_i$.

4.3.5. Global edge stiffness matrix

The off-diagonal terms $[\mathbf{K}_{i,j}]$ of the stiffness matrix correspond to edge stiffness matrices that are the sum of local edge stiffness matrices. The edge direction $(\mathbf{p}_j - \mathbf{p}_i)/\|\mathbf{p}_j - \mathbf{p}_i\|$ is an eigenvector of $[\mathbf{K}_{i,j}]$ associated with the eigenvalue $k_{i,j}$:

$$k_{i,j} = \sum_{\mathcal{T} \in \mathcal{S}(i,j)} \frac{\mu_{\mathcal{T}}(\mathbf{m}_i \cdot \mathbf{m}_j)}{36V(\mathcal{T})},$$

where $\mathcal{S}(i, j)$ is the set of tetrahedra adjacent to that edge (its shell). The tetrahedron volume $V(\mathcal{T})$ can be written as a function of triangle areas,

$$V(\mathcal{T}) = \frac{2}{3l_{i,j}^{\mathrm{opp}}} A_i A_j \sin\theta_{i,j},$$

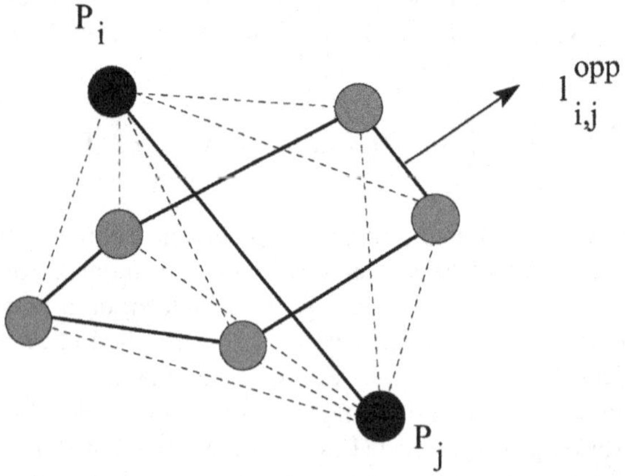

FIG. 4.5. Shell of an edge linking vertices $\mathbf{p}_i$ and $\mathbf{p}_j$ in a tetrahedral mesh: the adjacent tetrahedra are drawn with dashed lines whereas opposite edges to that edge are drawn with solid lines. One of the eigenvalues of edge stiffness matrix depends on the weighted sum of the lengths $l_{i,j}^{\mathrm{opp}}$.

where $l_{i,j}^{\text{opp}}$ is the length of the opposite edge in tetrahedron $\mathcal{T}$ (see Fig. 4.5).

$$k_{i,j} = \frac{1}{6} \sum_{\mathcal{T} \in \mathcal{S}(i,j)} \mu_{\mathcal{T}} l_{i,j}^{\text{opp}} \cot \theta_{i,j}.$$

4.4. Physical interpretation of isotropic stiffness matrix

Eq. (4.7) describes the local stiffness matrices and it can be interpreted as the sum of discrete differential operators. Indeed, the isotropic elastic energy related to the first invariant of the infinitesimal strain tensors, can be written as a quadratic functional of the displacement vector:

$$W(\mathbf{X}) = \frac{\lambda}{2}(\operatorname{tr} \mathbf{E}_{\mathrm{L}})^2 + \mu \operatorname{tr} \mathbf{E}_{\mathrm{L}}^2 = \frac{\lambda}{2}(\operatorname{div} \mathbf{u})^2 + \mu \operatorname{tr}\left(\nabla \mathbf{u}^{\mathrm{T}} \nabla \mathbf{u}\right) - \frac{\mu}{2} \| \mathbf{curl}\, \mathbf{u} \|^2.$$

The first variation of the elastic force $-\delta W$ can be interpreted as the density of linear elastic force per unit volume, and is given by the Lamé equation:

$$-\delta W = (\lambda + \mu) \nabla (\operatorname{div} \mathbf{u}) + \mu \Delta \mathbf{u}.$$

It is therefore natural to compare the Lamé equation with the expression of the discrete elastic force $\mathbf{F}_i(\mathcal{T})$ acting on a vertex i of tetrahedron $\mathcal{T}$:

$$\mathbf{F}_i(\mathcal{T}) = -\sum_{j=0}^{3} \left[\mathbf{B}_{ij}^{\mathcal{T}}\right] \mathbf{u}_j = \frac{-1}{36 V(\mathcal{T})} \sum_{j=0}^{3} \left[\lambda(\mathbf{m}_i \otimes \mathbf{m}_j) + \mu(\mathbf{m}_j \otimes \mathbf{m}_i) + \mu(\mathbf{m}_i \cdot \mathbf{m}_j)\mathbf{I}_3\right] \mathbf{u}_j.$$

The three terms of the local rigidity matrix may be interpreted as follows:

$$\mathbf{F}_i(\mathcal{T}) = \frac{-1}{36 V(\mathcal{T})} \sum_{j=0}^{3} \left[\underbrace{\lambda(\mathbf{m}_i \otimes \mathbf{m}_j)}_{\substack{\mathrm{T}_1 \\ \nabla(\operatorname{div} \mathbf{u}) \\ \text{operator}}} + \underbrace{\mu(\mathbf{m}_j \otimes \mathbf{m}_i)}_{\substack{\mathrm{T}_2 \\ \nabla(\operatorname{div} \mathbf{u}) \\ \text{pseudo-operator}}} + \underbrace{\mu(\mathbf{m}_i \cdot \mathbf{m}_j)\mathbf{I}_3}_{\substack{\mathrm{T}_3 \\ \Delta \mathbf{u} \\ \text{operator}}} \right] \mathbf{u}_j.$$

The first term of the local rigidity matrix corresponds to the integral of the operator $\nabla(\operatorname{div} \mathbf{u})$ over a subdomain of tetrahedron $\mathcal{T}$. Indeed, through Green's second formulae (BRONSHTEIN and SEMENDYAYEV [1985]), the integral over a domain D of that operator can be evaluated along its boundary ∂D:

$$\int_D \nabla(\operatorname{div} \mathbf{u}) = \int_{\partial D} (\operatorname{div} \mathbf{u}) \mathbf{n} \, \mathrm{d}S.$$

However, the divergence operator is actually constant over $\mathcal{T}$ and is equal to

$$\frac{-1}{6 V(\mathcal{T})} \sum_{j=0}^{3} \mathbf{m}_j \cdot \mathbf{u}_j.$$

Furthermore, since the integral $\int \mathbf{n}\,\mathrm{d}S$ over each triangle of $\mathcal{T}$ is equal to $\frac{1}{2}\mathbf{m}_i$, T_1 is equal to one third of the flux of $(\operatorname{div}\mathbf{un})$ through the face of $\mathcal{T}$ opposite to vertex $\mathbf{p}_i$:

$$T_1 = \frac{-1}{36V(\mathcal{T})}\sum_{j=0}^{3}\left[(\mathbf{m}_i \otimes \mathbf{m}_j)\right]\mathbf{u}_j = \frac{1}{3}\,\frac{1}{6V(\mathcal{T})}\left(\sum_{j=0}^{3}\mathbf{m}_j \cdot \mathbf{u}_j\right)\frac{\mathbf{m}_i}{2}.$$

Thus, we can provide a straightforward interpretation on the first term of the local rigidity tensor: it corresponds to the integration of the $\nabla(\operatorname{div}\mathbf{u})$ operator over a subdomain $\mathcal{D}_i$ for which

$$\int_{\mathcal{T}\mathcal{D}_i} \mathbf{n}\,\mathrm{d}S = \frac{\mathbf{m}_i}{6}.$$

A natural choice for this subdomain is to consider the shell $\mathcal{S}_i$ of vertex $\mathbf{p}_i$ homothetically scaled down by a ratio of $1/\sqrt{3}$. This subdomain is sketched in Fig. 4.6(a) and (b): the vertices of the subdomain are located at distance of $1/\sqrt{3}$ times the original edge length from vertex $\mathbf{p}_i$. Unfortunately, since $1/\sqrt{3} > 0.5$, the subdomain of two neighboring vertices overlap.

To obtain a non-overlapping subdomain $\mathcal{D}_i$, one should consider the subdomain defined by the middle of each edge, the barycenters of each triangles and the barycenter of the tetrahedron, as proposed by PUTTI and CORDES [1998]. More precisely, as

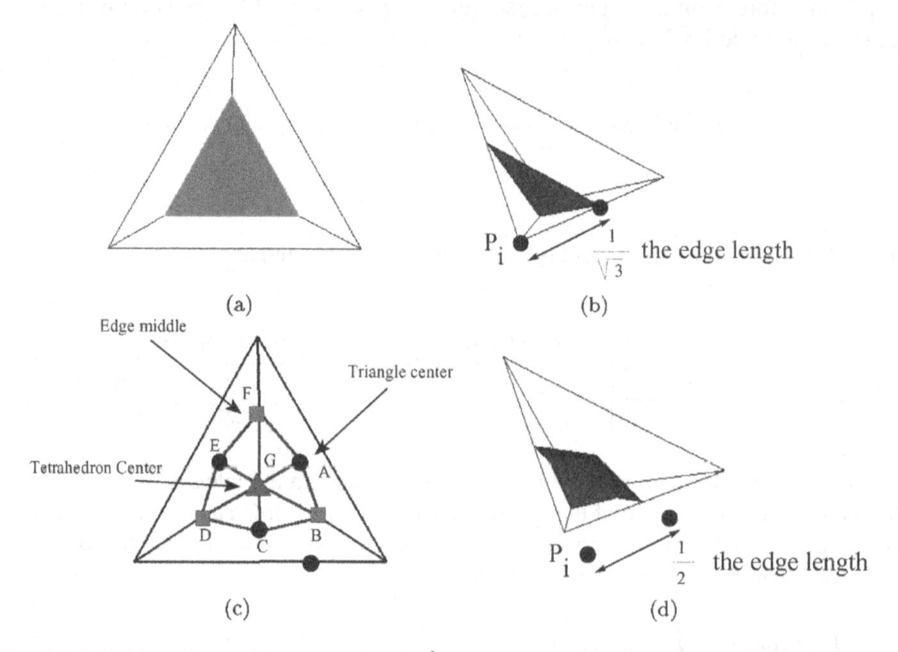

FIG. 4.6. Definitions of two subdomains for which $\int \mathbf{n}\,\mathrm{d}S$ is equal to one third the value through triangle Δ_i, opposite of vertex $\mathbf{p}_i$ in tetrahedron $\mathcal{T}$; (a) and (b): front and side view of the first subdomain consisting of a single triangle corresponding to the homothety Δ_i with a ratio of $1/\sqrt{3}$; (c) and (d): front and side view of the second non-overlapping subdomain consisting of 6 triangles linking the edge middles, triangle centers and tetrahedron center.

shown in Fig. 4.6(c), the subdomain consists in the six triangles $(FAG, GAB, BGC, CGD, DGE, EGF)$ where A, C, E are the centers of the three triangles adjacent to $\mathbf{p}_i$, B, D, F are the centers of the three adjacent edges and G is the tetrahedron barycenter. This subdomain is called *the barycentric dual cell* in Cosmi [2001].

Indeed, the flux over the six triangles may be written as a sum of cross products,

$$\int_{\mathcal{D}_i} \mathbf{n}\,\mathrm{d}S = A \times B + B \times C + C \times D + D \times E + E \times F.$$

Since A, B, C, D, E, F, G are simple barycentric coordinates of the four tetrahedron vertices $\mathbf{p}_i, \mathbf{p}_j, \mathbf{p}_k, \mathbf{p}_l$, it can be simply evaluated as a function of these vertices,

$$\int_{\mathcal{D}_i} \mathbf{n}\,\mathrm{d}S = \frac{1}{6}(\mathbf{p}_j \times \mathbf{p}_k + \mathbf{p}_k \times \mathbf{p}_l + \mathbf{p}_l \times \mathbf{p}_j) = \frac{\mathbf{m}_i}{6}.$$

Thus, to summarize, we have proved so far that term T_1 is the integral of the $\nabla(\operatorname{div}\mathbf{u})$ operator over a non-overlapping subdomain centered on $\mathbf{p}_i$.

The second term T_2 of the local rigidity matrix is the transposed matrix of the first term T_1 but cannot be interpreted in terms of a linear differential operator. In fact, if we write T_2 as $\sum(\mathbf{m}_i \cdot \mathbf{u}_j)\mathbf{m}_j$ we can state that T_2 corresponds to the flux of a scalar field equal to $\frac{1}{12V(\mathcal{T})}(\mathbf{m}_i \cdot \mathbf{u}_j)$ over each face of the subdomain $\mathcal{D}_i$. It should be noticed that T_2 has no equivalent in the continuous formulation (the Lamé equation) and is produced by the evaluation of $\|\mathbf{curl}\,\mathbf{u}\|^2$.

The third term T_3 corresponds to the discrete Laplacian operator and its expression originates from the evaluation of $\frac{1}{2}\operatorname{tr}(\nabla\mathbf{u}^{\mathrm{T}}\nabla\mathbf{u})$. The same approach as for the $\nabla(\operatorname{div}\mathbf{u})$ can be applied. First, the integral of the Laplacian operator is integrated over a domain D using the integral Gauss theorem. For the x component u^x of the displacement field, it gives

$$\int_D \Delta u^x\,\mathrm{d}V = \int_D \nabla \cdot (\nabla u^x)\,\mathrm{d}V = \int_{\partial D} (\nabla u^x) \cdot \mathbf{n}\,\mathrm{d}S.$$

If the domain D is included inside a tetrahedron, then the gradient of the displacement field is a constant vector,

$$\nabla u^x = \frac{-1}{6V(\mathcal{T})} \sum_{j=0}^{3} \mathbf{m}_j u_j^x.$$

If we suppose that the domain boundary coincides with triangle Δ_i, opposite to $\mathbf{p}_i$ in tetrahedron $\mathcal{T}$, then we get

$$\int_D \Delta u^x\,\mathrm{d}V = \frac{-1}{6V(\mathcal{T})} \sum_{j=0}^{3} \mathbf{m}_j \frac{1}{6V(\mathcal{T})} \sum_{j=0}^{3} \mathbf{m}_j u_j^x \cdot \frac{\mathbf{m}_i}{2}.$$

Therefore, T_3 corresponds to the integral of the Laplacian operator over a domain $\mathcal{D}$ for which

$$\int_{\partial\mathcal{D}\mathcal{T}} \mathbf{n}\,\mathrm{d}S = \frac{\mathbf{m}_i}{2},$$

for instance the subdomain defined in Fig. 4.6(c), corresponding to the barycentric dual cell of vertex $\mathbf{p}_i$ in tetrahedron $\mathcal{T}$. The Finite Element approximation of the Laplacian operator on tetrahedra was previously studied by PUTTI and CORDES [1998], DEBUNNE, DESBRUN, CANI and BARR [2001] and COSMI [2001].

To summarize, we have proved that the *variational formulation* of linear elasticity over tetrahedral meshes is not completely equivalent to the Finite Difference and Finite Volume methods. Indeed, the latter methods are equivalent to the *differential formulation* of Finite Element method which leads to the following equation of the elastic force:

$$\mathbf{F}_i(\mathcal{T}) = \frac{-1}{36V(\mathcal{T})} \sum_{j=0}^{3} \left[(\lambda + \mu)(\mathbf{m}_i \otimes \mathbf{m}_j) + \mu(\mathbf{m}_i \cdot \mathbf{m}_j)\mathbf{I}_3\right] \mathbf{u}_j. \tag{4.11}$$

The variational formulation of the FEM creates the stiffness matrix from the expression of the elastic energy whereas the differential formulation of the FEM is based on the Lamé differential equation.

4.5. *Computation of stiffness matrix: transversally anisotropic case*

From Section 3.2.4, the density of elastic energy for a transversally anisotropic material can be derived from the isotropic case by adding a corrective term:

$$W(\mathbf{X})_{\text{Transv.Ani}} = W(\mathbf{X}) + \Delta W_{\text{Ani}}(\mathbf{X}),$$
$$W(\mathbf{X})_{\text{Transv.Ani}} = W(\mathbf{X}) + \Delta\lambda I_1 I_4 + 2\Delta\mu\, I_5 - \left(\frac{\Delta\lambda}{2} + \Delta\mu\right) I_4^2,$$

where $\Delta\lambda$ and $\Delta\mu$ are the variation of Lamé coefficient in the direction of anisotropy $\mathbf{a}_0$ and where I_4 and I_5 are the constants defined in Eqs. (3.7) and (3.8). The evaluation of I_4 and I_5 with the linear tetrahedron finite element gives:

$$I_4 = \frac{-1}{6V(\mathcal{T})} \sum_{i=0}^{3} (\mathbf{a}_0 \cdot \mathbf{m}_i)(\mathbf{a}_0 \cdot \mathbf{u}_i),$$
$$(\operatorname{tr} \mathbf{E}_\text{L}) I_4 = \frac{1}{72V(\mathcal{T})^2} \mathbf{u}_i^\text{T} \left[(\mathbf{a}_0 \cdot \mathbf{m}_j)(\mathbf{m}_i \otimes \mathbf{a}_0)\right] \mathbf{u}_j,$$
$$I_4^2 = \frac{1}{36V(\mathcal{T})^2} \sum_{i,j=0}^{3} \mathbf{u}_i^\text{T} \left[(\mathbf{a}_0 \cdot \mathbf{m}_i)(\mathbf{a}_0 \cdot \mathbf{m}_j)(\mathbf{a}_0 \otimes \mathbf{a}_0)\right] \mathbf{u}_j.$$

Similarly for I_5:

$$I_5 = \frac{1}{144V(\mathcal{T})^2} \sum_{i,j=0}^{3} \mathbf{u}_i^\text{T} \left[(\mathbf{a}_0 \cdot \mathbf{m}_j)(\mathbf{a}_0 \otimes \mathbf{m}_i) + (\mathbf{a}_0 \cdot \mathbf{m}_i)(\mathbf{m}_j \otimes \mathbf{a}_0) \right.$$
$$\left. + (\mathbf{m}_i \cdot \mathbf{m}_j)(\mathbf{a}_0 \otimes \mathbf{a}_0) + (\mathbf{a}_0 \cdot \mathbf{m}_i)(\mathbf{a}_0 \cdot \mathbf{m}_j)\mathbf{I}_3\right] \mathbf{u}_j.$$

Thus the additional elastic energy term $\Delta W(\mathcal{T})_{\text{Ani}}$ due to transversal anisotropy can also be written as a bilinear function of vertex displacements,

$$\Delta W(\mathcal{T})_{\text{Ani}} = \frac{1}{2}\sum_{i,j=0}^{3} \mathbf{u}_i^{\text{T}}[\mathbf{A}_{ij}^{\mathcal{T}}]\mathbf{u}_j \tag{4.12}$$

with the local 3×3 matrix $[\mathbf{A}_{ij}^{\mathcal{T}}]$ being defined as

$$\begin{aligned}
[\mathbf{A}_{ij}^{\mathcal{T}}] = \frac{1}{144V(\mathcal{T})}\big(&\Delta\lambda(\mathbf{a}_0\cdot\mathbf{m}_j)(\mathbf{m}_i\otimes\mathbf{a}_0)\\
&-(\Delta\lambda+2\Delta\mu)(\mathbf{a}_0\cdot\mathbf{m}_i)(\mathbf{a}_0\cdot\mathbf{m}_j)(\mathbf{a}_0\otimes\mathbf{a}_0)\\
&+\Delta\mu(\mathbf{a}_0\cdot\mathbf{m}_j)(\mathbf{a}_0\otimes\mathbf{m}_i)+\Delta\mu(\mathbf{a}_0\cdot\mathbf{m}_i)(\mathbf{m}_j\otimes\mathbf{a}_0)\\
&+\Delta\mu(\mathbf{m}_i\cdot\mathbf{m}_j)(\mathbf{a}_0\otimes\mathbf{a}_0)+\Delta\mu(\mathbf{a}_0\cdot\mathbf{m}_i)(\mathbf{a}_0\cdot\mathbf{m}_j)\mathbf{I}_3\big).
\end{aligned}$$

4.5.1. Local vertex stiffness matrix

When $i = j$, the vertex stiffness matrix is written as

$$\begin{aligned}
[\mathbf{A}_{ii}^{\mathcal{T}}] = \frac{1}{144V(\mathcal{T})}\Big[&(\Delta\lambda+\Delta\mu)(\mathbf{a}_0\cdot\mathbf{m}_i)(\mathbf{m}_i\otimes\mathbf{a}_0).\\
&-(\Delta\lambda+2\Delta\mu)(\mathbf{a}_0\cdot\mathbf{m}_i)^2(\mathbf{a}_0\otimes\mathbf{a}_0)\\
&+\Delta\mu(\mathbf{a}_0\cdot\mathbf{m}_i)(\mathbf{a}_0\otimes\mathbf{m}_i)+\Delta\mu\|\mathbf{m}_i\|^2(\mathbf{a}_0\otimes\mathbf{a}_0)\\
&+\Delta\mu(\mathbf{a}_0\cdot\mathbf{m}_i)^2\mathbf{I}_3\Big].
\end{aligned}$$

This matrix has $\mathbf{c}_0$, the unit vector orthogonal to both $\mathbf{a}_0$ and $\mathbf{m}_i$ as first eigenvector with eigenvalue $\Delta\mu(\mathbf{a}_0\cdot\mathbf{m}_i)^2$. The existence of the other two eigenvectors, in the plane defined by $\mathbf{a}_0$ and $\mathbf{m}_i$, depends on the sign of $(2\Delta\mu+\Delta\lambda)(\mathbf{a}_0\cdot\mathbf{m}_i)^2 - \Delta\lambda\|\mathbf{m}_i\|^2$.

4.5.2. Global stiffness matrix

For a transversally anisotropic material, the global stiffness matrix $\mathbf{K}$ is assembled as the sum of local isotropic and anisotropic stiffness matrices:

$$[\mathbf{K}_{i,j}] = \sum_{\mathcal{T}\in\mathcal{S}(i,j)} [\mathbf{B}_{i,j}] + [\mathbf{A}_{i,j}]. \tag{4.13}$$

One should note that the global matrix $[\mathbf{K}_{i,j}]$ contains non-null values only if vertices i and j are linked by an edge of the tetrahedral mesh.

4.6. Work of gravity forces

We calculate the potential energy of gravity forces when a displacement field $\mathbf{u}(\mathbf{X})$ is applied on the body $\mathcal{M}_{\text{def}}$. If we write $\mathbf{g}$ the gravity vector ($\|\mathbf{g}\| = 9.8$ m/s^2) and ρ the density of the material (assumed constant for the whole body), then the potential energy of a tetrahedron $\mathcal{T}$ is a simple function of the center of mass $\mathcal{T}$:

$$W_g(\mathcal{T}) = \int_{\mathcal{T}} \rho\mathbf{X}\cdot\mathbf{g}\,\mathrm{d}V = \rho\int_{\mathcal{T}}\mathbf{X}\,\mathrm{d}V\cdot\mathbf{g} = \rho V(\mathcal{T})\frac{\mathbf{q}_0+\mathbf{q}_1+\mathbf{q}_2+\mathbf{q}_3}{4}\cdot\mathbf{g}.$$

If we drop the constant part of this energy, which is equivalent to consider the work of gravity forces when a displacement field $\mathbf{u}(\mathbf{X})$ is applied, then we get

$$W_g(T) = \rho V(T) \frac{\mathbf{u}_0 + \mathbf{u}_1 + \mathbf{u}_2 + \mathbf{u}_3}{4} \cdot \mathbf{g}$$
$$= \frac{\rho V(T)\mathbf{g}}{4} \begin{bmatrix} \mathbf{u}_0 & \mathbf{u}_1 & \mathbf{u}_2 & \mathbf{u}_3 \end{bmatrix} \begin{bmatrix} 1 \\ 1 \\ 1 \\ 1 \end{bmatrix}.$$

The potential energy of the whole model $\mathcal{M}_{\text{def}}$ is the dot product of the following two vectors:

$$W_g = \sum_T W_g(T) = \mathbf{U}^{\mathrm{T}} \mathbf{R}^g = \mathbf{U}^{\mathrm{T}} \begin{bmatrix} \cdots \\ \mathbf{r}_i^g \\ \cdots \end{bmatrix},$$

where $\mathbf{R}^g$ is a vector of size $3N$. More precisely, the sub-vector $\mathbf{r}_i^g$ of $\mathbf{R}^g$ associated with vertex i is proportional to the gravity vector, the coefficient being the volume of its neighboring tetrahedra:

$$\mathbf{r}_i^g = \rho \left(\sum_{T \in \mathcal{S}(i)} \frac{V(T)}{4} \right) \mathbf{g}. \tag{4.14}$$

4.7. *Work of external surface pressure*

Among external forces acting on deformable soft tissue models, we include a pressure force $\mathbf{f}_p$ which is applied on a part of its surface. We consider that such pressure force has a constant intensity $\|\mathbf{f}_p\| = p$ but its direction may be either constant (contact with a stream of gas) or directed along the surface normal (contact with a solid, fluid or gas at low speed). In the latter case, the force applied on a triangle T is

$$\mathbf{f}_p(T) = p\, \mathbf{n}(T).$$

For a tetrahedral mesh, we consider that such constant pressure $\mathbf{f}_p$ is applied on a set $\mathcal{C}$ of surface triangles. If we consider a triangle $T \in \mathcal{C}$ consisting of vertices $(\mathbf{p}_i, \mathbf{p}_j, \mathbf{p}_k)$, the work of $\mathbf{f}_p$ on this triangle is

$$W_p(T) = \int_T \mathbf{f}_p \cdot \mathbf{u}(\mathbf{X})\, \mathrm{d}A = A(T)\, \mathbf{f}_p \cdot \left(\frac{\mathbf{u}_i + \mathbf{u}_j + \mathbf{u}_k}{3} \right).$$

The work of external surface pressure on the whole model $\mathcal{M}_{\text{def}}$ is then

$$W_g = \mathbf{U}^{\mathrm{T}} \mathbf{R}^p = \mathbf{U}^{\mathrm{T}} \begin{bmatrix} \cdots \\ \mathbf{r}_i^p \\ \cdots \end{bmatrix} \tag{4.15}$$

where $\mathbf{r}_i^p$ is null if vertex $\mathbf{p}_i$ is not adjacent to any triangles in $\mathcal{C}$ and is proportional the sum of triangles area otherwise:

$$\mathbf{r}_i^p = \sum_{\mathbf{p}_i \in T,\, T \in \mathcal{C}} \frac{A(T)\mathbf{f}_p(T)}{3}.$$

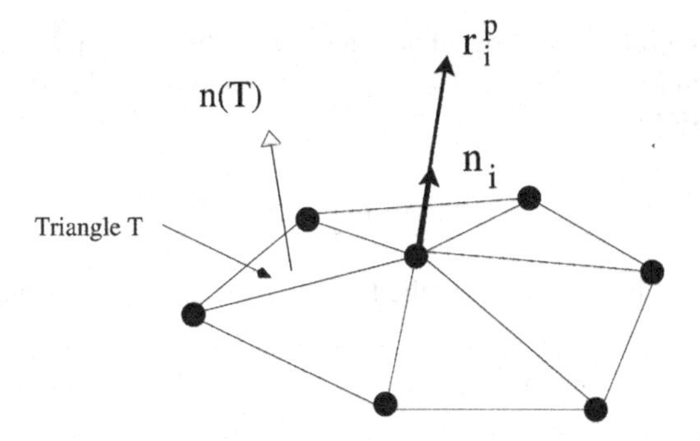

FIG. 4.7. The pressure applied on neighboring triangles results in a force directed along the surface normal at a vertex and proportional to the sum of neighboring triangle area. The vertex surface normal $\mathbf{n}_i$ is computed as the weighted average of triangle normals.

If the pressure force is applied along the surface normal, then vector $\mathbf{r}_i^p$ has an intuitive formulation. The nodal force, resulting from the pressure applied on neighboring triangle, is proportional to the area sum of surrounding triangles and is directed along the surface normal $\mathbf{n}_i$ at vertex $\mathbf{p}_i$ (see Fig. 4.7):

$$\mathbf{r}_i^p = \frac{p}{3}\left(\sum_{\mathbf{p}_i \in T,\, T \in \mathcal{C}} A(T)\right)\mathbf{n}_i,$$

where $\mathbf{n}_i$ is computed as the average of surrounding triangle normals $\mathbf{n}(T)$ weighted by their area,

$$\mathbf{n}_i = \frac{\sum_{T \in \mathcal{C}} \mathbf{n}(T)A(T)}{\sum_{T \in \mathcal{C}} A(T)}.$$

4.8. *Mass matrix*

The mass matrix is derived from the evaluation of the kinetic energy $\mathcal{E}(\mathcal{M}_{\text{def}})$ on the whole body $\mathcal{M}_{\text{def}}$. The density of kinetic energy $w(\mathbf{X}) = \rho(\dot{\mathbf{u}}(\mathbf{X}))^2$ where $\dot{\mathbf{u}}(\mathbf{X}) = \mathrm{d}\mathbf{u}/\mathrm{d}t$ is the speed of the material point $\mathbf{X}$. It follows that the kinetic energy of tetrahedron $\mathcal{T}$ is a bilinear form of the speed of nodal vertices $\dot{\mathbf{u}}_i$:

$$\mathcal{E}(\mathcal{T}) = \begin{bmatrix} \dot{\mathbf{U}}_0 \\ \dot{\mathbf{U}}_1 \\ \dot{\mathbf{U}}_2 \\ \dot{\mathbf{U}}_3 \end{bmatrix}^{\mathrm{T}} \begin{bmatrix} \mathbf{M}_{0,0}^{\mathcal{T}} & \mathbf{M}_{0,1}^{\mathcal{T}} & \mathbf{M}_{0,2}^{\mathcal{T}} & \mathbf{M}_{0,3}^{\mathcal{T}} \\ \mathbf{M}_{1,0}^{\mathcal{T}} & \mathbf{M}_{1,1}^{\mathcal{T}} & \mathbf{M}_{1,2}^{\mathcal{T}} & \mathbf{M}_{1,3}^{\mathcal{T}} \\ \mathbf{M}_{2,0}^{\mathcal{T}} & \mathbf{M}_{2,1}^{\mathcal{T}} & \mathbf{M}_{2,2}^{\mathcal{T}} & \mathbf{M}_{2,3}^{\mathcal{T}} \\ \mathbf{M}_{3,0}^{\mathcal{T}} & \mathbf{M}_{3,1}^{\mathcal{T}} & \mathbf{M}_{3,2}^{\mathcal{T}} & \mathbf{M}_{3,3}^{\mathcal{T}} \end{bmatrix} \begin{bmatrix} \dot{\mathbf{U}}_0 \\ \dot{\mathbf{U}}_1 \\ \dot{\mathbf{U}}_2 \\ \dot{\mathbf{U}}_3 \end{bmatrix}.$$

This tetrahedron mass matrix has size 12×12 and is composed of 4×4 local mass matrix between vertex i and j, $\mathbf{M}_{i,j}^{\mathcal{T}}$ that are 3×3 diagonal matrices,

$$\mathbf{M}_{i,j}^{\mathcal{T}} = \rho\left(\int_{\mathcal{T}} h_i(\mathbf{X})h_j(\mathbf{X})\,\mathrm{d}V\right)\mathbf{I}_3.$$

To evaluate the integral, we use the 3 barycentric coordinates (h_0, h_1, h_2) as integration variables. Based on Eqs. (4.3) and (4.4), the determinant of the Jacobian matrix is equal to the inverse of $6V(\mathcal{T})$,

$$\left|\frac{\partial h_0}{\partial \mathbf{X}} \frac{\partial h_1}{\partial \mathbf{X}} \frac{\partial h_2}{\partial \mathbf{X}}\right| = \frac{1}{216V(\mathcal{T})^3}|\mathbf{m}_0\ \mathbf{m}_1\ \mathbf{m}_2| = \frac{1}{6V(\mathcal{T})}.$$

Thus the integral can be computed explicitly using the expression below:

$$\begin{aligned}\int_{\mathcal{T}} h_i(\mathbf{X})h_j(\mathbf{X})\,\mathrm{d}V &= 6V(\mathcal{T})\int_0^1\int_0^{1-h_0}\int_0^{1-h_0-h_1} h_i\,h_j\,\mathrm{d}h_0\,\mathrm{d}h_1\,\mathrm{d}h_2 \\ &= \frac{V}{10} \quad \text{if } i = j \\ &= \frac{V}{20} \quad \text{if } i \neq j.\end{aligned}$$

Thus the local mass matrix $\mathbf{M}_{i,j}^{\mathcal{T}}$ is equal to $\frac{\rho V(\mathcal{T})}{10}\mathbf{I}_3$ if $i = j$ and to $\frac{\rho V(\mathcal{T})}{20}\mathbf{I}_3$, otherwise. If we perform *mass lumping* by considering only diagonal elements equal to the sum of row values, then we naturally get $\frac{\rho V(\mathcal{T})}{4}\mathbf{I}_3$, as if the tetrahedron mass is evenly spread over its four vertices.

The kinetic energy of the whole body can be written as a function of the global mass matrix built by assembling the local matrices $\mathbf{M}_{i,j}^{\mathcal{T}}$,

$$\mathcal{E}(\mathcal{M}_{\text{def}}) = \frac{1}{2}\dot{\mathbf{U}}^{\mathrm{T}}\mathbf{M}\dot{\mathbf{U}} = \frac{1}{2}\dot{\mathbf{U}}^{\mathrm{T}}[\mathbf{M}_{i,j}]\dot{\mathbf{U}},$$

where $\mathbf{M}_{i,j}$, the global 3×3 mass matrix between vertex i and j, depends on the volumes of tetrahedra adjacent to vertex i (if $i = j$) or tetrahedra adjacent to edge (i, j) if $i \neq j$:

$$\mathbf{M}_{i,i} = \rho \sum_{\mathcal{T}\in\mathcal{S}(i)} \frac{V(\mathcal{T})}{10}\mathbf{I}_3, \tag{4.16}$$

$$\mathbf{M}_{i,j} = \rho \sum_{\mathcal{T}\in\mathcal{S}(i,j)} \frac{V(\mathcal{T})}{20}\mathbf{I}_3 \quad \text{if } i \neq j. \tag{4.17}$$

If we perform *mass lumping* to get a diagonal mass matrix $\mathbf{M}$ (and therefore easily invertible), then the vertex mass is equal to one fourth of the mass of its adjacent tetrahedra:

$$(\mathbf{M}_{i,i})_{\text{lumping}} = \rho \sum_{\mathcal{T}\in\mathcal{S}(i)} \frac{V(\mathcal{T})}{4}\mathbf{I}_3.$$

4.9. Boundary conditions

In a surgical simulator, the boundary conditions of a soft tissue model are related to the existence of contacts with either its neighboring anatomical structures or with surgical tools (Fig. 4.8).

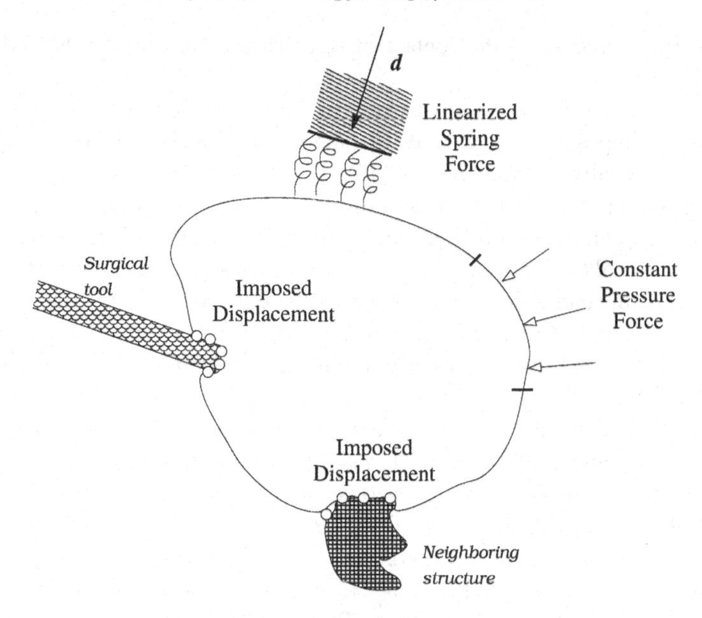

FIG. 4.8. The three different boundary conditions resulting from interaction with neighboring structures or with surgical tools.

We simplify the interaction with other physical material by considering that such an interaction can be represented either in terms of imposed displacements or elastic forces or surface pressure forces. If the material is stiff, or if it is significantly stiffer than the material of interest, we model the contact by imposing given displacements on a set of vertices. For instance, in the case of the liver model, we consider that vertices located near the vena cava (a stiff vessel) are stable (zero displacement).

If neighboring materials are as stiff (or less) than the material of interest, then we model the interaction as a linearized spring force. More precisely, for a boundary vertex $\mathbf{p}_i$, the applied force $\mathbf{r}_i^e$ is directed along a given direction $\mathbf{d}$, with stiffness k_e and rest displacement $\mathbf{u}_i^e$:

$$\mathbf{r}_i^e = -k_e\big((\mathbf{u}_i - \mathbf{u}_i^e)\cdot\mathbf{d}\big)\,\mathbf{d} = -k_e(\mathbf{d}\otimes\mathbf{d})\,\big(\mathbf{u}_i - \mathbf{u}_i^e\big). \tag{4.18}$$

Using a linearized spring allows to compute the static equilibrium by solving a linear system of equation. Indeed, the stiffness caused by the spring $k_e(\mathbf{d}\otimes\mathbf{d})$ can be added to the global stiffness matrix while the residual force $k_e(\mathbf{d}\otimes\mathbf{d})\mathbf{u}_i^e$ is added to the nodal load at node i. Furthermore, since the stiffness k_e is lower than the Young modulus of the material, the condition number of the updated stiffness matrix is not significantly modified.

In the sequel, we do not consider linearized spring boundary conditions explicitly. Instead, we modify the global stiffness matrix $\mathbf{K}$ into $\mathbf{K}^\star$, and we consider that a nodal force $\mathbf{r}_i^b$ is applied to vertex $\mathbf{p}_i$,

$$\big[\mathbf{K}_{i,i}^\star\big] = \big[\mathbf{K}_{i,i} + k_e(\mathbf{d}\otimes\mathbf{d})\big], \quad \mathbf{r}_i^b = k_e(\mathbf{d}\otimes\mathbf{d})\mathbf{u}_i^e.$$

When a soft tissue model is in contact with some fluids (bile, water, blood, ...) or gas (carbon dioxide, air, ...), we make the hypothesis that a constant pressure is applied along the normal direction of the contact surface. The computation of the nodal forces is detailed in Section 4.7.

Finally, the contact between surgical tools and a soft tissue model may be posed, in theory, either as imposed displacements (geometric method (BATHE [1982])) or as prescribed forces (penalty method (BATHE [1982])). However, in practice, the motion of surgical tools is controlled by the end-user through a force-feedback device. To decrease their cost, these devices are force-controlled and follow a simple open loop: the positions of surgical tools can be sent to a computer while they receive the force level that should be felt by the end-user. In other words, despite the low speed of a surgeon hands the position of a surgical tool varies significantly between two iterations ($dt = 20\,\text{ms}$) and therefore we found that the penalty method was not suited for deforming a soft tissue model.

Thus, after detecting the collision between soft tissue models and surgical tools, a set of imposed displacements at the collision nodes is computed. This computation is obviously ill-posed since it relies only on geometry (surface–volume intersection) rather than physical principles (Coulomb friction, for instance). Furthermore, a major challenge is to design a stable contact algorithm where a small displacement of the tool entails a small variation of node position. The geometric contact algorithm used in our hepatic surgery simulator, can be found in PICINBONO, LOMBARDO, DELINGETTE and AYACHE [2002].

To summarize, we consider only 2 types of boundary conditions in the remainder:

(1) *Imposed displacement*. We write $\mathcal{V}_d$ the set of vertices $\mathbf{p}_i$ for which the displacement $\mathbf{u}_i^b$ is known. In the scope of surgery simulation, these vertices are always lying on the surface of the mesh.
(2) *Applied nodal forces*. We write $\mathcal{V}_f$ the set of vertices $\mathbf{p}_i$ where an external force r_i^b is applied. Again, we make the hypothesis that applied forces may exist only on surface nodes.

4.10. *Equilibrium equations*

We apply the principle of virtual displacements described in Section 3.2.5 to obtain the finite element formulation of equilibrium equations. In a first stage, we only consider the static equilibrium by neglecting inertial forces. Thus, based on Eq. (3.10), we can state that the virtual elastic energy is equal to the sum of the virtual work of gravity and boundary forces,

$$\frac{1}{2}\widehat{\mathbf{U}}^{\mathrm{T}}\mathbf{K}\widehat{\mathbf{U}} = \widehat{\mathbf{U}}^{\mathrm{T}}\mathbf{R}^{g} + \widehat{\mathbf{U}}^{\mathrm{T}}\mathbf{R}^{b}.$$

Since this equation must hold for any set of compatible displacements, the static equation of equilibrium becomes

$$\mathbf{K}\mathbf{U} = \mathbf{R}^{g} + \mathbf{R}^{b}. \tag{4.19}$$

It is important to note that Eq. (4.19) is written for all nodes including the $\mathcal{V}_d$ nodes where the displacement is imposed. Therefore, in order to compute the unknown displacement vectors (where no displacement is imposed), it is important to write

Eq. (4.19) with a distinction between free nodes (subscript f) and constrained nodes (subscript c):

$$\begin{bmatrix} \mathbf{K}_{ff} & \mathbf{K}_{fc} \\ \mathbf{K}_{cf} & \mathbf{K}_{cc} \end{bmatrix} \begin{bmatrix} \mathbf{U}_f \\ \mathbf{U}_c \end{bmatrix} = \begin{bmatrix} \mathbf{R}_f^g + \mathbf{R}_f^b \\ \mathbf{R}_c^g + \mathbf{R}_c^b \end{bmatrix}$$

thus leading to

$$\mathbf{K}_{ff}\mathbf{U}_f = \mathbf{R}_f^g + \mathbf{R}_f^b - \mathbf{K}_{fc}\mathbf{U}_c. \tag{4.20}$$

In the case of a linear tetrahedron finite element, $\mathbf{K}_{fc}\mathbf{U}_c$ is non-zero only for free nodes that are neighbors to fixed nodes. In the remainder, we used simplified notations by dropping the subscript f for the stiffness matrix and displacement vector and by gathering all applied nodes into a single vector:

$$\mathbf{KU} = \mathbf{R}. \tag{4.21}$$

To get the dynamic law of motion, the work of inertial forces $-\frac{1}{2}\dot{\mathbf{U}}^{\mathrm{T}}\mathbf{M}\dot{\mathbf{U}}$ should be added to the work of body forces. By adding the work of damping forces, the following classical equation is obtained:

$$\mathbf{M}\ddot{\mathbf{U}} + \mathbf{C}\dot{\mathbf{U}} + \mathbf{KU} = \mathbf{R}, \tag{4.22}$$

where $\mathbf{C}$ is the damping matrix. In general, we assume that $\mathbf{C}$ follows Rayleigh damping,

$$\mathbf{C} = \gamma_1\mathbf{M} + \gamma_2\mathbf{K}. \tag{4.23}$$

This assumption is important for performing modal analysis but also for ensuring that the damping matrix, as the stiffness matrix, is also sparse.

4.11. Solution of equilibrium equations

The static equilibrium given by Eq. (4.21) is a linear system of equations with a symmetric positive definite stiffness matrix. Since this matrix is sparse, the classical method to solve this equation is to use the conjugated gradient algorithm [SAAD, 1996].

More precisely, when solving the complete system $\mathbf{KU} = \mathbf{R}$, we perform the following steps:

- *Node renumbering* by using the reverse cutting McKee algorithm (SAAD [1996]) in order to decrease the bandwidth of the stiffness matrix.
- *Matrix preconditioning* based on Cholesky factorization or incomplete LU decomposition (SAAD [1996]).
- *Application of the conjugated gradient algorithm* for solving the linear system of equation. We rely on the *Matrix Template Library* (LUMSDAINE and SIEK [1998]) for an efficient implementation of these algorithms in C++. When the stiffness matrix is poorly conditioned, for instance, for nearly incompressible materials, it is possible that the conjugated gradient algorithm fails. In which case, we resort to using direct methods for solving the system of equation, such as Gauss pivoting (SAAD [1996]).

Despite optimizing the bandwidth and the condition number of the stiffness matrix, the time required for solving the static equation is still too large for real-time computation. For instance, with a liver model composed of a mesh consisting of 1313 vertices, the solution of the linear system of size 3939×3939 requires 9 s on a PC Pentium II (450 MHz) with 140 iterations of the preconditioned conjugated gradient in order to reach an accuracy of 0.001 mm.

Therefore, solving directly the static equation with the conjugated gradient algorithm does not satisfy the real-time constraints mentioned in Section 1.3.2 since $T_c > T_{relaxation}$. As an alternative, we propose in the next sections, three soft tissue models that satisfy either hard or soft real-time constraints.

5. Quasi-static precomputed linear elastic model

5.1. Introduction

Since the complete solution of the static equilibrium equation is too computationally expensive for real-time constraint, a straightforward solution is to perform only few iterations of the conjugated gradient at each time step in order to increase the update rate. This approach, proposed by BARAFF and WITKIN [1998] is well-suited in the context of computer animation but is not applicable for surgery simulation where boundary conditions are constantly changing and are formulated in terms of imposed displacements. Indeed, using a conjugated gradient method would require to modify the stiffness matrix frequently as well as its preconditioning which would considerably reduce its efficiency.

Instead, we propose a *quasi-static precomputed linear elastic model* (COTIN, DELINGETTE and AYACHE [1999a]) that is based on a simple concept which consists in partially inverting the stiffness matrix in a precomputation stage before the simulation.

This model has the following characteristics:

- It is computationally very efficient: the computation complexity during the simulation is proportional to the cube of the number of imposed displacements.
- Only the position of surface nodes is updated during the simulation. In fact, only the data structure of the triangulated surface corresponding to the shell of the tetrahedral mesh is needed online.
- During the simulation the reaction forces at the nodes where the virtual instruments collide are also computed.
- The model is quasi-static, i.e., it computes the static equilibrium position at each iteration.

However, it relies on the following hypotheses:

- The mesh topology is not modified during the simulation. Thus, no simulation of cutting or suturing can be performed on this model.
- The interaction with neighboring tissues or with instruments is translated into modified boundary conditions (displacements or forces) only on surface nodes but not on the boundary conditions of internal nodes.

Therefore, the main limitation of this precomputed model comes from the first hypothesis which states that it is not suited for the simulation of tissue cutting.

5.2. Overview of the algorithm

One important feature of the model consists in making a distinction between surface and interior nodes. Thus, for the sake of clarity, we decompose the displacement and load vectors as well as the stiffness matrix according to surface and interior nodes with the s and i subscripts:

$$\begin{bmatrix} \mathbf{K}_{ss} & \mathbf{K}_{si} \\ \mathbf{K}_{is} & \mathbf{K}_{ii} \end{bmatrix} \begin{bmatrix} \mathbf{U}_s \\ \mathbf{U}_i \end{bmatrix} = \begin{bmatrix} \mathbf{R}_s \\ \mathbf{R}_i \end{bmatrix}.$$

It is important to note that only free vertices appear in this matrix as discussed in Section 4.10.

The solution of static equation can be obtained by multiplying the compliance matrix [**G**], corresponding to the inverse of the stiffness matrix [**K**], with the load vector. This compliance matrix can also be decomposed into surface and interior nodes,

$$\begin{bmatrix} \mathbf{U}_s \\ \mathbf{U}_i \end{bmatrix} = \begin{bmatrix} \mathbf{G}_{ss} & \mathbf{G}_{si} \\ \mathbf{G}_{is} & \mathbf{G}_{ii} \end{bmatrix} \begin{bmatrix} \mathbf{R}_s \\ \mathbf{R}_i \end{bmatrix}. \tag{5.1}$$

The load vector $\mathbf{R}_s$ that applies on free surface nodes can be decomposed into two parts. A first part $\mathbf{R}_s^0$, corresponds to loads that will not evolve during the simulation for instance gravity forces (see Section 4.6), constant pressure forces (see Section 4.7), applied nodal forces (see Section 4.9) or the presence of a non-zero imposed displacement vertex in its neighborhood (see Eq. (4.20)). The second part $\mathbf{R}_s^C$ corresponds to loads that are created by the contact of the soft tissue with surgical tools.

The principle of this soft tissue model is to compute the surface node positions $\mathbf{U}_s$ directly from the contact loads $\mathbf{R}_s^C$ by multiplying this vector with the compliance matrix $\mathbf{G}_{ss}$:

$$\begin{aligned} \mathbf{U}_s &= \mathbf{G}_{ss}\mathbf{R}_s^C + \mathbf{U}_s^0, \\ \mathbf{U}_s^0 &= \mathbf{G}_{ss}\mathbf{R}_s^0 + \mathbf{G}_{si}\mathbf{R}_i. \end{aligned} \tag{5.2}$$

Since the loads on interior nodes $\mathbf{R}_i$ do not evolve during the simulation, $\mathbf{U}_s^0$ is a displacement offset that is computed as the displacement of surface nodes when no contact loads are applied: $\mathbf{R}_s^C = \mathbf{0}$.

The goal of the precomputation stage is to compute the compliance matrix $[\mathbf{G}_{ss}]$.

5.3. Precomputation stage

5.3.1. Description of the algorithm

In the remainder, we write $[\mathbf{G}_{ss}^{ij}]$ the 3×3 submatrix of $\mathbf{G}_{ss}$ associated to vertex i and j. More precisely, a force $\mathbf{R}_s^j$ applied on vertex j entails an additional displacement of vertex i equal to $[\mathbf{G}_{ss}^{ij}]\mathbf{R}_s^j$.

The algorithm for computing the compliance matrix $\mathbf{G}_{ss}$ is described as Algorithm 1. It consists in solving $3 \times N_s$ times the linear system of equations $\mathbf{KU} = \mathbf{R}$, where N_s is the number of surface vertices. Note that the size of the stiffness matrix $\mathbf{K}$ is $N = N_s + N_i$ whereas the size of the compliance matrix $\mathbf{G}_{ss}$ is $3N_s \times 3N_s$.

```
1:  Set R_i = 0
2:  for all Surface Vertex i do
3:  for all j such that 0 ⩽ j ⩽ 2 do
4:      Set R_s = 0
5:      Set to 1.0 the jth component of the load R_s^i applied to vertex i
6:      Solve the static equilibrium equation KU = R
7:      for all Surface Vertex k do
8:          Store the computed displacement U_k of vertex k into the jth column
            of matrix [G_ss^ki]
9:      end for
10: end for
11: end for
```

ALGORITHM 1. Computation of the compliance matrix $\mathbf{G}_{ss}$.

The solution of equation $\mathbf{KU} = \mathbf{R}$ is performed using the steps described in Section 4.11 including node renumbering and matrix preconditioning. Since the rigidity matrix $\mathbf{K}$ is the same for all $3 \times N_s$ systems of equations, these two steps are performed only once, which significantly speeds-up the computation. Each time a linear system of equation is solved, the displacement of all surface nodes U_s corresponds to a column of matrix $\mathbf{G}_{ss}$. The storage of matrix $\mathbf{G}_{ss}$ requires only $(8 \times 9(N_s)^2)/2$ bytes (each element being stored as a double), since it is a symmetric matrix, as the inverse of a symmetric matrix.

Algorithm 1 can be slightly improved in the following way:

- Applying a unitary force successively along the X, Y and Z directions may cause a loss of accuracy in computing the compliance matrix, because the resulting displacement may be very large or very small depending on the size of the mesh. To obtain meaningful displacements, it is possible to apply a force f_{ref} and then divide the resulting displacement by f_{ref} to compute $\mathbf{G}_{ss}$. A good choice for f_{ref} is $\|[\mathbf{K}_{i,i}]\| * 0.1 * l$, where $[\mathbf{K}_{i,i}]$ is the block diagonal stiffness matrix of vertex i, and l is the estimated diameter of the object. This choice of force scale, produces displacements which are roughly equal to 10% of the diameter.
- It is sometimes necessary to obtain the displacement of some interior nodes during the simulation. This is the case, for instance, when vessels or tumors, located inside an organ, need to be displayed during the simulation. In this case, it is possible in the final loop of the algorithm (lines 7, 8 and 9 of Algorithm 1) to add these inside vertices to the list of surface vertices. Thus, it does not entail the solution of any additional system of equations, but only an additional storage requirement since the compliance matrix becomes a rectangular matrix of size $3N_s \times 3(N_s + N_i^{\star})$ where $N_i^{\star}$ is the number of additional interior nodes.

This precomputation stage is quite computationally expensive and requires between a few minutes up to several hours depending on the number of the mesh vertices and the stiffness of the material. For instance, the liver model presented in Fig. 5.1 is composed of 1394 vertices, 8347 edges and 6342 tetrahedra. Its triangulated surface is composed of 1224 triangles and 614 vertices which is enough to produce a smooth

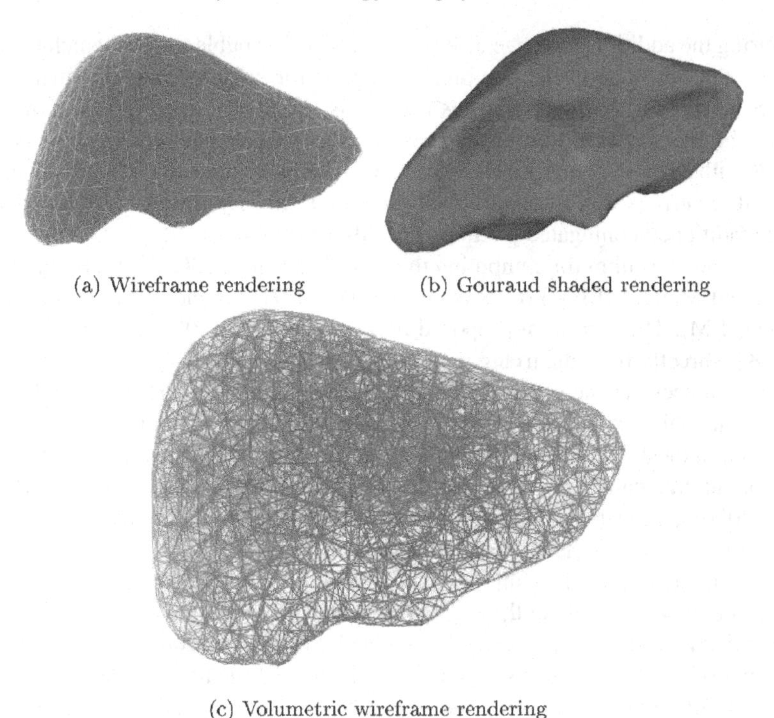

(a) Wireframe rendering
(b) Gouraud shaded rendering
(c) Volumetric wireframe rendering

FIG. 5.1. Visualization of a liver model with 1394 vertices and 6342 tetrahedra.

visual rendering. The Poisson ratio of the material is set to 0.45 while its Young modulus is $E = 1000$ kPa. In this case, the precomputation time required nearly 4 h on a Pentium PII 450 MHz, while the compliance matrix is stored in a file of size 13 Mb.

5.3.2. Other methods for computing the compliance matrix

At least two alternative methods have been proposed in the literature to compute the compliance matrix $\mathbf{G}_{ss}$. The first one, proposed by BRO-NIELSEN and COTIN [1996] is based on matrix condensation (MACMILLAN [1955]). More precisely, the compliance matrix $\mathbf{G}_{ss}$ can be directly obtained from the inversion of the stiffness matrix $\mathbf{K}_{ii}$ of interior nodes. From Eq. (5.1), we can derive the following equations:

$$\begin{aligned}
&\mathbf{K}_{ii}\mathbf{U}_i = \mathbf{R}_i - \mathbf{K}_{is}\mathbf{U}_s,\\
&\mathbf{K}_{ss}\mathbf{U}_s + \mathbf{K}_{si}\left(\mathbf{K}_{ii}^{-1}\mathbf{R}_i - \mathbf{K}_{ii}^{-1}\mathbf{K}_{is}\mathbf{U}_s\right) = \mathbf{R}_s,\\
&\left(\mathbf{K}_{ss} - \mathbf{K}_{si}\mathbf{K}_{ii}^{-1}\mathbf{K}_{is}\right)\mathbf{U}_s = \mathbf{R}_s - \mathbf{K}_{si}\mathbf{K}_{ii}^{-1}\mathbf{R}_i.
\end{aligned} \tag{5.3}$$

From Eq. (5.3), we can deduce the expression of the compliance matrix,

$$\mathbf{G}_{ss} = \left(\mathbf{K}_{ss}^{\star}\right)^{-1} = \left(\mathbf{K}_{ss} - \mathbf{K}_{si}\mathbf{K}_{ii}^{-1}\mathbf{K}_{is}\right)^{-1}. \tag{5.4}$$

Therefore, the computation of $\mathbf{G}_{ss}$ requires the inversion of two matrices: the first one of size $3N_i \times 3N_i$ and the second one of size $3N_s \times 3N_s$. This method has the disadvantage of requiring the additional storage of $9(N_i)^2$ numbers in double format, which in general is greater than the size of the compliance matrix: for large meshes, this method may become unpractical. Furthermore, this method is slightly more complex to implement whereas the method proposed in the previous section only requires to solve equation $\mathbf{KU} = \mathbf{R}$ with a sparse matrix $\mathbf{K}$. However, the condensation method is well-suited when the rigidity matrix is very ill-conditioned (Poisson ratio very close to 0.5) in which case the preconditioned conjugated gradient algorithm may fail.

The second algorithm for computing the compliance matrix $\mathbf{G}_{ss}$ is to use the Boundary Element Method (BEM) (CANAS and PARIS [1997]) instead of the Finite Element Method (FEM). The algorithm proposed by JAMES and PAI [1999] creates the stiffness matrix $\mathbf{K}^{\star}_{ss}$ directly from the triangulated surface of the object.

The differences between BEM and FEM are well-understood (HUNTER and PULLAN [1997]). The main advantage of BEM techniques is that they do not require a volumetric tetrahedral mesh but only its triangulated surface. While there exist several free software[6] for automatically creating tetrahedral meshes from triangulated surfaces (SIMAIL, OWEN [2000], JOE [1991]), having a control over the final number of vertices and the quality of tetrahedral elements is still an issue.

On the other hand, BEM techniques have several disadvantages over FEM. First, they make strong hypotheses about the nature of the elastic material: only homogeneous and isotropic linear elastic materials can be modeled. Second, the computation of the compliance matrix, and above all its diagonal elements, is difficult to implement and often numerically unstable because singular integrals must be evaluated over each triangle. The quality of the triangle geometry can influence the stability of this computation. Third, this method cannot compute the displacement of any interior point, which can be a limitation when the displacement of internal structures (vessels, tumors, ...) is needed. Finally, the BEM presented in JAMES and PAI [1999] uses centroid collocation to compute the stiffness matrix. Thus, this matrix allows to compute the displacements of the centroids of all triangles but not the displacements of the triangulation vertices. Therefore, the mesh being deformed is not the original triangulated mesh but its dual mesh which is called a *simplex mesh* (DELINGETTE [1999]). Mapping the displacements of triangle centroids into the displacements of vertices is not trivial since the duality between triangulation and simplex meshes is not a one-to-one mapping from the geometrical standpoint (DELINGETTE [1999]).

To conclude, the algorithm proposed by James et al. is more difficult to implement than our method and it is only suitable for simple material. However, when there is no software program for creating tetrahedral meshes from triangulations, this approach should be used.

5.4. On-line computation

5.4.1. Data structure

Before starting the simulation, the compliance matrix $\mathbf{G}_{ss}$, previously stored into a file as described in Section 5.3.1, is loaded into a specific data structure. Indeed, this

[6] A list of available software can be found at the following two URLs: http://www-users.informatik.rwth-aachen.de/~roberts/meshgeneration.html and http://www.andrew.cmu.edu/user/sowen/softsurv.html.

data structure only describes the triangulated surface shell of the volumetric tetrahedral mesh with a list of surface vertices and a list of surface triangles. Note that the number of surface vertices is usually greater than N_s because some surface vertices have an imposed displacement. For display purposes, the triangulated data structure may contain additional information such as 2D or 3D texture coordinates as well as parameters describing the rendered material. Finally, the data structure contains a list of imposed displacements and applied nodal forces as a storage of boundary conditions.

For each free vertex of index i, an array of 3×3 matrices $[\mathbf{G}_{ss}^{ji}]$, for all $j \in \{0, \ldots, N_s - 1\}$, is stored inside the vertex data structure. These N_s matrices $[\mathbf{G}_{ss}^{ji}]$ allow to compute the displacement of all surface vertex j, once a force is applied on vertex i.

The data structure optimizes the computation time of deformation but at the cost of being less efficient in terms of memory requirement. Indeed, the compliance matrix $\mathbf{G}_{ss}$ is a symmetric matrix, but it is stored as a non-symmetric matrix in this data structure. To optimize memory at a small additional computational cost, one could alternatively store the symmetric matrix as a double array of 3×3 compliance matrices $[\mathbf{G}_{ss}^{ji}]$ which is filled only if $i < j$.

5.4.2. *Algorithm description and collision processing*

The sketch of the algorithm is given in Algorithm 2 and includes two independent parts. The first part, between lines 1 and 8, consists in detecting and computing the contact between the soft tissue model and each virtual surgical instrument. In Fig. 5.2, we present an example of contact between a liver model and a tool. The collision detection algorithm (LOMBARDO, CANI and NEYRET [1999]) makes the assumption that the handle and the tool extremity can be approximated by a set of cylinders with rectangular section. Its efficiency depends on the availability of graphics cards since it relies on the

1: Reset the list of imposed displacement $l_{\text{displacement}}$ to the empty list
2: Reset the list of applied forces l_{force} to the empty list
3: Reset the position of free surface vertices to their rest position $+ \mathbf{U}_{ss}^0$
4: **for all** Surface Tools ST_i **do**
5: **if** collision between the soft tissue model and ST_i **then**
6:     Add imposed displacement to the list $l_{\text{displacement}}$
7: **end if**
8: **end for**
9: **if** $l_{\text{displacement}}$ is not empty **then**
10: Compute the list of applied forces l_{force} from $l_{\text{displacement}}$
11: **for all** Applied forces $\mathbf{F}_j^\star$ on vertex j in l_{force} **do**
12:     **for all** Free surface vertex k **do**
13:         Add to current position of vertex k, the displacement $[\mathbf{G}_{ss}^{kj}]\mathbf{F}_j^\star$
14:     **end for**
15: **end for**
16: **end if**

ALGORITHM 2. On-line computation of mesh deformation.

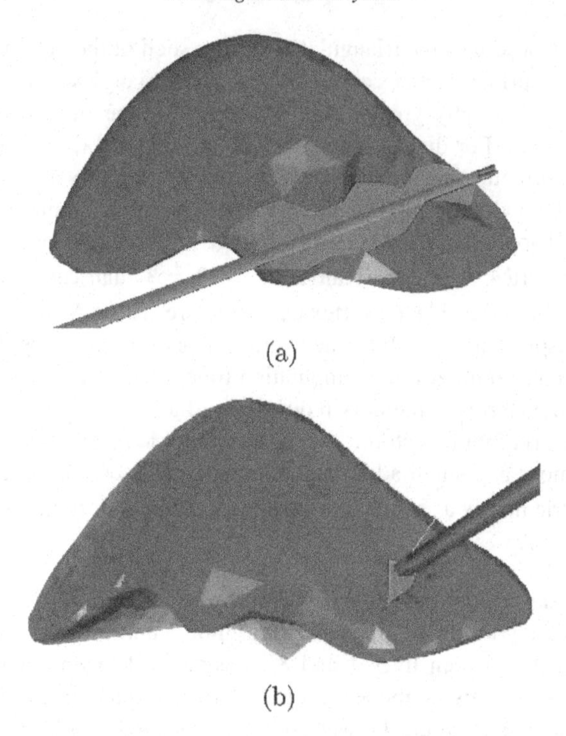

FIG. 5.2. Example of collision computation between the handle (a) and the extremity (b) of a surgical tool and a liver soft tissue model (PICINBONO, LOMBARDO, DELINGETTE and AYACHE [2002]). The position of triangles displayed in light gray have been displaced such that the tool is tangent to the liver surface.

OpenGL (WOO, NEIDER and DAVIS [1997]) library. Once a collision has been detected, the collided triangles must be moved such that the tissue model is no longer in contact with the surgical tool. This computation turns out to be quite complex since it not only depends on the tool position but also on its trajectory. The algorithm is described in PICINBONO, LOMBARDO, DELINGETTE and AYACHE [2002]. The outcome of this computation is a list $l_{\text{displacement}}$ of imposed displacements that should apply on each vertex of the collided triangles.

5.4.3. Imposing displacements

The second part of Algorithm 2, between lines 9 and 16 computes the position of all surface vertices, given the list of imposed displacements.

The first task corresponding to line 10 consists in computing the set of forces $\{\mathbf{F}_j^\star\}$ that should be applied to each vertex j of $l_{\text{displacement}}$ in order to bring the displacement of these vertices to $\mathbf{U}_j^b$.

To be more didactic, we first consider that only one vertex displacement $\mathbf{U}_j^b$ is imposed on a vertex of index j. Without any collision with a surgical tool, this vertex has a displacement $\mathbf{U}_j^0$ under the application of the *normal* boundary conditions (gravity forces, pressure forces, ... described in Section 4.9). Because the material is

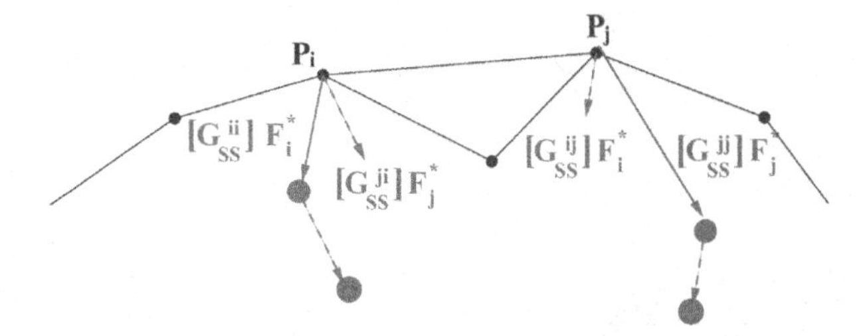

FIG. 5.3. Principle of superposition when applying two forces $\mathbf{F}_i^\star$ and $\mathbf{F}_j^\star$ to the two nodes i and j.

linear elastic, it follows the superposition principle: the displacements resulting from the application of two sets of nodal forces is the sum of the displacements resulting from the application of each set of forces. Thus, the force $\mathbf{F}_j^\star$ to be computed is the force that should be applied on vertex j in order to create a displacement of that vertex equal to $\mathbf{U}_j^b - \mathbf{U}_j^0$. Because the quantity $[\mathbf{G}_{ss}^{jj}]\mathbf{F}_j^\star$ gives the additional displacement of vertex j resulting from the application of force $\mathbf{F}_j^\star$, the force $\mathbf{F}_j^\star$ is given by

$$\mathbf{F}_j^\star = [\mathbf{G}_{ss}^{jj}]^{-1}\left(\mathbf{U}_j^b - \mathbf{U}_j^0\right).$$

When the displacements of two vertices i and j are imposed, the problem is slightly more complex. Indeed, the application of force $\mathbf{F}_i^\star$ on vertex i not only displaces vertex i of the amount $[\mathbf{G}_{ss}^{ii}]\mathbf{F}_i^\star$, but it also moves vertex j by the amount $[\mathbf{G}_{ss}^{ij}]\mathbf{F}_i^\star$ (see Fig. 5.3). Since $\mathbf{F}_j^\star$ also displaces vertex i of $[\mathbf{G}_{ss}^{ij}]\mathbf{F}_j^\star$, to compute the applied force, a 6×6 symmetric linear system of equations needs to be solved,

$$\begin{cases} [\mathbf{G}_{ss}^{ii}]\mathbf{F}_i^\star + [\mathbf{G}_{ss}^{ij}]\mathbf{F}_j^\star = \mathbf{U}_i^b - \mathbf{U}_i^0, \\ [\mathbf{G}_{ss}^{ji}]\mathbf{F}_i^\star + [\mathbf{G}_{ss}^{jj}]\mathbf{F}_j^\star = \mathbf{U}_j^b - \mathbf{U}_j^0. \end{cases}$$

Similarly, when the list of imposed displacements $l_{\text{displacement}}$ contains p elements, then a symmetric linear system of equations of size $3p \times 3p$ needs to be solved to find the set of nodal forces. If we use the set of indices i_j, $j \in [1, \ldots, p]$ to denote the set of vertices where a displacement $\mathbf{U}_{i_j}$ is imposed, then this linear system of equations can be written as

$$\begin{bmatrix} [\mathbf{G}_{ss}^{i_1,i_1}] & [\mathbf{G}_{ss}^{i_1,i_2}] & \cdots & [\mathbf{G}_{ss}^{i_1,i_p}] \\ [\mathbf{G}_{ss}^{i_2,i_1}] & [\mathbf{G}_{ss}^{i_2,i_2}] & \cdots & \vdots \\ \vdots & \vdots & \ddots & \vdots \\ [\mathbf{G}_{ss}^{i_p,i_1}] & \cdots & \cdots & [\mathbf{G}_{ss}^{i_p,i_p}] \end{bmatrix} \begin{bmatrix} \mathbf{F}_{i_1}^\star \\ \vdots \\ \vdots \\ \mathbf{F}_{i_p}^\star \end{bmatrix} = \begin{bmatrix} \mathbf{U}_{i_1}^b - \mathbf{U}_{i_1}^0 \\ \vdots \\ \vdots \\ \mathbf{U}_{i_p}^b - \mathbf{U}_{i_p}^0 \end{bmatrix}. \tag{5.5}$$

In Fig. 5.4, we show an example of a mesh where the same displacement is imposed on three vertices. In this particular case, the direction of computed forces departs strongly from the direction of the prescribed displacement.

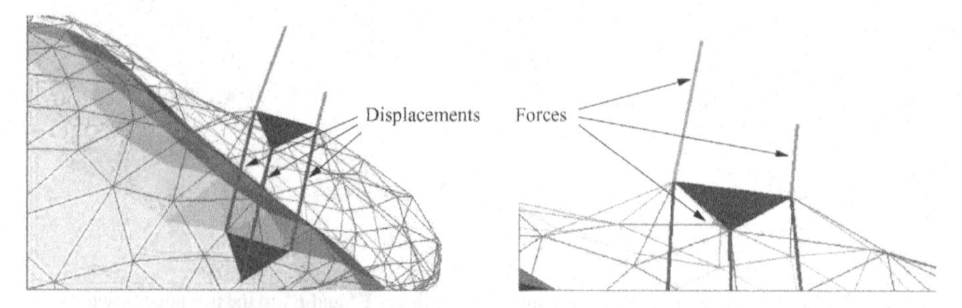

FIG. 5.4. (Right) The same displacement is imposed on the three vertices of a triangle; (left) from Eq. (5.5) we compute, the three forces that should be applied on these three vertices to move them of the given displacement.

5.4.4. *Results*

Once the set of nodal forces is computed, the additional displacement on all surface (and potentially internal) nodes are computed as described in lines 11 to 15 of Algorithm 2. The number of matrix–vector operations is $p \times N_s$ for p applied forces. In general, p, the number of vertices collided with the surgical tools, is small (from 3 to 20) when compared to N_s (see Fig. 5.5). This is why we chose to store the N_s array of compliance matrix $[\mathbf{G}_{ss}^{ji}]$ at vertex j, in order to optimize the inner loop (lines 12 to 14).

The computational efficiency of this quasi-static precomputed model on the liver mesh shown in Fig. 5.1 is presented in Table 5.1. These performances, measured on three different hardware platforms, correspond to the frequency update that can be achieved when running Algorithm 2 in a loop without any computation for visual and haptic rendering.

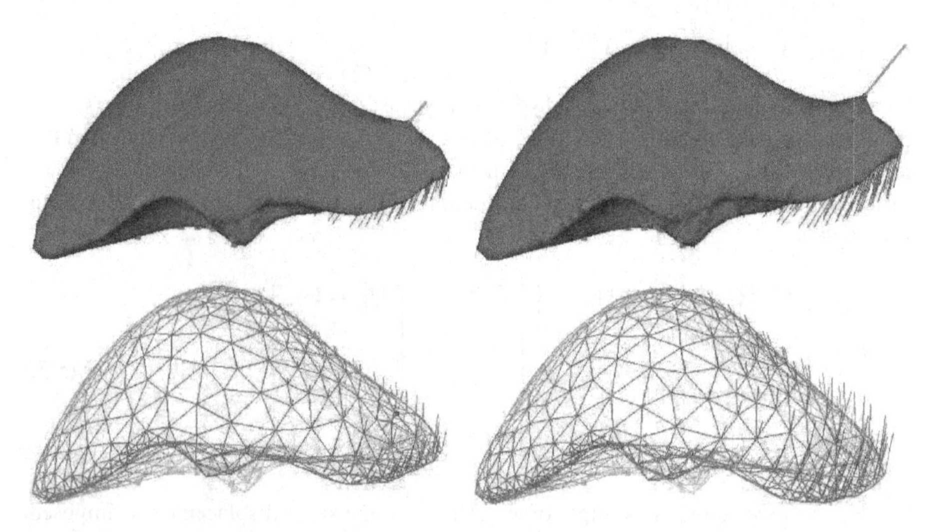

FIG. 5.5. Liver deformation based on a linear elastic pre-computed model (COTIN, DELINGETTE and AYACHE [1999b]). Solid lines indicate the imposed displacements.

TABLE 5.1
Computation efficiency of quasi-static precomputed linear elastic model for different boundary conditions: either when applying nodal forces or when imposing displacements

Simulation frequency (liver model with 614 surface nodes)		Pentium PIII 600 MHz
Force applied on 1 node		3772 Hz
Force applied on 5 nodes		754 Hz
Force applied on 10 nodes		377 Hz
Force applied on 20 nodes		188 Hz
Imposed displacements on	1 node	3759 Hz
	5 nodes	561 Hz
	10 nodes	185 Hz
	20 nodes	40 Hz

When applying one nodal force, corresponding to the execution of lines 12 to 14 in Algorithm 2, the computation time is nearly equal to 0.3 ms. The time required to compute the mesh deformation when applying p forces is strictly proportional to this value: $0.3 \times p$ ms.

When imposing p displacements, which is what occurs in practice in a surgical simulator, the additional computation is the solution of a $3p \times 3p$ linear symmetric system of equations. For $p = 1$, the overhead is very small and hardly perturbs the simulation frequency. However, for larger value of p, the overhead becomes dominant. For 20 vertices, for instance, solving the system of equations of size 60×60 is 3 times more costly than computing the $20 * 614 = 1280$ matrix–vector products and additions.

5.4.5. *Discussion*

As a whole, the proposed method is "very efficient", since it allows real-time visual rendering even for large meshes. When the material is soft enough and when the number of collided vertices remains small (typically less than 15), this model can also be compatible with real-time haptic rendering. In fact, it is one of the few algorithms which are suitable for the first software architecture described in Section 1.3.2 (see also Fig. 1.6(a)) consisting of one synchronous loop including visual and haptic rendering. Furthermore, our approach has one major advantage for haptic rendering computation: it already provides the nodal reaction forces through the algorithm described in Section 5.4.3. Indeed, the set of forces $\mathbf{F}^{\star}_{i_j}$ corresponds to the set of physical forces that have been applied on each node of index i_j in order to deform the soft tissue model: thus, $-\mathbf{F}^{\star}_{i_j}$ corresponds to the nodal reaction force. From this set of forces, one can easily compute the reaction force along the direction of the tool, as well as the torque at the extremity of the tool.

Using the terminology introduced in Section 1.3.2, we can also state that the quasi-static precomputed linear elastic model has a *very low relaxation time* (or equivalently that it has a high speed of convergence). Indeed, each time Algorithm 2 is run, the soft tissue is deformed to its static equilibrium position. Because this algorithm can be run at a high frequency, as seen in Table 5.1, this implies that the relaxation time is very

1: **for all** Free surface vertex k **do**
2: **if** $k \notin l_{\text{displacement}}$ **then**
3: Let $\mathbf{p}_k$ be the position of vertex k after Algorithm 2
4: Let $\mathbf{p}_k^{\text{previous}}$ be the position of vertex k at the previous iteration.
5: $\mathbf{p}_k \Leftarrow \gamma \mathbf{p}_k + (1-\gamma)\mathbf{p}_k^{\text{previous}}$
6: **end if**
7: $\mathbf{p}_k^{\text{previous}} \Leftarrow \mathbf{p}_k$
8: **end for**

ALGORITHM 3. Additional part of Algorithm 2 that adds a visco-elastic behavior controlled by delay parameter γ.

low. In fact, for some soft tissue, this time is too low and degrades the visual realism of the simulation. This is the case, for instance, when the operator grasps and displaces some soft tissue and suddenly ceases the grasping. Because the model has no longer any displacements imposed on its surface, it returns in one iteration to its rest position, while in reality, it takes several milliseconds.

To add some visco-elastic behavior, one can increase the relaxation time artificially by using a delay function. This approach is described in Algorithm 3. For vertices which are not colliding with a surgical tool, the final vertex position is a weighted sum between the position computed by Algorithm 2 and the vertex position at the previous iteration. The weight parameter $0 \leqslant \gamma \leqslant 1$ controls the damping of the material deformation: for $\gamma = 1$, the deformation is not damped (quasi-static motion) while for $\gamma = 0$, the motion is infinitely damped (no motion). Any intermediate value of γ modifies the relaxation time of the material. Note that this damping is not applied to vertices colliding with tools because the collision would otherwise appear visually unrealistic. Algorithm 3 assumes that the model has a damping matrix $\mathbf{C}$ which is proportional to the identity matrix: more sophisticated hypotheses (but often more computationally intensive) could be proposed.

6. Dynamic linear elastic model

In this section, we describe two different soft tissue models that are able to address with the limitation of the previous model: the simulation of tissue cutting. Using the terminology defined in Section 1.3.2, these two methods can be qualified as "Explicit Iterative Methods" sharing the advantage of requiring a small computation time for each iteration but with the drawback of having a low speed of convergence.

The main difference between these two models is that the first can model the visco-elastic behavior of the soft tissue properly whereas the second does not require the evaluation of any time step and is unconditionally stable.

Finally, we propose in Section 6.3 a *hybrid model* which combines any of the two previous models with the precomputed linear elastic model seen in Section 5.

6.1. Tensor–mass model

6.1.1. Introduction

The *tensor–mass model* is based on the dynamic law of motion described in Eq. (4.22):

$$\mathbf{M}\ddot{\mathbf{U}} + \mathbf{C}\dot{\mathbf{U}} + \mathbf{K}\mathbf{U} = \mathbf{R}.$$

This second order differential equation couples the motion of tissue under the influence of inertia $\mathbf{M}\ddot{\mathbf{U}}$, of visco-elasticity $\mathbf{C}\dot{\mathbf{U}}$, elasticity $\mathbf{K}\mathbf{U}$ and external loads $\mathbf{R}$.

The most efficient way to solve the equation above is by far to use modal analysis (BATHE [1982]). By making simple assumptions about the damping matrix $\mathbf{C}$, it is possible to simplify the above PDE into a small set of ordinary differential equations with an appropriate change of basis. The proper basis is given by the eigenvectors associated to the largest eigenvalues of the generalized eigenproblem $\mathbf{K}\phi = \omega^2\mathbf{M}\phi$.

However, the eigenproblem must be solved each time the rigidity matrix is modified. Therefore, this approach is not suitable for simulating tissue cutting, since the computation cost to solve the eigenproblem is very high.

Instead, a classical method to solve Eq. (4.22), is to use integration methods: the time dimension is uniformly discretized with a time step Δt, and each term of that equation is supposed to be constant during each time interval. There is an important distinction between *implicit integration schemes* and *explicit integration schemes* depending whether the position of the model at time $t + \Delta t$ requires the solution or not of a global linear systems of equations (see also the discussion in Section 1.3.2).

Implicit schemes are *unconditionally stable* which allows the use of large time steps. In structural analysis, the Houbolt method (HOUBOLT [1950], BATHE [1982]) and the Newmark method (NEWMARK [1959], BATHE [1982]) are the most commonly used. However, these schemes require either to inverse a sparse matrix or to solve at each iteration a linear system of equations. Considering the time required to solve such a linear system (a few seconds for a small-size mesh), these implicit schemes cannot be used for real-time interaction.

Instead, we chose to use *explicit integration schemes* which have several nice properties (ease of implementation, low computational cost) compared to implicit schemes but with the drawback of being *conditionally stable*: the time step must be smaller than a critical time step $\Delta t_{\text{critical}}$. Therefore, smaller time step Δt must be used for explicit schemes which yields a larger relaxation time and a longer time for convergence.

6.1.2. Mass matrix

Regarding the mass matrix, a common choice consists in replacing the symmetric positive definite matrix $\mathbf{M}$ with a diagonal matrix, where each diagonal element is the sum of all row elements in the original matrix: this lumped mass matrix is detailed in Section 4.8.

In order to keep the time step Δt large enough during the simulation, we propose a further simplification of the mass matrix $\mathbf{M}$ by considering that the nodal mass is constant for all nodes, which makes $\mathbf{M}$ proportional to the identity matrix,

$$\mathbf{M} = m_0\mathbf{I}_3,$$

where m_0 is the average mass per node computed as the total mass of the tissue divided by the number of nodes in the initial mesh.

Indeed, the critical time step Δt of the iterative scheme is inversely proportional to the highest eigenvalue of the matrix $\mathbf{M}^{-1}\mathbf{K}$, while the speed of convergence is related to the ratio between the largest to the smallest eigenvalues of the same matrix, also called the *condition number* of that matrix.

From the equation of the nodal stiffness matrix $[\mathbf{K}_{i,i}]$, we can state that the nodal stiffness is proportional to the size (for instance, the largest foot height) of all the tetrahedra surrounding each node:

$$[\mathbf{K}_{i,i}] = \sum_{T \in S(i)} \frac{1}{36V(T)} \Big((\lambda_T + \mu_T)(\mathbf{m}_i \otimes \mathbf{m}_i) + \mu_T A_i^2 \mathbf{I}_3 \Big).$$

Thus, the largest eigenvalue of $\mathbf{K}$ is determined by the largest tetrahedra while the condition number is given by the size ratio between the largest and smallest tetrahedra. On the other hand, when performing mass lumping, as in BRO-NIELSEN [1998], the nodal mass of $\mathbf{M}^{-1}$ is inversely proportional to the volume of tetrahedra surrounding each node. Therefore, the power spectrum of $\mathbf{M}^{-1}\mathbf{K}$ largely differs from that of $\mathbf{K}$: the largest eigenvalue of $\mathbf{M}^{-1}\mathbf{K}$ now becomes related to the tetrahedron of smallest size while the condition number is related to the square ratio between the largest and smallest tetrahedra. These properties of $\mathbf{M}^{-1}\mathbf{K}$ have two consequences for the simulation of tissue cutting: both the speed of convergence and the time step Δt decrease as tetrahedra of small size are created.

By choosing a mass matrix proportional to the identity matrix, we keep the spectral properties of the rigidity matrix: the creation of small tetrahedra does not entail any decrease of the time step and limits the decrease of the speed of convergence. However, this choice is a gross approximation of physics since the total mass of the tissue increases as the number of elements increases. As claimed in Section 3.1.5, we prefer to satisfy real-time constraints of the simulation (by keeping a large value of Δt) at the expense of coarse approximations of the tissue dynamic behavior.

6.1.3. Numerical integration

Several explicit iterative schemes can be proposed from Eq. (4.22) depending on the choice of damping matrix and discretization of time derivatives. Below, we propose three explicit schemes that are of interest in the context of surgery simulation. In the remainder, we write ${}^t\mathbf{U}$ the displacement vector at time t.

Euler method. This method uses central finite differences to estimate acceleration but right finite difference to estimate speed. Furthermore, sophisticated damping matrix such as Rayleigh damping can be employed in this scheme:

$$\frac{m_0}{\Delta t^2}\left({}^{t-\Delta t}\mathbf{U} - 2\,{}^t\mathbf{U} + {}^{t+\Delta t}\mathbf{U}\right) + \frac{1}{\Delta t}(\gamma_1 m_0 \mathbf{I}_3 + \gamma_2 \mathbf{K})\left({}^t\mathbf{U} - {}^{t-\Delta t}\mathbf{U}\right) + \mathbf{K}\,{}^t\mathbf{U} = {}^t\mathbf{R}.$$

The displacement at time $t + \Delta t$ can be computed through the recurrent equation:

$${}^{t+\Delta t}\mathbf{U} = {}^t\mathbf{U} + (1 - \Delta t \gamma_1)\left({}^t\mathbf{U} - {}^{t-\Delta t}\mathbf{U}\right)$$

$$-\mathbf{K}\left(\frac{\Delta t^2}{m_0}{}^t\mathbf{U}+\frac{\gamma_2\Delta t}{m_0}\left({}^t\mathbf{U}-{}^{t-\Delta t}\mathbf{U}\right)\right)+\frac{\Delta t^2}{m_0}{}^t\mathbf{R}.$$

Euler method with central finite difference. In this case, central finite differences are used to estimate both acceleration and speed, while constant damping is used $\gamma_2=0$:

$$\frac{m_0}{\Delta t^2}\left({}^{t-\Delta t}\mathbf{U}-2{}^t\mathbf{U}+{}^{t+\Delta t}\mathbf{U}\right)+\frac{\gamma_1 m_0}{2\Delta t}\left({}^{t+\Delta t}\mathbf{U}-{}^{t-\Delta t}\mathbf{U}\right)+\mathbf{K}{}^t\mathbf{U}={}^t\mathbf{R},$$

which leads to the following update equation:

$${}^{t+\Delta t}\mathbf{U}={}^t\mathbf{U}+\frac{2-\gamma_1\Delta t}{2+\gamma_1\Delta t}\left({}^t\mathbf{U}-{}^{t-\Delta t}\mathbf{U}\right)-\frac{2\Delta t^2}{m_0(2+\gamma_1\Delta t)}\left(\mathbf{K}{}^t\mathbf{U}-{}^t\mathbf{R}\right). \tag{6.1}$$

Runge–Kutta method of order 4. The Runge–Kutta method (PRESS, FLANNERY, TEUKOLSKY and VETTERLING [1991]) is an integration method of fourth order of accuracy, but which requires four evaluations of the Euler recurrent equation. To describe this method, it is necessary to write the original equation as a first order differential equation,

$$\frac{\mathrm{d}}{\mathrm{d}t}\begin{bmatrix}\dot{\mathbf{U}}\\ \mathbf{U}\end{bmatrix}=\begin{bmatrix}\ddot{\mathbf{U}}\\ \dot{\mathbf{U}}\end{bmatrix}=\begin{bmatrix}-\frac{\mathbf{C}}{m_0} & -\frac{\mathbf{K}}{m_0}\\ 1 & 0\end{bmatrix}\begin{bmatrix}\dot{\mathbf{U}}\\ \mathbf{U}\end{bmatrix}+\begin{bmatrix}\frac{\mathbf{R}}{m_0}\\ 0\end{bmatrix}.$$

Now, the state of a soft tissue model at time t is described by two vectors: displacement vector ${}^t\mathbf{U}$ and the velocity vector ${}^t\dot{\mathbf{U}}$. Applying the simple Euler method on this system gives the following relation:

$$\begin{bmatrix}{}^{t+\Delta t}\dot{\mathbf{U}}\\ {}^{t+\Delta t}\mathbf{U}\end{bmatrix}-\begin{bmatrix}{}^t\dot{\mathbf{U}}\\ {}^t\mathbf{U}\end{bmatrix}=\Delta t\begin{bmatrix}\frac{1}{m_0}\left(-\mathbf{C}{}^t\dot{\mathbf{U}}-\mathbf{K}{}^t\mathbf{U}+{}^t\mathbf{R}\right)\\ {}^t\dot{\mathbf{U}}\end{bmatrix}=\begin{bmatrix}\delta v\left({}^t\mathbf{U},{}^t\dot{\mathbf{U}}\right)\\ \delta u\left({}^t\mathbf{U},{}^t\dot{\mathbf{U}}\right)\end{bmatrix}.$$

The fourth order Runge–Kutta method requires to compute the following eight incremental displacement and velocity vectors:

$$\begin{aligned}
&\delta v_1=\delta v\left({}^t\mathbf{U},{}^t\dot{\mathbf{U}}\right), && \delta u_1=\delta u\left({}^t\mathbf{U},{}^t\dot{\mathbf{U}}\right),\\
&\delta v_2=\delta v\left({}^t\mathbf{U}+\frac{\delta u_1}{2},{}^t\dot{\mathbf{U}}+\frac{\delta v_1}{2}\right), && \delta u_2=\delta u\left({}^t\mathbf{U}+\frac{\delta u_1}{2},{}^t\dot{\mathbf{U}}+\frac{\delta v_1}{2}\right),\\
&\delta v_3=\delta v\left({}^t\mathbf{U}+\frac{\delta u_2}{2},{}^t\dot{\mathbf{U}}+\frac{\delta v_2}{2}\right), && \delta u_3=\delta u\left({}^t\mathbf{U}+\frac{\delta u_2}{2},{}^t\dot{\mathbf{U}}+\frac{\delta v_2}{2}\right),\\
&\delta v_4=\delta v\left({}^t\mathbf{U}+\frac{\delta u_3}{2},{}^t\dot{\mathbf{U}}+\frac{\delta v_3}{2}\right), && \delta u_4=\delta u\left({}^t\mathbf{U}+\frac{\delta u_3}{2},{}^t\dot{\mathbf{U}}+\frac{\delta v_3}{2}\right).
\end{aligned}$$

Finally, the velocity and displacement for the next time step are given by the following equation:

$$\begin{bmatrix}{}^{t+\Delta t}\dot{\mathbf{U}}\\ {}^{t+\Delta t}\mathbf{U}\end{bmatrix}=\begin{bmatrix}{}^t\dot{\mathbf{U}}\\ {}^t\mathbf{U}\end{bmatrix}+\frac{1}{6}\begin{bmatrix}\delta v_1\\ \delta u_1\end{bmatrix}+\frac{1}{3}\begin{bmatrix}\delta v_2\\ \delta u_2\end{bmatrix}+\frac{1}{3}\begin{bmatrix}\delta v_3\\ \delta u_3\end{bmatrix}+\frac{1}{6}\begin{bmatrix}\delta v_4\\ \delta u_4\end{bmatrix}.$$

TABLE 6.1
Comparison between three explicit integration methods for soft tissue modeling

	Euler method	Euler central finite differences	Runge–Kutta method
Computation time	low	low	high
Damping	Rayleigh	basic	basic
Time step	small	medium	high

Comparison between the three methods. We summarized in Table 6.1 the properties of the three methods described above. Three qualitative criteria were proposed to outline the advantages and drawbacks of each method. In terms of computation time required to update the position of a model, the first two Euler methods are equivalent while the Runge–Kutta method is at least four times slower. As far as damping is concerned, only the first Euler method allows to use Rayleigh damping while the two other methods can only use diagonal damping matrices. Having a non-diagonal damping matrix helps in keeping a continuous field of velocity throughout the model which improves the visual realism of the simulation. Finally, the Runge–Kutta method is more stable than the Euler method and our experience showed that a tenfold increase of the time step can be observed in the former case. The Euler method with central finite differences allows larger time steps than the Euler method because the velocity computation leaps over position computation by one time step.

6.1.4. Data structure

With explicit schemes, the update of the mesh position can be performed locally, at the vertex level, without creating any global matrix. Indeed, for each free vertex of index i, we can take advantage of the sparse nature of the rigidity matrix $\mathbf{K}$, in order to compute the matrix–vector product $\mathbf{KU}$. More precisely, from Eq. (4.13), it is clear that the off-diagonal stiffness matrices $[K_{i,j}]$ are non-null matrices only when there is an edge connecting vertices i and j in the tetrahedral mesh. Therefore, only the set $\mathcal{N}(i)$ of vertices connected to vertex i by an edge is involved when computing the elastic force $\mathbf{F}_i$ applied on vertex i. For instance, the update Eq. (6.1) can be computed for a vertex i as

$$
\begin{aligned}
{}^{t+\Delta t}\mathbf{u}_i = {}^{t}\mathbf{u}_i + \frac{2-\gamma_1\Delta t}{2+\gamma_1\Delta t}\left({}^{t}\mathbf{u}_i - {}^{t-\Delta t}\mathbf{u}_i\right) \\
- \frac{2\Delta t^2}{m_0(2+\gamma_1\Delta t)}\left(\sum_{j\in\mathcal{N}(i)} [\mathbf{K}_{i,j}]^{t}\mathbf{u}_j + [\mathbf{K}_{i,i}]^{t}\mathbf{u}_i - {}^{t}\mathbf{R}_i\right).
\end{aligned}
$$

The data structure that is suitable for performing this computation follows the data structure required for storing a tetrahedral mesh. The basic structure consists in a double-linked list of vertices, edges and tetrahedra. For each vertex, we store its current position ${}^{t}\mathbf{q}_i$, its rest position $\mathbf{p}_i$ and the symmetric tensor $[\mathbf{K}_{i,i}]$. For each edge, we store its two adjacent vertices (vertex i and vertex j) as well as the tensor $[\mathbf{K}_{i,j}]$, as sketched in Fig. 6.1. We therefore take advantage of the symmetric nature of the stiffness matrix by storing the off-diagonal rigidity matrix only once.

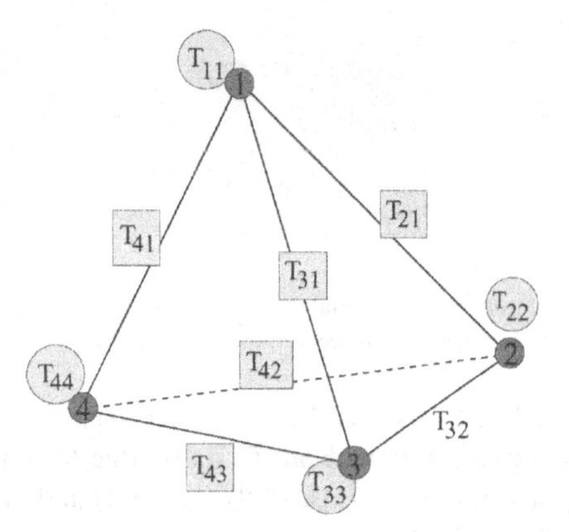

FIG. 6.1. Representation of the data structure of a tensor–mass model. The 3×3 rigidity matrices are stored at each edge and each vertex. The symmetry of the rigidity matrix enables to store only one tensor per edge.

Finally, for each tetrahedron, we store its four vertices and its six edges as well as the Lamé coefficients λ_i, μ_i, the area vectors $\mathbf{m}_i$ and if required the direction of anisotropy $\mathbf{a}_0$.

6.1.5. *Cutting and refinement algorithms*

One of the basic tasks in surgery simulation consists in cutting and tearing soft tissue. With the dynamic linear elastic model, these tasks can be achieved efficiently.

To perform an hepatectomy (partial resection of the liver), the use of scalpel instruments is not appropriate because of the important vascularization of the liver. Instead, surgeons usually proceed with a set of pliers that smash hepatic cells or with a cavitron device that destroys the hepatic parenchyma with ultrasound energy: in both cases, the resection is performed by removing soft tissue. It is therefore important to simulate the removal of bits of soft tissue located at the vicinity of a surgical tool. To perform this simulation, two basic meshing techniques must be implemented: removal of tetrahedra and local refinement.

At first sight, removing a single tetrahedron from a tetrahedral mesh is straightforward. However, in order to obtain a visually realistic simulation, one should avoid to produce isolated or self-intersecting tetrahedra or even tetrahedra connected through a single vertex. A proper way to keep "visually appealing" meshes is to constrain the mesh to be a *manifold* mesh in addition to being a *conformal* mesh. Indeed, in a manifold mesh, the shell of a vertex located on the mesh surface is homeomorphic a half-sphere (the shell is a sphere for interior vertices) which allows to define unambiguously a surface normal at that vertex. However, by adding this topological constraint, even removing a single tetrahedron is not straightforward as discussed in FOREST, DELINGETTE and AYACHE [2002b]. The detailed description of the topological issues relevant to the operation of tetrahedron removal falls outside the scope of this chapter; instead we present briefly the algorithms related to the computation of soft tissue deformation.

Once a collision between a surgical tool and a set of tetrahedra has been detected, each tetrahedron of the set is removed one after the other. After updating the topological

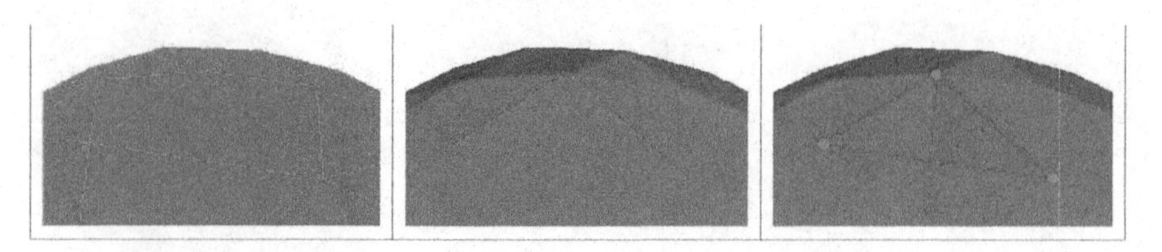

FIG. 6.2. To remove the tetrahedron whose external triangle has been selected (dark gray), it is necessary to update the local rigidity matrices stored at the vertices and edges of that tetrahedron.

structure of the mesh, the local vertex and edge stiffness matrices must also be updated (see Fig. 6.2). When removing tetrahedron $\mathcal{T}$, its 6 edge tensors $[\mathbf{B}^{\mathcal{T}}_{i,j}]$ and 4 vertex tensors $[\mathbf{B}^{\mathcal{T}}_{i,i}]$ are computed based on Eqs. (4.8) and (4.9) and are subtracted from the current edge and vertex local rigidity matrices:

$$[\mathbf{K}_{i,i}] = [\mathbf{K}_{i,i}] - [\mathbf{B}^{\mathcal{T}}_{ii}], \qquad [\mathbf{K}_{i,j}] = [\mathbf{K}_{i,j}] - [\mathbf{B}^{\mathcal{T}}_{ij}].$$

These ten local operations are performed efficiently because of the specific data structure associated with a tetrahedron.

The second meshing technique, local refinement, can be used in two cases. First, it can be used offline (before the simulation), to increase the mesh resolution at places of high curvature or near structures of interest (tumors, gall blader, ...). Second, it is often necessary to refine the mesh locally during the removal of soft tissue when the tetrahedra to be removed arc too large. In the former case, sophisticated meshing techniques can be employed while in the latter case, real-time constraints allow the application of only basic refinement algorithms. An example of such a basic algorithm consists in adding a vertex at the middle of an edge and then splitting all tetrahedra adjacent to that edge into two tetrahedra (see Fig. 6.3). In this case, the edge and vertex tensors of all tetrahedra adjacent to that edge are first removed and the contributions from all newly created tetrahedra are then added. A more sophisticated refinement algorithm can be found in FOREST, DELINGETTE and AYACHE [2002b].

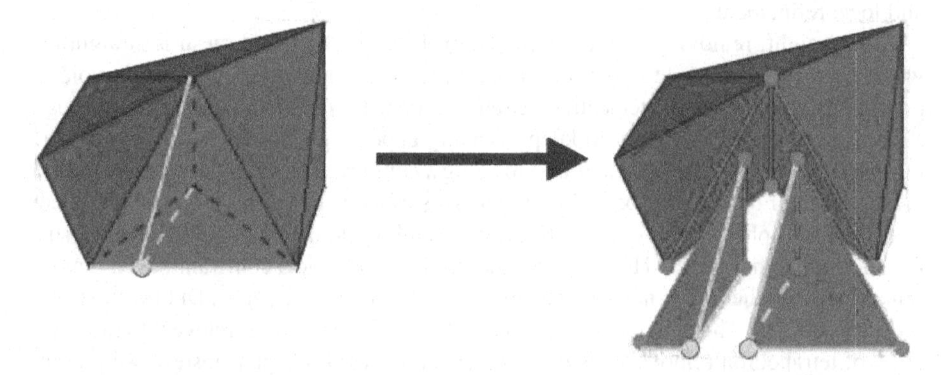

FIG. 6.3. Local refinement of a tetrahedral mesh. An edge is split into two edges by inserting a vertex. The rigidity matrices must be updated for vertices and edges that already existed (drawn in dark grey) while these matrices must be computed for newly created vertices and edges (drawn in light grey).

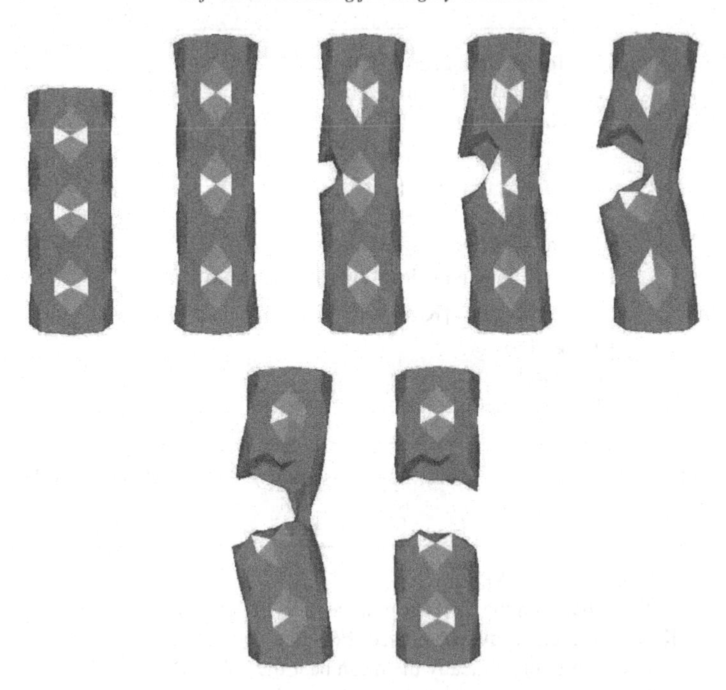

FIG. 6.4. Deformation of a cylinder subject to gravity forces: some tetrahedra are progressively being removed at its center leading to a separation into independent solids.

The proper adjustment of stiffness matrices during the removal of soft tissue reinforces the visual realism of the simulation significantly: this is especially the case when the tissue is cut while being stretched. For instance, in Fig. 6.4, we show the deformation of a cylinder being cut: the cylinder is fixed at its upper part and is under the influence of gravity forces along its main axis.

6.1.6. Algorithm description

Before describing the deformation algorithm for a tensor–mass model, we shortly describe the initialization stages in Algorithm 4. Once the vertex and edge stiffness matrices have been assembled, it is necessary to estimate a time step Δt that allow the stability of the iterative schemes described in Section 6.1.3. Finding the critical time step (i.e., the highest possible time step) is actually a difficult task because of the lack of a closed-form expression. However, a practical approach is to estimate the critical time step as a product of an unknown constant with the time step given by the Courant–Friedrich–Levy condition (PRESS, FLANNERY, TEUKOLSKY and VETTERLING [1992]):

$$(\Delta t)_{\text{Courant}} = l_{\max}\sqrt{\frac{\rho}{\lambda + 2\mu}}.$$

Algorithm 5 presents the different loops required to update a tensor–mass model. Unlike the precomputed quasi-static model, it is not necessary to maintain an explicit list of vertices that are displaced by the collision with a surgical tool: it is sufficient

1: **for all** Tetrahedron $\mathcal{T}$ **do**
2: Compute the 4 area vectors $\mathbf{m}_i$
3: **for all** Vertex i of $\mathcal{T}$ **do**
4: Compute the local rigidity matrix $[\mathbf{B}_{ii}^{\mathcal{T}}]$
5: $[\mathbf{K}_{i,i}] \Leftarrow [\mathbf{K}_{i,i}] + [\mathbf{B}_{ii}^{\mathcal{T}}]$
6: **end for**
7: **for all** Edge between vertices i and j of $\mathcal{T}$ **do**
8: Compute the local rigidity matrix $[\mathbf{B}_{ij}^{\mathcal{T}}]$
9: $[\mathbf{K}_{i,j}] \Leftarrow [\mathbf{K}_{i,j}] + [\mathbf{B}_{ij}^{\mathcal{T}}]$
10: **end for**
11: **end for**
12: Estimate time step Δt.

ALGORITHM 4. Matrix assembly for the tensor–mass model performed before any simulation.

1: **for all** Surface tools ST_i **do**
2: **if** collision between the soft tissue model and ST_i **then**
3: **if** ST_i represent a cavitron device **then**
4: Eventually refine locally the mesh near the collision
5: Remove tetrahedra located near the extremity of ST_i
6: **end if**
7: Impose displacements on vertices near the contact zone and raise a flag on these vertices
8: **end if**
9: **end for**
10: **for all** edge e connecting vertex i and j **do**
11: add elastic force $[\mathbf{K}_{i,j}]^t\mathbf{u}_i$ to vertex i
12: elastic force $[\mathbf{K}_{i,j}]^{\mathrm{T}t}\mathbf{u}_j$ to vertex j
13: **end for**
14: **for all** vertex i **do**
15: **if** vertex i is free (flag not raised) **then**
16: compute elastic force $[\mathbf{K}_{i,i}]^t\mathbf{u}_i$
17: update vertex position ${}^t\mathbf{p}_i$ based on one of the three iterative schemes described in Section 6.1.3
18: **else**
19: reset flag
20: **end if**
21: **end for**

ALGORITHM 5. On-line computation of tensor–mass model.

(see line 7) to raise a flag stating that these vertices are not free vertices. A second important feature of this algorithm is the existence of a loop on the mesh edges in order to compute the matrix–vector products $\sum_{j\in\mathcal{N}(i)}[\mathbf{K}_{i,j}]^t\mathbf{u}_j$. This approach is more efficient than scanning iteratively the neighbors $\mathcal{N}(i)$ for each vertex i. When using the fourth order Runge–Kutta algorithm, the algorithm from lines 10 to 21 must be modified since it is then necessary to scan four times the edges and vertices of the mesh. For the Euler method, only lines 11 and 12 must be modified in order to compute

$$\mathbf{K}\left(\frac{\Delta t^2}{m_0}{}^t\mathbf{U} + \frac{\gamma_2 \Delta t}{m_0}\left({}^t\mathbf{U} - {}^{t-\Delta t}\mathbf{U}\right)\right)$$

instead of $\mathbf{K}^t\mathbf{U}$.

6.1.7. Comparison between spring–mass and tensor–mass models

We have used the word "tensor–mass model" to designate a finite element model based on Newtonian dynamics and discretized with an explicit scheme. This word has been chosen in order to stress the similarity between a "tensor–mass model" and a "spring–mass model". In particular, it is the purpose of this section to oppose to the widely spread belief stating that "finite element models are slower and more complex to implement than spring–mass models".

A spring–mass model (BARAFF and WITKIN [1998]) consists of a set of masses and a set of springs connecting these masses. The force applied to a point $\mathbf{p}_i$ in a spring–mass system, is given by the relation

$$\mathbf{F}_i = \sum_{j \in \mathcal{N}(i)} k_{ij}\left(\|\mathbf{p}_i\mathbf{p}_j\| - l_{ij}^0\right)\frac{\mathbf{p}_i\mathbf{p}_j}{\|\mathbf{p}_i\mathbf{p}_j\|}, \tag{6.2}$$

where k_{ij} is the stiffness coefficient between vertices i and j, l_{ij}^0 is the length at rest.

Similarly, on a tensor–mass model, the elastic force applied on vertex i is given by

$$\mathbf{F}_i = [\mathbf{K}_{i,i}]\mathbf{u}_i + \sum_{j \in \mathcal{N}(i)} [\mathbf{K}_{i,j}]\mathbf{u}_j. \tag{6.3}$$

By comparing Eqs. (6.2) and (6.3), it is clear that both dynamic models have the same computational complexity which is linear in the number of edges. In practice, we have observed a slight computational advantage for the tensor–mass model, mostly because it does not include any square root evaluation.

However, both approaches differ substantially in terms of biomechanical modeling. Spring–mass systems constitute a discrete representation of an object and their behavior strongly depends on the topology of the spring network. Adding or removing a spring may change the elastic behavior of the whole system drastically. Conversely, a finite element model is a continuous representation of the object and its behavior is independent of the mesh topology (it mostly depends on the mesh resolution). This is an advantage when mesh cutting is performed since it produces continuous and natural deformations.

Because all biomechanical data related to biological soft tissue are formulated as parameters found in continuum mechanics (such as Young's modulus or Poisson coefficients), it is *a priori* difficult to model realistic soft tissue deformations with a spring–mass system. However, several authors (LOUCHET, PROVOT and CROCHEMORE [1995], DEUSSEN, KOBBELT and TUCKE [1995]) have developed genetic or simulated annealing

TABLE 6.2
Comparison between the three soft tissue models: pre-computed quasi-static, tensor–mass and spring–mass models

	Pre-computed	Tensor–mass	Spring–mass
Computational efficiency	+++	+	+
Biomechanical realism	+	+	−
Cutting simulation	−	++	+
Large displacements	−	−	+

algorithms to identify spring parameters (stiffness and damping) from a set of known deformations of an object.

Finally, as previously mentioned, the tensor–mass model is only valid for small displacements. This model is invariant under the application of a global translation, but if a global rotation is applied to the rest shape $\mathcal{M}_{\text{rest}}$, then the forces applied to all vertices will not be null. On the opposite, a spring–mass model under the same displacement would not deform, since the length of the springs are preserved under a rigid transformation. The difference between these three soft tissue models is summarized in Table 6.2.

6.2. *Relaxation-based elastic models*

6.2.1. *Introduction*

In this section, we introduce an alternative algorithm to the tensor–mass model. This algorithm is based on Gauss–Seidel relaxation and has the following properties:

- Its iterative scheme is unconditionally stable. It does not require the estimation of any critical time step.
- The relaxation algorithm is fairly efficient (small computation time required for one iteration) but it is slightly less efficient than a tensor–mass model.
- The algorithm is based on static equilibrium equations whereas tensor–mass models are based on the dynamic law of motion.
- The position of each vertex is updated asynchronously, one vertex after the other.

However, when compared to tensor–mass models, relaxation-based elastic models have two drawbacks. First, their implementation requires the following property for the mesh data structure: each vertex should be able to access efficiently its adjacent edges. This topological "vertex–edge" relationship can be stored in two ways inside a data structure. In a first approach, a list of edges can be stored explicitly at each vertex. After removing or adding tetrahedra, the edge list must be updated for all vertices belonging to these tetrahedra. To achieve this update, each edge must have a list of adjacent tetrahedra which should also be explicitly updated upon the removal or addition of tetrahedra.

In a second approach, the list of edges adjacent to a vertex is recovered through the knowledge of a single tetrahedron adjacent to this vertex. This approach is only applicable if we constrain the tetrahedral mesh to be a manifold mesh (see FOREST, DELINGETTE and AYACHE [2002b] for more details). Indeed, in such case, the neighborhood of a vertex is homeomorphic to a topological sphere or half-sphere. By marching around a vertex from a given tetrahedron, it is possible to obtain all tetrahedra adjacent to a given vertex and consequently the list of all adjacent edges. In this case as in the former case, we do store a list of adjacent edges for each vertex in order to avoid duplicating the search algorithm. However, when a tetrahedron is removed or added, this topological list is reseted and the pointer to the adjacent tetrahedron is eventually updated.

The second drawback of relaxation algorithms is that they require in average 3 times more storage than the tensor–mass model. Indeed, in addition to the symmetric stiffness matrix, a non-symmetric stiffness matrix must be stored.

6.2.2. *Overview of the algorithm*

Following the notations of Eq. (6.3) the static problem $\mathbf{KU} = \mathbf{R}$ can be written at the level of each vertex i as

$$[\mathbf{K}_{i,i}]\mathbf{u}_i + \sum_{j\in\mathcal{N}(i)} [\mathbf{K}_{i,j}]\mathbf{u}_j = \mathbf{R}_i. \tag{6.4}$$

For relaxation algorithms, the displacement of a vertex $\mathbf{u}_i$ is updated independently from other vertices. Therefore, the notation ${}^{t+\Delta t}\mathbf{u}_i$ to describe the position of vertex i at the next time step cannot be used, since formally there is no temporal evolution (and no temporal variable t) in relaxation algorithms. Thus, we note ${}^{+}\mathbf{u}_i$ the next position of vertex i and $\mathbf{u}_i$ its current position.

The principle of relaxation algorithms is quite straightforward: each vertex is moved in order to locally solve Eq. (6.4). Thus, the displacement ${}^{+}\mathbf{u}_i$ is given by

$${}^{+}\mathbf{u}_i = -\sum_{j\in\mathcal{N}(i)} [\mathbf{K}_{i,i}]^{-1}[\mathbf{K}_{i,j}]\mathbf{u}_j + [\mathbf{K}_{i,i}]^{-1}\mathbf{R}_i. \tag{6.5}$$

This is equivalent to minimizing the total mechanical energy by successively optimizing each variable $\mathbf{u}_i$. It is therefore similar to the Iterative Conditional Mode (ICM) algorithm (BESAG [1986]) used in statistical analysis.

If all displacements $\{\mathbf{u}_i\}$ are successively updated according to Eq. (6.4), then this method is equivalent to the Gauss–Seidel relaxation method (SAAD [1996]). More precisely, we can decompose the stiffness matrix $\mathbf{K}$ as the sum of three terms: $\mathbf{K}_D$ a 3×3 block diagonal matrix, $\mathbf{K}_C$ the lower triangle matrix of $\mathbf{K}$ and $\mathbf{K}_C^{\mathrm{T}}$ the upper triangle matrix of $\mathbf{K}$:

$$\mathbf{K} = \underbrace{\begin{bmatrix} [\mathbf{K}_{1,1}] & \mathbf{0} & \cdots & \mathbf{0} \\ \mathbf{0} & [\mathbf{K}_{2,2}] & \ddots & \mathbf{0} \\ \vdots & \ddots & \ddots & \vdots \\ \mathbf{0} & \mathbf{0} & \cdots & [\mathbf{K}_{N,N}] \end{bmatrix}}_{\mathbf{K}_D} + \underbrace{\begin{bmatrix} \mathbf{0} & \mathbf{0} & \cdots & \mathbf{0} \\ [\mathbf{K}_{2,1}] & \mathbf{0} & \ddots & \mathbf{0} \\ \vdots & \ddots & \ddots & \vdots \\ [\mathbf{K}_{N,1}] & [\mathbf{K}_{N,2}] & \cdots & \mathbf{0} \end{bmatrix}}_{\mathbf{K}_C} + \mathbf{K}_C^{\mathrm{T}}.$$

With this notation, the Gauss–Seidel relaxation consists in the application of an iterative equation

$${}^{k+1}\mathbf{U} = (\mathbf{K}_D + \mathbf{K}_C)^{-1}\left(-\mathbf{K}_C^{\mathrm{T}}\,{}^{k}\mathbf{U} + \mathbf{R}\right), \tag{6.6}$$

where ${}^{k}\mathbf{U}$ is the displacement vector at iteration k.

To speed-up convergence, we use over-relaxation (known as the Simultaneous Over-Relaxation algorithm (SAAD [1996])) that consists in anticipating future correction with an overrelaxation parameter ω,

$$ {}^{k+1}\mathbf{U} = (\mathbf{K}_D + \omega\mathbf{K}_C)^{-1}\left(-\omega\mathbf{K}_C^{\mathrm{T}}\,{}^{k}\mathbf{U} + (1-\omega)\mathbf{K}_D\,{}^{k}\mathbf{U} + \omega\mathbf{R}\right). \tag{6.7} $$

This equation translates at the vertex level with the recursion

$$ {}^{+}\mathbf{u}_i = (1-\omega)\mathbf{u}_i - \omega \sum_{j\in\mathcal{N}(i)} [\mathbf{K}_{i,i}]^{-1}[\mathbf{K}_{i,j}]\mathbf{u}_j + \omega[\mathbf{K}_{i,i}]^{-1}\mathbf{R}_i. \tag{6.8} $$

If $\omega = 1$, then the SOR algorithm is equivalent to the Gauss–Seidel relaxation. Convergence is guaranteed for values of ω comprised between 1 and 2, while fastest convergence is obtained for a critical value

$$ \omega_{\text{optimal}} = \frac{2}{1+\sqrt{1-\rho_{GS}}}, $$

where ρ_{GS} is the spectral radius (the modulus of the largest eigenvalue) of the matrix $(\mathbf{K}_D + \omega\mathbf{K}_C)^{-1}\mathbf{K}_C^{\mathrm{T}}$.

The overrelaxation parameter ω controls the dynamics of the soft tissue model. With $\omega \equiv 2$, the model tends to overshoot around the solution whereas with $\omega \equiv 1$, the motion is very damped. In practise, we chose a value of $\omega = 1.2$ as a trade-off between these two behaviors.

6.2.3. *Algorithm description*

The application of the SOR recursive Eq. (6.8) requires the computation of matrices $[\mathbf{K}_{i,i}]^{-1}[\mathbf{K}_{i,j}]$ and $[\mathbf{K}_{i,i}]^{-1}$. For speed-up purposes, these matrices are stored respectively at each vertex and edge. Because the matrix $\mathbf{K}_D^{-1}\mathbf{K}$ is no longer symmetric, at each edge linking vertices i and j, we store the two 3×3 matrices $[\mathbf{K}_{i,i}]^{-1}[\mathbf{K}_{i,j}]$ and $[\mathbf{K}_{j,j}]^{-1}[\mathbf{K}_{i,j}]^{\mathrm{T}}$.

The algorithm of the relaxation-based elastic model is presented as Algorithm 6. A large part is dedicated to the update of these additional matrices each time a topological change of the mesh occurs. A flag is positioned at each vertex and edge in order to indicate whether matrices $[\mathbf{K}_{i,i}]^{-1}[\mathbf{K}_{i,j}]$ and $[\mathbf{K}_{i,i}]^{-1}$ are up-to-date or not. This flag is raised each time a topological change takes place at a vertex or edge level and it is lowered once these matrices are updated.

6.3. *Hybrid models*

6.3.1. *Motivation*

We have previously described two types of *linear elastic models*:

(1) a *quasi-static* pre-computed elastic model which is computationally efficient but that does not allow any change of topology (cutting, tearing) (see Section 5).
(2) two *dynamic* elastic models (tensor–mass and relaxation-based models) that have lower convergence speed but that allow topology changes (see Sections 6.1

1: **for all** Surface tools ST_i **do**
2: **if** collision between the soft tissue model and ST_i **then**
3:   **if** ST_i represents a cavitron device **then**
4:     Possibly refine locally the mesh near the collision
5:     Remove tetrahedra located near the extremity of ST_i
6:   **end if**
7:   Impose displacements on vertices near the contact zone
8: **end if**
9: **end for**
10: **for all** free vertex i **do**
11: **if** flag raised at vertex i **then**
12:   compute and store $[\mathbf{K}_{i,i}]^{-1}$
13:   lower flag at vertex i
14: **end if**
15: $\mathbf{u}_i^\star \Leftarrow (1-\omega)\mathbf{u}_i + \omega[\mathbf{K}_{i,i}]^{-1}\mathbf{R}_i$
16: **for all** edge e connecting vertex i and j **do**
17:   **if** flag raised at edge e **then**
18:     **if** flag raised at vertex j **then**
19:       compute and store $[\mathbf{K}_{j,j}]^{-1}$
20:       lower flag at vertex j
21:     **end if**
22:     compute and store $[\mathbf{K}_{i,i}]^{-1}[\mathbf{K}_{i,j}]$ and $[\mathbf{K}_{j,j}]^{-1}\,[\mathbf{K}_{i,j}]^{\mathrm{T}}$
23:     lower flag at edge e
24:   **end if**
25:   $\mathbf{u}_i^\star \Leftarrow \mathbf{u}_i^\star - \omega[\mathbf{K}_{i,i}]^{-1}\,[\mathbf{K}_{i,j}]\mathbf{u}_j$
26: **end for**
27: $\mathbf{u}_i \Leftarrow \mathbf{u}_i^\star$
28: **end for**

ALGORITHM 6. On-line computation of the relaxation-based model.

and 6.2). In the remainder, we use tensor–mass models as the method for deforming.

To combine these two approaches, we make a distinction between two types of anatomical structures that usually appear in a surgical simulation:

- Anatomical structures which are the target of the surgical procedure. On these structures, tearing and cutting need to be simulated. In many cases, they correspond to pathological structures and only represent a small subset of the anatomy that needs to be visualized during the simulation.
- Anatomical structures which only need to be visualized or eventually deformed but which are not submitted to any surgical action.

Thus, in a hybrid model, we propose to model the former type of anatomical structures as tensor–mass models whereas the latter type of structures should be modeled as a pre-computed linear model. However, this method is only efficient if the number of tensor–mass elements is kept as low as possible.

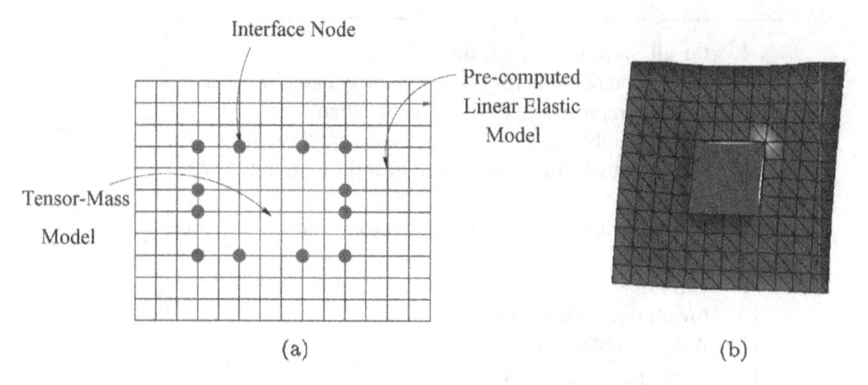

FIG. 6.5. (a) Definition of the interface nodes in a hybrid elastic model; (b) hybrid elastic model with eight interface nodes (COTIN, DELINGETTE and AYACHE [2000]).

6.3.2. *Description*

A hybrid elastic model $\mathcal{M}_{\text{hybrid}}$ is composed of two different types of elements: let $\mathcal{M}_{\text{dynamic}}$ be the set of tensor–mass elements and let $\mathcal{M}_{\text{quasi-static}}$ be the set of pre-computed linear elastic elements. The model $\mathcal{M}_{\text{dynamic}}$ is connected to $\mathcal{M}_{\text{quasi-static}}$ by a set of common vertices called *interface nodes*. These interface nodes define additional boundary conditions for each model. As seen in Fig. 6.5, the two models may not be completely connected along their entire boundaries. In fact, a way to reduce the number of tensor–mass elements, is to associate a fine pre-computed elastic model with a coarse tensor–mass model. As shown in Fig. 6.5(b), this incomplete interface causes some visual artifacts due to the non-continuity between two neighboring parts. However, if the interface zone between the two elastic models is not an important visual cue, a different mesh resolution can be used.

Since both linear elastic models follow the same physical law, their combination should behave exactly as a global linear elastic model. Thus, the additional boundary conditions imposed at the interface nodes must be consistent with responding terms of forces and displacements for both models.

Fig. 6.6 summarizes the computation loop of a hybrid model. Since the pre-computed model $\mathcal{M}_{\text{quasi-static}}$ is more efficient with force boundary conditions than with imposed displacements (see Section 5.4.3), its update is based on forces applied at interface nodes by $\mathcal{M}_{\text{dynamic}}$ but also on imposed displacements resulting from the contact with surgical tools. The applied forces originating from $\mathcal{M}_{\text{dynamic}}$ are computed as reaction forces (opposite of elastic force) at interface nodes. At this stage, the displacement of all surface nodes of $\mathcal{M}_{\text{quasi-static}}$ is computed and the position of interface nodes becomes new displacement constraints for $\mathcal{M}_{\text{dynamic}}$. After $\mathcal{M}_{\text{quasi-static}}$, $\mathcal{M}_{\text{dynamic}}$ is updated based on displacements imposed at the interface nodes by $\mathcal{M}_{\text{quasi-static}}$ and the displacements imposed by the user interaction.

6.3.3. *Examples*

In Fig. 6.7, we present an example of a hybrid cylinder model undergoing deformation caused by gravity forces. The different stages of the deformation process are shown.

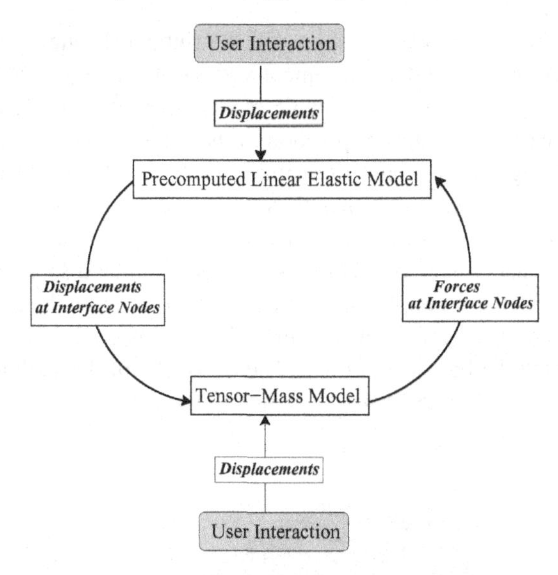

FIG. 6.6. Interaction loop for a hybrid elastic model. Both models are updated alternatively while allowing for user interaction.

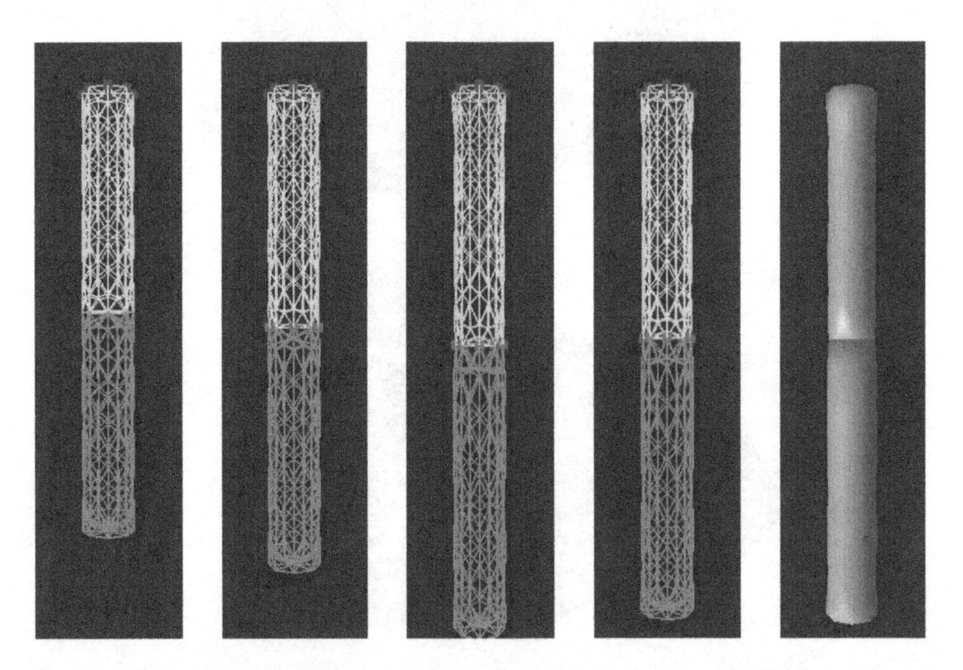

FIG. 6.7. Deformation of a hybrid elastic model under a gravity force: the upper cylinder consists of a precomputed linear elastic model whereas the lower part is a tensor–mass model. The leftmost figure corresponds to the initial position of the mesh and the rightmost figure to the equilibrium state.

When the equilibrium is reached, as shown in the rightmost figure, forces applied at the interface nodes are null and displacement vectors stabilize to a constant value. In this example, both quasi-static and dynamic models have the same elastic properties and we verified that the equilibrium position is the same as the one that would have been reached by a single quasi-static or dynamic elastic model. Furthermore, this hybrid model converges significantly faster than the corresponding dynamic elastic model.

The second example is related to the simulation of hepatectomy, i.e., the removal of one of the eight anatomical segments – known as Couinaud segments (COUINAUD [1957]) – of a liver. In this example the segment number six has to be removed. A tetrahedral mesh of a liver has been created from a CT scan image. It is composed of 1537 vertices and 7039 tetrahedra – see Fig. 6.8. The tetrahedra of the sixth anatomical segment, which represent 18% (280 vertices and 1260 tetrahedra) of the global mesh, are

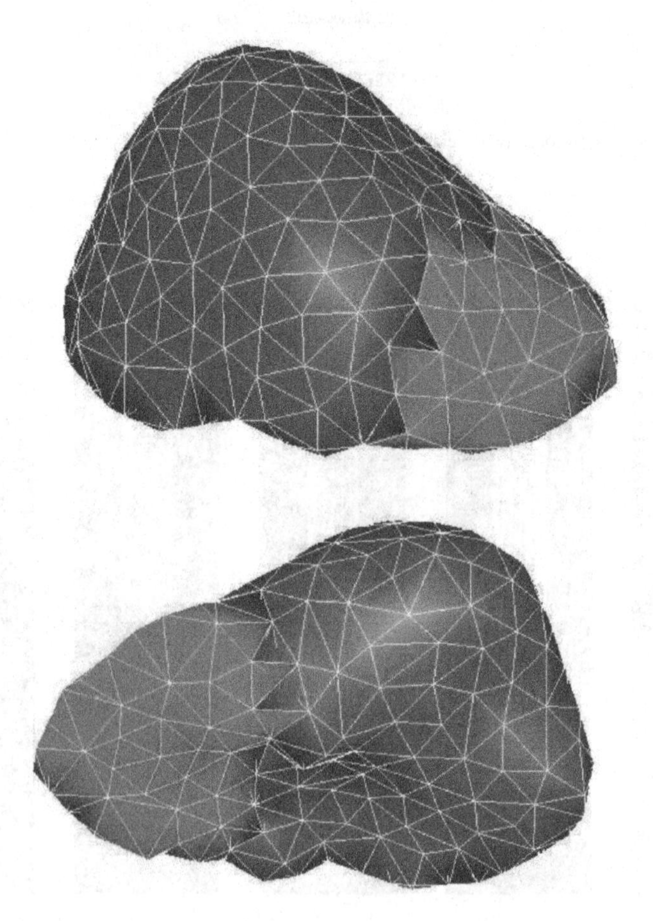

FIG. 6.8. Display of a hybrid liver model. The part displayed in blue corresponds to the pre-computed quasi-static elastic model whereas the red part corresponds to the tensor–mass model. The interface nodes ensure the visual continuity between the two elastic models.

modeled with a tensor–mass model and the remaining tetrahedra with a pre-computed linear elastic model.

In Fig. 6.9, we show different stages of the hepatectomy simulation. The first six pictures show the deformation of the model when the tool collides with the dynamic model. Since both models have the same elastic characteristics, it is not possible to visually distinguish the interface between the two different elastic models.

The last six pictures show the cutting of the liver segment by removing additional tetrahedra. The cutting occurs for the tetrahedron being collided by the tool. One can notice that each part of the hybrid model deforms naturally itself during the resection simulation.

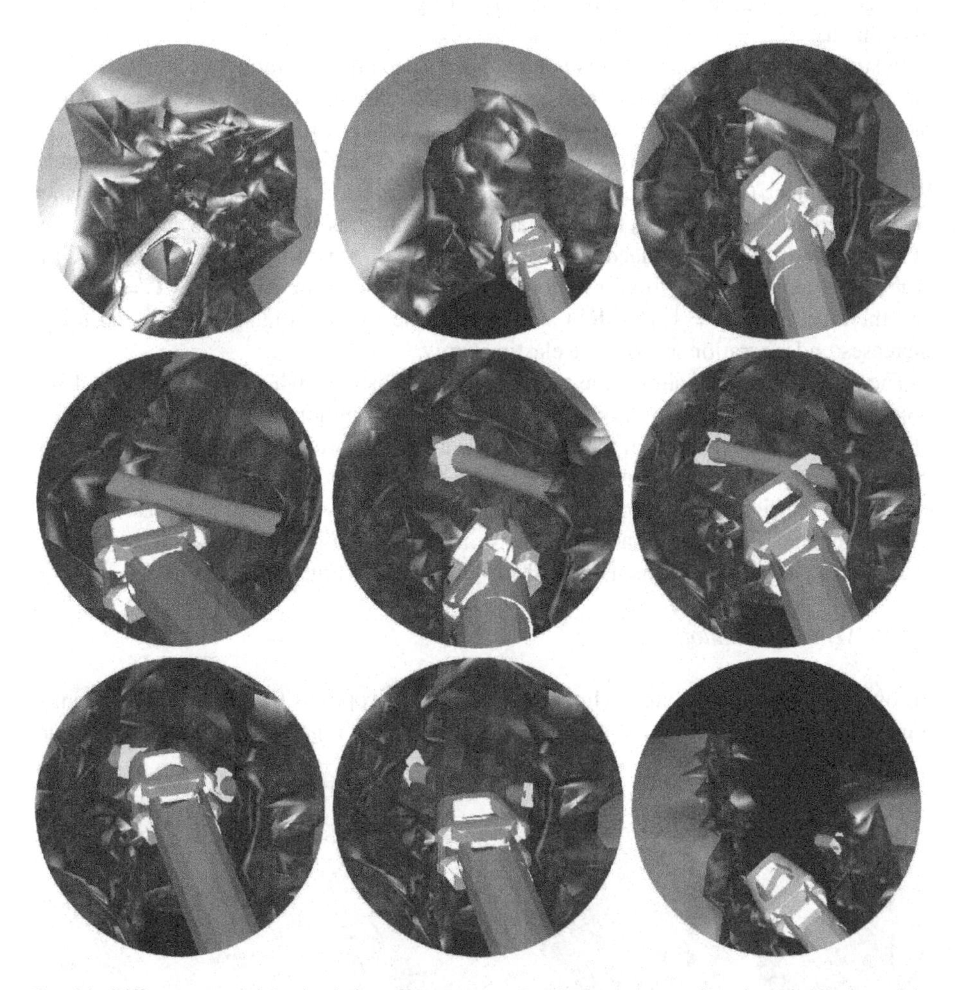

FIG. 6.9. Different stages of the simulation of hepatectomy. In this simulation, we have included lineic models of the main bifurcations of the portal vein (FOREST, DELINGETTE and AYACHE [2002b]). The simulation consists in removing some hepatic parenchyma but also to clamp and cut each vessel.

7. Large displacement non-linear elastic model

7.1. Shortcomings of linear elasticity

The physical behavior of a soft tissue may be considered as linear elastic for small displacements and small deformations (FUNG [1993], MAUREL, WU, MAGNENAT THALMANN and THALMANN [1998]). The hypothesis of small displacements corresponds to displacements that are typically less than 10% of the mesh size.

In the context of surgery simulation, this hypothesis is often violated. For instance, the lobes of the liver are often folded to access underlying structures such as the gall bladder. Also during the resection of a soft tissue, it is common that pieces being cut undergo large rotations either under the action of gravity or under the action of surgical instruments.

In such cases, linear elasticity is not an appropriate physical model because it makes the assumption of infinitesimal strain instead of finite strain. To exhibit the shortcomings of linear elasticity, we produced two examples pictured in Figs. 7.1 and 7.2.

In a first example, we illustrate the action of a global rotation on a linear elastic model. When an object (an icosahedron in Fig. 7.1) undergoes a global rotation, its elastic energy increases, leading to a large variation of volume (as seen in the wireframe mesh of the rightmost figures). Indeed, the infinitesimal strain tensor $\mathbf{E}_{\mathrm{L}}(\mathbf{X}) = \frac{1}{2}(\nabla\mathbf{U} + \nabla\mathbf{U}^{\mathrm{T}})$ is not invariant when a global rotation $\mathbf{R}$ is applied since in this case $\nabla\mathbf{U} = \mathbf{R} - \mathbf{I}_3$ and therefore $\mathbf{E}_{\mathrm{L}}(\mathbf{X}) = \frac{1}{2}(\mathbf{R} + \mathbf{R}^{\mathrm{T}}) - \mathbf{I}_3 \neq [\mathbf{0}]$. The two invariants $(\operatorname{tr}\mathbf{E}_{\mathrm{L}})^2$ and $\operatorname{tr}\mathbf{E}_{\mathrm{L}}^2$ increases under rotation as does the elastic energy.

The second example shows the effect of linear elasticity when only one part of an object undergoes a large rotation (which is the most common case). The cylinder pictured in Fig. 7.2 has its bottom face fixed while a force is being applied at the central top vertex. The arrows correspond to the trajectories of some vertices: because of the linear elastic hypothesis, these trajectories are straight lines. This results in unrealistic distortions of the mesh. Moreover, abnormal deformations are not equivalent in all directions since the object only deforms itself in the rotation plane (Fig. 7.2(c) and (d)).

7.2. St Venant–Kirchhoff elasticity

To overcome the limitations of linear elasticity, we proposed to adopt the St Venant–Kirchhoff elasticity. The St Venant–Kirchhoff model is a generalization of the linear

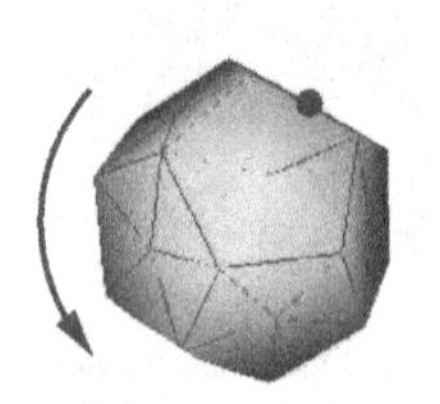
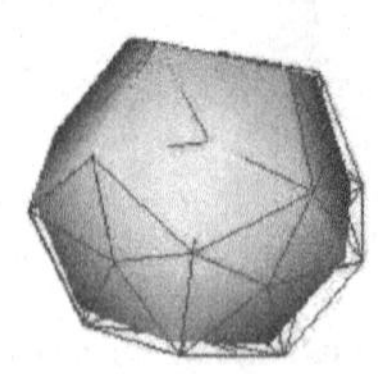
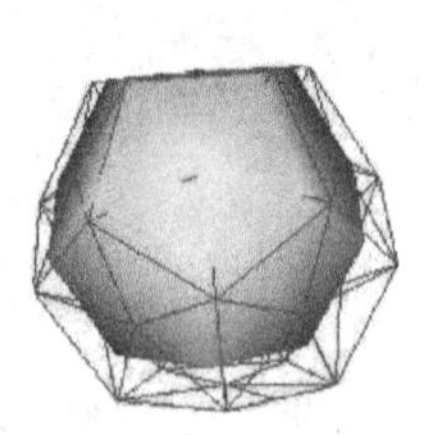
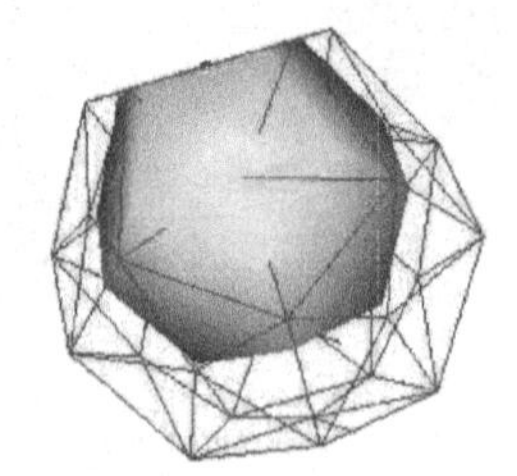

FIG. 7.1. Global rotation of the linear elastic model (wireframe).

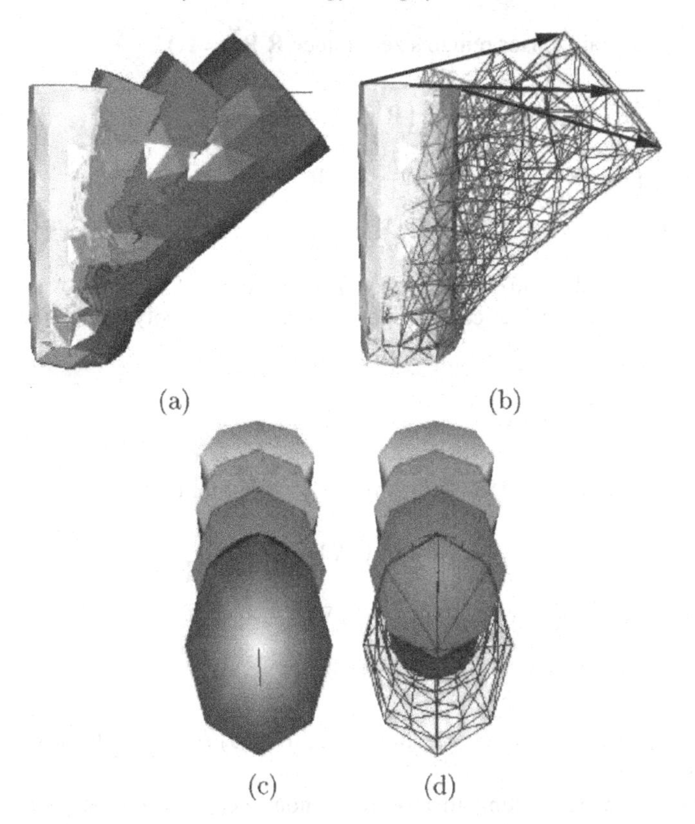

FIG. 7.2. Successive deformations of a linear elastic cylinder (PICINBONO, DELINGETTE and AYACHE [2001]). (a) and (b): side view; (c) and (d): top view.

model for large displacements, and is a particular case of hyperelastic materials. It has been used to model various materials (Table 3.8.4 of CIARLET [1987] provides the constants for materials like steel, glass, lead or rubber) including facial soft tissue (GLADILIN [2002]) and trabecular bone (BAYRAKTAR, ADAMS, GUPTA, PAPADOPOULOS and KEAVENY [2003]). A St Venant–Kirchhoff material relies on the Hooke's law as the definition of elastic energy (see Eq. (3.5) in Section 3.2.3) but the linearized strain tensor $\mathbf{E}_{\mathrm{L}}$ is replaced by the Green–Lagrange strain tensor $\mathbf{E}$:

$$\mathbf{E}(\mathbf{X}) = \frac{1}{2}\Big(\nabla\mathbf{U} + \nabla\mathbf{U}^{\mathrm{T}} + \nabla\mathbf{U}^{\mathrm{T}}\nabla\mathbf{U}\Big), \tag{7.1}$$

$$W_{\mathrm{NL}}(\mathbf{X}) = \frac{\lambda}{2}(\operatorname{tr}\mathbf{E})^2 + \mu \operatorname{tr}\mathbf{E}^2. \tag{7.2}$$

The Green–Lagrange strain tensor $\mathbf{E}$ is no longer a linear function of the displacement field. A first property is that the elastic energy becomes invariant under the application of rotations. Indeed, when a rigid transformation (with rotation matrix $\mathbf{R}$) is applied to an object, the gradient of the displacement field is $\nabla\mathbf{U} = \mathbf{R} - \mathbf{I}_3$ and therefore the

Green–Lagrange strain tensor remains zero (since $\mathbf{R}\,\mathbf{R}^{\mathrm{T}} = \mathbf{I}_3$),

$$\begin{aligned}\mathbf{E}(\mathbf{X}) &= \frac{1}{2}\left(\mathbf{R} - \mathbf{I}_3 + \mathbf{R}^{\mathrm{T}} - \mathbf{I}_3 + \left(\mathbf{R}^{\mathrm{T}} - \mathbf{I}_3\right)(\mathbf{R} - \mathbf{I}_3)\right)\\ &= \frac{1}{2}\left(\mathbf{R} + \mathbf{R}^{\mathrm{T}} - 2\mathbf{I}_3 + \mathbf{R}^{\mathrm{T}}\mathbf{R} - \mathbf{R} - \mathbf{R}^{\mathrm{T}} + \mathbf{I}_3\right)\\ &= [\mathbf{0}].\end{aligned}$$

A second property is that the elastic energy W_{NL} (Section 3.2.3), which was a quadratic function of $\nabla\mathbf{U}$ in the linear case, is now a fourth-order polynomial function with respect to $\mathbf{U}$:

$$\begin{aligned}W_{\mathrm{NL}} &= \frac{\lambda}{2}(\operatorname{tr}\mathbf{E})^2 + \mu \operatorname{tr}\mathbf{E}^2\\ &= \frac{\lambda}{2}\left[(\operatorname{div}\mathbf{U}) + \frac{1}{2}\|\nabla\mathbf{U}\|^2\right]^2 + \mu\|\nabla\mathbf{U}\|^2 - \frac{\mu}{2}\|\operatorname{rot}\mathbf{U}\|^2\\ &\quad + \mu(\nabla\mathbf{U} : \nabla\mathbf{U}^t\nabla\mathbf{U}) + \frac{\mu}{4}\left\|\nabla\mathbf{U}^t\nabla\mathbf{U}\right\|^2,\\ W_{\mathrm{NL}} &= W_{\mathrm{Linear}} + \frac{\lambda}{2}(\operatorname{div}\mathbf{U})\|\nabla\mathbf{U}\|^2 + \frac{\lambda}{8}\|\nabla\mathbf{U}\|^4\\ &\quad + \mu\left(\nabla\mathbf{U} : \nabla\mathbf{U}^t\nabla\mathbf{U}\right) + \frac{\mu}{4}\left\|\nabla\mathbf{U}^t\nabla\mathbf{U}\right\|^2,\end{aligned} \tag{7.3}$$

where W_{Linear} is given by Eq. (3.5) and $A : B = \operatorname{tr}(A^t B) = \sum_{i,j} a_{ij}b_{ij}$ is the dot product of two matrices.

Furthermore, we can extend this isotropic non-linear elastic energy to take into account "transversally isotropic" materials as performed in Section 3.2.4 for the linear elastic model. In fact, Eq. (3.9), which defines the additional anisotropic term, still holds for St Venant–Kirchhoff elasticity. However, for the sake of clarity, we chose to keep only the anisotropic contribution which penalizes the material stretch in the direction given by unit vector $\mathbf{a}_0$:

$$W_{\mathrm{Trans\_iso}} = W_{\mathrm{NL}} + \left(-\frac{\Delta\lambda}{2} + \Delta\mu\right)\left(\mathbf{a}_0^t\mathbf{E}\mathbf{a}_0\right)^2,$$

where $\Delta\lambda$ and Δ are the variations of Lamé coefficients along the direction of anisotropy.

7.3. Finite element modeling

By adopting the same methodology as the one presented in Section 4.3, we provide a closed form expression of the elastic energy of a linear tetrahedron finite element,

$$\begin{aligned}W_{\mathrm{NL}}(\mathcal{T}) &= \frac{1}{2}\sum_{j,k}\mathbf{U}_j^t\left[\mathcal{B}_{jk}^{\mathcal{T}}\right]\mathbf{U}_k + \frac{1}{2}\sum_{j,k,l}\left(\mathbf{U}_j.\mathcal{C}_{jkl}^{\mathcal{T}}\right)(\mathbf{U}_k.\mathbf{U}_l)\\ &\quad + \frac{1}{2}\sum_{j,k,l,m}\mathcal{D}_{jklm}^{\mathcal{T}}(\mathbf{U}_j.\mathbf{U}_k)(\mathbf{U}_l.\mathbf{U}_m),\end{aligned} \tag{7.4}$$

where the terms $\mathcal{B}_{jk}^{\mathcal{T}}$, $\mathcal{C}_{jkl}^{\mathcal{T}}$ and $\mathcal{D}_{jklm}^{\mathcal{T}}$, called "stiffness parameters", are given by

- $\mathcal{B}_{jk}^{\mathcal{T}}$ is a (3×3) symmetric matrix (which corresponds to the linear component of the energy),

$$36V(\mathcal{T})\mathcal{B}_{jk}^{\mathcal{T}} = \lambda(\mathbf{m}_j \otimes \mathbf{m}_k) + \mu\left[(\mathbf{m}_k \otimes \mathbf{m}_j) + (\mathbf{m}_j.\mathbf{m}_k)\mathbf{I}_3\right] + \left(-\frac{\Delta\lambda}{2} + \Delta\mu\right)(\mathbf{a}_0 \otimes \mathbf{a}_0)(\mathbf{m}_j \otimes \mathbf{m}_k)(\mathbf{a}_0 \otimes \mathbf{a}_0),$$

- $\mathcal{C}_{jkl}^{\mathcal{T}}$ is a vector,

$$216\,(V(\mathcal{T}))^2\,\mathcal{C}_{jkl}^{\mathcal{T}} = \frac{\lambda}{2}\mathbf{m}_j(\mathbf{m}_k.\mathbf{m}_l) + \frac{\mu}{2}\left[\mathbf{m}_l(\mathbf{m}_j.\mathbf{m}_k) + \mathbf{m}_k(\mathbf{m}_j.\mathbf{m}_l)\right] + \left(-\frac{\Delta\lambda}{2} + \Delta\mu\right)(\mathbf{a}_0 \otimes \mathbf{a}_0)(\mathbf{m}_j \otimes \mathbf{m}_k)(\mathbf{a}_0 \otimes \mathbf{a}_0)\mathbf{m}_l,$$

- and $\mathcal{D}_{jklm}^{\mathcal{T}}$ is a scalar,

$$1296\,(V(\mathcal{T}))^3\,\mathcal{D}_{jklm}^{\mathcal{T}} = \frac{\lambda}{8}(\mathbf{m}_j.\mathbf{m}_k)(\mathbf{m}_l.\mathbf{m}_m) + \frac{\mu}{4}(\mathbf{m}_j.\mathbf{m}_m)(\mathbf{m}_k.\mathbf{m}_l) + \frac{1}{4}\left(-\frac{\Delta\lambda}{2} + \Delta\mu\right)(\mathbf{a}_0.\mathbf{m}_j)(\mathbf{a}_0.\mathbf{m}_k)(\mathbf{a}_0.\mathbf{m}_l)(\mathbf{a}_0.\mathbf{m}_m).$$

- The last term of each stiffness parameter models the anisotropic behavior of the material.

The elastic force applied at each vertex $\mathbf{p}_i$ of tetrahedron $\mathcal{T}$ is obtained as the derivation of the elastic energy $W_{\mathrm{NL}}(\mathcal{T})$ with respect to the displacement $\mathbf{p}_i$,

$$\mathbf{F}_i(\mathcal{T}) = \underbrace{\sum_j \left[\mathcal{B}_{ij}^{\mathcal{T}}\right]\mathbf{U}_j}_{\mathbf{F}_i^1(\mathcal{T})} + \underbrace{\sum_{j,k}(\mathbf{U}_k \otimes \mathbf{U}_j)\mathcal{C}_{jki}^{\mathcal{T}} + \frac{1}{2}(\mathbf{U}_j.\mathbf{U}_k)\mathcal{C}_{ijk}^{\mathcal{T}}}_{\mathbf{F}_i^2(\mathcal{T})} + \underbrace{2\sum_{j,k,l}\mathcal{D}_{jkli}^{\mathcal{T}}\mathbf{U}_l\mathbf{U}_k^t\mathbf{U}_j}_{\mathbf{F}_i^3(\mathcal{T})}\,. \tag{7.5}$$

The first term of the elastic force ($\mathbf{F}_i^1(\mathcal{T})$) corresponds to the linear elastic case presented in Section 4.4.

7.4. Non-linear tensor–mass model

In this section, we generalize the tensor–mass model introduced in Section 6.1 to the case of large displacement elasticity. The only changes in the tensor–mass algorithm are related to the computation of the elastic force $\mathbf{F}_i$ applied at vertex i.

In the case of linear elasticity, this force was computed by a first scan of all edges to compute the terms $[\mathbf{K}_{ij}]\mathbf{u}_j$ followed by a scan of all vertices to add the terms $[\mathbf{K}_{ii}]\mathbf{u}_i$.

TABLE 7.1
Storage of the stiffness parameters on the mesh

Stiffness parameters distribution	Tensors	Vectors		Scalars		
Vertex p	$\mathcal{B}^{pp}$	$\mathcal{C}^{ppp}$			$\mathcal{D}^{pppp}$	
Edge (p, j)	$\mathcal{B}^{pj}$	$\mathcal{C}^{ppj}$	$\mathcal{C}^{jpp}$	$\mathcal{D}^{jppp}$	$\mathcal{D}^{jjjp}$	$\mathcal{D}^{jpjp}$
		$\mathcal{C}^{jjp}$	$\mathcal{C}^{pjj}$	$\mathcal{D}^{pjjp}$	$\mathcal{D}^{jjpp}$	
Triangle (p, j, k)		$\mathcal{C}^{jkp}$		$\mathcal{D}^{jkpp}$	$\mathcal{D}^{jpkp}$	$\mathcal{D}^{pjkp}$
		$\mathcal{C}^{kjp}$		$\mathcal{D}^{jjkp}$	$\mathcal{D}^{jkjp}$	$\mathcal{D}^{kjjp}$
		$\mathcal{C}^{pjk}$		$\mathcal{D}^{kkjp}$	$\mathcal{D}^{kjkp}$	$\mathcal{D}^{jkkp}$
Tetrahedron (p, j, k, l)				$\mathcal{D}^{jklp}$	$\mathcal{D}^{jlkp}$	$\mathcal{D}^{kjlp}$
				$\mathcal{D}^{kljp}$	$\mathcal{D}^{ljkp}$	$\mathcal{D}^{lkjp}$

We proposed to apply the same principle to the quadratic term ($\mathbf{F}_2^p(\mathcal{T})$ of Eq. (7.5)) and the cubic term ($\mathbf{F}_3^p(\mathcal{T})$). The former requires *stiffness vectors* for vertices, edges and triangles, and the latter requires *stiffness scalars* for vertices, edges, triangles and tetrahedra.

The task of assembling global stiffness parameters is slightly more time consuming than in the linear case, since 31 parameters must be assembled instead of 2; these parameters are presented in Table 7.1.

For vertex, edge and triangle parameters, one needs to add the contributions of all neighboring tetrahedra. For instance, the vertex rigidity vector $\mathcal{C}^{ppp}$ is computed at vertex p as

$$\mathcal{C}^{ppp} = \sum_{\mathcal{T}\in\mathcal{S}(p)} \mathcal{C}^{\mathcal{T}}_{ppp}.$$

For the 6 scalar parameters $\mathcal{D}^{jklp}$ stored at each tetrahedron, no assembly is required since there is no other contribution originating from another tetrahedron.

The computation of the elastic force is performed by successively scanning tetrahedra, triangles, edges and vertices of the mesh. When scanning triangles for instance, the contributions from the three triangles are computed and added to the elastic force of each of its three vertices. The contribution for each element is summarized in Eq. (7.5).

$$\mathbf{F}_i = \mathbf{F}_i^{\text{vertex}} + \mathbf{F}_i^{\text{edge}} + \mathbf{F}_i^{\text{triangle}} + \mathbf{F}_i^{\text{tetrahedron}} \tag{7.6}$$

with

$$\mathbf{F}_i^{\text{vertex}} = \boxed{\begin{array}{l} \text{Vertex contribution} \\ \hline [\mathcal{B}^{pp}]\mathbf{U}_p \\ \quad +\left[(\mathbf{U}_p \otimes \mathbf{U}_p) + \tfrac{1}{2}(\mathbf{U}_p.\mathbf{U}_p)\mathbf{I}_3\right]\mathcal{C}^{ppp} \\ +2\mathcal{D}^{pppp}\mathbf{U}_p\mathbf{U}_p^t\mathbf{U}_p \end{array}},$$

$$\mathbf{F}_i^{\text{edge}} = \sum_{\text{edges}(p,j)} \boxed{\begin{array}{l} \text{Edge contribution} \\ \hline \left[\mathcal{B}^{pj}\right]\mathbf{U}_j \\ \quad + \left[(\mathbf{U}_j \otimes \mathbf{U}_p) + (\mathbf{U}_j.\mathbf{U}_p)\mathbf{I}_3\right]\mathcal{C}^{ppj} + (\mathbf{U}_p \otimes \mathbf{U}_j)\mathcal{C}^{jpp} \\ \quad + (\mathbf{U}_j \otimes \mathbf{U}_j)\mathcal{C}^{jjp} + \frac{1}{2}(\mathbf{U}_j.\mathbf{U}_j)\mathcal{C}^{pjj} \\ \quad + 2\Big[\mathcal{D}^{jppp}\left(2\mathbf{U}_p\mathbf{U}_p^t\mathbf{U}_j + \mathbf{U}_j\mathbf{U}_p^t\mathbf{U}_p\right) + \mathcal{D}^{jjpp}\mathbf{U}_p\mathbf{U}_j^t\mathbf{U}_j \\ \qquad + \left(\mathcal{D}^{jpjp} + \mathcal{D}^{pjjp}\right)\mathbf{U}_j\mathbf{U}_j^t\mathbf{U}_p + \mathcal{D}^{jjjp}\mathbf{U}_j\mathbf{U}_j^t\mathbf{U}_j\Big] \end{array}},$$

$$\mathbf{F}_i^{\text{triangle}} = \sum_{\text{faces}(p,j,k)} \boxed{\begin{array}{l} \text{Triangle contribution} \\ \hline \left[(\mathbf{U}_k \otimes \mathbf{U}_j)\mathcal{C}^{jkp} + (\mathbf{U}_j \otimes \mathbf{U}_k)\mathcal{C}^{kjp} + (\mathbf{U}_j.\mathbf{U}_k)\mathcal{C}^{pjk}\right] \\ \quad + 2\Big[\left(\mathcal{D}^{pjkp} + \mathcal{D}^{jpkp}\right)\left(\mathbf{U}_j\mathbf{U}_k^t\mathbf{U}_p + \mathbf{U}_k\mathbf{U}_j^t\mathbf{U}_p\right) \\ \quad + 2\mathcal{D}^{jkpp}\mathbf{U}_p\mathbf{U}_j^t\mathbf{U}_k \\ \quad + \left(\mathcal{D}^{kjjp} + \mathcal{D}^{jkjp}\right)\mathbf{U}_j\mathbf{U}_j^t\mathbf{U}_k + \mathcal{D}^{jjkp}\mathbf{U}_k\mathbf{U}_j^t\mathbf{U}_j \\ \quad + \left(\mathcal{D}^{jkkp} + \mathcal{D}^{kjkp}\right)\mathbf{U}_k\mathbf{U}_k^t\mathbf{U}_j + \mathcal{D}^{kkjp}\mathbf{U}_j\mathbf{U}_k^t\mathbf{U}_k\Big] \end{array}},$$

$$\mathbf{F}_i^{\text{tetrahedron}} = \sum_{\text{tetra}(p,j,k,l)} \boxed{\begin{array}{l} \text{Tetrahedron contribution} \\ \hline 2\Big[\left(\mathcal{D}^{jklp} + \mathcal{D}^{kjlp}\right)\mathbf{U}_l\mathbf{U}_j^t\mathbf{U}_k \\ \quad + \left(\mathcal{D}^{jlkp} + \mathcal{D}^{ljkp}\right)\mathbf{U}_k\mathbf{U}_j^t\mathbf{U}_l \\ \quad + \left(\mathcal{D}^{kljp} + \mathcal{D}^{lkjp}\right)\mathbf{U}_j\mathbf{U}_k^t\mathbf{U}_l\Big] \end{array}}.$$

In terms of data structure, the non-linear tensor–mass model requires the addition of triangles in the mesh topological description. In our case, we chose to store triangles in a hash table which is hashed by the three indices of its vertices in lexicographic order. Furthermore, each tetrahedron owns pointers towards its four triangles and reversely, each triangle owns pointers towards its two neighboring tetrahedra.

During the simulation of resection, tetrahedra are iteratively removed near the extremities of virtual cavitron instruments. When removing a single tetrahedron, 280 floating point numbers are updated to suppress the tetrahedron contributions to the stiffness parameters of the surrounding vertices, edges and triangles:

$$\begin{aligned} &4 * (1 \text{ tensor} + 1 \text{ vector} + 1 \text{ scalar}) \\ &\qquad + 6 * (1 \text{ tensor} + 4 \text{ vectors} + 5 \text{ scalars}) \\ &\qquad + 4 * (3 \text{ vectors} + 9 \text{ scalars}) \\ &\quad = 280 \text{ real numbers}. \end{aligned}$$

By locally updating stiffness parameters, the tissue has exactly the same properties as if the corresponding tetrahedron had been removed at its rest position. Because of the volumetric continuity of finite element modeling, the tissue deformation remains realistic during cutting.

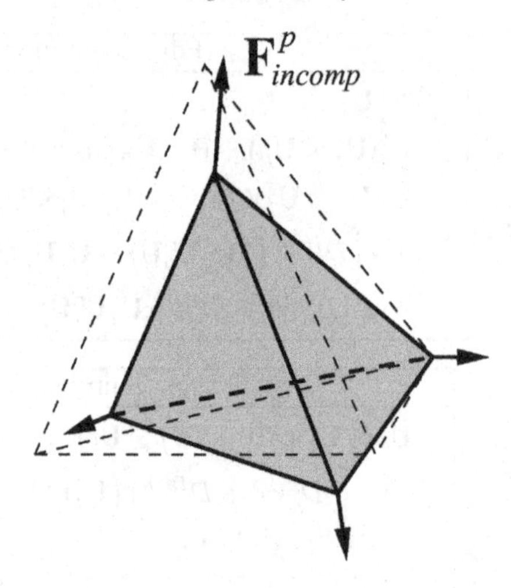

FIG. 7.3. Penalization of volume variation.

7.5. *Incompressibility constraint*

Living tissue, which is made essentially of water is almost incompressible, a property which is difficult to model and which, in most cases, leads to instability problems. This is the case with the St Venant–Kirchhoff model: the material remains incompressible when the Lamé constant λ tends towards infinity. Taking a large value for λ would impose to decrease the time step and therefore to increase the computation time. Another reason to add an external incompressibility constraint to the model is intrinsic to the model itself: the St Venant–Kirchhoff model relies on the Green–Lagrange strain tensor E which is invariant with respect to rotations. But it is also invariant with respect to symmetries, which could lead to the reversal of some tetrahedra under strong constraints.

We chose to penalize volume variation by applying to each vertex of the tetrahedron a force directed along the normal of the opposite face $\mathbf{N}_p$ (see Fig. 7.3), the norm of the force being proportional to the square of the relative volume variation,

$$\mathbf{F}^p_{\text{incomp}} = \operatorname{sign}(V - V_0)\left(\frac{V - V_0}{V_0}\right)^2 \vec{\mathbf{N}}_p. \tag{7.7}$$

Since the volume V is proportional to the height of each vertex facing its opposite triangle, when V is greater than V_0 then the force $\mathbf{F}^p_{\text{incomp}}$ tends to decrease V by moving each vertex along the normal of the triangle facing it. These forces act as an artificial pressure inside each tetrahedron. This method is closely related to Lagrange multipliers, which are often used to solve problem of energy minimization under constraints.

7.6. *Results*

In a first experiment, we wish to highlight the contributions of our new deformable model in the case of partial rotations. Fig. 7.4 shows the same experience as the one presented for linear elasticity (Section 7.1, Fig. 7.2). On the left we can see that the cylinder vertices are now able to follow non-straight trajectories (Fig. 7.4(a)), leading

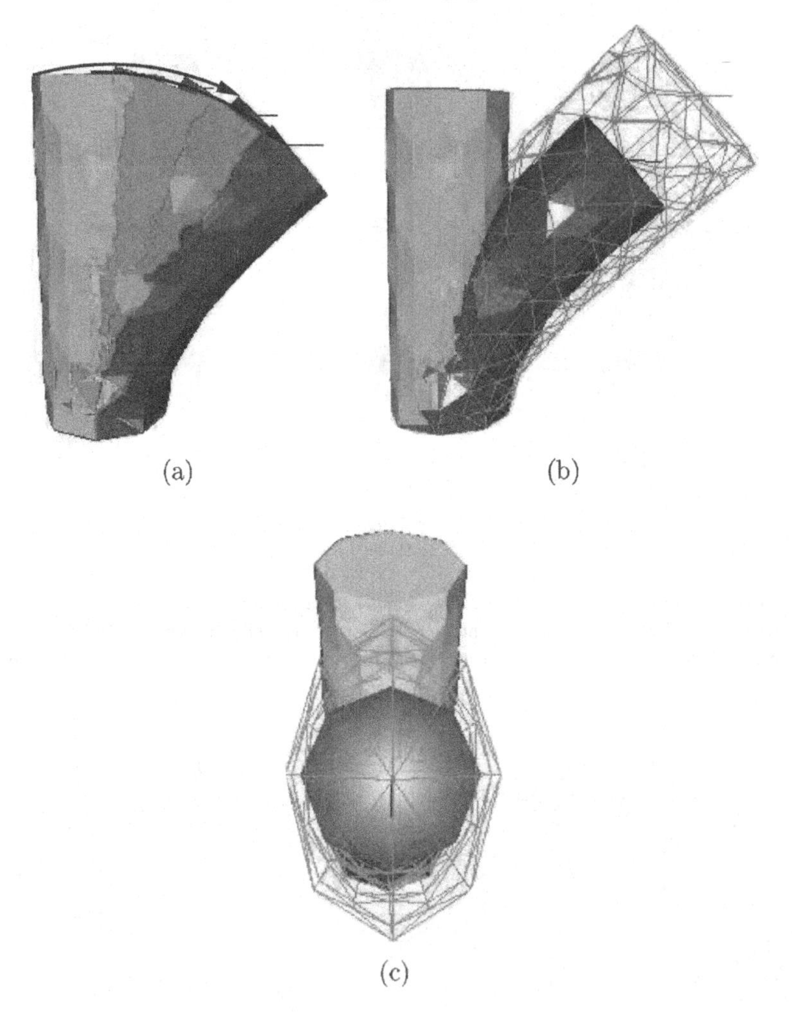

FIG. 7.4. (a) Successive deformations of the non-linear model (PICINBONO, DELINGETTE and AYACHE [2003]). Side (b) and top (c) view of the comparison between linear (wireframe) and non-linear model (solid rendering).

to much more realistic deformations than in the linear (wireframe) case (Fig. 7.4(b) and (c)).

The second example presents the differences between isotropic and anisotropic materials. The three cylinders of Fig. 7.5 have their top and bottom faces fixed, and are submitted to the same forces. While the isotropic model on the left undergoes a "snake-like" deformation, the last two, which are anisotropic along their height, stiffen in order to minimize their stretch in the anisotropic direction. The rightmost model, being twice as stiff as the middle one in the anisotropic direction, starts to squeeze in the plane of isotropy because it cannot stretch anymore.

In the third example (Fig. 7.6), we apply a force to the right lobe of the liver (the liver is fixed in a region near the center of its back side, and Lamé coefficients are:

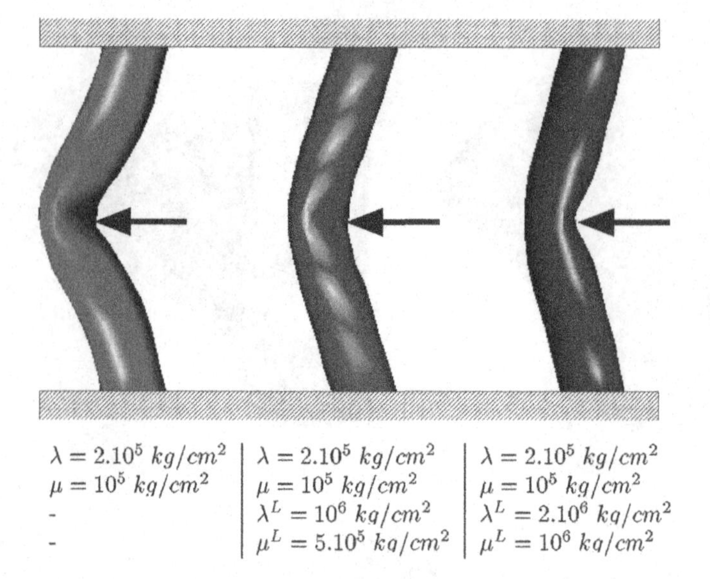

FIG. 7.5. Shearing deformation of tubular structures under the action of the force indicated by the arrow. The leftmost figure corresponds to an isotropic non-linear material while the center and rightmost figures correspond to a non-linear anisotropic material, the direction of anisotropy being the cylinder axis.

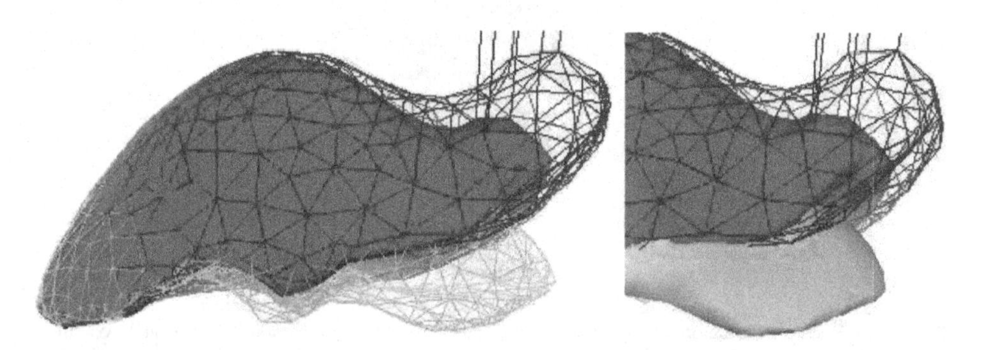

FIG. 7.6. Linear (upper mesh in wireframe), non-linear (Gauraud shaded) liver models and rest shape (lower mesh in wireframe). In both cases, the same forces showed in solid lines are applied to three surface nodes lying on the left lobe (PICINBONO, DELINGETTE and AYACHE [2003]).

$\lambda = 40$ kPa and $\mu = 10$ kPa). Using the linear elastic model, the right part of the liver undergoes a large (and unrealistic) volume increase, whereas with non-linear elasticity, the right lobe is able to rotate partially, while adopting a more realistic deformation.

Adding the incompressibility constraint on the same examples decreases the volume variation even more (see Table 7.2), and also stabilizes the behavior of the deformable models in highly constrained areas.

The last example is the simulation of a typical laparoscopic surgical gesture on the liver. One tool is pulling the edge of the liver sideways while a bipolar cautery device

TABLE 7.2
Volume variation results. For the cylinder: left, middle and right stand for the different deformations of Figs. 7.5 and 7.6

Volume variation (%)	Linear	Non-linear	Non-linear incomp.
Cylinder (left – middle – right)	7 – 28 – 63	0.3 – 1 – 2	0.2 – 0.5 – 1
Liver	9	1.5	0.7

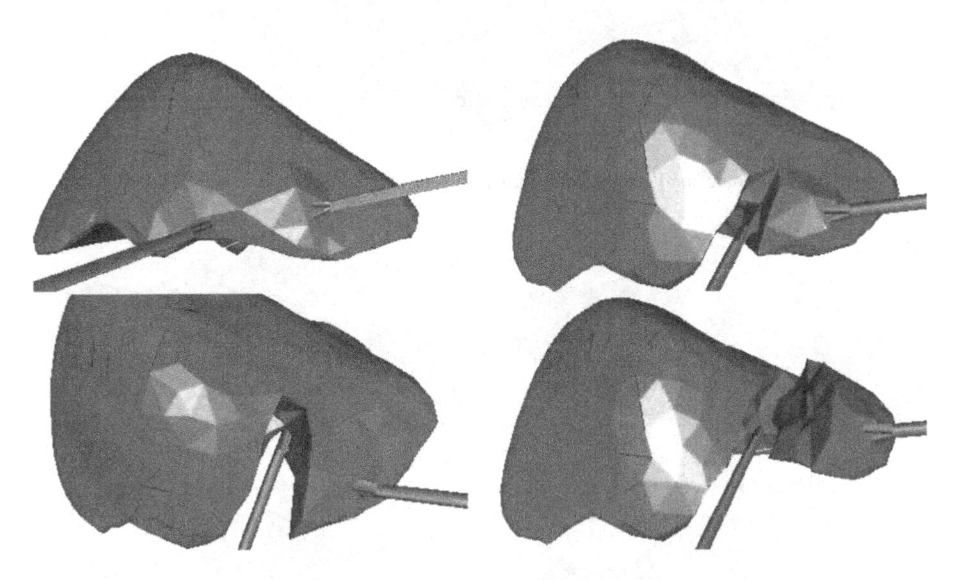

FIG. 7.7. Simulation of laparoscopic liver surgery.

cuts it. During the cutting, the surgeon pulls away the part of the liver he wants to remove. This piece of liver undergoes large displacements and the deformation appears fairly realistic with this new non-linear deformable model (Fig. 7.7).

Obviously, the computation time of this model is larger than for the linear model because the force equation is more complex (Eq. (7.5) in Section 7.3 to be compared with Eq. (6.3) in Section 6.1.7). With our current implementation, the simulation refresh rate is five times slower than with the linear model. Nevertheless, with this non-linear model, we can reach an update cycle of 25 Hz on meshes made of about 2000 tetrahedra (on a PC Pentium PIII 500 MHz). This is enough to achieve real-time visual feedback with quite complex objects, and even to provide a realistic haptic feedback using force extrapolation as described in PICINBONO, LOMBARDO, DELINGETTE and AYACHE [2000].

7.7. *Optimization of non-linear deformations*

We showed that non-linear elasticity allows to simulate much more realistic deformations than linear elasticity when the model undergoes large displacements. However,

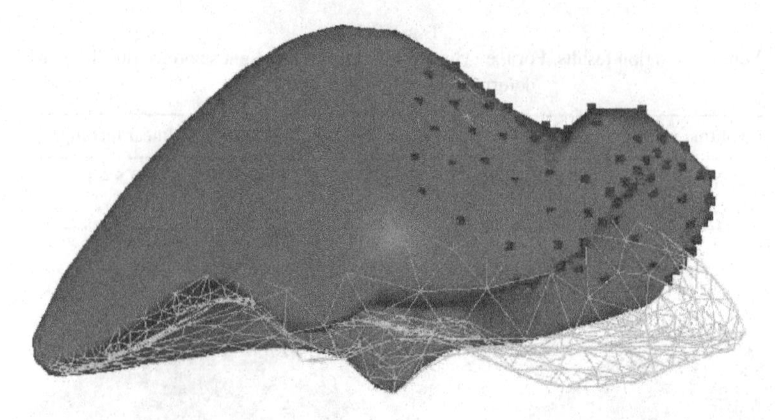

FIG. 7.8. Adaptable non-linear model deformation compared to its rest position (wireframe).

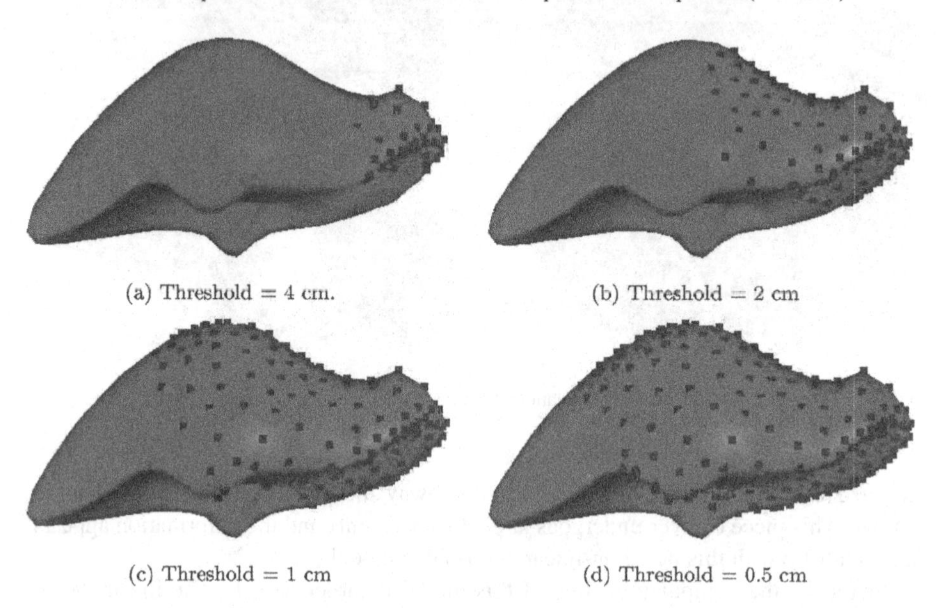

(a) Threshold = 4 cm. (b) Threshold = 2 cm

(c) Threshold = 1 cm (d) Threshold = 0.5 cm

FIG. 7.9. Deformation of the adaptive non-linear model for several values of the threshold.

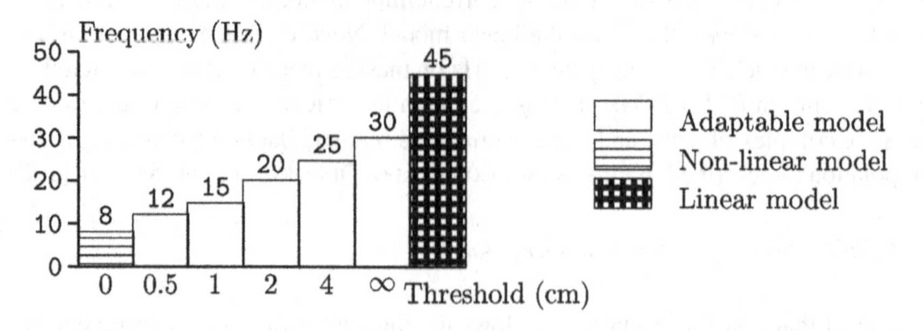

FIG. 7.10. Updating frequencies of the adaptable model for several values of the threshold.

non-linear elasticity is more computationally expensive than linear elasticity. Since non-linear elastic forces tend to linear elastic forces as the maximum vertex displacement decreases to zero, we propose to use non-linear elasticity only at parts of the mesh where displacements are larger than a given threshold, the remaining part using linear elasticity. Thus, we modified the force computation algorithm in the following manner: for each vertex, we first compute the linear part of the force, and we add the non-linear part only if its displacement is larger than a threshold. Fig. 7.8 shows a deformation computed with this optimization (same model as in Fig. 7.6). This liver model is made of 6342 tetrahedra and 1394 vertices. The threshold is set to 2 cm while the mesh is about 30 cm long. The points drawn on the surface identify vertices using non-linear elasticity. With this method, we reach an update frequency of 20 Hz instead of 8 Hz for a fully non-linear model. The same deformation is presented on Fig. 7.9 for different values of the threshold. With this method, we can choose a trade-off between the bio-mechanical realism of the deformation and the update frequency of the simulation. The diagram on Fig. 7.10 shows the update frequencies reached for each value of the threshold, in comparison with the fully linear and the fully non-linear models. Even when this threshold tends towards infinity, the adaptable model is slower than the linear model, because the computation algorithm of the non-linear force is more complex. Indeed, the computation of non-linear forces requires to visit all vertices, edges, triangles and tetrahedra of the mesh, whereas only vertices and edges need to be visited for the

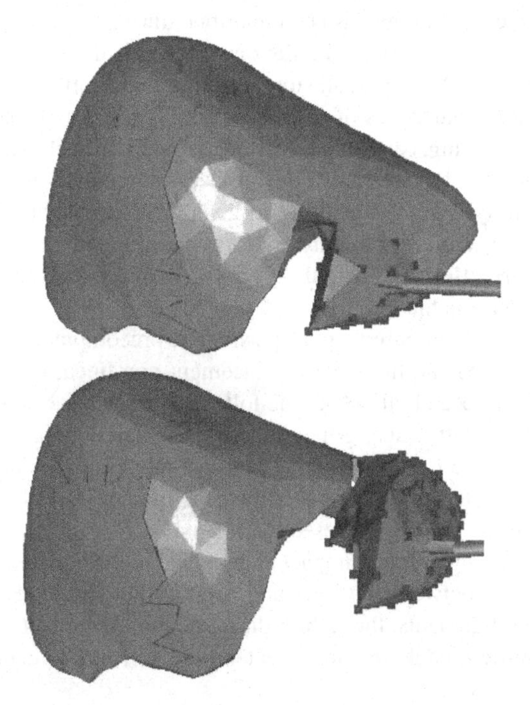

FIG. 7.11. Simulation of hepatectomy based on a non-linear adaptable elastic model. Non-linear elastic force are applied on vertices outlined with a box.

linear model. For the simulation example of Fig. 7.7, this optimization leads to update frequencies varying between 50 and 80 Hz, depending on the number of points modeling non-linear elasticity (Fig. 7.11). The minimal frequency of 50 Hz is reached at the end of the simulation, when all vertices of the resected part of the liver are using large displacement elasticity (on the right of Fig. 7.11).

In general, two strategies can be used to set the value of this threshold. In the first strategy, the threshold is increased until a given update frequency is matched as demonstrated previously. The second strategy is physically-motivated and sets the threshold to 10% of the typical size of the mesh since it corresponds to the extent of displacement for which linear elasticity remains a valid constitutive law.

8. Conclusion

In this chapter, we have presented several algorithms for computing in real-time the deformation of soft tissues in a surgical simulator. We wish to stress two important aspects of these algorithms. First of all, using linear tetrahedra as finite elements helped us to write closed-form expressions of the elastic energy and its derivatives, even in the case of large displacement elasticity. These expressions nicely decouple the physical parameters (Lamé coefficients) from the geometry of each tetrahedron both in its rest position (direction of anisotropy, rest volume, area vectors) and in its deformed state (displacement vectors). Furthermore, it enables to quickly assemble local and global stiffness matrices when the mesh topology has been modified during a cutting simulation.

Second, in the context of surgery simulation, soft tissue deformation algorithms are closely tied with the visualization, collision detection and haptic rendering algorithms. Furthermore, the traditional stages of matrix assembly, matrix preconditioning, system solution and post-processing, cannot be easily decoupled like in classical software packages available in structural mechanics. This implies that the data structure and the flow chart must be carefully designed in order to achieve a reasonable trade-off between these performances. Therefore, building a successful simulator can only be achieved by a multidisciplinary effort covering the fields of biomechanics, numerical analysis, robotics and computer graphics.

An hepatectomy simulator based on the quasi-static precomputed linear elastic model (introduced in Section 5) and the large displacement non-linear elastic models (introduced in Section 7) has been built where the following three basic surgical gestures can be rehearsed: touching soft tissue, gripping soft tissue and cutting parenchyma with a cavitron. Furthermore, we recently added a physical model of the portal vein (Forest, Delingette and Ayache [2003]), which allows the user to simulate the clamping and cutting of vessels during the hepatic resection.

However, to increase the training impact and realism of the simulation, it is important to simulate the contact between the liver and neighboring structures such as the gallbladder, the different ligaments, the right kidney, the peritoneum, etc. These additional surface and volumetric models require to extend the soft tissue models introduced in this chapter in two ways.

First, it is necessary to extend the precomputed linear elastic model to include large-displacement non-linear elasticity. Indeed, the linear domain of biological soft tissue is usually rather small, and therefore many surgical gestures can only be simulated by using large-displacement elasticity (like rotating the lobe of the liver or resecting the gall-bladder). The precomputation of non-linear elastic material is not a trivial task since it implies solving a complex third-order algebraic equation in the case of St Venant–Kirchhoff elasticity (see Section 7.2). Instead, it may be possible to find suitable approximations which can be computed efficiently.

Second, it is necessary to extend the concept of hybrid models (introduced in Section 6.3) in order to cope with the deformation of models including several tens of thousands of vertices. Ideally, we would like to provide accurate but computationally expensive soft tissue models in the center of the surgical field where the user performs complex gestures and at the same time to provide less expensive models but potentially less accurate, away from the center of the surgical field. Of course, during surgery, the focus of the surgeon may switch from the gall-bladder to the hepatic parenchyma which implies that those tissue models should evolve dynamically from one level of accuracy to the other. Achieving this level of scalability with the constraint that the topology of these models may change over time, is the main challenge of soft tissue modeling for surgery simulation.

Finally, we would likc to stress the importance of validating the different components of a surgical simulator. Concerning soft tissue models, there are at least three levels of validation that need to be achieved. A first validation consists in comparing the soft tissue deformation algorithms that rely on strong hypotheses against well-known finite element packages in order to evaluate the range of approximations that are performed. In the second level of validation, the biomechanical behavior of each anatomical structure must be compared to experimental dataset. Ideally, one would like to validate both boundary conditions and the constitutive law of each biological tissue. However, in practice, this validation is made difficult by the lack of quantitative experimental information. The third level of validation consists in evaluating the dynamic behavior of each soft tissue during the simulation since some models that appear too soft or too stiff. Finally, and most importantly, it is required to validate the whole simulation system by assessing its ability to succeed in training young residents to perform a given surgical task.

Despite these remaining issues to be solved, we believe that practical surgery simulators will be fully operational and actually part of the surgical studies in the near future.

List of mathematical symbols

f_{u} Update frequency of the soft tissue model
t Discrete or continuous time variable
X_t Position of the model at time t
$T_{\mathrm{relaxation}}$ Relaxation time
T_{c} Computation time
$T_{\mathrm{interaction}}$ Latency caused by the software and hardware architecture
Δt Time step used in the discretization of temporal derivatives
$\mathbf{F}$ Global force vector
$\mathbf{K}$ Global stiffness matrix

$\mathbf{U}$ Global displacement vector
$\mathbf{M}$ Global mass matrix
$\mathbf{C}$ Global damping matrix
$\dot{\mathbf{U}}$ Global speed vector
$\mathcal{M}_{\text{rest}}$ Soft tissue model at its rest position
$\mathcal{M}_{\text{def}}$ Soft tissue model at its deformed position
Ω Region of space for the rest configuration
$\boldsymbol{\Phi}(x, y, z)$ Deformation function that maps point (x, y, z) from the rest configuration to the deformed configuration
$\mathbf{X}$ Point in the rest configuration
$\mathbf{U}(\mathbf{X})$ Displacement function
$\mathbf{C}(\mathbf{X})$ Right Cauchy–Green strain tensor
$\mathbf{E}(\mathbf{X})$ Green–Lagrange strain tensor
$\mathbf{I}_3$ 3×3 identity matrix
$\mathbf{E}_{\text{L}}$ Linearized strain tensor
e_{ij} Element of the linearized strain tensor
$\mathbf{T}(\mathbf{X})$ Cauchy stress tensor
$W(\mathbf{X})$ Density of elastic energy
λ, μ Isotropic Lamé coefficients
E, ν Isotropic Young modulus and Poisson ratio
$\mathbf{a}_0$ Unit vector along the direction of anisotropy for transversally isotropic materials
$\lambda^{\mathbf{a}_0}, \mu^{\mathbf{a}_0}$ Lamé coefficients along the direction of anisotropy
$\Delta\lambda, \Delta\mu$ Difference between the Lamé coefficients along the direction of anisotropy and those in the orthogonal plane
$\Delta W_{\text{Ani}}(\mathbf{X})$ Additional term of the density of elastic energy caused by anisotropy
I_4, I_5 Deformation invariants estimated along the direction of anisotropy
$\mathbf{p}_i$ Point of a tetrahedron in its rest position
$\mathbf{q}_i$ Point of a tetrahedron in its deformed position
$\mathbf{u}_i$ Displacement vector of a vertex of a tetrahedron
$\mathcal{T}$ Tetrahedron as a linear finite element
$h_j(\mathbf{X})$ Shape functions associated with a linear tetrahedron
$\mathbf{P}$ 4×4 matrix describing the shape functions
$V(\mathcal{T})$ Volume of tetrahedron $\mathcal{T}$
$\mathbf{m}_i$ Area vector opposite to vertex i
V_i 6 times the volume of the tetrahedron made by the origin $\mathbf{o}$ and vertices $\mathbf{p}_{i+1}$, $\mathbf{p}_{i+2}$ and $\mathbf{p}_{i+3}$
T_i Triangle opposite to vertex i
$\mathbf{n}_i$ Normal vector at the triangle T_i opposite to vertex i in a tetrahedron
$\theta_{i,j}$ Angle between normal vectors of triangles T_i and T_j
A_i Area of triangle T_i
$l_{i,j}$ Length of the edge connecting vertices i and j
f_i Height of vertex o above triangle T_i
$\mathbf{B}_{i,j}^{\mathcal{T}}$ Element (i, j) of the 3×3 stiffness matrix for a tetrahedron $\mathcal{T}$ made of an isotropic material

$\mathbf{A}^{\mathcal{T}}_{i,j}$ Element (i, j) of the 3×3 stiffness matrix for a tetrahedron $\mathcal{T}$ made of an transversally isotropic material

$\mathbf{K}_{i,j}$ 3×3 global stiffness matrix between vertex i and j

$k_{i,j}$ Eigenvalue along the edge direction of matrix $\mathbf{K}_{i,j}$

$W_g(\mathcal{T})$ Work of gravity forces

$W_p(\mathcal{T})$ Work of external surface pressure

$\mathbf{M}_{i,j}$ 3×3 global mass matrix between vertex i and j

$\mathbf{K}^{\star}_{i,j}$ 3×3 global stiffness matrix between vertex i and j that includes spring boundary conditions

$\mathbf{R}^g$ Global vector of gravity forces

$\mathbf{R}^b$ Global vector of boundary forces

Acknowledgements

We thank Matthias Teschner, Denis Laurendeau and Jean-Marc Schwartz for their priceless comments and for proofreading this article.

The work presented in this paper is a joint work between the authors and mainly two former PhD students: Stéphanc Cotin and Guillaume Picinbono. Stéphane Cotin developed the precomputed linear elastic model of Section 5 as well as a first version of the tensor–mass model described in Section 6.1. Guillaume Picinbonno proposed the extension of the tensor–mass model to the case of large displacement elasticity (in Section 7). We also thank Clément Forest and Jean-Christophe Lombardo for their numerous contributions on force-feedback rendering, collision detection as well as mesh data structure. This work was fueled with the stimulating remarks and propositions from our INRIA colleagues who participated in the AISIM and CAESARE joint initiatives: Marie-Paule Cani, Marina Vidrascu, Marc Thiriet, Christian Laugier. Also, we are grateful to Prof. Marescaux, Prof. Leroy and Prof. Luc Soler from the IRCAD research center for their long-term vision and for sharing their expertise of abdominal surgery with us. Finally, we acknowledge the strong support we received from Gilles Khan, INRIA Vice-President for Research, during the different stages of this research work.

References

ACKERMAN, M.J. (1998). The visible human project. *Proc. IEEE: Special Issue on Surgery Simulation* **86** (3), 504–511.

AYACHE, N. (2003). Epidaure: a research project in medical image analysis, simulation and robotics at INRIA. *IEEE Trans. Medical Imaging*, Invited Editorial.

AYACHE, N., DELINGETTE, H. (eds.) (2003). *Int. Symp. on Surgery Simulation and Soft Tissue Modeling*, Juan-Les-Pins, France, June 1998. In: Lecture Notes in Comput. Sci. **2673** (Springer-Verlag, New York).

BARAFF, D., WITKIN, A. (1998). Large steps in cloth simulation. In: *Computer Graphics Proceedings*, SIGGRAPH'98, Orlando, USA, July 1998, pp. 43–54.

BASSINGTHWAIGHTE, J.B. (2000). Strategies for the physiome project. *Ann. Biomed. Engrg.* **28**, 1043–1058.

BATHE, K.-L. (1982). *Finite Element Procedures in Engineering Analysis* (Prentice Hall, New York).

BAYRAKTAR, H., ADAMS, M., GUPTA, A., PAPADOPOULOS, P., KEAVENY, T. (2003). The role of large deformations in trabecular bone mechanical behavior. In: *ASME Bioengineering Conference*, Key Biscayne, FL, USA, June 2003.

BENZLEY, S.E., PERRY, E., CLARK, B., MERKLEY, K., SJAARDEMA, G. (1995). Comparison of all-hexahedral and all-tetrahedral finite element meshes for elastic and elasto-plastic analysis. In: *Proc. 4th Int. Meshing Roundtable*, Sandia National Laboratories, October 1995, pp. 179–191.

BERCI, G., HUNTER, J.G., SACKIER, J.M. (1994). Training in laparoscopic cholecystectomy: Quantifying the learning curve. *J. Endoscopic Surgery* **8**, 28–31.

BESAG, J. (1986). On the statistical analysis of dirty pictures. *J. Roy. Statist. Soc.* **48** (3), 326–338.

BIELSER, D., GROSS, M.H. (2000). Interactive simulation of surgical cuts. In: *Proc. Pacific Graphics 2000*, Hong-Kong, October 2000 (IEEE Computer Society Press), pp. 116–125.

BOUX DE CASSON, F., LAUGIER, C. (1999). Modelling the dynamics of a human liver for a minimally invasive simulator. In: *Proc. Int. Conf. on Medical Image Computer-Assisted Intervention*, Cambridge, UK, September 1999.

BRO-NIELSEN, M. (1998). Finite element modeling in surgery simulation. *Proc. IEEE: Special Issue on Surgery Simulation* **86** (3), 490–503.

BRO-NIELSEN, M., COTIN, S. (1996). Real-time volumetric deformable models for surgery simulation using finite elements and condensation. In: *Eurographics'96, vol. 3*. pp. 57–66.

BRONSHTEIN, I.N., SEMENDYAYEV, K.A. (1985). *Handbook of Mathematics* (Van Nostrand–Reinhold, New York).

BROWN, J.D., ROSEN, J., KIM, Y., CHANG, L., SINANAN, M., HANNAFORD, B. (2003). In-vivo and in-situ compressive properties of porcine abdominal soft tissue. In: *Medicine Meets Virtual Reality*, MMVR'03, Newport Beach, USA, January 2003.

CANAS, J., PARIS, F. (1997). *Boundary Element Method: Fundamentals and Application* (Oxford Univ. Press, London).

CARTER, F.J. (1998). Biomechanical testing of intra-abdominal soft tissue. In: *Int. Workshop on Soft Tissue Deformation and Tissue Palpation*, Cambridge, MA, October 1998.

CIARLET, P.G. (1987). *Mathematical Elasticity, vol. 1: Three-dimensional Elasticity* (North-Holland Amsterdam). ISBN 0-444-70259-8.

COSMI, F. (2001). Numerical solution of plane elasticity problems with the cell method. *Comput. Methods Engrg. Sci.* **2** (3).

COSTA, I.F., BALANIUK, R. (2001). Lem – an approach for real time physically based soft tissue simulation. In: *Int. Conf. Automation and Robotics*, ICRA'2001, Seoul, May 2001.

Cotin, S., Delingette, H., Ayache, N. (1999a). Real-time elastic deformations of soft tissues for surgery simulation. *IEEE Trans. Visual. Comput. Graph.* **5** (1), 62–73.

Cotin, S., Delingette, H., Ayache, N. (1999b). Real-time elastic deformations of soft tissues for surgery simulation. *IEEE Trans. Visual. Comput. Graph.* **5** (1), 62–73.

Cotin, S., Delingette, H., Ayache, N. (2000). A hybrid elastic model allowing real-time cutting, deformations and force-feedback for surgery training and simulation. *The Visual Computer* **16** (8), 437–452.

Cotin, S., Delingette, H., Clement, J.-M., Tassetti, V., Marescaux, J., Ayache, N. (1996). Volumetric deformable models for simulation of laparoscopic surgery. In: *Proc. Int. Symp. on Computer and Communication Systems for Image Guided Diagnosis and Therapy, Computer Assisted Radiology*, CAR'96. In: Int. Congr. Ser. **1124** (Elsevier, Amsterdam).

Couinaud (1957). *Le foie, Études anatomiques et chirurgicales* (Masson, Paris).

Cover, S.A., Ezquerra, N.F., O'Brien, J.F. (1993). Interactively deformable models for surgery simulation. *IEEE Comput. Graph. Appl.* **13**, 68–75.

Dan, D. (1999). Caractérisation mécanique du foie humain en situation de choc, PhD thesis, Université Paris 7.

Debunne, G., Desbrun, M., Cani, M.-P., Barr, A.H. (2001). Dynamic real-time deformations using space and time adaptive sampling. In: *Computer Graphics Proceedings*, SIGGRAPH'01, August 2001.

Decoret, X., Schaufler, G., Sillion, F., Dorsey, J. (1999). Multi-layered impostors for accelerated rendering. *Computer Graphics Forum (Eurographics'99)* **18**, 61–73.

Delingette, H. (1998). Towards realistic soft tissue modeling in medical simulation. *Proc. IEEE: Special Issue on Surgery Simulation* **86**, 512–523.

Delingette, H. (1999). General object reconstruction based on simplex meshes. *Int. J. Comput. Vision* **32** (2), 111–146.

Delingette, H., Montagnat, J. (2001). Shape and topology constraints on parametric active contours. *J. Comput. Vision and Image Understanding* **83**, 140–171.

Desbrun, M., Gascuel, M.-P. (1995). Animating soft substances with implicit surfaces. In: *Computer Graphics*, SIGGRAPH'95, Los Angeles.

Deussen, O., Kobbelt, L., Tucke, P. (1995). Using simulated annealing to obtain a good approximation of deformable bodies. In: *Proc. Eurographics Workshop on Animation and Simulation*, Maastricht, Netherlands, September 1995 (Springer-Verlag, Berlin).

Duncan, J., Ayache, N. (2000). Medical image analysis: Progress over two decades and the challenges ahead. *IEEE Trans. on Pattern Analysis and Machine Intelligence* **22** (1), 85–106.

Forest, C., Delingette, H., Ayache, N. (2002a). Cutting simulation of manifold volumetric meshes. In: *Modelling and Simulation for Computer-aided Medicine and Surgery*, MS4CMS'02.

Forest, C., Delingette, H., Ayache, N. (2002b). Removing tetrahedra from a manifold mesh. In: *Computer Animation*, CA'02, Geneva, Switzerland, June 2002 (IEEE Computer Society), pp. 225–229.

Forest, C., Delingette, H., Ayache, N. (2003). Simulation of surgical cutting in a manifold mesh by removing tetrahedra, *Medical Image Analysis*, submitted for publication.

France, L., Angelidis, A., Meseure, P., Cani, M.-P., Lenoir, J., Faure, F., Chaillou, C. (2002). Implicit representations of the human intestines for surgery simulation. In: *Conf. on Modeling and Simulation for Computer-aided Medicine and Surgery*, MS4CMS'02, Rocquencourt, November 2002.

France, L., Lenoir, J., Meseure, P., Chaillou, C. (2002). Simulation of minimally invasive surgery of intestines. In: Richir, S. (ed.), *Fourth Virtual Reality International Conference*, VRIC'2002, pp. 21–27. ISBN 2-9515730.

Fung, Y.C. (1993). *Biomechanics – Mechanical Properties of Living Tissues*, second ed. (Springer-Verlag, Berlin).

Gibson, S., Samosky, J., Mor, A., Fyock, C., Grimson, E., Kanade, T., Kikinis, R., Lauer, H., McKenzie, N. (1997). Simulating arthroscopic knee surgery using volumetric object representations, real-time volume rendering and haptic feedback. In: Troccaz, J., Grimson, E., Mosges, R. (eds.), *Proc. First Joint Conf. CVRMed-MRCAS'97*. In: Lecture Notes in Comput. Sci. **1205**, pp. 369–378.

Gladilin, E. (2002). Biomechanical modeling of soft tissue and facial expressions for craniofacial surgery planning, PhD thesis, Freie Univerisität Berlin, Germany.

Gottschalk, S., Lin, M., Manocha, D. (1996). Obb-tree: A hierarchical structure for rapid interference detection. In: *Proc. SIGGRAPH 96*, New Orleans, LA, pp. 171–180. ISBN 0-201-94800-1.

HODGSKINSON, R., CURREY, J.D. (1992). Young modulus, density and material properties in cancellous bone over a large density range. *J. Materials Science: Materials in Medicine* **3**, 377–381.

HOUBOLT, J.C. (1950). A recurrence matrix solution for the dynamic response of elastic aircraft. *J. Aeronautical Sci.* **17**, 540–550.

HUMPHREY, J.D., STRUMPF, R.K., YIN, F.C.P. (1990). Determination of a constitutive relation for passive myocardium: I. A new functional form. *ASME J. Biomech. Engrg.* **112**, 333–339.

HUMPHREY, J.D., YIN, F.C.P. (1987). On constitutive relations and finite deformations of passive cardiac tissue: I. A pseudostrain-energy function. *ASME J. Biomech. Engrg.* **109**, 298–304.

HUNTER, P., PULLAN, A. (1997). *FEM/BEM Notes* (University of Auckland, New-Zeland). Available at http://www1.esc.auckland.ac.nz/Academic/Texts/fembemnotes.pdf.

JAMES, D.L., PAI, D.K. (1999). Artdefo accurate real time deformable objects. In: *Computer Graphics, SIGGRAPH'99*, pp. 65–72.

JOE, B. (1991). Geompack – a software package for the generation of meshes using geometric algorithms. *J. Advanced Eng. Software* **13**, 325–331.

KAISS, M., LE TALLEC, P. (1996). La modélisation numérique du contact œil-trépan. *Revue Européenne des éléments Finis* **5** (3), 375–408.

KASS, M., WITKIN, A., TERZOPOULOS, D. (1988). Snakes: Active contour models. *Int. J. Comput. Vision* **1**, 321–331.

KAUER, M., VUSKOVIC, V., DUAL, J., SZÉKELY, G., BAJKA, M. (2001). Inverse finite element characterization of soft tissues. In: *Proc. 4th Int. Conf. on Medical Image Computing and Computer-Assisted Intervention*, MICCAI'01, Utrecht, October 2001. In: Lecture Notes in Comput. Sci. **2208**, pp. 128–136.

KAYE, J., PRIMIANO, F., METAXAS, D. (1997). A 3d virtual environment for modeling mechanical cardiopulmonary interactions. *Medical Image Analysis (Media)* **2** (2), 1–26.

KNOTT, D., PAI, D. (2003). Collision and interference detection in real-time using graphics hardware. In: *Proc. Graphics Interface*, Halifax, Canada, June 2003.

KUHN, CH., KÜHNAPFEL, U., KRUMM, H.-G., NEISIUS, B. (1996). A 'virtual reality' based training system for minimally invasive surgery. In: *Proc. Computer Assisted Radiology*, CAR'96, Paris, June 1996, pp. 764–769.

KUHNAPFEL, U., AKMAK, H., MAA, H. (2000). Endoscopic surgery training using virtual reality and deformable tissue simulation. *Computers and Graphics* **24**, 671–682.

LIU, Z., BILSTON, L.E. (2000). On the viscoelastic character of liver tissue: experiments and modelling of the linear behaviour. *Biorheology* **37**, 191–201.

LIU, Z., BILSTON, L.E. (2002). Large deformation shear properties of liver tissue. *Biorheology* **39**, 735–742.

LOMBARDO, J.-C., CANI, M.-P., NEYRET, F. (1999). Real-time collision detection for virtual surgery. In: *Computer Animation*, Geneva, Switzerland, May 1999, pp. 82–89.

LORENSEN, W., CLINE, H.E. (1987). Marching cubes: a high resolution 3d surface construction algorithm. *ACM Computer Graphics (SIGGRAPH'87)* **21**, 163–169.

LOUCHET, J., PROVOT, X., CROCHEMORE, D. (1995). Evolutionary identification of cloth animation model. In: *Workshop on Computer Animation and Simulation*, Eurographics'95, pp. 44–54.

LUMSDAINE, A., SIEK, J. (1998). The Matrix Template Library. http://www.lsc.nd.edu/research/mtl/.

MACMILLAN, R.H. (1955). A new method for the numerical evaluation of determinants. *J. Roy. Aeronaut. Soc.* **59** (772).

MANDUCA, A., MUTHUPILLAI, R., ROSSMAN, P., GREENLEAF, J., EHMAN, L. (1996). Visualization of tissue elasticity by magnetic resonance elastography. In: *Proc. of Visualization in Biomedical Imaging*, VBC'96, Hamburg, Germany, pp. 63–68.

MARCUS, B. (1996). Hands on: Haptic feedback in surgical simulation. In: *Proc. of Medicine Meets Virtual Reality IV*, MMVR IV, San Diego, CA, January 1996, pp. 134–139.

MARK, W., RANDOLPH, S., FINCH, M., VAN VERTH, J., TAYLOR II R.M. (1996). Adding force feedback to graphics systems: Issues and solutions. In: Rushmeier, H. (ed.), *ACM SIGGRAPH Computer Graphics Annual Conference*, SIGGRAPH'96 (Addison–Wesley, Reading, MA), pp. 447–452.

MAUREL, W., WU, Y., MAGNENAT THALMANN, N., THALMANN, D. (1998). *Biomechanical Models for Soft Tissue Simulation*, ESPRIT Basic Research Series (Springer-Verlag, Berlin).

SERRANO, C.M., LAUGIER, C. (2001). Realistic haptic rendering for highly deformable virtual objects. In: *Proc. Int. Conf. on Virtual Reality*, Yokohama, Japan, March 2001.

MILLER, K. (2000). Constitutive modelling of abdominal organs. *J. Biomech.* **33** (3), 367–373.

MONTAGNAT, J., DELINGETTE, H. (1998). Globally constrained deformable models for 3d object reconstruction. *Signal Processing*, 173–186.

NAVA, A., MAZZA, E., KLEINERMANN, F., AVIS, N., MCCLURE, J. (2003). Determination of the mechanical properties of soft human tissues through aspiration experiments. In: *Proc. Conf. on Medical Robotics, Imaging And Computer Assisted Surgery*, MICCAI 2003, Montreal, Canada, November 2003, In: Lecture Notes in Comput. Sci.

NEWMARK, N.M. (1959). A method of computation for structural dynamics. *J. Engrg. Mech. Division* **85**, 67–94.

O'MAHONY, A., WILLIAMS, J., KATZ, J. (1999). Anisotropic elastic properties of cancellous bone from a human edentulous mandible. In: *Proc. ASME Bioengineering'99 Conference*.

OWEN, S. (2000). A survey of unstructured mesh generation technology. Technical report, Department of Civil and Environmental Engineering, Carnegie Mellon University

PAPADEMETRIS, X., SHI, P., DIONE, D.P., SINUSAS, A.J., CONSTABLE, R.T., DUNCAN, J.S. (1999). Recovery of soft tissue object deformation from 3d image sequences using biomechanical models. In: *XVIth Int. Conf. on Information Processing In Medical Imaging*, IPMI'99, Visegrád, Hungary, June 28–July 2, 1999, pp. 352–357.

PARTHASARATHY, V.N., GRAICHEN, C.M., HATHAWAY, A.F. (1993). A comparison of tetrahedron quality measures. *Finite Elements in Analysis and Design* **15**, 255–261.

PICINBONO, G., DELINGETTE, H., AYACHE, N. (2001). Non-linear and anisotropic elastic soft tissue models for medical simulation. In: *IEEE Int. Conf. Robotics and Automation*, ICRA'2001, Seoul, Korea, May 2001. Best conference paper award.

PICINBONO, G., DELINGETTE, H., AYACHE, N. (2003). Non-linear anisotropic elasticity for real-time surgery simulation. *Graphical Models* **65** (5), 305–321.

PICINBONO, G., LOMBARDO, J.-C., DELINGETTE, H., AYACHE, N. (2000). Anisotropic elasticity and forces extrapolation to improve realism of surgery simulation. In: *IEEE Int. Conf. Robotics and Automation*, ICRA'2000, San Francisco, USA, April 2000, pp. 596–602.

PICINBONO, G., LOMBARDO, J.-C., DELINGETTE, H., AYACHE, N. (2002). Improving realism of a surgery simulator: linear anisotropic elasticity, complex interactions and force extrapolation. *J. Visual. Comput. Animation* **13** (3), 147–167.

PRESS, W.H., FLANNERY, B.P., TEUKOLSKY, S.A., VETTERLING, W.T. (1991). *Numerical Recipes in C* (Cambridge Univ. Press, Cambridge, UK).

PRESS, W.H., FLANNERY, B.P., TEUKOLSKY, S.A., VETTERLING, W.T. (1992). *Numerical Recipes in FORTRAN: The Art of Scientific Computing*, second ed. (Cambridge Univ. Press Cambridge, UK).

PUSO, M.A., WEISS, J.A. (1998). Finite element implementation of anisotropic quasi-linear viscoelasticity using a discrete spectrum approximation. *ASME J. Biomech. Engrg.* **120** (1).

PUTTI, M., CORDES, C. (1998). Finite element approximation of the diffusion operator on tetrahedra. *SIAM J. Scientific Comput.* **19** (4), 1154–1168.

QUARTERONI, A., TUVERI, M., VENEZIANI, A. (2000). Computational vascular fluid dynamics: problems, In: models and methods. *Computing and Visualization in Science* **2**, 163–197.

RADETZKY, A. (1998). The simulation of elastic tissues in virtual medicine using neuro-fuzzy systems. In: *Medical Imaging'98: Image Display*, San Diego, CA, February 1998.

SAAD, Y. (1996). *Iterative Methods for Sparse Linear Systems* (WPS).

SAKUMA, I., NISHIMURA, Y., KONG CHUI, C., KOBAYASHI, E., INADA, H., CHEN, X., HISADA, T. (2003) In vitro measurement of mechanical properties of liver tissue under compression and elongation using a new test piece holding method with surgical glue. In: *Int. Symp. on Surgery Simulation and Soft Tissue Modeling*, Juan-Les-Pins, France, June 2003. In: Lecture Notes in Comput. Sci. **2673** (Springer-Verlag, Berlin), pp. 284–292.

SATAVA, R. (1994). Medicine 2001: The King Is Dead. In: *Proc. Conf. Virtual Reality in Medicine*.

SATAVA, R. (1996). Medical virtual reality: The current status of the future. In: *Proc. 4th Conf. Medicine Meets Virtual Reality*, MMVR IV, pp. 100–106.

SCHROEDER, W.J., ZARGE, J., LORENSEN, W. (1992). Decimation of triangles meshes. *Computer Graphics (SIGGRAPH'92)* **26**.

SERMESANT, M., COUDIÈRE, Y., DELINGETTE, H., AYACHE, N. (2002). Progress towards an electro-mechanical model of the heart for cardiac image analysis. In: *IEEE Int. Symp. Biomedical Imaging*, ISBI'02, pp. 10–14.

Sermesant, M., Faris, O., Evans, F., McVeigh, E., Coudière, Y., Delingette, H., Ayache, N. (2003). Preliminary validation using in vivo measures of a macroscopic electrical model of the heart. In: Ayache, N., Delingette, H. (eds.), *Int. Symp. Surgery Simulation and Soft Tissue Modeling*, IS4TM'03. In: Lecture Notes in Comput. Sci. **2673** (Springer-Verlag, Heidelberg).

Sillion, F.X., Drettakis, G., Bodelet, B. (1997). Efficient impostor manipulation for real-time visualization of urban scenery. In: *Proc. Eurographics'97*, Budapest, Hungary, September 1997.

Simail: product of Simulog S.A. – 1, rue James Joule, 78286 Guyancourt cedex, France, http://www.simulog.fr.

Soferman, Z., Blythe, D., John, N. (1998). Advanced graphics behind medical virtual reality: Evolution of algorithms, hardware and software interfaces. *Proc. IEEE: Special Issue on Surgery Simulation* **86** (3), 531–554.

Soler, L., Delingette, H., Malandain, G., Montagnat, J., Ayache, N., Clément, J.-M., Koehl, C., Dourthe, O., Mutter, D., Marescaux, J. (2000). Fully automatic anatomical, pathological and functional segmentation from ct-scans for hepatic surgery. In: *Medical Imaging 2000*, San Diego, February 2000.

Soler, L., Malandain, G., Delingette, H. (1998). Segmentation automatique: application aux angioscanners 3d du foie. *Traitement du signal* **15** (5), 411–431. (in French).

Spencer, A.J.M. (1972). *Deformations of Fibre-Reinforced Materials* (Clarendon Press, Oxford).

Spencer, A.J.M. (1984). *Continuum Theory of Fiber-Reinforced Composites* (Springer-Verlag, New York).

Szekely, G., Bajka, M., Brechbuhler, C. (1999). Virtual reality based simulation for endoscopic gynaecology. In: *Proc. Medicine Meets Virtual Reality*, MMVR'99, San Francisco, USA, pp. 351–357.

Teschner, M., Heidelberger, B., Muller, M., Pomeranets, D., Gross, M. (2003). Optimized spatial hashing for collision detection of deformable objects. In: *Proc. Vision, Modeling, Visualization*, VMV'03, Munich, Germany, November 2003.

Vidrascu, M., Delingette, H., Ayache, N. (2001). Finite element modeling for surgery simulation. In: *First MIT Conf. on Computational Fluid and Solid Mechanics*.

Vlachos, A., Peters, J., Boyd, C., Mitchell, J.L. (2001). Curved pn triangles. In: *2001 ACM Symp. on Interactive 3D Graphics*.

Weiss, J.A., Gardiner, J.C., Quapp, K.M. (1995). Material models for the study of tissues mechanics. In: *Proc. Int. Conf. on Pelvic and Lower Extremity Injuries*, Washington, DC, December 1995, pp. 249–261.

Woo, M., Neider, J., Davis, T. (1997). *OpenGL Programing Guide* (Addison–Wesley, Reading, MA).

Yamashita, Y., Kubota, M. (1994). Ultrasonic characterization of tissue hardness in the in-vivo human liver. In: *Proc. IEEE Ultrasonics Symposium*, pp. 1449–1453.

Zorin, D., Schroeder, P., Sweldens, W. (1996). Interpolating subdivision for meshes with arbitrary topology. In: *Proc. 23rd Annual Conf. on Computer Graphics and Interactive Techniques* (ACM Press), pp. 189–192.

Subject Index

Printed in the United States
By Bookmasters